COMPUTER ARITHMETIC:
PRINCIPLES, ARCHITECTURE,
AND DESIGN

Computer Arithmetic

PRINCIPLES, ARCHITECTURE, AND DESIGN

Kai Hwang

School of Electrical Engineering
Purdue University

John Wiley & Sons,

New York • Chichester • Brisbane • Toronto • Singapore

Library of Congress Cataloging in Publication Data:

Hwang, Kai.
Computer arithmetic: principles, architecture, and design.

Includes bibliographical references and indexes.
1. Electronic digital computers. 2. Computer
arithmetic and logic units. 3. Title.
TK7888.3.H9 621.3819'58'2 78-18922
ISBN 0-471-03496-7

Printed in the United States of America

10 9 8 7

To my wife

and our

parents and sons

for their

love and encouragement

Preface

This book is designed to be used in an introductory course for upper-division and first-year graduate students in electrical, mechanical, and industrial engineering, and in computer and mathematical sciences. It is also suitable as a reference book for practicing engineers and computer scientists who are involved in the design, research, or development of digital computing machines. A minimum background in basic logic design is assumed, but no background in electronics is needed nor is a detailed knowledge of the interior circuitry of IC chips required. The book presents machine arithmetic theories and algorithms, processor architectures, design methodologies, and hardware arithmetic functions associated with modern digital computer systems. Arithmetic algorithms and their efficient implementation schemes are illustrated with flowcharts, arithmetic and Boolean equations, schematic logic circuit diagrams, numerical tabulations, control matrices, and case studies of existing high-performance arithmetic processors. Up-to-date literature guides are also provided at the end of each chapter.

Chapters 1 and 2 present the basic concepts of machine arithmetic, internal number representations, and an introduction to modern digital electronic devices that are frequently used as building blocks in digital system design. Chapters 3 to 8 are devoted to the theory and design of standard arithmetic units that perform fixed-point addition, subtraction, multiplication, or division. Emphases are on the designs of high-speed and cost-effective adders/subtractors, multipliers, and dividers, and on some new and unconventional approaches, such as signed-digit arithmetic, cellular array arithmetic, multioperand addition, multiplier recoding, SRT, convergence, and high-radix division methods. Chapters 9 and 10 are devoted to floating-point arithmetic design, starting with the fundamental issues and leading to some advanced topics on floating-point singularities, error analyses, axiomatic rounding theory, and multiple-precision FLP arithmetic. Chapter 11 deals with the evaluation of elementary functions, pipelined arithmetic design, and error control techniques in arithmetic processors. The arithmetic designs in **IBM** System/**360** Model **91**, **ILLIAC III**, Texas Instruments **Advanced Scientific Computer**, and Control Data **STAR-100** Computer systems are illustrated in case study examples. Bibliographic notes, reference lists, and problems are provided at the end of each chapter.

This book grew out of a sequence of computer design courses that I taught at the University of Miami, and Wayne State, and Purdue Universities. It is appropriate for three different course offerings. First, it can be used as the text for a senior course with a suggested title "Computer Arithmetic Theory and Design," if Chapters 1 to 5, 7, and 9 are covered in one semester. Second, it can be offered for use in a one-semester graduate course under the title "Advanced Computer Arithmetic," if all chapters except Chapters 1 to 3 are covered. Third, it can be offered as the text for a sequence of two quarter courses for dual-level students (a mixture of seniors and graduates), covering Chapters 1 to 3, 5, 7, and 9 in the first quarter and the remaining five chapters in the second quarter. A Solutions Manual for homework problems is available and may be obtained from John Wiley.

I gratefully acknowledge the technical guidance I have received from Professors H. C. Torng of Cornell University, Arthur Gill of the University of California at Berkeley, and Norman Abramson of the University of Hawaii. The inspiring and encouraging atmosphere created by my colleagues at Purdue University has been of immeasurable value in this project. In particular, I am indebted to Professors Clarence Coates, King-Sun Fu, George Saridis, Richard Schwartz, William Hayt, Robert Pierret, Fred Mowle, Steven Bass, Pen-Min Lin, Janak Patel, Mr. Lionel M. Ni, and Mr. T. P. Chang for their valuable comments, stimulating discussions and assistance. The special help from Professors M. Ghausi of Oakland University, Francis Yu of Wayne State University, Daniel Atkins of the University of Michigan, and V. C. Hamacher of the University of Toronto is greatly appreciated. Thanks are also due to numerous authors for permission to include their published results in this book. Linda Stovall and Jackie Ganster have typed the manuscript, and to them I owe my deep gratitude. Comments and suggestions from readers are invited and will be considered for future editions of the book.

Kai Hwang

Lafayette, Indiana
1979

Contents

COMPUTER ARITHMETIC: PRINCIPLES, ARCHITECTURE, AND DESIGN

Chapter

1

Computer Arithmetic and Number Systems

1.1 An Overview of Computer Arithmetic

Arithmetic has played important roles in human civilization, especially in the areas of science, engineering, and technology. Machine arithmetic can be traced back as early as 500 BC in the form of the *abacus* used in China. Throughout the history of computing machines, arithmetic processors are the major working forces of computers. Arithmetic processors are used to execute arithmetic operations and to generate numerical solutions to computational problems. Since the advent of Medium/Large Scale Integrated (MSI/LSI) electronic circuits, more and more sophisticated arithmetic processors have become standard hardware features for today's high-performance digital computer systems. A modern digital computer may be equipped with a number of hardware arithmetic processors, which manipulate data of different formats or solve special arithmetic functions for general and dedicated applications. These hardware arithmetic processors provide faster computing speed than the conventional means of computation. The choice of an appropriate arithmetic system underlines both computer architectural design and programming applications.

Let **R** be the set of all real numbers. A *real arithmetic* can be defined as an algebraic mapping

$$\mathbf{f}:\mathbf{R} \times \mathbf{R} \rightarrow \mathbf{R} \tag{1.1}$$

Typical examples of **f** are the standard two-operand arithmetic operations $+$, $-$, $\times$, $\div$. Let **M** be the set of machine representable numbers. Each number in set **M** can have only finite number of digits in it. Set **M** must be a proper finite subset of **R**, that is, $\mathbf{M} \subset \mathbf{R}$. A *machine arithmetic* can be modeled by the following mapping:

$$\mathbf{g}:\mathbf{M} \times \mathbf{M} \rightarrow \mathbf{M} \tag{1.2}$$

The machine arithmetic differs from the real arithmetic in the fundamental issue of number precision. Only *finite-precision* computations can be performed by arithmetic processors, whereas real arithmetic may produce *arbitrary-precision* results with no restriction in length.

In other words, we can consider machine arithmetic an approximated real arithmetic subject to appropriate rounding controls. In terms of functional mappings, the above machine arithmetic function g can be related to the real arithmetic function **f** by the following composite function

$$\mathbf{g} = \mathbf{h} \circ \boldsymbol{\rho} \tag{1.3}$$

The mapping **h** is defined by

$$\mathbf{h} = \mathbf{f}\ |\mathbf{M} \times \mathbf{M} \rightarrow \mathbf{R} \tag{1.4}$$

as the real function **f** restricted to the machine domain **M** × **M**. The mapping defined by

$$\boldsymbol{\rho}: \mathbf{R} \rightarrow \mathbf{M} \tag{1.5}$$

is the selected rounding scheme, which will trim the arbitrary long result to a machine representable number. Note that digital computers in contrast to conventional arithmetic units can perform computations with arbitrary precision, given a sufficient storage capacity. Finite-precision results are expected only for machines with restricted word length (in registers or adders) and limited memory capacity.

Digital computer arithmetic is a field evolved from system architecture and logic design of digital computers. Arithmetic design therefore includes the development of arithmetic algorithm and its logic implementation. In the computer architecture domain it interacts with the efficiency of implementing arithmetic operations. In the numerical analysis domain it concerns the accuracy of approximated real arithmetic. The interactions affecting the design of computer arithmetic systems with other computer-related disciplines are illustrated in Fig. 1.1. It can be said that the design of arithmetic systems involves the major areas of computer engineering, which is a mixed field of computer science and digital engineering.

This text describes the principles and design techniques of four major areas of computer arithmetic systems: *radix arithmetic*, *signed-digit arithmetic*, *elementary function evaluation*, and *pipelined arithmetic*. Chapters 1 and 2 cover number systems and digital circuit technology, and provide the groundwork for the remaining chapters. The radix arithmetic is divided into fixed-point and floating-point systems as encountered in most computers. Both standard and unconventional add/subtract, multiply and divide algorithms and their implementations are presented with more emphasis on new approaches, such as multioperand carry-save adders, multiplier recoding, convergence and SRT divisions, cellular array multipliers and dividers, signed-digit arithmetic, and so on. Floating-point arithmetic is presented from elementary to advanced levels. The final chapter discusses elementary functions and pipelined arithmetic design.

Interested readers are encouraged to check the bibliographic notes and the reference list at the end of each chapter for additional sources of information. The majority of this material is selected from original journal or conference publications.

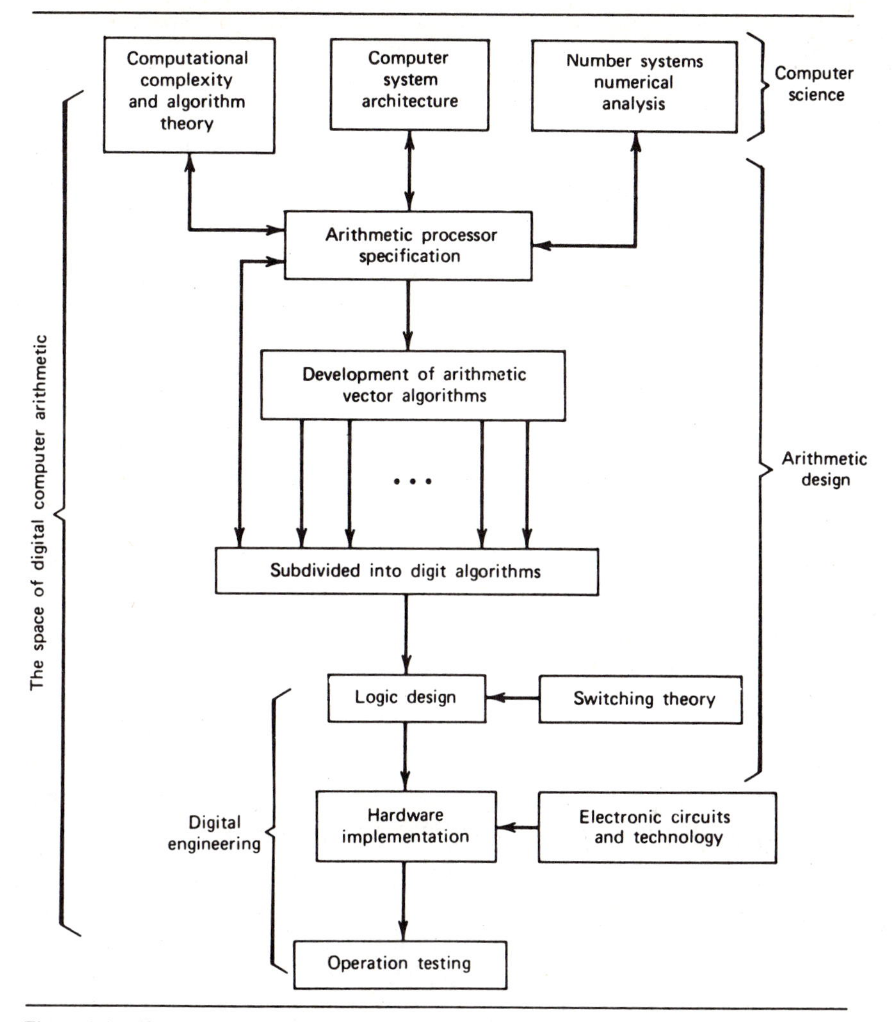

Figure 1.1 The interactions of arithmetic design with other disciplines of computer science and digital engineering.

Exercises are designed to help students digest the essential points in the book and, hopefully, to stimulate new and improved techniques of their own.

1.2 Machine Arithmetic Number Systems

Implementation of arithmetic algorithms in a digital computer depends largely on how numerical data are stored in memory and in registers. Different internal number representation may result in different arithmetic hardware design. The choice of an appropriate number system impacts simultaneously on computer architecture from the designer's viewpoint as well as on numerical analysis methods applied by the users.

Because only finite-precision arithmetic is implemented in digital computers, all the allowable numeric representations must be restricted to finite length. The real arithmetic operations can be realized only by machines with finite accuracy as restricted by the word length. A good choice of arithmetic system and internal number representations affects both efficient implementation of the machine operations and the accuracy of approximated real arithmetic.

The arithmetic designer must be aware of the architectural realities of *time* and *space* efficiency and the numerical analytic reality of providing sufficient *accuracy* for a wide range of application problems. Generally, machine arithmetic number systems can be divided into the following five categories.

A. Conventional Radix Number System

Conventional computers use the *fixed radix* arithmetic system with a radix $r \geq 2$ and a digit set

$$\{0, 1, \ldots, r-1\} \tag{1.6}$$

All the digits of a number are positively weighted and each number is uniquely represented. Some special systems may use *mixed radix* numbers, in which different digits of a number assume different base values. We shall formally describe fixed radix number systems in section 1.4.

B. Signed-Digit Number System

This system can be considered as an extended case of the fixed-radix system, in which both positive and negative weighted digits are allowed in a digit set

$$\{-\alpha, \ldots, -1, 0, 1, \ldots, \alpha\} \tag{1.7}$$

where α is a bounded positive integer. For a given numeric value, the signed-digit representation may not be unique. Because there may be more than one signed-digit representation of a given number, such a number system is considered *redundant*. Redundant signed-digit number systems will be treated in detail in section 1.5.

C. Residue Number System

No weighting factor is assigned to each digit of a residue number; therefore, the order of the digits is immaterial in determining the value of the number. Also, mixed radices are assigned to different digits. A *residue number* **x** is a integer represented by an n-tuple

$$\mathbf{x} = \{r_1, r_2, \ldots, r_n\}_{\mathbf{m}} \tag{1.8}$$

with respect to another n-tuple

$$\mathbf{m} = \{m_1, m_2, \ldots, m_n\} \tag{1.9}$$

Each r_i is called the *residue* of **x** modulo m_i, where all the n *moduli* $\{m_i | i = 1, 2, \ldots, n\}$ are pairwise relatively prime. All the n residue digits r_i for $i = 1, 2, \ldots, n$ can be independently processed. Because of this, addition and multiplication are inherently carry-free in residue arithmetic. Residue arithmetic was originally proposed by Svoboda [16] to combat arithmetic errors in the design of reliable computing systems. Readers are advised to read Szabo and Tanaka [15] for the details of residue arithmetic and its applications.

D. Rational Number System

This arithmetic system represents such numeric quantities as fractions in terms of numerator-denominator integer pairs. Addition, subtraction, multiplication, and division of such rational numbers always yield rational numbers, so these operations may be closed without resort to infinite precision, such as the number $\frac{1}{3} = 0.333\ldots$. A practical limitation does make such an ideal situation intractable, however, in that the numerators and denominators can become large very early in only a moderate sized computation.

Investigation of the design implications of approximated arithmetic has been suggested in a limited precision rational system, where the precision limitation is realized via bounding the numerator and denominator sizes for allowed fractions. The finite precision bounds may be made separately or jointly, limiting the number of digits in the numerator and denominator, yielding systems termed *fixed-slash* or *floating-slash* rational number systems. The rational arithmetic system is still in its postulate stage as far as realization is concerned. Interested readers may refer to the work of Matula [12] and of Hwang and Chang [7] for the potential development.

E. Logarithmic Number System

This system employs a real number $\boldsymbol{\mu} > 1$ as the *base*. The set of real numbers is defined by the following logarithmic space $\mathbf{L}_\mu$

$$\mathbf{L}_\mu = \{\mathbf{x} \,\|\, \mathbf{x}| = \boldsymbol{\mu}^i, \text{ i an integer}\} \cup \{0\} \tag{1.10}$$

The notion of applying exponential expressions was suggested to enable geometric rounding rather than arithmetic rounding in order to enhance the number accuracy. The details of realizing logarithmic arithmetic via integer arithmetic and log-antilog subroutines can be found in Marasa [10].

Arithmetic designs described in this book are based mainly on the conventional radix number system or on the redundant signed-digit number system. Details of residue, rational, and logarithmic number systems are not within the scope of this text.

1.3 Classification of Arithmetic Operations

Arithmetic instructions in modern digital computers can generally be classified into three classes from both the user and designer's viewpoints.

A. Standard Arithmetic Operations

This class includes mainly the four primitive arithmetic functions: **add**, **subtract**, **multiply**, and **divide** in either fixed-point or floating-point modes. All other mathematical functions can be expressed in terms of these four standard operations. We briefly distinguish the two operation modes below. Detailed implications will appear in subsequent sections.

(1) Fixed-Point (FXP) Arithmetic is commonly used on problem data with *fixed*-radix point, such as those encountered in business or statistical calculations. FXP operations can be further divided into two subclasses according to the apparent position for the radix point. In *integer arithmetic* all results are lined up at the right end of the registers, as if there were a radix point at the extreme right. In *fraction arithmetic* all results, regardless of length, are lined up at the left end of the registers.
(2) Floating-Point (FLP) Arithmetic is used mainly for scientific and engineering computations, in which frequent magnitude scaling is required. FLP operations can also be divided into two subclasses according to their data format. When normalized data operands are enforced, we have the so-called *normalized FLP arithmetic*. When operands are not required to be normalized during intermediate and final steps, we have the *unnormalized FLP arithmetic*. Most computers in use adopt the normalized operations.

B. Elementary Arithmetic Functions

Elementary functions refer to those special arithmetic operations frequently used in mathematical computations. These include *exponential*, *square root*, *logarithm*, *trigonometric*, *hyperbolic functions*, and so on. Not all computers include these functions as standard hardware features. At present most computers implement these elementary functions with software or firmware routines. Special-purpose hardware function evaluators for generating elementary functions are becoming more and more popular as hardware cost decreases.

C. Pseudoarithmetic Operations

This class of operations requires a certain degree of arithmetic calculations, but mainly for dedicated purposes in the execution of computer programs. Two subclasses of pseudoarithmetic operations are described below.

(1) Address Arithmetic is performed primarily for effective memory address calculation, such as *indexing*, *indirect*, *relative*, or *offset* addressing schemes.
(2) Data Editing Arithmetic includes logical alphanumeric and data transfer operations such as *compare*, *complement*, *load*, *store*, *pack*, *unpack*, *shifting*, *normalize*, and so on. These operations are used for transforming data from one format to another, checked for consistency with a source format, or tested for controlling the program sequence.

We are mainly interested in the standard FXP or FLP arithmetics and some elementary function evaluations. The address arithmetic and data editing operations are not our main concerns. The theory and design techniques presented in this text focus on the first two classes of arithmetic operations. In many computers, the arithmetic instructions are subdivided according to the data precision enforced. Multiple-precision arithmetic applies to both FXP and FLP operations.

(1) Single-Precision (SP) Arithmetic refers to those operations defined over standard data operands with word length equal to that of one memory word.
(2) Double-Precision (DP) Arithmetic uses twice the word length per each operand. *Triple-precision* or higher precision operation can be similarly defined.

Some computing machines offer separate arithmetic hardware for handling *binary* and *decimal* data such as that found in IBM System/360. In these machines, operations can be defined directly on decimal operands. Usually, code conversion and pack or unpack instructions are needed in order to manipulate data directly in decimal format.

Table 1.1 summarizes the above arithmetic operation classifications. For example, an ADD operation can be specified under 16 categories as shown below

$$\left\{\begin{array}{l}\text{SP}\\ \text{DP}\end{array}\right\}\left\{\begin{array}{l}\text{FXP}\left\{\begin{array}{c}\text{Integer}\\ \text{Fractional}\end{array}\right\}\\ \text{FLP}\left\{\begin{array}{c}\text{Normalized}\\ \text{Unnormalized}\end{array}\right\}\end{array}\right.\quad\left\{\begin{array}{l}\text{binary}\\ \text{decimal}\end{array}\right\}\text{ADD}$$

The above 16 specifications of the ADD operation in a digital computer are determined by the number precision, integer or fraction FXP, normalized or unnormalized FLP arithmetic over either binary or decimal data.

If there is no limit on hardware investment and the applications do require an extensive amount of hardware arithmetic functions for enhancing the processing speed, one can design an arithmetic unit to handle both *real* and *complex* data, or some dedicated hardware machines for special arithmetic calculations such as a *Super Fast-Fourier Transformer* (SFFT), a *matrix manipulator*, and so on. Such

Table 1.1 Classification of Arithmetic Operations

Standard Arithmetic Operations	**Elemenatry Arithmetic Functions**	**Pseudoarithmetic Operations**
ADD, SUBTRACT, MULTIPLY, and DIVIDE operated with:	Exponential	**Data editing arithmetic:**
Single precision	Logarithm	Compare, Range
Double precision	Square root	Complement
FXP integer	Cubic root	Negation
FXP fractional	Squaring	Load, Store
Normalized FLP	Cubic power	Pack, Unpack
Unnormalized FLP	High-Order powers or roots	Increment
Binary arithmetic	Trigonometric and hyperbolic functions	Decrement
Decimal arithmetic or	Special polynomial	Shift, Rotate
High-Radix arithmetic	Transcendental function	Normalize
Negative operand		**Address Arithmetic:**
		Relative addressing,
		Indexing
		Base addressing
		Offset addressing

high-level special-purpose arithmetic machines are indeed feasible as the technology grows. The design techniques contained in this text provide the knowledge necessary to build either dedicated or general-purpose computing machines.

1.4 Conventional Radix Number System

The foundation of most existing arithmetic processors is the conventional radix number system. Only a handful of machines realized the redundant signed-digit arithmetic [14]. The remaining three classes of arithmetic systems based on residue, rational, and logarithmic number systems have not been widely accepted yet.

A radix-r number $\mathbf{X}$ is represented in a digital computer by a *digital vector* of $(n + k)$-tuples

$$\mathbf{X} = (x_{n-1}, \ldots, x_0 \,.\, x_{-1}, \ldots, x_{-k})_r \tag{1.11}$$

where each component x_i for $-k \leq i \leq n - 1$ is called the ith *digit* of the vector $\mathbf{X}$. Each digit can assume r distinct values

$$\{0, 1, \ldots, r - 1\} \tag{1.12}$$

where r is the *radix* of the number system. In a *fixed-radix number system*, all the digits assume the same radix value. The conventional decimal number representation belongs to this category with $r = 10$.

The first n digits $(x_{n-1}, \ldots, x_1, x_0)$ form the *integer* portion of the number $\mathbf{X}$ and the remaining k negatively indexed digits $(x_{-1}, x_{-2}, \ldots, x_{-k})$ form the *fraction* portion of the number $\mathbf{X}$. A *radix point* is used to divide these two portions. In an actual computer circuit, the radix point is implicit; that is, it does not occupy a physical position in a storage device. We shall discuss only radix number systems with positive integer radix $r \geq 2$. Number systems with negative, fractional, and complex radices are beyond the scope of this book.

Mixed-radix numbers are those assuming different radix values in different digit positions. The way we count time (hours, minutes, seconds) follows mixed radices of (24, 60, 60). In a *weighted* radix number system, we associate with each digital vector $\mathbf{X}$ a unique *value* denoted as

$$\mathbf{X}_v = \sum_{i=-k}^{n-1} x_i \cdot \omega_i \tag{1.13}$$

where each ω_i is called a *weighting factor* for the ith digit. The $n + k$ weighting factors form a *weight vector* denoted as

$$\mathbf{W} = (\omega_{n-1}, \ldots, \omega_0, \omega_{-1}, \ldots, \omega_{-k}) \tag{1.14}$$

The value of the number $\mathbf{X}$ can be obtained by $\mathbf{X} \cdot \mathbf{W}$, the *dot product* of the two vectors $\mathbf{X}$ and $\mathbf{W}$. In particular, the weight vector

$$\mathbf{W} = (r^{n-1}, \ldots, r^0, r^{-1}, \ldots, r^{-k}) \tag{1.15}$$

leads to the conventional *positional representation* of a radix-r number **X** with a value

$$\mathbf{X}_v = \mathbf{X} \cdot \mathbf{W}$$

$$= \sum_{i=-k}^{n-1} x_i \cdot \omega_i$$

$$= \sum_{i=-k}^{n-1} x_i \cdot r^i \qquad \textbf{(1.16)}$$

In practice, there are four radix number systems that are frequently used corresponding to radix values $r = 2, 8, 10$, and 16 for *binary*, *octal*, *decimal*, and *hexadecimal* number systems, respectively.

The higher the radix value r, the more *binary digits* (bits) are required to encode each radix-r digit. In general, at least k bits are required to encode a radix-r digit, where

$$k = \lceil \log_2 r \rceil \qquad \textbf{(1.17)}$$

The notation $\lceil x \rceil$ refers to the least integer that is not less than the real number x. For example, we may wish to encode each decimal digit ($r = 10$) with $k = \lceil \log_2 10 \rceil = 4$ bits. Several 4-bit binary codes for representing decimal digits are given in Table 1.2, of which the BCD and Gray code are the most popular ones.

Table 1.2 Several Binary Codes for Decimal Digits

Decimal Digit	BCD Code (8-4-2-1 Code)	Gray Code	5-2-1-1 Code	Excess-3 Code
0	0000	0010	0000	0011
1	0001	0110	0001	0100
2	0010	0111	0011	0101
3	0011	0101	0101	0110
4	0100	0100	0111	0111
5	0101	1100	1000	1000
6	0110	1101	1010	1001
7	0111	1111	1100	1010
8	1000	1110	1110	1011
9	1001	1010	1111	1100

Human beings are accustomed to the decimal system, whereas in the digital computer all data are binary coded. The popular choice of internal binary encoding is based on the *efficiency* of representation, *facility* for arithmetic design, and *reliability* of operation.

The designer should be aware that internal radix conversion is required for nonbinary machines. However, the user need not be aware of the necessary code conversion. The computers may appear to the user as octal, decimal, hexadecimal, or other radix machines. Higher-radix methods have been suggested for the design of high-speed hardware multiply and divide units, such as radix $r = 256$ used for multiply and divide in the ILLIAC III computer [6].

There are essentially two different ways of representing positive and negative conventional radix numbers in a digital computer. The expression shown in Eq. 1.11 corresponds to the so-called *fixed-point* representation. Another way of storing radix numbers is the floating-point representation. We shall describe both radix number representations in detail in sections 1.8 and 1.9.

1.5 Signed-Digit Number Systems

Redundant number systems may not be convenient for manual computation, but they have been useful in designing high-speed arithmetic machines. *Signed-digit* (**SD**) number representations allow redundancy to exist. Each signed digit may need more than one bit to represent it (including the sign bit of each **SD**), and this may increase the data storage space and require wider data buses. However, the gain in speed can far justify its merit. We shall illustrate the use of **SD** representations to design totally parallel adders in Chapter 4, recoded multipliers in Chapter 5, and high-radix dividers in Chapter 7.

SD numbers are formally defined as follows: Given a radix r, each digit of an **SD** number can assume the following $2\alpha + 1$ values

$$\Sigma_r = \{-\alpha, \ldots, -1, 0, 1, \ldots, \alpha\} \tag{1.18}$$

where the maximum digit magnitude α must be within the following region:

$$\left\lceil \frac{r-1}{2} \right\rceil \leq \alpha \leq r - 1 \tag{1.19}$$

Because integer $\alpha \geq 1$, $r \geq 2$ must be assumed. In order to yield *minimum redundancy* in the balanced digit set Σ_r, one can choose the following value for the maximum magnitude

$$\alpha = \left\lfloor \frac{r}{2} \right\rfloor \tag{1.20}$$

where $\lfloor x \rfloor$ stands for the largest integer that is less than or equal to the real number x. Therefore, $\alpha = r_e/2$ where r_e is even and $\alpha = (r_o - 1)/2$ when r_o is odd. This implies that adjacent odd-even radix values may result in the same digit set. The digit set corresponding to this choice of α is shown below in two different forms, but is actually the same set when $r_o = r_e + 1$.

$$\Sigma_{r_o} = \left\{-\frac{r_o - 1}{2}, \ldots, -1, 0, 1, \ldots, \frac{r_o - 1}{2}\right\}$$

$$\Sigma_{r_e} = \left\{-\frac{r_e}{2}, \ldots, -1, 0, 1, \ldots, \frac{r_e}{2}\right\} \tag{1.21}$$

For example, radix-2 **SD** system has the digit set $\Sigma = \{-1, 0, 1\}$, radix-4 **SD** system has the digit set $\Sigma_4 = \{-2, -1, 0, 1, 2\}$, and so on. The algebraic value, $\mathbf{Y}_v$, of an **SD** number

$$\mathbf{Y} = (y_{n-1} \cdots y_o y_{-1} \cdots y_{-k})_r \tag{1.22}$$

can be evaluated by

$$\mathbf{Y}_v = \sum_{i=-k}^{n-1} y_i \cdot r^i \tag{1.23}$$

This is similar to Eq. 1.16 except, now, **Y** can be itself positive or negative with no need for an *explicit sign*. The zero has a unique expression if and only if $y_i = 0$ for all i in Eq. 1.22. The negation $-\mathbf{Y}$ of an **SD** number **Y** is derived directly by changing the sign of all nonzero digits in **Y**. Note that "zero" digits are not considered signed. In practice, only powers-of-two radix values $r = 2^k$ have been used.

The original motivation of using **SD** number system is to eliminate carry propagation chains in addition (or subtraction). To break the carry chain, the lower bound on α should be made tighter as

$$\left\lceil \frac{r+1}{2} \right\rceil \le \alpha \le r - 1 \tag{1.24}$$

We shall consider **SD** addition using α within the above interval in Chapter 4. For division employing **SD** representation, the bound given in Eq. 1.19 is sufficiently tight.

Let us consider several numerical examples. Given an explicit value $\mathbf{Y}_v = -3$ with word length $n = 4$, $k = 0$, radix-2 **SD** system is considered with a digit set $\{\bar{1}, 0, 1\}$. For clarity, the negative digit -1 is denoted by *overbar* $\bar{1}$. There are five legitimate **SD** representations that yield value -3.

$$\begin{aligned}
\mathbf{Y} &= (0\ 0\ \bar{1}\ \bar{1})_2 = -2 - 1 \\
&= (0\ \bar{1}\ 0\ 1)_2 = -4 + 1 \\
&= (\bar{1}\ 1\ 0\ 1)_2 = -8 + 4 + 1 \\
&= (0\ \bar{1}\ 1\ \bar{1})_2 = -4 + 2 - 1 \\
&= (\bar{1}\ 1\ 1\ \bar{1})_2 = -8 + 4 + 2 - 1
\end{aligned}$$

These representations have different zero and nonzero digit distribution. The number of nonzero digits in an n-digit **SD** vector **Y** with value $\mathbf{Y}_v$ is called the *weight* $\omega(n, \mathbf{Y}_v)$. The above number has weights ranging from 2 to 4. In general, the weight of a binary n-digit **SD** vector is defined below with $|y_i| = 1$ if $y_i \neq 0$

$$\omega(n, \mathbf{Y}_v) = \sum_{i=0}^{n-1} |y_i| \tag{1.25}$$

The **SD** vector with the *minimal weight* is called a *minimal* **SD** *representation* with respect to given values of n and $\mathbf{Y}_v$. In the above example, there are two **SD** vectors that are minimal, namely, $(0\ 0\ \bar{1}\ \bar{1})_2$ and $(0\ \bar{1}\ 0\ 1)_2$. The minimal **SD** representation is of particular interest for multiplier recoding to be discussed in Chapter 5.

The conventional radix-r radix number can be easily converted to an equivalent **SD** form as follows:

Let $\mathbf{X} = (x_{n-1}, \ldots, x_1 x_0)_r$ be a conventional radix-r number and $\mathbf{Y} = (y_{n-1}, \ldots, y_1 y_0)_r$ be the equivalent **SD** number. Equivalence implies that they represent the same algebraic value. For every conventional digit x_i we generate an *interim difference* digit d_i by

$$d_i = x_i - r \cdot b_{i+1} \tag{1.26}$$

where the *borrow digit*,

$$b_{i+1} = \begin{cases} 0, & \text{if } x_i < \alpha \\ 1, & \text{if } x_i \geq \alpha \end{cases} \tag{1.27}$$

The ith **SD** digit, y_i, is then obtained by adding d_i and b_i

$$y_i = d_i + b_i \tag{1.28}$$

For example, given $\mathbf{X} = (0648)_{10}$ with $r = 10$, $n = 4$, and $\alpha = 6$ for the given **SD** set $\{\bar{6}, \ldots, \bar{1}, 0, 1, \ldots, 6\}$. We obtain $\mathbf{Y} = (1\ \bar{4}\ 5\ \bar{2})_{10}$ by going through the conversion sequence shown in Table 1.3. It is interesting to note that the **SD** generation can be started from either end; in fact there is no borrow propagation. All the signed digits are independently generated, as seen from the above example.

Table 1.3 Conversion of a Conventional Decimal Number to an Equivalent **SD** Form

Digit Position i	4	3	2	1	0	Comment
x_i		0	6	4	8	Conventional decimal
d_i		0	$\bar{4}$	4	$\bar{2}$	Interim difference
b_i	0	1	0	1		Borrow digit
y_i		1	$\bar{4}$	5	$\bar{2}$	SD form

The conversion of an **SD** number **Y** back to the conventional radix form **X** is done by adding two numbers, $\mathbf{Y}^+$ and $\mathbf{Y}^-$, formed from the positive digits and negative digits in **Y**, respectively. The above example shows that

$$\mathbf{Y} = \mathbf{Y}^+ + \mathbf{Y}^- = (1\ \bar{4}\ 5\ \bar{2})_{10}$$

where

$$\begin{array}{rl} \mathbf{Y}^+ & = +(1\ 0\ 5\ 0)_{10} \\ +)\ \mathbf{Y}^- & = -(0\ 4\ 0\ 2)_{10} \\ \hline \mathbf{X} & = \quad (0\ 6\ 4\ 8)_{10} \end{array}$$

Note that carry propagation exists in the above reverse conversion process because ordinary subtraction is performed.

1.6 Arithmetic Algorithms and Their Implementations

An *arithmetic algorithm* is a set of procedures which is sequenced or looped in a certain fashion to process legitimate input data and to produce meaningful results in prespecified data formats. The procedures must be machine executable in discrete time domain. No infinite loops are allowed, because algorithms are required to be self-halting at the end of computation.

Any arithmetic design project can be divided into two phases. The first is to develop *efficient algorithms* and the second is to develop their *logic implementations.* Both phases require extensive design efforts. Algorithm development is more challenging and creative, whereas the implementation requires more engineering background. However, it is fair to say that the two phases support each other. One cannot achieve an efficient design without either. Arithmetic algorithms can be classified into two general classes according to the data sizes.

(1) Digit Algorithms handle individual digits of the operands separately and generate the result on a local basis. The arithmetic rules governing the operation of an 1-bit full adder provide a typical example of digit algorithm. Usually, digit algorithms can be implemented with small- or moderate-size combinational or sequential logic circuits with limited inputs/outputs. In the case of nonbinary systems, each digit may be internally represented by a binary encoding. The complexity of digit algorithm does not necessarily increase with the radix value. The circuit complexity however increases rapidly, with higher radix systems. Table 1.4 lists some common digit algorithms.

(2) Vector Algorithms handle the entire operands as vector quantities. In a sense, one can organize the local digit algorithms to form a vector algorithm. One can view the digit algorithm as a microscopic level of processing, which is closer to the level of the logic design in terms of Boolean equations or lookup tables. The vector algorithms process numeric data at the macroscopic level in terms of vector-valued arithmetic equations. The result may appear in *digit-vector* form after normal execution of the standard or elementary arithmetic operation; it may also appear as a *condition-vector.* For example, a one-bit condition vector may be the result of algorithms for singularity detection such as overflow and zero detection, whereas a two-bit condition vector may be the outputs of a magnitude comparison, range estimate algorithms. Table 1.5 lists some typical vector algorithms with single or double operands.

Arithmetic algorithms may be implemented in modern digital computer by three approaches: *software*, *hardware*, and *firmware*, or a combination of these approaches. Early computers had only basic sets of hardware functions such as ADD/SUBTRACT, COMPLEMENT, and SHIFT. Other standard operations such as MULTIPLY and DIVIDE, and elementary functions such as LOGARITHM, SQUARE ROOT and so on, were implemented with software routines. Later, firmware emerged to replace the stored software routines by microprograms stored in the read-only memories. Now, almost all computers have some hardware implemented standard arithmetic functions and some elementary functions.

Table 1.4 Typical Arithmetic Digit Algorithms (Avizienis [3]).

Number System[a]	Name of Digit Algorithm	Arithmetic or Logic Specification[b]
CR, SD	Digit Transfer	$D_i \leftarrow S_i$
CR, SD	Shift right k digital positions	$D_{i-k} \leftarrow S_i$
CR, SD	Shift left k digital positions	$D_{i+k} \leftarrow S_i$
CR	$(r-1)$'s complement	$D_i \leftarrow (r-1) - S_i$
SD	Additive Inverse	$D_i \leftarrow -S_i$
CR	Carry	$C_{i+1} = \lfloor (X_i + Y_i + C_i)/r \rfloor$ or $C_{i+1} = X_i Y_i \vee Y_i C_i \vee X_i C_i$ (Binary)
CR	Sum	$S_i \leftarrow (X_i + Y_i + C_i) \text{Mod } r$ or $S_i = X_i \oplus Y_i \oplus C_i$(Binary)
SD	Carry (Transfer)	$t_{i+1} = \text{Sign}((X_i + Y_i) \times \lfloor (X_i + Y_i)/\alpha \rfloor)$
SD	Sum	$S_i \leftarrow X_i + Y_i + t_i - t_{i+1} \times r$
CR	Digit Product Residue	$P_{ij} = (X_i \times Y_j) \text{Mod } r$
CR	Digit Product Carry	$C_{ij} = \lfloor (X_i \times Y_j)/r \rfloor$
CR	Product-and-Sum	$Z_{ij} = ((X_i \times Y_j) + U_{ij} + V_{ij}) \text{Mod } r$

[a] CR means *Conventional Radix* number system and SD means *Signed-Digit* number system.

[b] $\lfloor x \rfloor$ = The largest integer $\leq \mathbf{x}$;

$$\text{Sign}(\mathbf{x}) = \begin{cases} 1 & \text{if } \mathbf{x} > 0, \\ 0 & \text{if } \mathbf{x} = 0, \\ -1 & \text{if } \mathbf{x} < 0, \end{cases}$$

and α is the maximum digit magnitude allowed in an SD number system.

The software approach is time-consuming and occupies large memory space for the routine statements, although it was cheaper in the early days of computer industry. With rapidly decreasing hardware cost, fewer computers use software for evaluating frequently used functions. The hardware approach provides faster execution speed and occupies minimum memory space, as little as one or two words for the instruction itself. Firmware offers a systematic method to sequence the micro-operations of an arithmetic algorithm. This control memory offers much faster execution speed than that of the software routines. Horizontal microprogramming using long control words exploits the maximum hardware parallelism, which contributes to the faster speed. In recent years, the mature development of microelectronics has led to a feasible utilization of large-scale, cellular, iterative, or pipelined logic arrays to implement arithmetic operations. This ultimate hardware approach may offer super computing speed.

Table 1.5 Typical Radix Arithmetic Vector Algorithms (Avizienis [3])

Vector Algorithm Name and Specification[a]	Fixed-Point Notation Used[b]	Subdivided Digit Algorithms for Specified Positions	Remarks
Direct transfer $D \leftarrow S$	RC, DC	$D_i \leftarrow S_i$ for $n-1 \geq i \geq 0$	$\lvert D \rvert = \lvert S \rvert = n$
Sign extension m positions $D \leftarrow \text{Ext}(S)$	RC,	$D_i \leftarrow S_{n-1} = \text{Sign}(S)$ for $n \leq i \leq n+m-1$	$\lvert D \rvert = n+m$
	DC	$D_i \leftarrow S_i$ for $n-1 \geq i \geq 0$	$\lvert S \rvert = n$
Arithmetic left shift m positions $D \leftarrow S \times r^m$	RC, DC	$D_{i+k} \leftarrow S_i$ for $n-1 \geq i \geq m$	Overflow may occur at the left end when significant digits are lost.
	RC	$D_i \leftarrow 0$ for $m-1 \geq i \geq 0$	
	DC	$D_i \leftarrow \begin{cases} 0, & \text{if } S > 0 \\ r-1, & \text{if } S < 0 \end{cases}$ for $m-1 \geq i \geq 0$	
Sum $D \leftarrow S + T$ or Difference $D \leftarrow S - T$	RC, DC	$D_i \leftarrow (S_i \pm T_i \pm C_i)\text{Mod } r$ for $n-1 \geq i \geq 0$	Addition overflow or subtraction underflow may occur.
	RC	$C_0 \leftarrow 0$ (initial carry or borrow)	
	DC	$C_0 \leftarrow C_n$ (end-around carry or borrow)	
Round off m positions (1) Truncation	RC, DC	$D_i \leftarrow S_i$ for $n-1 \geq i \geq m$ $D_i \leftarrow 0$ for $m-1 \geq i \geq 0$	The least significant m positions are chopped
(2) Local Adjustment Rounding	RC, DC	$D_i \leftarrow S_i$ for $n-1 \geq i \geq m+1$, $D_m \leftarrow S_m + \left\lfloor \frac{S_m}{r/2} \right\rfloor$, if $S_m \leq r-2$. $D_i \leftarrow 0$ for $m-1 \geq i \geq 0$	
(3) Sum Adjustment rounding	RC, DC	$D_i \leftarrow (S_i + C_i)\text{Mod } r$, for $n-1 \geq i \geq m+1$, where C_i is carry-in due to $D_{m+1} \leftarrow S_m + 1$ if $S_{m-1} = 1$, $D_i \leftarrow 0$ for $m-1 \geq i \geq 0$	Overflow may occur
Zero Test $F_z \leftarrow S = 0$	RC	$F_z \leftarrow 1$, if $S_i = 0 \ \forall 0 \leq i \leq n-1$	Clean Zero
	DC	$F_z = 1$, if either $S_i = 0$ for all $n-1 \geq i \geq 0$ or $S_i = r-1$ for all i	Both $+0$ and -0 are used

[a] $D = D_{n-1} \cdots D_1 D_0$ and $S = S_{n-1} \cdots S_1 S_0$.
[b] RC means Radix-Complement FXP notation.
DC means Diminished-Radix-Complement FXP notation.

We will describe both the algorithms and their implementation methods associated with various arithmetic operations. Vector algorithms will be illustrated by flow charts, and digit algorithms, if needed, by Boolean equations. The hardware realization of arithmetic algorithms utilizing MSI/LSI devices will be presented. The machine hosts are described in terms of functional block diagrams instead of numerous low-level logic design details.

1.7 Specification of Arithmetic Processors

Arithmetic processors are the sections of a digital computer that actually perform arithmetic duties. A general specification of digital arithmetic processors is given below. The formulation is described from the designer's viewpoint. The designer must consider the guiding arithmetic algorithms, the system architecture, and logic implementation of the processor. The designer must separate the arithmetic design from logic design in two sequential stages. This two-stage approach makes the algorithmic description separable and independent of the rapidly changing device technology.

Figure 1.2 depicts a digital arithmetic processor which can be described by a 13-tuple

$$\mathbf{A} = \langle \mathbf{I}, \mathbf{R}, \mathbf{O}, \mathbf{F}, \mathbf{C}, \mathbf{S}, \mathbf{T}, \boldsymbol{\Delta}, \mathbf{f}, \mathbf{g}, \mathbf{h}, \mathbf{p}, \mathbf{q} \rangle \tag{1.29}$$

where

1. $\mathbf{I} = \{I_1, I_2, \ldots, I_k\}$ is a set of *input operands*, with $k = 1$ for single-operand operation and $k = 2$ for double-operand operation, and so on. Each operand I_i is a digital vector with a prespecified number representation satisfying the following machine properties:

   1.a Each $I_i \in \mathbf{M}$, where

   $$\mathbf{M} = \{m \mid \mathbf{m}_{\min} \leq m \leq \mathbf{m}_{\max}\} \tag{1.30}$$

   is a finite set of finite-precision machine representable numbers.

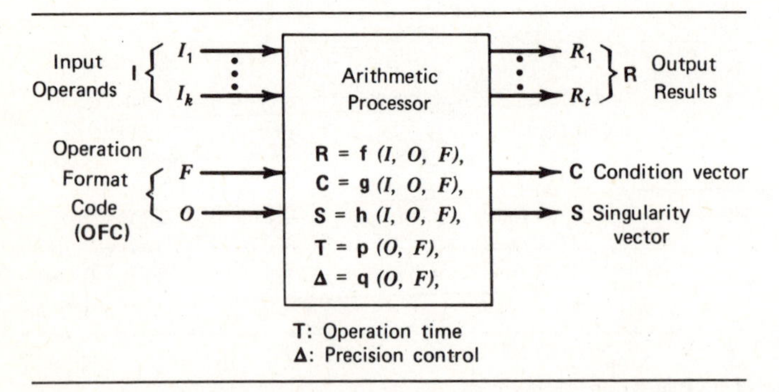

Figure 1.2 The general specification of a digital arithmetic processor.

1.b Each I_i is within a known precision

$$I_i - \Delta_L \leq \mathbf{I}_i \leq I_i + \Delta_H \tag{1.31}$$

where $\Delta_L, \Delta_H \in \mathbf{\Delta}$ will be defined in part (7) below.

2. $\mathbf{R} = \{R_1, R_2, \ldots, R_t\}$ is a set of *resultant output digital vectors* satisfying (1.a) and (1.b) but allowed to have different prespecified representation.

3. The *operation-format code* (**OFC**) is a vector consisting of a number of operation-format pairs (**O**, **F**)

$$\mathbf{OFC} = (O_1, F_1; O_2, F_2; ,,, ; O_r, F_r) \tag{1.32}$$

where $O_i \in \mathbf{O}$, is the set of allowable *operators*, and $F_i \in \mathbf{F}$, is the set of allowable data *formats*.

4. **C** is the *condition-vector* and **S** is the *singularity-vector* identifying specific conditions associated with the result (such as the *sign* of the result, and so on) or indicating the singularity conditions when legitimate results are not generated. Singularities may include *overflow*, *underflow*, and so forth.

5. **f** is a vector-valued mapping relating the inputs and outputs

$$\mathbf{f}: \mathbf{I} \times \mathbf{O} \times \mathbf{F} \rightarrow \mathbf{R}; \tag{1.33}$$

g and **h** are vector-valued mappings relating inputs and condition and singularity vectors respectively

$$\begin{aligned} \mathbf{g}: \mathbf{I} \times \mathbf{O} \times \mathbf{F} \rightarrow \mathbf{C}; \\ \mathbf{h}: \mathbf{I} \times \mathbf{O} \times \mathbf{F} \rightarrow \mathbf{S} \end{aligned} \tag{1.34}$$

6. **p** is a time function which determines the expected execution time of the OFC (**T** is the time domain).

$$\mathbf{p}: \mathbf{O} \times \mathbf{F} \rightarrow \mathbf{T} \tag{1.35}$$

7. $\mathbf{\Delta}$ is the set of precision control parameters and **q** specifies the precision-control mechanism

$$\mathbf{q}: \mathbf{O} \times \mathbf{F} \rightarrow \mathbf{\Delta} \tag{1.36}$$

The preceding mathematical formulation of a digital arithmetic processor is realizable through the two-stage arithmetic design process described earlier. The details include the development of effective vector algorithms, the host machine system architecture, arithmetic processor specification in terms of the above mappings, and the digit-level algorithmic logic design. After an effective algorithm is found, the major undertaking of an arithmetic designer is to establish these mappings in a form which can be easily translated into logic design tasks.

1.8 Fixed-Point Number Representations

A signed radix number must be either positive or negative but not both. Usually, the leftmost digit of an n-digit positional number is reserved as the sign indicator. Consider the following number with radix r

$$\mathscr{A} = (a_{n-1}a_{n-2} \ldots a_1 a_0)_r \tag{1.37}$$

where the sign digit a_{n-1} assumes the value

$$a_{n-1} = \begin{cases} 0, & \text{if } \mathscr{A} \geq 0 \\ r-1, & \text{if } \mathscr{A} < 0 \end{cases} \tag{1.38}$$

The remaining digits in $\mathscr{A}$ specify either the *true magnitude* or one of two *complemented magnitudes of* $\mathscr{A}$. Because $\mathscr{A}$ is a positional number, we should place a radix point in it to distinguish the integer portion from the fraction. If a fixed position is chosen for the radix point, we call it a *fixed-point (FXP) number*. Theoretically, the radix point could be located between any two adjacent digits in the given representation.

There are two commonly accepted conventions of FXP number representations. The two conventions are interchangable. Every n-digit integer can be considered as a fraction multiplied by a constant factor r^n, and every k-digit fraction can be considered as an integer multiplied by a constant factor r^{-k}, where r is the fixed radix. It is rather easy to convert between integers and fractions.

Therefore, the two most commonly chosen positions for the radix point are either at the left extreme or at the right extreme of the magnitude portion of the number. In the former case, the radix point lies between the sign digit a_{n-1} and the most significant magnitude digit a_{n-2}. This makes all FXP numbers fractions strictly less than one. In the latter case, the radix point is located at the right of the least significant digit a_o, which renders all FXP integers. The two conventions are essentially equivalent. One can easily convert one to the other either by multiplying or dividing by a constant factor r^{n-1}. In most of the discussions, we assume the integer convention unless otherwise noted. The fixed-radix point is self-implied; there is no need to mark it explicitly.

For *positive* FXP numbers, the sign digit $a_{n-1} = 0$ and the remaining digits $a_{n-2}, \ldots, a_1, a_o$ always display the *true magnitude*. We denote the true or absolute magnitude of a number $\mathscr{A}$ as

$$|\mathscr{A}| = (m_{n-2}m_{n-3} \cdots m_1 m_0)_r \tag{1.39}$$

where $m_i = a_i$ for $i = n-2, \ldots, 1, 0$ when $a_{n-1} = 0$ or $\mathscr{A} \geq 0$. To sum up, we represent a *positive* FXP number as

$$\begin{aligned} \mathscr{A} &= (0a_{n-2} \cdots a_1 a_0)_r \\ &= (0m_{n-2} \cdots m_1 m_0)_r \end{aligned} \tag{1.40}$$

This means that the magnitude equals

$$|\mathscr{A}| = \sum_{i=0}^{n-1} a_i \cdot r^i = \sum_{i=0}^{n-2} m_i \cdot r^i \tag{1.41}$$

There are three different notations for representing a *negative* FXP number. Let $\bar{\mathscr{A}}$ be the negative version of a positive number $\mathscr{A}$ as defined in Eq. 1.40. Then $\bar{\mathscr{A}}$ has a sign digit with value $r - 1$. Three distinct FXP representations of negative numbers are given below:

Sign-Magnitude Representation

$$\bar{\mathscr{A}} = ((r - 1)m_{n-2} \cdots m_1 m_0)_r \tag{1.42}$$

where m_i for $n - 2 \geq i \geq 0$ are the true magnitude digits. The positive version $\mathscr{A}$ differs from the negative version $\bar{\mathscr{A}}$ only in the sign digit in this notation.

Diminished-Radix Complement Representation

$$\bar{\mathscr{A}} = ((r - 1)\bar{m}_{n-2} \cdots \bar{m}_1 \bar{m}_0)_r \tag{1.43}$$

where $\bar{m}_i = (r - 1) - m_i$ for $n - 2 \geq i \geq 0$. In this notation, $\bar{\mathscr{A}} = r^n - 1 - \mathscr{A}$. It is also called the $(r - 1)$'s *complement representation.*

Radix Complement Representation

$$\bar{\mathscr{A}} = (((r - 1)\bar{m}_{n-2} \cdots \bar{m}_1 \bar{m}_0) + 1)_r \tag{1.44}$$

In this notation, $\bar{\mathscr{A}} = r^n - \mathscr{A}$. It is also called *r's complement representation.*

The above representations are summarized in Table 1.6 for two special radix systems, the binary ($r = 2$) and the decimal ($r = 10$) numbers. A practical example is

Table 1.6 Fixed-Point Representations of Negative Binary or Decimal Numbers

Representation System	**Binary** ($r = 2$)	**Decimal** ($r = 10$)
Sign-magnitude	$(1m_{n-2} \cdots m_1 m_0)_2$	$(9m_{n-2} \cdots m_1 m_0)_{10}$
Diminished radix complement	1's complement $(1\bar{m}_{n-2} \cdots \bar{m}_1 \bar{m}_0)_2$	9's complement $(9\bar{m}_{n-2} \cdots \bar{m}_1 \bar{m}_0)_{10}$
Radix complement	2's complement $(1\bar{m}_{n-2} \cdots \bar{m}_1 \bar{m}_0 + 1)_2$	10's complement $(9\bar{m}_{n-2} \cdots \bar{m}_1 \bar{m}_0 + 1)_{10}$

used to clarify the various binary notations as shown in Table 1.7. We store in a 16-bit minicomputer the positive and negative versions of a signed FXP number with true magnitude $\mathbf{A} = (547)_{10} = (1000100011)_2$. Note that in order to fill up the full 16-bit word, leading *zeros* are filled in for positive numbers in all notations and

Table 1.7 Binary Fixed-Point Representation for ±547

FXP Representation	Binary Number +547	Binary Number −547
Sign-magnitude	0000001000100011	1000001000100011
One's complement	0000001000100011	1111110111011100
Two's complement	0000001000100011	1111110111011101

for negative numbers in sign-magnitude notation. On the other hand, leading *ones* are filled in for either one's complement or two's complement negative numbers. We call this a *sign extension* for the sign-complement number system.

The range of integers associated with each of the above three FXP representation systems is determined by the word length n. In all three notations, the upper bound is determined by the largest positive integer fitting an n-bit word including the sign. The binary pattern of this upper bound is $(011 \cdots 1)_2$, which is equal to $2^{n-1} - 1$ in decimal. For sign-magnitude and one's complement numbers, the lower bounds are $(111 \cdots 1)_2 = -(2^{n-1} - 1)_{10}$ and $(100 \cdots 0)_2 = -(2^{n-1} - 1)_{10}$, respectively. In two's complement arithmetic, the most negative number is $-2^{n-1} = (100 \cdots 0)_2$. *Overflow* occurs when a positive number exceeds the upper bound. Similarly, *underflow* occurs when a negative number exceeds the lower bound. The range of binary integers for an n-bit machine are summarized in Table 1.8.

All FXP numbers have unique representations except the *zero* in sign-magnitude or in one's complement notations. Positive and negative zeros have different representations in these two number representations. For two's complement arithmetic, there is a unique zero representation $000 \cdots 0$. We call this unique zero vector a *clean zero* in contrast to the *dirty zero* for the other two FXP number representations.

Table 1.8 Integer Ranges and Zero Representations in Fixed-Point Binary Arithmetic System Having Word Length of n Bits.

Representation System	Integer range	Zero Representations: Positive	Zero Representations: Negative
Sign-magnitude	$-(2^{n-1} - 1) \leq \mathbf{A} \leq 2^{n-1} - 1$ $111 \ldots 1 \leq \mathbf{A} \leq 011 \ldots 1$	$000 \ldots 0$	$100 \ldots 0$
One's complement	$-(2^{n-1} - 1) \leq \mathbf{A} \leq 2^{n-1} - 1$ $100 \ldots 0 \leq \mathbf{A} \leq 011 \ldots 1$	$000 \ldots 0$	$111 \ldots 1$
Two's complement	$-2^{n-1} \leq \mathbf{A} \leq 2^{n-1} - 1$ $100 \ldots 0 \leq \mathbf{A} \leq 011 \ldots 1$	$000 \ldots 0$	$000 \ldots 0$

1.9 Floating-Point Number Representations

The fixed-point notations are convenient for representing small radix numbers with bounded orders of magnitude. Consider a binary computer with 32-bit words. The range of fixed-point integers which this computer can handle is restricted to $\pm(2^{31} - 1)$, which equals roughly $\pm 10^{11}$. This range may be inadequate in engineering and scientific applications. In order to represent numbers in a much wider range, we employ a two-part representation

$$\mathbf{f} = (m, e) \tag{1.45}$$

to express a real number

$$\mathbf{f} = m \times r^e \tag{1.46}$$

where m and e are each a signed, fixed-point number and r is a given *radix* (*base*). One possible representation format of the 32-bit operand is to reserve the 22 leftmost bits for the m field and the remaining 10 bits for the e field with an implied base $r = 2$.

The two components m and e are called the *mantissa* and *exponent* of the number **f**, respectively. In general, the mantissa m may assume any one of the three fixed-point notations as described in previous section. For example, one can consider the following convention: m is a fraction in sign-magnitude notation and the exponent e is a two's complement integer with either *unbiased* or *biased* value (to be described shortly). The radix r of the number **f** does not appear in the representation (m, e) because it is self-implied.

It is interesting to observe that the radix point in mantissa m can float around by adjusting the magnitude of the exponent e. For this reason, we call the notation (m, e) a *floating-point representation* of the number f, which itself is called a *floating-point* (*FLP*) *number*.

Consider a numerical example in a decimal FLP system

$$\mathbf{f} = -0.000005078125 \times 10^{+3} \tag{1.47}$$

It can be represented by the following two FLP codes

$$\begin{aligned} \mathbf{f}_1 &= (-0.000005078125, +3) \\ \mathbf{f}_2 &= (-0.507812500000, -2) \end{aligned} \tag{1.48}$$

One can shift the mantissa k places to the left (right) and simultaneously decrement (increment) the value of the exponent by k without changing the real value of the number being represented. In the above example, we have shifted the mantissa to the left 5 digits and decremented the exponent by 5 accordingly. The shifting of the mantissa and scaling of the exponent are frequently performed in normalized FLP arithmetic operations.

An FLP number is said to be *normalized* if the most significant position of the mantissa contains a nonzero digit such as that shown in Eq. 1.48. Normalizing an FLP number requires shifting the mantissa to the left, pushing off the redundant leading 0's in the more significant digits, and decreasing the exponent accordingly until a nonzero digit appears in the most significant position. In normalized arithmetic, all the FLP numbers must be prenormalized before they can be manipulated. Therefore, after every intermediate computation step, one must apply postnormalization procedures to ensure the integrity of the normalized form.

Normalized mantissa m may assume absolute magnitude within the range

$$\frac{1}{r} \leq |m| < 1 \tag{1.49}$$

where r is the implied radix. For binary numbers ($r = 2$), this refers to the range $0.5 \leq |m| < 1$. In other words, all the normalized nonzero binary numbers should have a mantissa of no less than one-half. The representation of an FLP zero is the only exception.

The exponent e can be either a positive or a negative integer. One has to compare and equalize the exponents of two FLP numbers before they can be added or subtracted. We wish to simplify the comparison operation without involving the signs of the exponents. One way to bypass the sign is to convert all the exponents to positive integers by adding a positive constant to each exponent, forming a so-called *biased-exponent*. A popular choice of this *bias constant* has magnitude equal to that of the most negative exponent. Consider an FLP representation with a q-bit exponent field. An unbiased exponent in two's complement notation may assume values within the range

$$-2^{q-1} \leq e_{\text{unbiased}} \leq 2^{q-1} - 1 \tag{1.50}$$

We may choose the bias constant $\mathbf{b} = 2^{q-1}$ to convert all the exponents to positive integers in the following range

$$0 \leq e_{\text{biased}} \leq 2^q - 1 \tag{1.51}$$

One should be aware of the fact that the true exponent (unbiased exponent) is retrievable from the biased exponent by applying the following equation:

$$e_{\text{unbiased}} = e_{\text{biased}} - 2^{q-1} \tag{1.52}$$

Consider the storage of a normalized FLP number $\mathbf{f} = -0.5078125 \times 2^{-2}$ in a digital computer with 32 bits per word. The following data format is assumed.

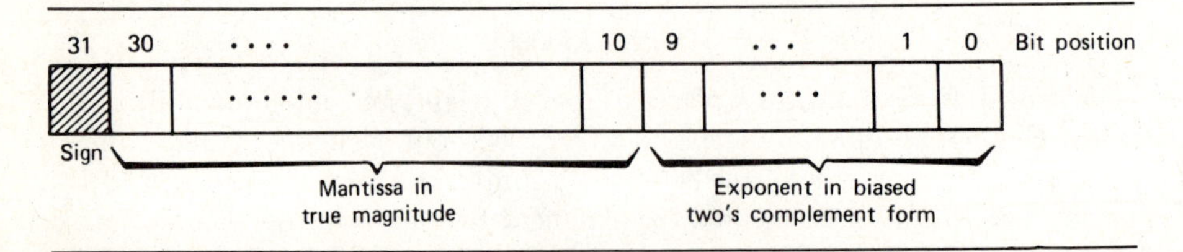

Bit 31 is reserved as the sign indicator, "0" for positive and "1" for negative numbers. The implied binary point is located between bit 31 and bit 30. Because $m = (-0.5078125)_{10} = (1100000100000000000000)_2$ in sign-magnitude form and $e_{\text{biased}} = (-2 + 512)_{10} = (510)_{10} = (0111111110)_2$, the complete FLP representation of this number is obtained as

110000010000000000000	0111111110
↑ Sign / Normalized Mantissa	Biased Exponent

The bias constant used is $2^{q-1} = 2^9 = (512)_{10} = (1000000000)_2$. Three particular values of the exponent are listed below for comparison purpose.

e_{unbiased}	$(-512)_{10} =$ $(1000000000)_2$	Zero = $(0000000000)_2$	$(+511)_{10} =$ $(0111111111)_2$
e_{biased}	$(0000000000)_2$	$(1000000000)_2$	$(1111111111)_2$

Theoretically, there may exist infinitely many zero representations in FLP notation, provided the mantissa equals zero regardless of the value of the exponent. In computer design, we wish to have a unique zero representation for both FXP and FLP numbers. A "clean" FXP zero is composed of a sequence of 0's. We can also define such a "clean zero" for FLP arithmetic by using the biased exponent. What we must do is assigning a zero mantissa and the most negative exponent in biased form to represent an FLP zero. The "dirty" zeros containing nonzero exponent are not considered legal representation. With the biased exponent we can avoid this "dirty zero" turmoil.

It is the mantissa length that determines the *precision* of an FLP number. Means of increasing the precision are to be discussed next. It is the exponent e that displays the order of magnitude of FLP number. In the above example, the exponents lie within $(-512, +511)$. We can, therefore, express positive normalized FLP numbers $\mathbf{F}^+$ within the range

$$0.5 \times 2^{-512} \leq \mathbf{F}^+ \leq (1 - 2^{-21}) \times 2^{+511} \qquad \textbf{(1.53)}$$

and negative normalized FLP number $\mathbf{F}^-$ within the range

$$-(1 - 2^{-21}) \times 2^{+511} \leq \mathbf{F}^- \leq -0.5 \times 2^{-512} \qquad \textbf{(1.54)}$$

The FLP ranges shown in the above inequalities are depicted in Fig. 1.3. There is a gap around the zero (origin) due to the normalization of the mantissas. Normalized

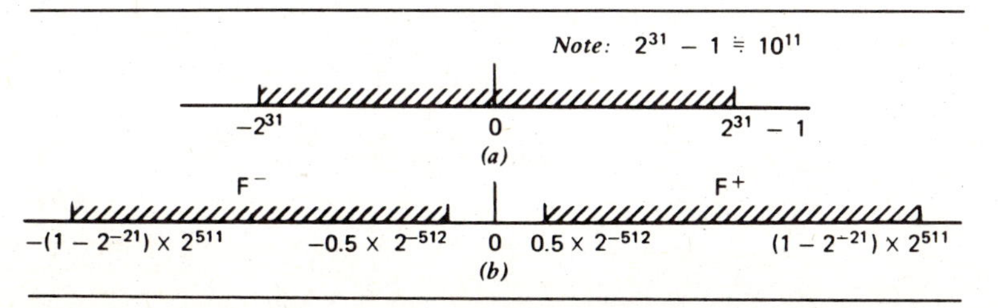

Figure 1.3 Ranges of 32-bit numbers in FXP and FLP representations. (*a*) Allowable range of 32-bit FXP numbers (integers) in two's complement representation. (b) Allowable range of 32-bit normalized FLP numbers with 21-bit mantissa and 10-bit exponent of base 2.

FLP arithmetic for general radix $r \geq 2$ and advanced topics on FLP arithmetic design will be studied in Chapters 9 and 10, respectively.

1.10 Multiple-Precision Arithmetic

Digital numbers presented in previous sections have *fixed-length* and *single-precision* format, in which each number occupies a single memory word. In some applications, *variable-length* or *multiple-precision* formats are needed to increase the number of significant digits in the number for better accuracy or wider operating range. In order to increase the number of significant digits contained in the mantissa, one can use two or more words to represent one floating-point number. For an example, one can use two 32-bit memory words for each floating-point number. The accuracy can be increased to $22 + 32 = 54$ bits for the mantissa if we maintain the remaining 10 bits in the first word as the exponent. Alternatively, we can use 48 bits for the mantissa and 16 bits for the exponent. There exists a tradeoff between these two field lengths.

Several standard arithmetic operations requiring multiple-precision data formats are described below:

1. After *adding* or *subtracting* two n-bit single-precision numbers, one may expect a single-precision *sum* and a 1-bit *carry*. This carry-out needs a special flip flop to hold it for subsequent computations.
2. After *multiplying* two n-bit binary numbers, one may expect a $2n$-bit *product* (or somtimes a $(2n - 1)$-bit product). If only a single-precision number is allowed, one has to round off the lower half of the product, which causes a rounding error. To maintain the full precision, a double-length product will be generated.
3. After *dividing* a double-length ($2n$ bits) integer by a single-length (n bits) integer, one may expect a single-length quotient and a single-length remainder. If the pair of results (quotient; remainder) is anticipated, a double-precision word is needed to hold it for subsequent use.

When multiple precision numbers are expected in a computation, provision has to be made to indicate whether the given number has single-, double-, or triple-precision, and so on. Several commonly used methods are described below:

1. Use two special symbols as the *end markers* to identify the beginning and the ending of a digital number.
2. Attach a separate word as the *length indicator* of the variable-length number. Usually, this length indicator shows the number of memory words needed for the multiple-precision number.
3. The use of an exponent to convert an FXP number into an FLP number with a *significance tag vector* or a *guard digit*.

Double-precision computations are almost always required for FLP arithmetic, except in statistical work where FXP double-precision arithmetic is more beneficial for calculating the sum of squares and cross products. FXP multiple-precision computation is much simpler than its FLP counterpart. There are also other means of controlling the precision of digital numbers, such as using rounding schemes or using guard registers to be discussed in later chapters.

1.11 Bibliographic Notes

This first chapter gives an overview of the spectrum of computer arithmetic, including theory, design, and application problems. Excellent surveys of number systems and arithmetic design methods have been given by Garner [5, 6] and Tung [18]. Matula [11, 13] has provided comprehensive number-theoretic treatments of radix arithmetic and its interactions with other disciplines of computer engineering. Knuth [8] presented a thorough historical account of machine arithmetic, especially on floating-point computations. Kulisch [9] has given an axiomatic treatment of the mathematical foundations of computer arithmetic, which summarizes number spaces, rounding schemes, and arithmetic structures in computers. The redundant signed-digit number systems were originated from the work of Robertson [14] and Avizienis [1, 2]. The material presented in section 1.5 is a relaxed version of the signed-digit number system defined in [2]. It includes the radix-2 redundant numbers as well, on which the theory of SRT division (Chapter 7) is based.

Residue arithmetic systems are discussed in detail in Szabo and Tanaka [15]. Rational number systems were studied by Matula [12] and Hwang and Chang [7] Logarithmic number systems were treated in depth in Marasa's thesis [10]. Classification of arithmetic operations may be compared with those found in existing arithmetic machines, such as the IBM 7030 (Stretch Project) in Buchholz [4], IBM System/360 in [6], and the CDC 6600 in Thornton [17]. The classification of arithmetic algorithms and processor specification are extended from the work of Avizienis [3]. Case studies of several fixed-point, floating-point, and multiple-precision data representations can be found in references [4, 6, 8, 17].

A series of four international triennial **Symposia on Computer Arithmetic** was held by the IEEE Computer Society in 1969, 1972, 1975, and 1978. The **Transactions on Computers** of the **Institute of Electrical Electronics Engineers** has thus far published three special issues on *Computer Arithmetic*, including the key papers presented at the first three Symposia. These special issues of IEEETC appeared in August 1970, June 1973, and July 1977, respectively. Recent results reported in these special issues are selectively included in this text. However, the readers are urged to study the original presentations for additional details.

References

[1] Avizienis, A., "A Study of Redundant Number Representations for Parallel Digital Computers," *Ph.D. dissertation*, University of Illinois, Digital Computer Laboratory, May 1960.

[2] Avizienis, A., "Signed-Digit Number Representations for Fast Parallel Arithmetic," *IRE Trans. on Elec. Computers*, Vol. EC-10, September 1961, pp. 389–400.

[3] Avizienis, A., "Digital Computer Arithmetic: A Unified Algorithmic Specification," *Proc. Symposium on Computers and Automata*, Polytechnic Institute of Brooklyn, 1971, pp. 509–525.

[4] Buchholz, W. (ed.), *Planning A Computer System: Project Stretch*, McGraw-Hill, New York, 1961.

[5] Garner, H. L., "Number Systems and Arithmetic," in *Advances in Computers*, Vol. 6, Academic Press, New York, 1965, pp. 131–194.

[6] Garner, H. L., "A Survey of Recent Contributions to Computer Arithmetic," *IEEE Trans. Comp.*, Vol. C-25, No. 12, December 1976, pp. 1277–1282.

[7] Hwang, K. and Chang, T. P., "A New Interleaved Rational/Radix Number System for High-Precision Arithmetic Computations", *Proc. of the Fourth Symposium on Computer Arithmetic*, October 1978.

[8] Knuth, D. E., *The Art of Computer Programming*, Vol. 2, *Seminumerical Algorithms*, Addison-Wesley, Reading, Mass. 1969, Chap. 4.

[9] Kulisch, U., "Mathematical Foundations of Computer Arithmetic," *IEEE Trans. Computers*, Vol. C-26, No. 7, July 1977, pp. 610–620.

[10] Marasa, J. D., "Accumulated Arithmetic Error in Floating-Point and Alternative Logarithmic Number Systems," *M.S. Thesis*, Seven Institute Technology Washington University, St. Louis, Mo., June 1970.

[11] Matula, D. W., "Number Theoretic Foundations for Finite-Precision Arithmetic," in *Applications of Numbers Theory to Numerical Analysis*, W. Zaremba (ed.), Academic Press, New York, 1972, pp. 479–489.

[12] Matula, D. W., "Fixed-Slash and Floating-Slash Rational Arithmetic," *Proc. of the Third Symposium on Computer Arithmetic*, IEEE Catalog No. 75 CH1017-3C, November 1975, pp. 90–91.

[13] Matula, D. W., "Radix Arithmetic: Digital Algorithms for Computer Architecture," in *Applied Computation Theory: Analysis, Design, Modeling*, R. T. Yeh (ed.), Prentice-Hall, Englewood Cliffs, N.J., 1976, Chap. 9.

[14] Robertson, J. E., "Redundant Number Systems for Digital Computer Arithmetic," *Notes for the Univ. of Michigan Engineering Summer Conference*, in "Topics in the Design of Digital Computing Machines," Ann Arbor, Mich., July 6–10, 1959.

[15] Szabo, N. S. and Tanaka, R. I., *Residue Arithmetic and Its Applications to Computer Technology*, McGraw-Hill, New York, 1967.

[16] Svoboda, A., *Digitale Informationswandler*, Braunschweig, Germany, Vieweg and Sohn, 1960.

[17] Thornton, J. E., *Design of A Computer The Control Data 6600*, Scott, Foresman, Glenview, Illinois., 1970, Chap. V.

[18] Tung, C., "Arithmetic," in *Computer Science* (A. F. Cardenas et al., ed.) Wiley-Interscience, 1972, Chap. 3.

Problems

Prob. 1.1 Distinguish the following pairs of terminologies in relation to digital computer arithmetic systems:

(a) Real arithmetic versus machine arithmetic.
(b) Digit algorithm versus vector algorithm.
(c) Fixed-radix versus mixed-radix number systems.
(d) Redundant versus nonredundant number representations.
(e) Fixed-point versus floating-point number representations.
(f) Integer versus fraction fixed-point arithmetics.
(g) Single-precision versus double-precision arithmetic operations.
(h) Normalized versus unnormalized floating-point operations.
(i) Diminished radix complement versus radix complement number systems.
(j) Biased versus unbiased exponents in *FLP* representations.
(k) Dirty versus clean zero in internal number representations.

Prob. 1.2 Given two algebraic values of $(39.25)_{10}$ and $(-39.25)_{10}$. Find the fixed-point representations in each of the following number systems with radix r. Each number consists of n integer digits and k fraction digits.

(a) Radix complement system with $(r, n, k) = (8, 4, 2)$.
(b) Diminished radix complement system with $(r, n, k) = (2, 8, 4)$.
(c) Show one possible signed-digit representation with $(r, n, k) = (4, 8, 8)$ over digit set $\{\bar{2}, \bar{1}, 0, 1, 2\}$. At least one negative digit must appear in the representation.

Prob. 1.3 Given an algebraic value $(-14)_{10}$, the word length $n = 6$ and a digit set $\{\bar{1}, 0, 1\}$. Find all the possible 6-digit signed-digit representations that assumes the value -14. Indicate the minimal sign-digit representation among all possible **SD** representations. Is this minimal representation unique?

Prob. 1.4 Convert the conventional octal number $(376)_8$ into an equivalent signed-digit number with $r = 8$, $n = 6$, and $\alpha = 5$ for the digit set

$$\{-5, -4, -3, -2, -1, 0, 1, 2, 3, 4, 5\},$$

where α is the maximum digit magnitude.

Prob. 1.5 Compare the advantages and disadvantages of implementing arithmetic algorithms with hardware, firmware, and software. Comment on design efforts, computing speed, and hardware requirements.

Prob. 1.6 Find the normalized internal machine representations of the following floating-point numbers in a 32-bit FLP arithmetic processor with the same FLP data format as given in section 1.9.

(a) -1023.75×2^{-7} (b) $+0.0025 \times 2^{+501}$
(c) $+932.875 \times 2^{+327}$ (d) $-0.007813 \times 2^{-490}$

Prob. 1.7 A certain computer has the following 36-bit normalized floating-point data format with both mantissa and exponent in two's complement form. Determine

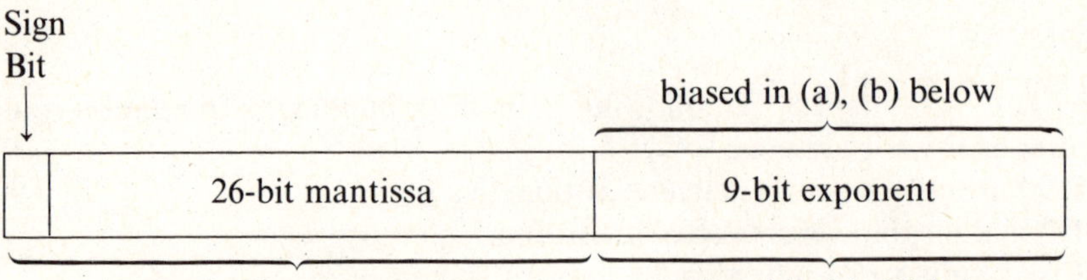

the bit patterns of the following extreme *FLP* numbers found in this 36-bit computer. Binary base ($r = 2$) is assumed for the exponent
(a) The largest positive number with biased exponent.
(b) The most negative number with biased exponent.
(c) Repeat (a) except with unbiased exponent field.
(d) Repeat (b) except with unbiased exponent field.

Prob. 1.8 Let **A** be a radix-r n-digit fixed-point integer. Let $\bar{\mathbf{A}}$ be obtained from **A** by digitwise complementing **A** with respect to the value $r - 1$.
(a) Prove that $\mathbf{A} + \bar{\mathbf{A}} = r^n - 1$, if **A** is in diminished-radix complement form.
(b) Prove that $\mathbf{A} + \bar{\mathbf{A}} + 1 = r^n$, if **A** is in radix-complement form.

Note that in part (a) $\bar{\mathbf{A}} = \bar{\mathscr{A}}$ as given in Eq. 1.43 and in part (b) $\bar{\mathbf{A}} + 1 = \bar{\mathscr{A}}$ as given in Eq. 1.44.

Prob. 1.9 Show how to normalize a *FLP* number if its mantissa is represented in two's complement form (instead of sign-magnitude representation given in the text). Repeat the question for the case of one's complement mantissa. The Boolean equations for either normalized conditions must be derived in terms of the mantissa bits and sign bit.

Prob. 1.10 Matula [12] has proposed a *fixed-slash rational number system.* The precision limitation of this system is imposed by separately bounding the numerator and the denominator size. Both the numerator and the denominator terms may be stored in fixed-length fields in a standard radix format as shown below. For simplicity, assume unsigned integers for both the numerator and the denominator.

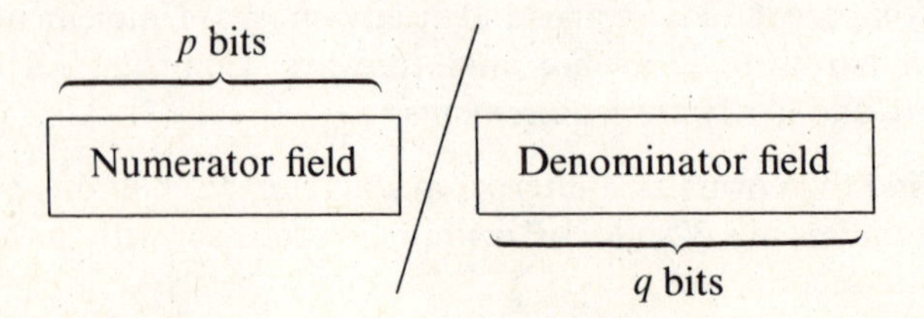

Answer the following questions associated with fixed-slash rational number system:

(a) What number precision range can be provided by this rational number representation system?

(b) Suppose that the sum $p + q = n$ is a constant (fixed size) and the implicit position of the slash can vary. Comment on the change of precision ranges when the slash is allowed to float to the right k places such that $1 \leq k \leq q - 1$.

(c) Repeat the same question for floating the slash to the left k places such that $1 \leq k \leq p - 1$.

In each of the above questions, show the number ranges and the smallest gaps between adjacent rational numbers within the ranges of interest.

Chapter 2

Integrated Circuits and Digital Devices

2.1 Electronic Technologies and Logic Families

Electronic components used in digital computers have experienced four generations of development. The first generation of digital computers (prior to 1952) utilized vacuum tubes as the basic circuit components. The second generation of computer electronics was marked (1958) by the use of discrete transistor and diode semiconductor components, which are relatively 20 times smaller in volume, 100 times faster in speed, and 100 times lesser failure rate than their vacuum tube predecessors. The third generation (1963) has been characterized by *Integrated Circuits* (ICs) in the forms of *Small Scale Integration* (SSI) and *Medium Scale Integration* (MSI) chips. Each SSI/MSI chip may contain up to 100 circuit components with an operating speed measured in terms of nanoseconds. The SSI/MSI devices require much reduced power consumption and packaged cost and even results in higher reliability. Since 1969, the *Large Scale Integration* (LSI) devices have become more and more popular, signaling the beginning of the fourth generation of electronic computers. An LSI functional device may contain over 1000 circuit components in one monolithic chip with further reduced cost, power, and failure rate and higher density and speed (over 2500 times faster than vacuum tubes).

In this text, we are considering digital designs with MSI/LSI component devices. A brief introduction to commonly used digital IC and memories devices is given in this chapter. The introduction is not restricted to standard MSI/LSI circuits. The trend of custom designed standard circuit functions is also presented. Schematic functional block diagrams and terminal characteristics of major categories of arithmetic IC building blocks are described. The description tends to be independent of the fabrication technologies, which are changing so rapidly. It is the functional behavior of these devices that concerns us most. Knowledge of these MSI/LSI devices is necessary to understand the design and hardware organization of various arithmetic processors in subsequent chapters. Apparently, a brief introduction will not cover all of the standard end-products available in the electronic market. To supplement this, we provide a general literature guide to help interested readers locate the major IC manuals, data books, and application notes.

Various solid-state electronic technologies and the corresponding logic families are briefly characterized below. In general, there are two classes of digital electronic devices that are frequently used in modern computer construction: the MOS and bipolar devices as listed below.

A. Metal-Oxide-Semiconductor (MOS) Technologies

1. The *PMOS* which uses *p*-channel MOS circuitry.
2. The *NMOS* which employs *n*-channel silicon gate processing.
3. The *CMOS* which combines both *p*-channel and *n*-channel transistors on the same substrate to form the *complementary* MOS.
4. The *CCD's*, Charge-Coupled Devices, which use the presence or absence of minority-carrier charges in the potential wells created at the silicon surface represent ones and zeros.
5. The *VMOS*, a V-groove MOS, which is capable of producing MOS transistors with V-shaped notched channels. The short channels result in high breakdown voltage and high packing density.

B. Bipolar Technologies and Logic Families

1. The TTL circuits, which use the multiple emitter *Transistor-Transistor Logic* circuitry.
2. The ECL circuits, which are made from nonsaturated *Emitter-Coupled Logic* circuitry.
3. The I^2L circuits, which employ the bipolar *Integrated Injection Logic* circuitry.
4. The *Schottky TTL* and *Schottky* I^2L are TTL and I^2L circuits with Schottky diodes attached in the multiple emitters and multiple collectors, respectively.

Both logic gate functions and memory arrays have been fabricated with the above technologies. The general trend in future technology development is to achieve higher switching speed, higher density (memory capacity), lower power dissipation, and lower cost requirements. In order to compare the speed of various arithmetic unit designs, we define Δ as the unit gate delay corresponding to one level of logic circuit. A good measure of the value of Δ would be the propagation delay of a NAND gate or of a NOR gate. Because every switching function can be implemented exclusively

Table 2.1 Logic Symbols and Time Delays of Typical Gate Functions

Gate Function	Logic Symbol (Positive logic)	Time Delay in Terms of Multiple of Δ
NAND	A, B → $\overline{A \cdot B}$	Δ
NOR	A, B → $\overline{A + B}$	Δ
NOT	A → $\overline{A}$	Δ
AND	A, B → $A \cdot B$	2Δ
OR	A, B → $A + B$	2Δ
XOR	A, B → $A \oplus B$	3Δ
XNOR	A, B → $A \odot B$	3Δ
AOI	A, B, C, D → $\overline{AB + CD}$	$\Delta + \Delta_{RC}$

with NAND or NOR gates, time delay in multilevel switching circuits can be measured by the NAND gate levels or numbers of Δ. Time delays of typical gate functions are listed in Table 2.1 in terms of Δ. Note that the AOI gate with wired logic has a delay of $\Delta + \Delta_{RC}$, where Δ_{RC} is determined by the RC constant of the load. For small Δ_{RC}, the wired gate function has an approximated delay of one Δ only.

2.2 Multiplexers and Demultiplexers

A 2^n-input *multiplexer* (MPX) has 2^n input lines, $x_{2^n-1}, \ldots, x_1, x_0$, and one output line f. The unit transfers the logic value on the ith input x_i to the output line under the control of n selection lines $s_{n-1}, \ldots, s_1, s_0$. That is,

$$f = x_i \quad \text{for} \quad i = (s_{n-1} \cdots s_1 s_0)_2 = \sum_{j=0}^{n-1} s_j \cdot 2^j \tag{2.1}$$

Digital multiplexers can be considered as the electronic equivalent of multiway rotary switches with external controls. In some applications it is called a *Data Selector*, because it selects one out of 2^n data inputs and transmits the information to the output. A 4-input multiplexer with two control lines selecting one out of the four inputs to appear on the output is shown in Fig. 2.1. The logic circuit diagram shows an additional output terminal which carries the complemented output denoted as $\bar{f}$. Furthermore, one or more strobe inputs may be available, which provides the chip enable capability. With the wired-logic circuit implementation the total time delay of a multiplexer should be only 2Δ from the data inputs to the output line.

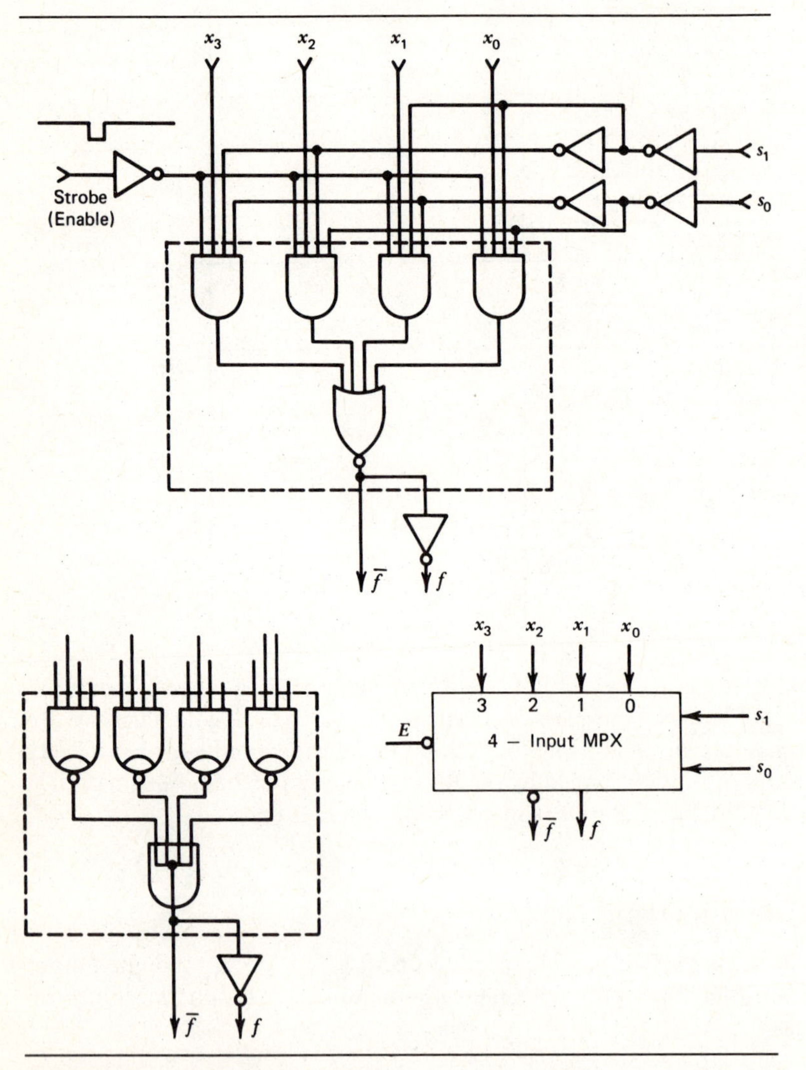

Figure 2.1 The schematic logic circuit of a 4-input multiplexer and its AOI wired circuit implementation (within the dash lines).

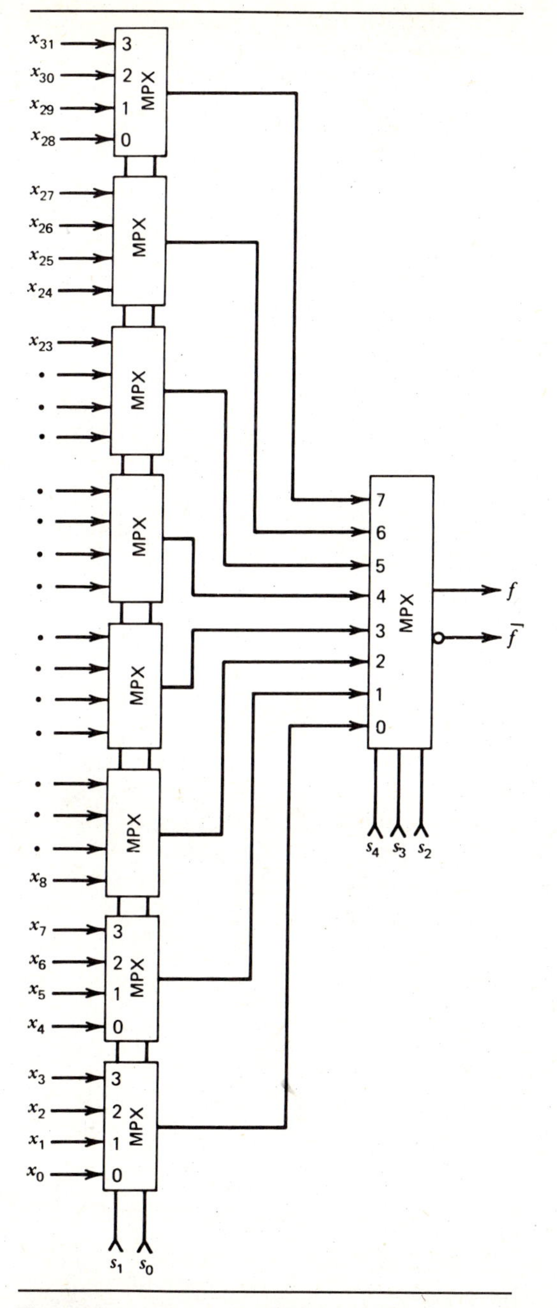

Figure 2.2 A 32-input multiplexer built with eight 4-input multiplexers and one 8-input multiplexer.

As constrained by present packaging technology, digital multiplexers come with *quad 2-input multiplexer* (four 2-input multiplexers), *dual 4-input multiplexer*, *dual 8-input multiplexer*, and *single 16-input multiplexer* on monolithic IC chips ranging from 14 to 24 pins. One can build a large-size multiplexer with a number of small multiplexers interconnected in the form of a multilevel tree. An example of a 32-input multiplexer constructed from connecting the outputs of four packages of a dual 4-input multiplexer to the inputs of an 8-input multiplexer is given in Fig. 2.2. The general rule that one should follow to construct such tree-structured multiplexers is to reduce the total number of IC packages used and to maintain as few levels as possible. A 2^n-input multiplexer can be used to implement any switching function of $n + 1$ variables with n variables on the selection lines, and each of the 2^n input lines is only a function of the remaining variable.

A *demultiplexer* is the opposite of a digital multiplexer. The circuit receives binary data on one input line and transfers it to one of 2^n possible output lines. The output line being selected is determined by the bit pattern of n selection lines. The logic circuit and logic symbol of a *4-output demultiplexer* is shown in Fig. 2.3.

Line decoders are a variation of a demultiplexer in which the dummy data input line is eliminated. A *3-by-8 line decoder* is shown in Fig. 2.4, in which the three selection lines a_2, a_1, a_0 are called *address lines*. The bit combination of these address lines

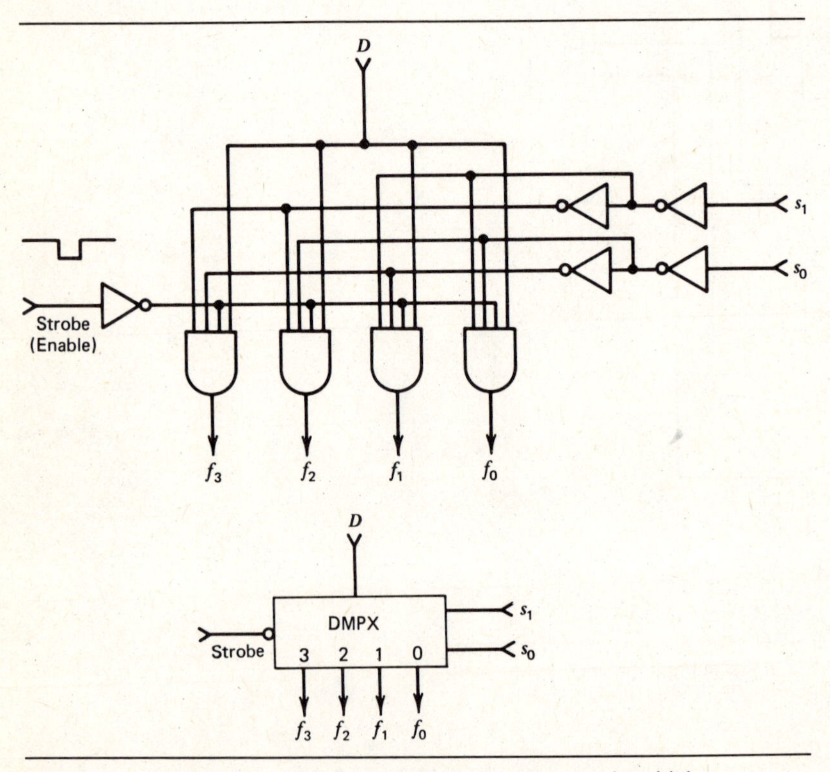

Figure 2.3 The schematic circuit and logic symbol of a 4-output demultiplexer.

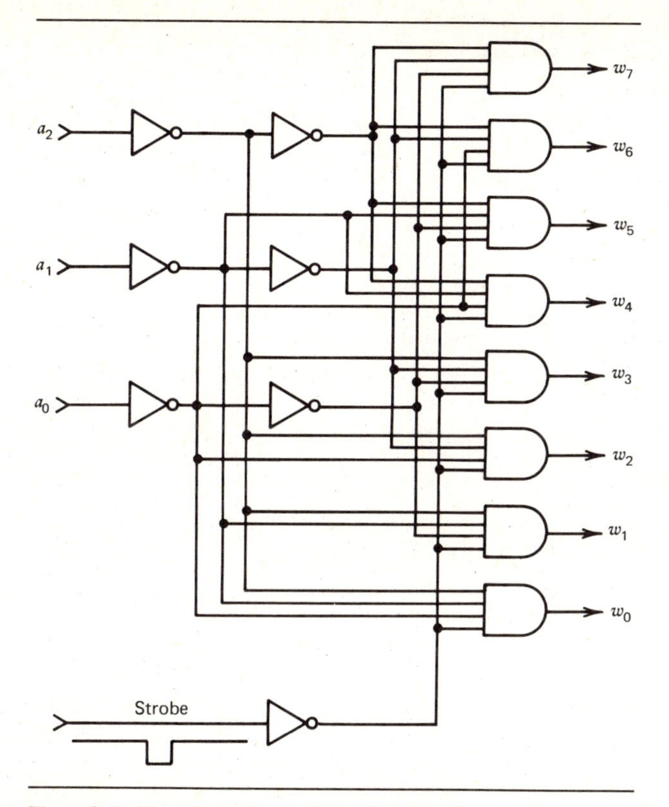

Figure 2.4 The schematic of a 3-to-8 line decoder.

determines which of the eight output lines should go "high." Again, the enable input can be used to expand the decoder. Line decoders are used in addressing memory systems, called *address decoder*, as well as in decoding vector output patterns. At present, *dual* 2-*by*-4 *line decoders*, 3-*by*-8 *decoders*, and 4-*by*-16 *line decoders* are available in monolithic packages with or without chip-enable inputs. Similarly, one can construct a large line decoder by interconnecting a number of small line decoders with enables. In Fig. 2.5, a 6-by-64 line decoder is constructed from using four 4-by-16 line decoders with two enable inputs each. The higher two bits in the address register are used to distinguish the four line decoders.

Combinational logic modules for converting binary data to its BCD equivalent and vice versa are frequently needed. The functional block diagram and truth-table description of a 6-bit *binary-to-BCD* converter is given in Fig. 2.6. This device converts a 6-bit binary number to a 7-bit two-decade BCD number. To convert longer binary numbers, one can use multiple numbers of this module. Figure 2.7 shows that three 6-bit binary-to-BCD converting modules can be interconnected to convert an 8-bit binary number to its three-decade BCD equivalent. Conversely, we can construct a standard 6-bit *BCD-to-binary converter* as shown in Fig. 2.8. Two such packages are needed to convert a two-decade BCD number to a 7-bit binary number. Longer-range

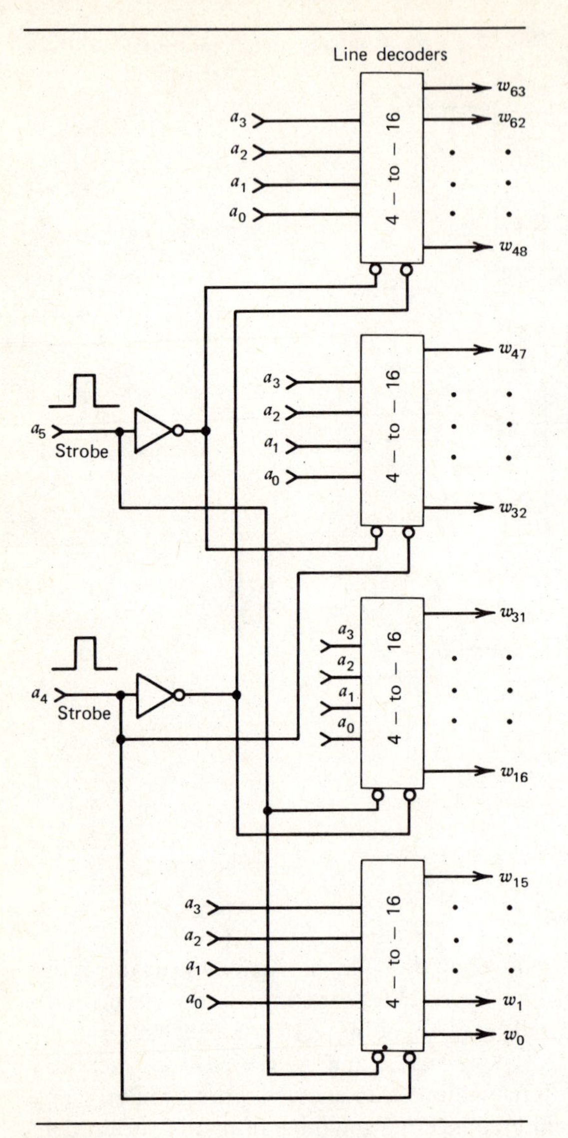

Figure 2.5 A 6-to-64 line decoder built with four 4-to-16 line decoders with two strobe enables each.

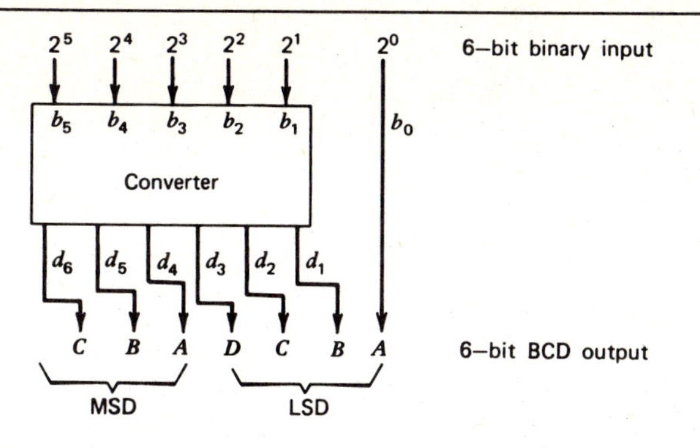

FUNCTION TABLE

Binary Words	Inputs $b_5\ b_4\ b_3\ b_2\ b_1$	Outputs $d_6\ d_5\ d_4\ d_3\ d_2\ d_1$	Binary Words	Inputs $b_5\ b_4\ b_3\ b_2\ b_1$	Outputs $d_6\ d_5\ d_4\ d_3\ d_2\ d_1$
0,1	0 0 0 0 0	0 0 0 0 0 0	32,33	1 0 0 0 0	0 1 1 0 0 1
2,3	0 0 0 0 1	0 0 0 0 0 1	34,35	1 0 0 0 0	0 1 1 0 1 0
4,5	0 0 0 1 0	0 0 0 0 1 0	36,37	1 0 0 1 0	0 1 1 0 1 1
6,7	0 0 0 1 1	0 0 0 0 1 1	38,39	1 0 0 1 1	0 1 1 1 0 0
8.9	0 0 1 0 0	0 0 0 1 0 0	40,41	1 0 1 0 0	1 0 0 0 0 0
10,11	0 0 1 0 1	0 0 1 0 0 0	42,43	1 0 1 0 1	1 0 0 0 0 1
12,13	0 0 1 1 0	0 0 1 0 0 1	44,45	1 0 1 1 0	1 0 0 0 1 0
14,15	0 0 1 1 1	0 0 1 0 1 0	46,47	1 0 1 1 1	1 0 0 0 1 1
16,17	0 1 0 0 0	0 0 1 0 1 1	48,49	1 1 0 0 0	1 0 0 1 0 0
18,19	0 1 0 0 1	0 0 1 1 0 0	50,51	1 1 0 0 1	1 0 1 0 0 0
20.21	0 1 0 1 0	0 1 0 0 0 0	52,53	1 1 0 1 0	1 0 1 0 0 1
22,23	0 1 0 1 1	0 1 0 0 0 1	54,55	1 1 0 1 1	1 0 1 0 1 0
24,25	0 1 1 0 0	0 1 0 0 1 0	56,57	1 1 1 0 0	1 0 1 0 1 1
26,27	0 1 1 0 1	0 1 0 0 1 1	58,59	1 1 1 0 1	1 0 1 1 0 0
28,29	0 1 1 1 0	0 1 0 1 0 0	60,61	1 1 1 1 0	1 1 0 0 0 0
30,31	0 1 1 1 1	0 1 1 0 0 0	62,63	1 1 1 1 1	1 1 0 0 0 1

Figure 2.6 The block diagram and function table of a 6-bit Binary-to BCD converter module.

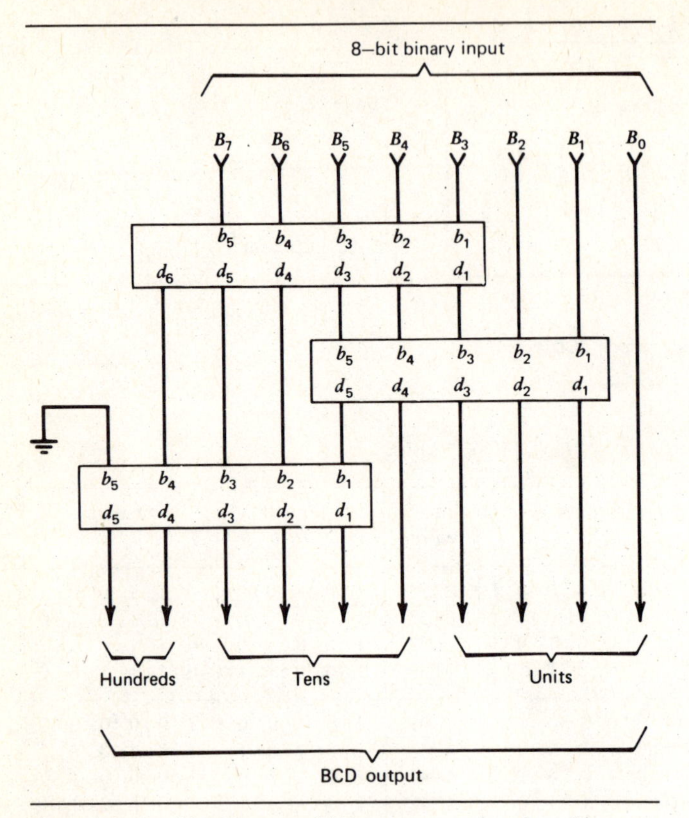

Figure 2.7 An 8-bit Binary-to-BCD converter using three Binary-to-BCD converting modules.

monolithic BCD-to-binary or binary-to-BCD converters can be fabricated with LSI circuits. The internal structure of the LSI should be similar to those presented in [2].

2.3 Basic Add/Subtract Logic

Basic MSI/LSI arithmetic functions, such as *adders*, *complementers*, *comparators*, and *multipliers*, are introduced in subsequent sections. Other handy arithmetic functions such as *Carry Look Ahead* (CLA) unit, *Arithmetic Logic Unit* (ALU), and so on, will be described in later chapters when they are used.

A one-bit *Full Adder* (FA) adds two binary digits A_i, B_i, and one carry-input C_i to produce a sum output S_i and a carry-output C_{i+1} as shown in Fig. 2.9. The two outputs are related to the three inputs by the following Boolean equations. Various possible ways of implementing full adders can be found in reference [17].

$$\begin{aligned} S_i &= A_i \oplus B_i \oplus C_i \\ C_{i+1} &= A_iB_i + B_iC_i + C_iA_i \end{aligned} \qquad \textbf{(2.2)}$$

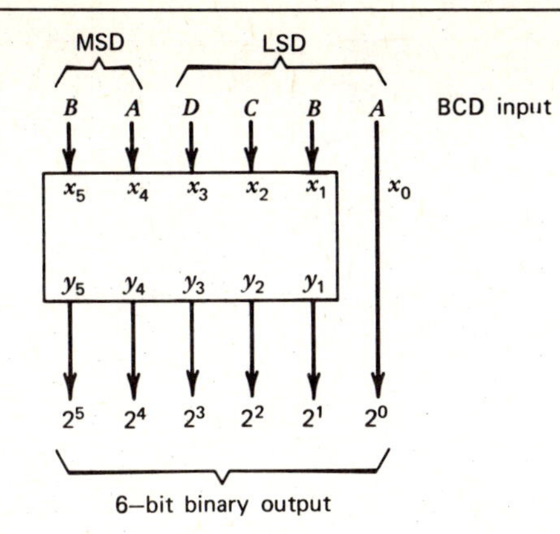

BCD Words	Inputs x_5	x_4	x_3	x_2	x_1	Outputs y_5	y_4	y_3	y_2	y_1
0–1	0	0	0	0	0	0	0	0	0	0
2–3	0	0	0	0	1	0	0	0	0	1
4–5	0	0	0	1	0	0	0	0	1	0
6–7	0	0	0	1	1	0	0	0	1	1
8–9	0	0	1	0	0	0	0	1	0	0
10–11	0	1	0	0	0	0	0	1	0	1
12–13	0	1	0	0	1	0	0	1	1	0
14–15	0	1	0	1	0	0	0	1	1	1
16–17	0	1	0	1	1	0	1	0	0	0
18–19	0	1	1	0	0	0	1	0	0	1
20–21	1	0	0	0	0	0	1	0	1	0
22–23	1	0	0	0	1	0	1	0	1	1
24–25	1	0	0	1	0	0	1	1	0	0
26–27	1	0	0	1	1	0	1	1	0	1
28–29	1	0	1	0	0	0	1	1	1	0
30–31	1	1	0	0	0	0	1	1	1	1
32–33	1	1	0	0	1	1	0	0	0	0
34–35	1	1	0	1	0	1	0	0	0	1
36–37	1	1	0	1	1	1	0	0	1	0
38–39	1	1	1	0	0	1	0	0	1	1

Figure 2.8 The block diagram and function table of a standard BCD-to-Binary Converting Module.

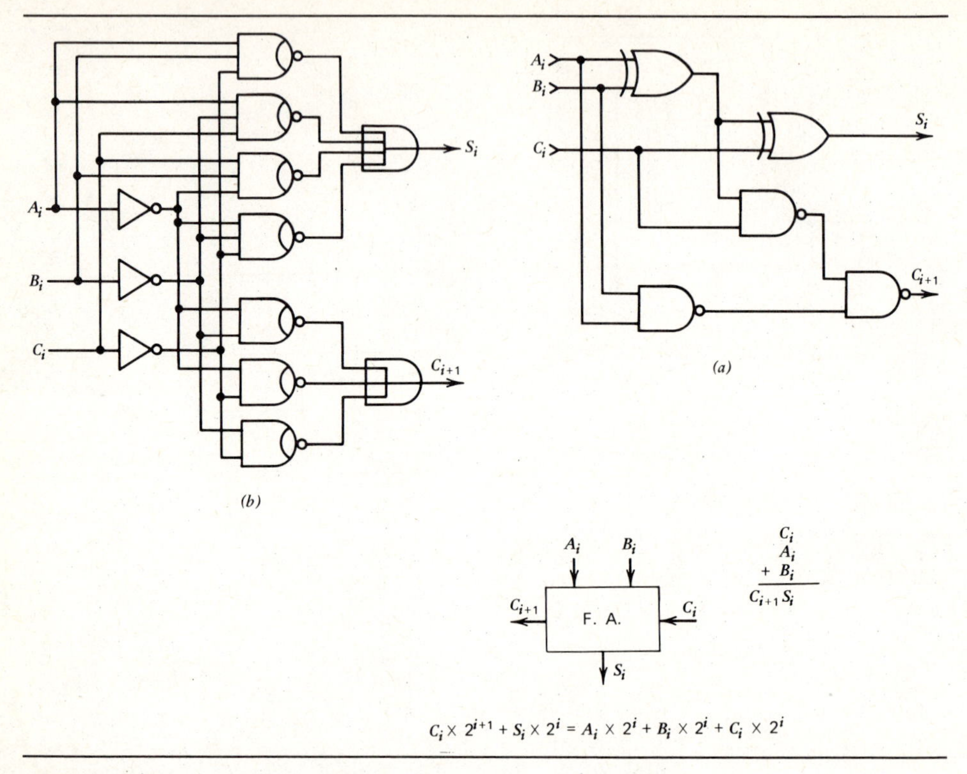

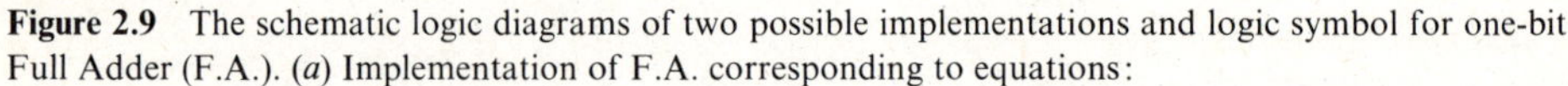

Figure 2.9 The schematic logic diagrams of two possible implementations and logic symbol for one-bit Full Adder (F.A.). (*a*) Implementation of F.A. corresponding to equations:

$$\begin{cases} S_i = A_i \oplus B_i \oplus C_i \\ C_{i+1} = A_iB_i + B_iC_i + A_iC_i \end{cases}$$

and with

$$\begin{cases} 6\Delta \text{ time delay for } S_i \\ 5\Delta \text{ time delay for } C_{i+1} \end{cases}$$

(*b*) Wired-logic implementation of F.A. with minimum-delay of 2Δ for both outputs

This basic adding cell can be modified to become a 4-input and 4-output *Controlled Add/Subtract cell* (CAS) as shown in Fig. 2.10. The additional input P is used to control the ADD ($P = 0$) or SUBTRACT ($P = 1$) operations. In case of subtraction, the C_i input is called *borrow-in* and the C_{i+1} output *borrow-out*. The input-output relationship of a CAS cell is specified by the following pair of Boolean equations.

$$S_i = A_i \oplus (B_i \oplus P) \oplus C_i$$
$$C_{i+1} = (A_i + C_i) \cdot (B_i \oplus P) + A_iC_i \tag{2.3}$$

Equation 2.3 is identical to Eq. 2.2 when $P = 0$. When $P = 1$, we have

$$S_i = A_i \oplus \overline{B}_i \oplus C_i$$

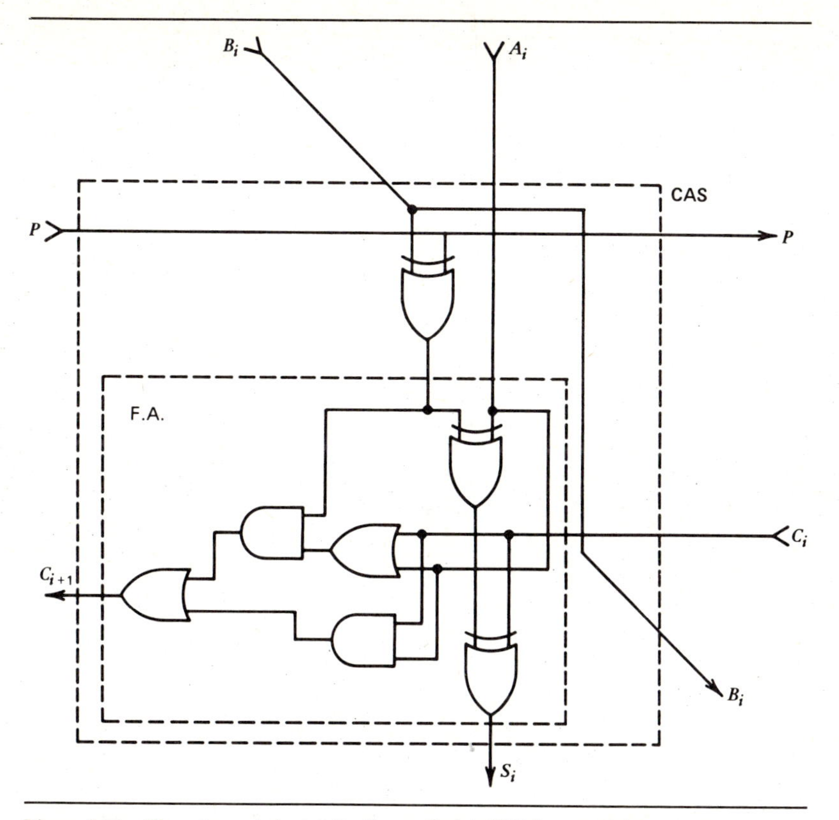

Figure 2.10 The schematic logic of a Controlled Add/Subtract (CAS) cell.

and $C_{i+1} = A_i\bar{B}_i + \bar{B}_iC_i + A_iC_i$. Such CAS cells are used extensively in constructing iterative arithmetic logic arrays in Chapters 6, 8, and 11.

One can connect n 1-bit full adders in cascade to form an n-bit ripple carry adder. Each 1-bit full adder has 2Δ time delay, assuming wired-logic implementation of the internal sum-of-products. At present, 4-bit or 8-bit monolithic adders generating the binary sum of two 4-bit or 8-bit numbers are available in MSI package form. The carry input and output lines enable it to be extended to any word length. Some of the n-bit adders are featured with full carry lookahead across the n-bits. Details of carry lookahead techniques in constructing high-speed adders will be discussed in Chapter 3.

2.4 Complementers and Magnitude Comparators

Two types of binary complementing circuits are frequently used in arithmetic unit design. The *one's complementer* shown in Fig. 2.11(a) performs bitwise complementation through the **eXclusive**-OR(XOR) gates, when the control line goes high.

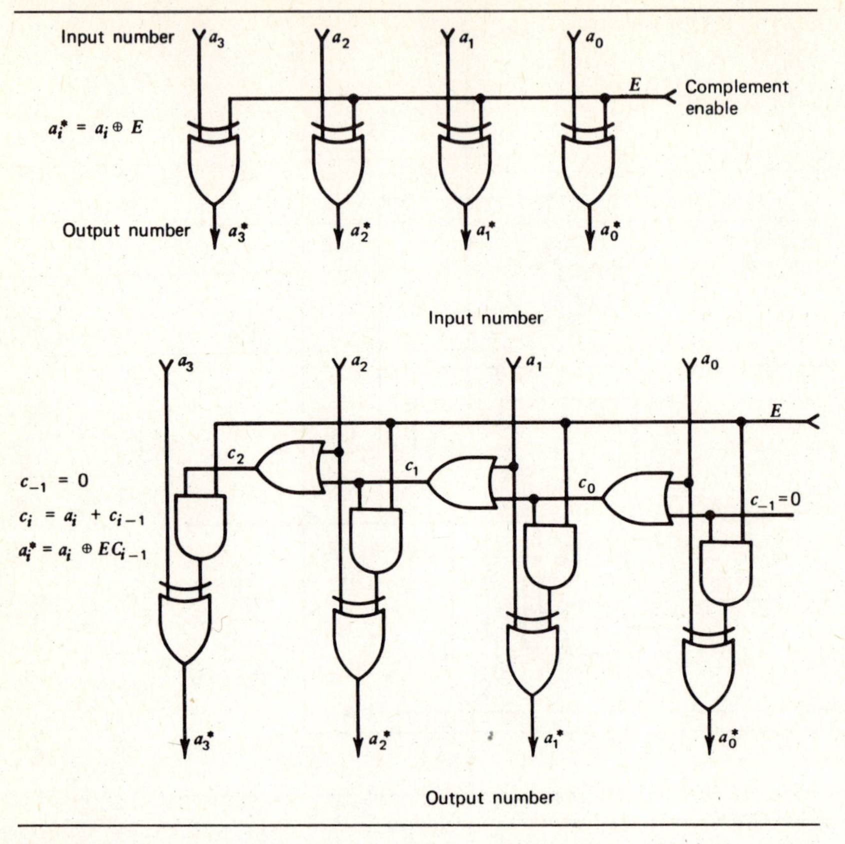

Figure 2.11 The circuit diagrams of two binary complementers with enable control. (*a*) A 4-bit one's complementer. (*b*) A 4-bit two's complementer.

These bitwise operations are independent and only 3Δ delays are required to complement any signed binary number to its one's complement form. To complement an $(n + 1)$-bit sign-magnitude number to one's complement form, the leading sign bit can be used as the control signal.

In the case of two's complement operation, a bit-scanning technique is used to perform the desired complementation. Let $\mathbf{A} = a_n \cdots a_1 a_0$ be the given $(n + 1)$-bit signed number whose two's complement form is desired. The bit-scanning starts from the right end, a_0, of the number and preceeds to the left until the "very" first "1" is found, say $a_i = 1$ for the smallest i such that $0 \le i \le n$. Then every input bit to the right of a_i, including a_i itself, is retained unchanged and every input bit to the left of a_i if flipped, that is, "1" becomes "0" and "0" becomes "1." Such a scanning *two's complementer* is implemented by the circuit shown in Fig. 2.11(b). The output bits from the right are identical with the input bits until the very first "1" is encountered in the input, and from that particular bit position on bitwise complementation is performed. The chain-output C_i of scanning stage-i is a "1." if and only if an input bit $a_i = 1$ at stage-i or the chain-input $C_{i-1} = 1$ from the chain-output of the preceding

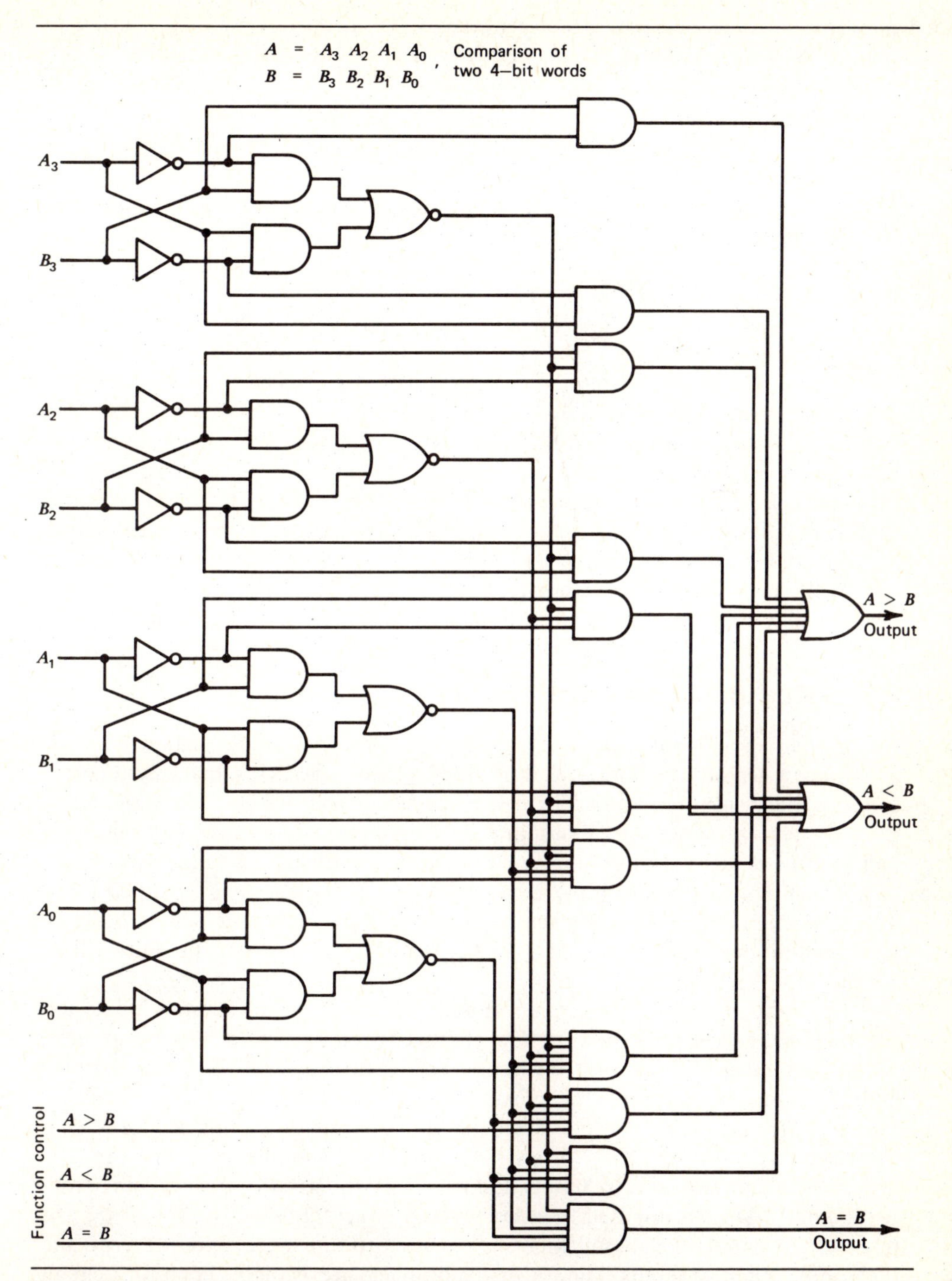

Figure 2.12 The schematic logic diagram of a 4-bit magnitude comparator.

stage $(i - 1)$ at the right. The initial chain-input C_{-1} at the right end must always set as zero. The control line enables the two's complement operation. The ouputs will be identical with the inputs when the control is "low." Again, one can use the sign bit as the control signal.

As an example, the output of a 4-bit two's complementer should be 0110 for an input number 1010, in which the second bit from right is the very first "1" encountered. The total time delay in converting an $(n + 1)$-bit signed number using such a combinational two's complementer is

$$\Delta_{TC} = n \cdot 2\Delta + 5\Delta = (2n + 5) \cdot \Delta \tag{2.4}$$

where each scanning stage contributes 2Δ delay and the 5Δ is due to the AND and XOR gates.

A 4-bit *magnitude comparator* is shown in Fig. 2.12. The unit compares two 4-bit binary numbers $\mathbf{A} = a_3a_2a_1a_0$ and $\mathbf{B} = b_3b_2b_1b_0$ and produces three outputs: $\mathbf{A} = \mathbf{B}$, $\mathbf{A} > \mathbf{B}$ and $\mathbf{A} < \mathbf{B}$. This device is fully expandable to any number of bits without external gates. Words of greater length can be compared by connecting the 4-bit comparators in cascade. The *cascading inputs* are used for this purpose. Figure 2.13 illustrates the interconnections of a 24-bit comparator implemented with six 4-bit comparators. At present, two 24-bit numbers can be compared in about 40 nsec, with the combinational comparator. Magnitude comparison can be also performed by using the adder in an arithmetic logic unit, to be described in Chapter 4.

2.5 Modular Array Multipliers

Modular combinational logic arrays, which perform fast multiplication on short or moderate-length operands, are introduced below. The array multiply modules can be divided into two types. The first type, called *Nonadditive Multiply Modules* (NMM), performs the local multiplication (summation of shifted multiplicands), independent of the results of other multiply modules. Figure 2.14 shows the design of a 4-by-4 nonadditive multiply module with 12 full adders and 16 AND gates. This device multiplies two 4-bit numbers, $\mathbf{A} = A_3A_2A_1A_0$ and $\mathbf{B} = B_3B_2B_1B_0$, to yield an 8-bit product $\mathbf{P} = \mathbf{A} \times \mathbf{B} = P_7P_6 \cdots P_1P_0$. An 18-pin IC package will be sufficient for this module.

This basic multiply module can be modified to receive some additive inputs and add them to the resulting product. The additive feature makes it possible to interconnect a number of small additive multiply cells to form an arbitrary n-by-n parallel multiplication array. Figure 2.15 depicts the functional structure of a 4-by-2 *Additive Multiply Module* (AMM), which evaluates the arithmetic function $\mathbf{P} = \mathbf{A} \times \mathbf{B} + \mathbf{C} + \mathbf{D}$. $\mathbf{A} = a_3a_2a_1a_0$ and $\mathbf{B} = b_1b_0$ are the multiplicand and multiplier. $\mathbf{C} = c_3c_2c_1c_0$ and $\mathbf{D} = d_1d_0$ are two additive inputs, and $P = P_5P_4P_3P_2P_1P_0$ is the resulting product. Eight AND gates are used to generate the product terms a_ib_j for $0 \leq i \leq 3$ and $0 \leq j \leq 1$. An array of eight full adders is necessary to sum up the product terms and the external additive inputs. These 4-by-2 multipliers are available in an MSI circuit with 20 external leads. Monolithic array multipliers up to 16-by-16 in size are now available [36]. We shall describe how these basic multiply modules can be used to

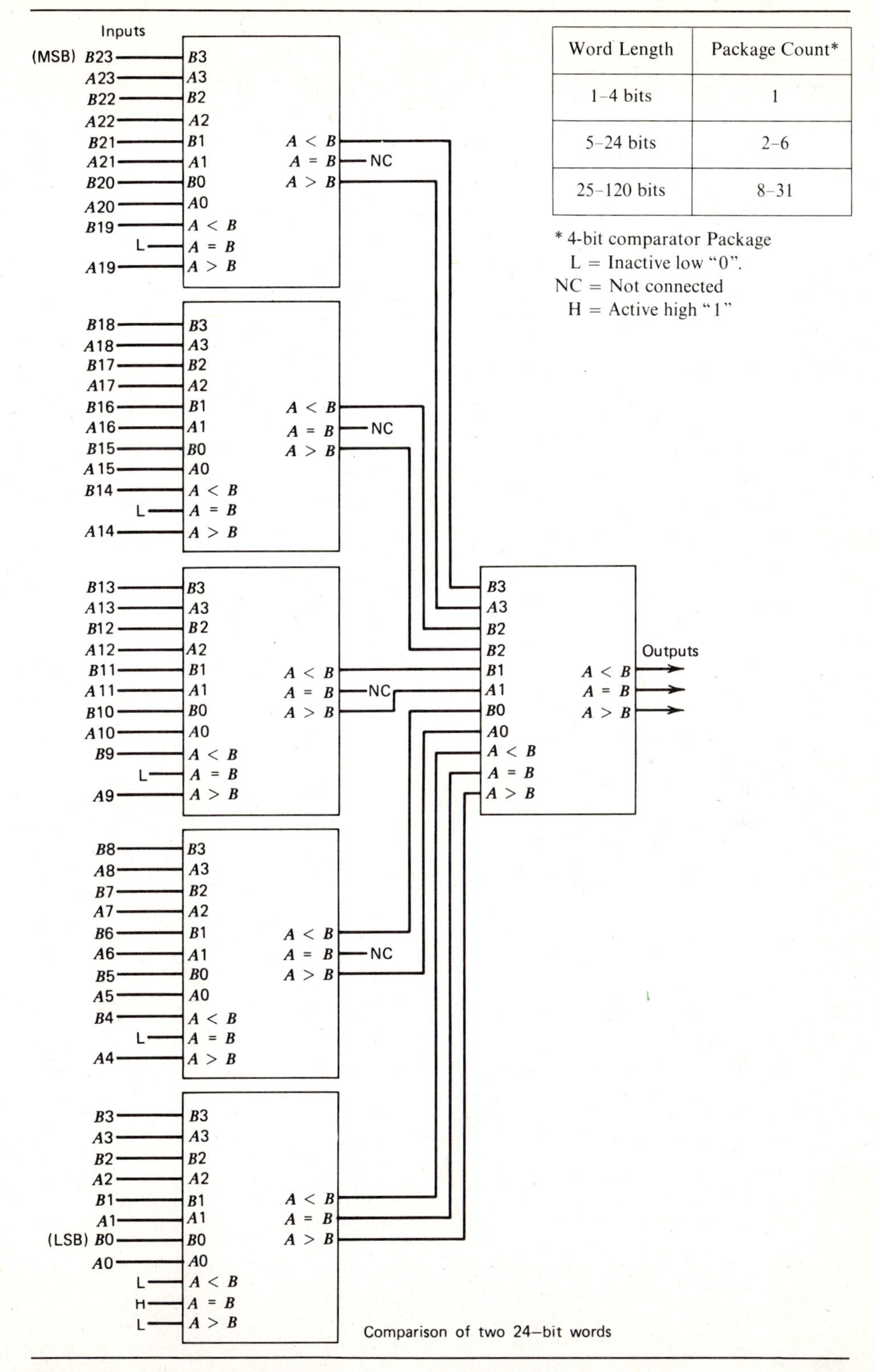

Word Length	Package Count*
1–4 bits	1
5–24 bits	2–6
25–120 bits	8–31

Figure 2.13 A 24-bit magnitude comparator built with six 4-bit comparators in cascade, and package counts of comparators of various word lengths.

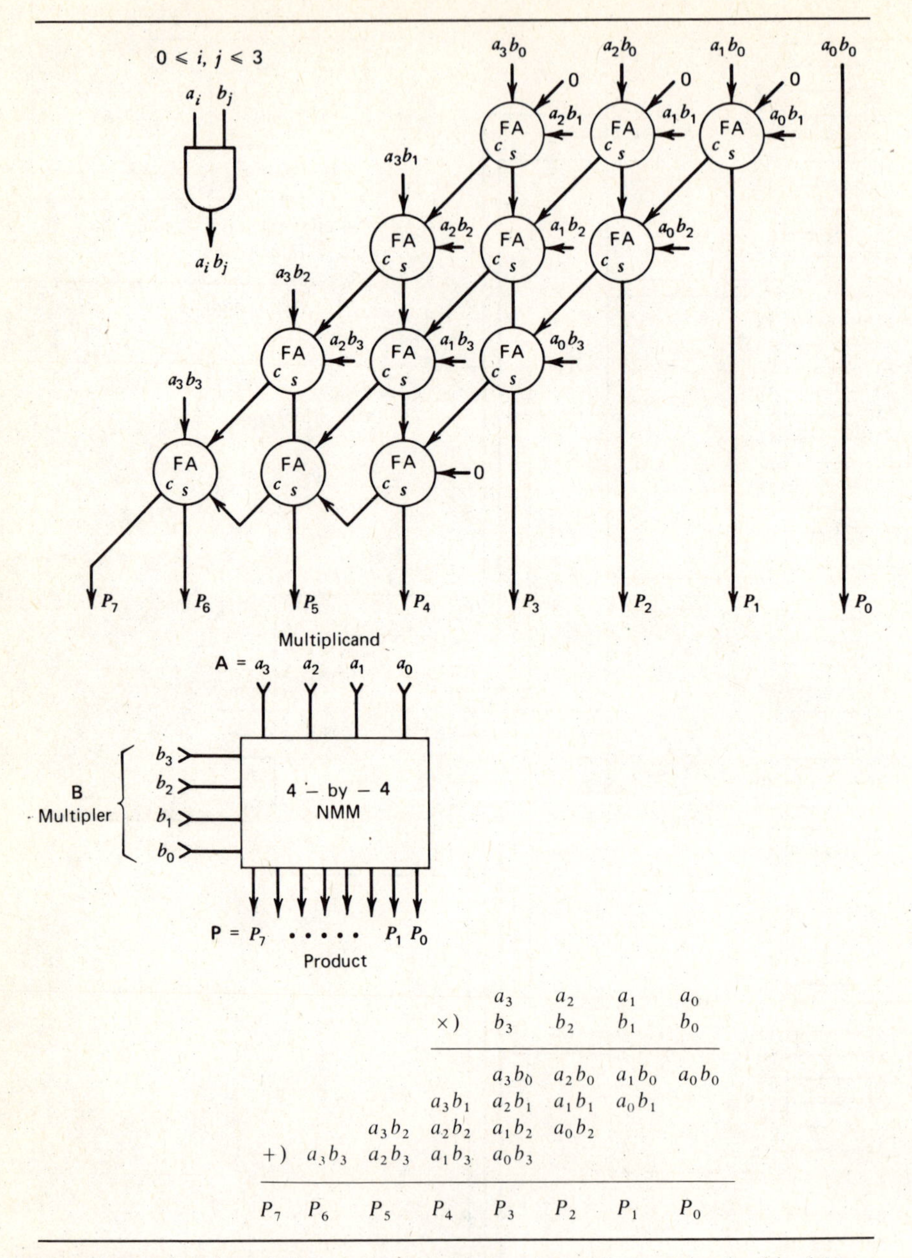

Figure 2.14 The schematic diagram, logic symbol, and operation matrix of a 4-by-4 Non-additive Multiply Module (NMM).

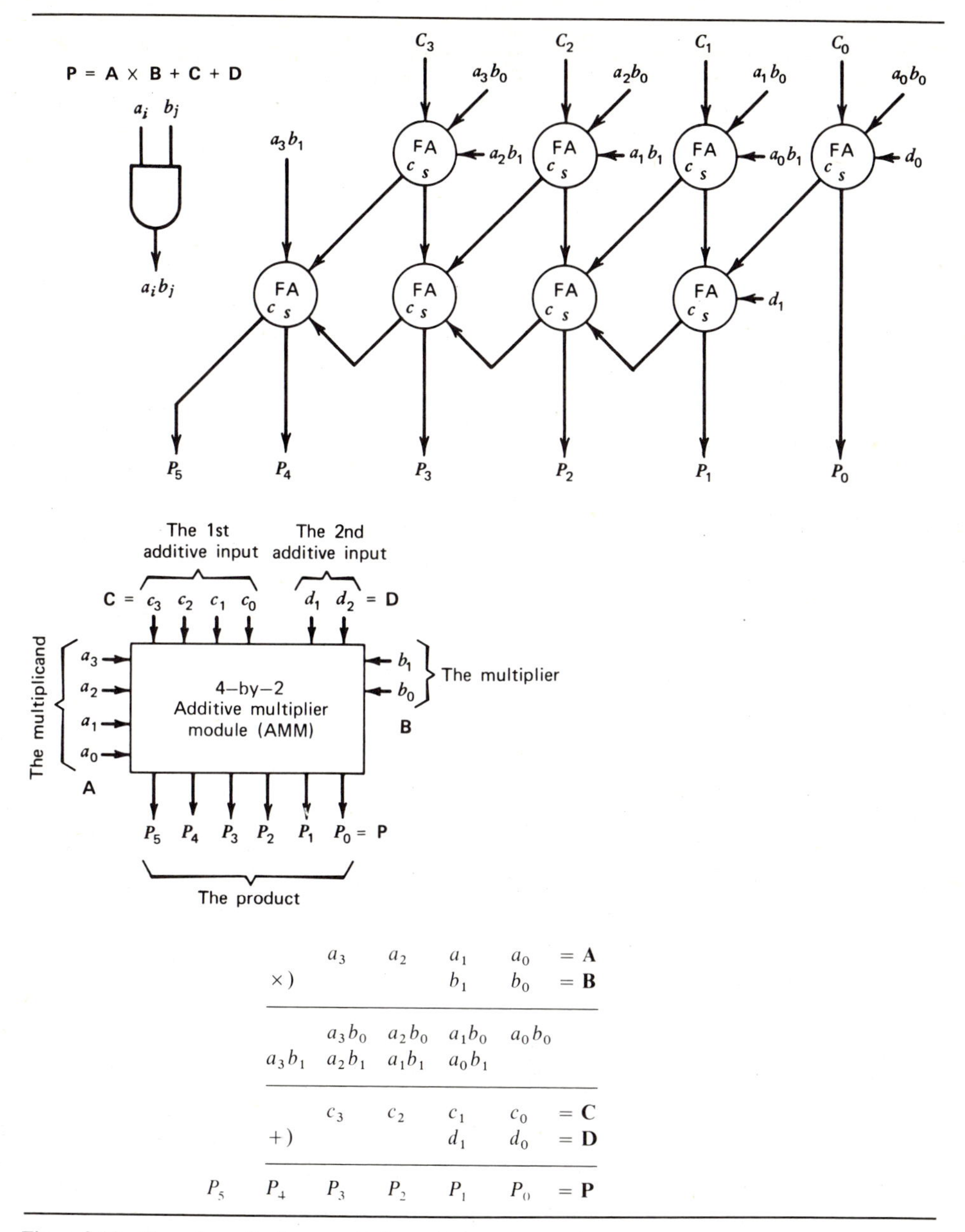

		a_3	a_2	a_1	a_0	= **A**
	×)			b_1	b_0	= **B**
		a_3b_0	a_2b_0	a_1b_0	a_0b_0	
	a_3b_1	a_2b_1	a_1b_1	a_0b_1		
		c_3	c_2	c_1	c_0	= **C**
	+)			d_1	d_0	= **D**
P_5	P_4	P_3	P_2	P_1	P_0	= **P**

Figure 2.15 The schematic diagram, logic symbol, and operation matrix of a 4-by-2 Additive Multiply Module (AMM).

Table 2.2 Application Parameters of Fundamental Array Multiply Modules, NMMs and AMMs, of Various Practical Sizes

Array Size $m \times n$	Nonadditive Multiply Module (NMM)		Additive Multiply Module (AMM)	
	No. of external leads[a]	Delay time[b]	No. of external leads[a]	Delay time[b]
2×2	10	4Δ	14	6Δ
4×2 or 2×4	14	8Δ	20	10Δ
4×4	18	12Δ	26	14Δ
8×4 or 4×8	26	20Δ	38	22Δ
8×8	34	28Δ	50	30Δ

[a] Including two extra leads needed for power and ground per each package.

[b] Excluding the 2Δ delay due to the AND gates for generating all the summand terms simultaneously.

construct large multiplication networks in Chapter 6. Table 2.2 summarizes the application parameters for both types of multiply modules of various practical sizes. To use the nonadditive multiply modules, additional summing devices such as the multiple-operand Wallace tree adders must be used. Details of these provisions and a universal approach to array multiplier construction will be given in Chapter 6.

2.6 Features of Multifunction Registers

Functional features associated with working registers in digital arithmetic processors are introduced in this section. These hardware features play key roles in the design of the control-state counters, shifters, complementers, and general-purpose registers frequently used in arithmetic processors.

We start with the design of a *counting register*, which, when used with a decoder, can control the state transition in a digital system. The block diagram of a 16-state *control counter* is shown in Fig. 2.16. The 4-bit counting register in this unit can perform four distinct functions—CLEAR, COUNT UP, COUNT DOWN, and PARALLEL LOAD—under the control of two function selection lines X and Y. A 4-by-16 line decoder is used to decode the 16 counter states $C_0, C_1, \ldots, C_{15}$.

The internal circuit of the 4-bit counting register is shown in Fig. 2.17. The excitation logic to each flip-flop in the counting register is implemented with a 4-input multiplexer. In order to expand the 4-bit register, we attach a count-enable input line

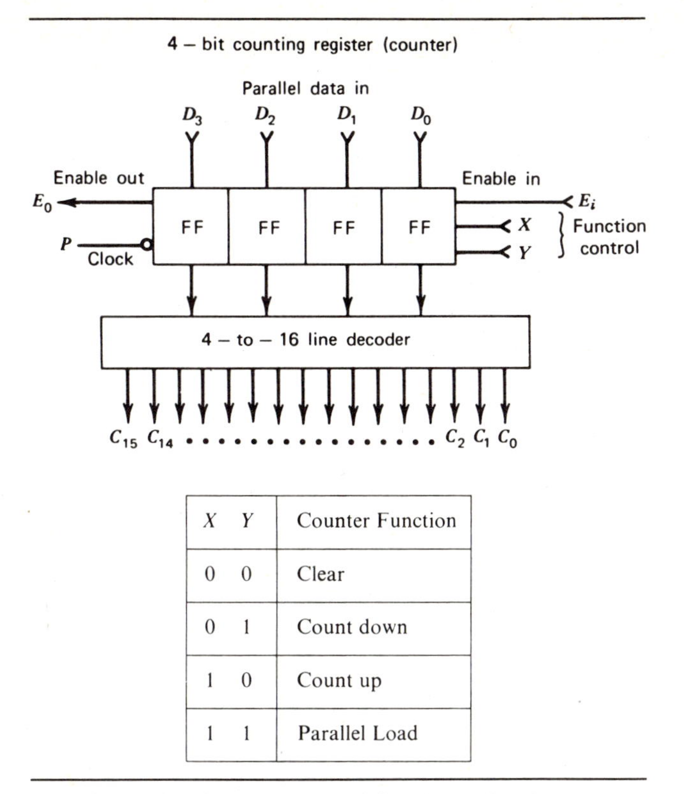

X	Y	Counter Function
0	0	Clear
0	1	Count down
1	0	Count up
1	1	Parallel Load

Figure 2.16 The block diagram and function table of a 16-state control counter with cycle decoder.

E_i and a count-enable output line E_0 to each 4-bit slice. For example, one can interconnect two 4-bit counting registers to form an 8-bit register which can count up to $2^8 - 1 = 255$ states. The output line E_0 of the lower stage is fed to the input line E_i of the higher stage. The input to the rightmost stage is set to a one.

Shifting registers or simply *shifters* are often used in arithmetic processor design. A 4-bit bidirectional parallel-load shift register is illustrated in Fig. 2.18. The external input into the rightmost flip-flop during LEFT SHIFT is denoted as SLI (Shift Left Input) and that into the leftmost flip-flop during RIGHT SHIFT is denoted as SRI (Shift Right Input). The four possible input selections to each flip-flop are SHIFT RIGHT, SHIFT LEFT, LOAD A, and LOAD B. The enable E is used to control the clocking. When E is low, no clock pulses can get through. The register will retain its old data when the clock is not enabled.

A register can be designed to complement itself under external control. In Fig. 2.19, a 4-bit four function *complementing register* is illustrated. The four functions performed are CL (Clear), OC (One's Complement), TC (Two's Complement), and PL (Parallel Load). The circuit is implemented with J-K flip-flops without using the multiplexer to select the inputs. Instead, the input excitation logic is implemented with random logic gates and a function decoder.

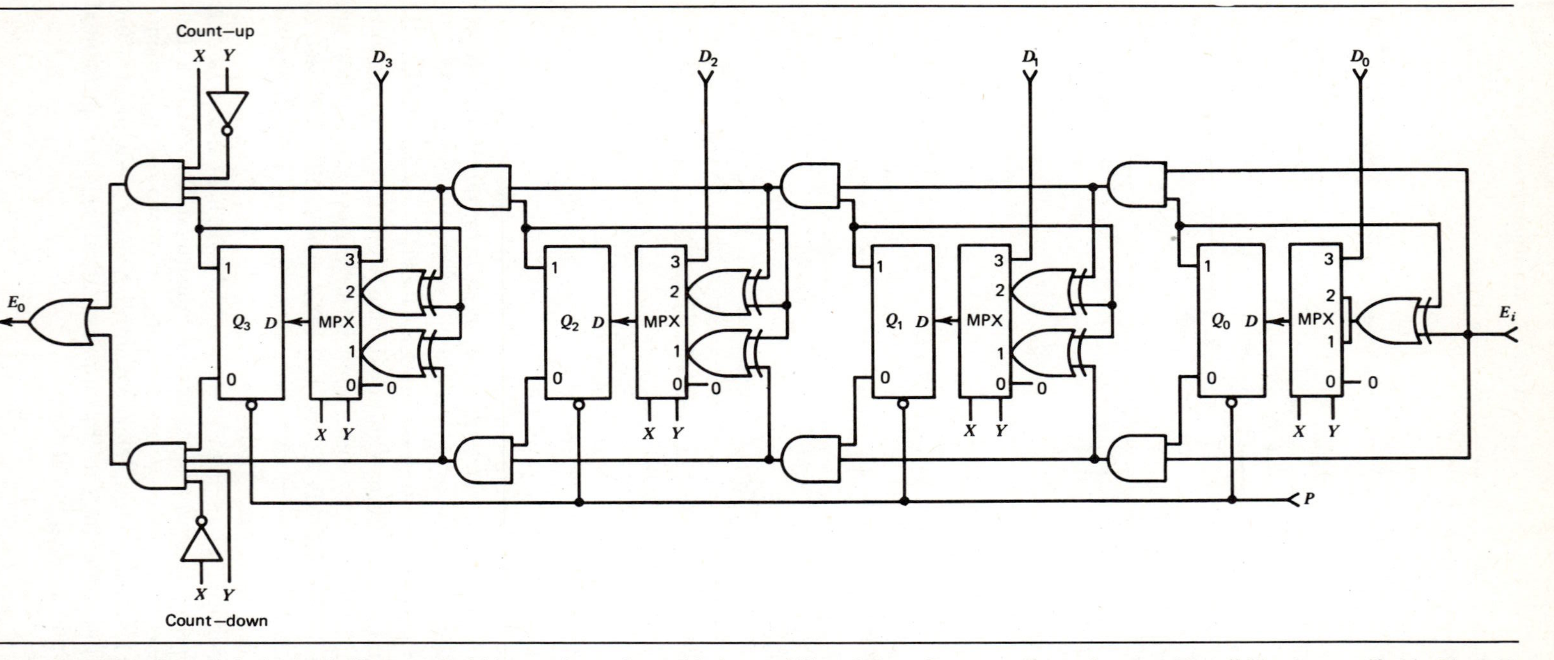

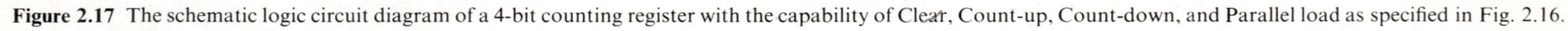

Figure 2.17 The schematic logic circuit diagram of a 4-bit counting register with the capability of Clear, Count-up, Count-down, and Parallel load as specified in Fig. 2.16.

X Y	0 0	0 1	1 0	1 1
Function	Shift right	Shift left	Load A	Load B

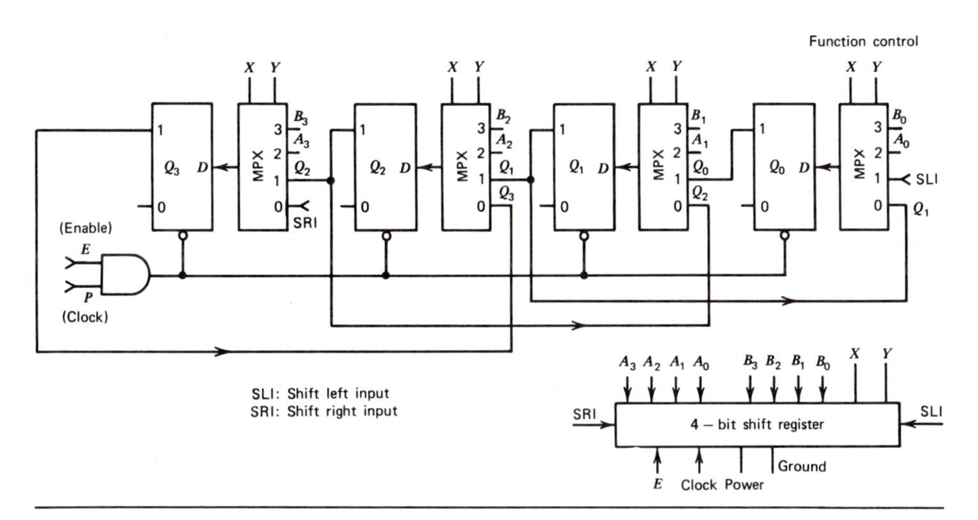

Figure 2.18 The schematic logic circuit diagram of a 4-bit bidirectional shift register with two parallel-load entries.

Bitwise complementation is performed in one's complement operation. Carries between bits may be generated in two's complement operation. Thus, a carry input line T_i and a carry output line T_0 are introduced to each 4-bit slice so that they can be connected in cascade to yield any word length. Figure 2.20 shows the interconnection of four 4-bit complementing registers to form a 16-bit multifunction complementer. The carry input T_i to the rightmost stage is always a zero, similar to that shown in Fig. 2.11(b).

2.7 General-Purpose Register Design

In addition to the *counting*, *shifting*, and *complementing* operations described above, one can establish additional logic operations, such as bitwise XOR, AND, OR, SET, and so on, in a general-purpose register. We choose the design of an eight-function register to illustrate two different approaches in constructing general-purpose registers. Table 2.3 specifies the eight functions to be performed by a register constructed with D-type flip-flops. The next-state equations of the flip-flops are shown below.

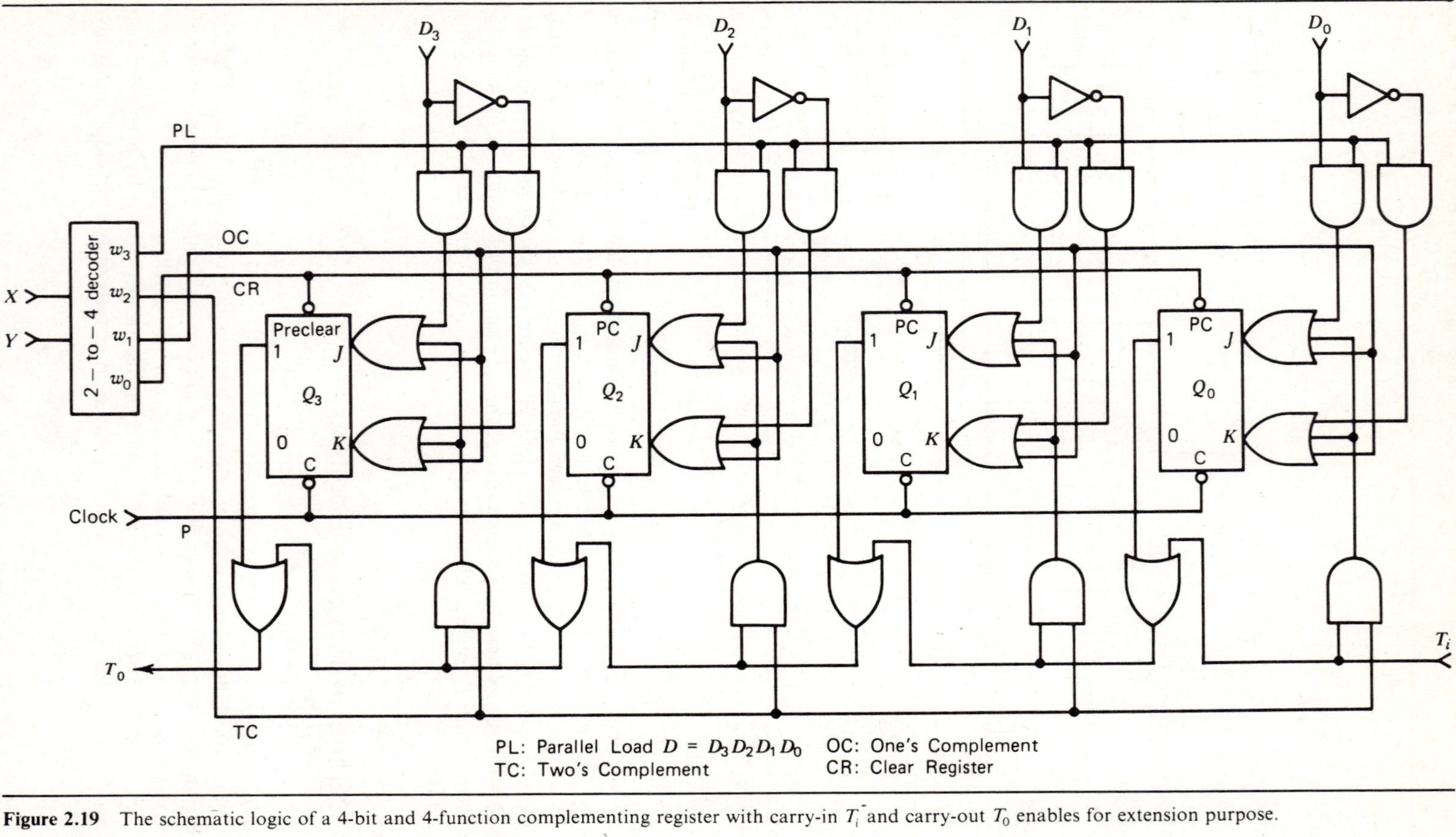

Figure 2.19 The schematic logic of a 4-bit and 4-function complementing register with carry-in T_i and carry-out T_0 enables for extension purpose.

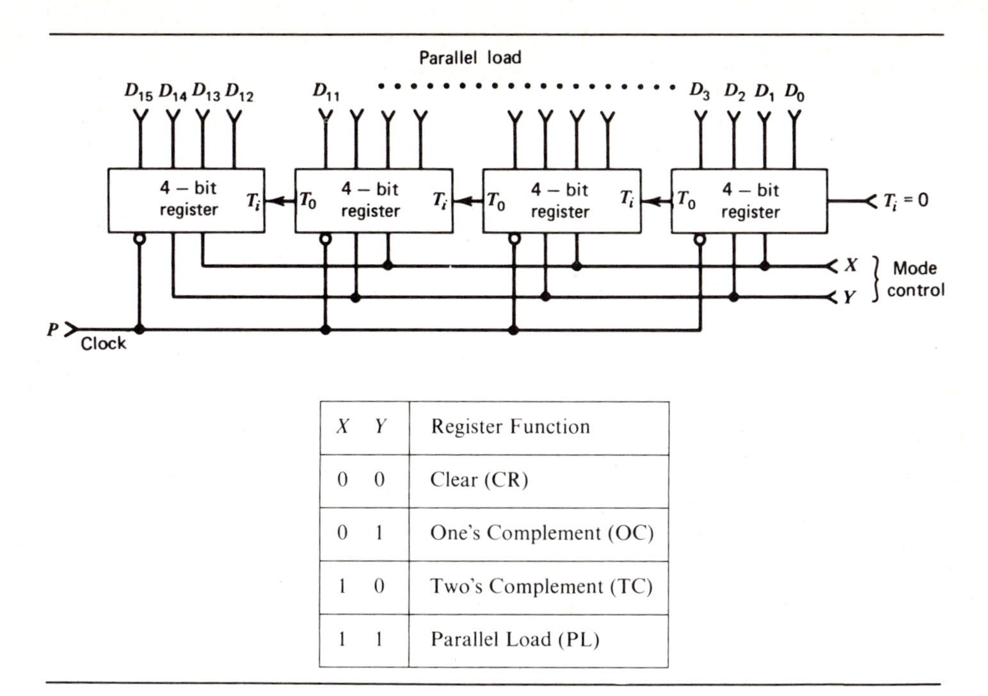

X	Y	Register Function
0	0	Clear (CR)
0	1	One's Complement (OC)
1	0	Two's Complement (TC)
1	1	Parallel Load (PL)

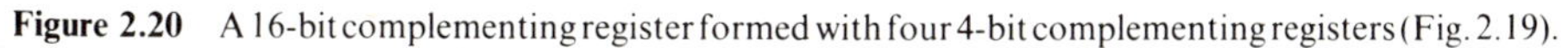

Figure 2.20 A 16-bit complementing register formed with four 4-bit complementing registers (Fig. 2.19).

Table 2.3 The Function Specification and Next States of an n-Bit General-Purpose Register

FNC Decode	Function Control XYZ	Register Function	Next States of F.F.'s $Q_{n-1}(t+1)$	$Q_i(t+1)$	$Q_0(t+1)$
CR	0 0 0	Clear	0	0	0
SR	0 0 1	Shift right	SRI	$Q_{i+1}(t)$	$Q_1(t)$
SL	0 1 0	Shift left	$Q_{n-2}(t)$	$Q_{i-1}(t)$	SLI
PL	0 1 1	Parallel load	D_{n-1}	D_i	D_0
XOR	1 0 0	XOR	$Q_{n-1}(t) \oplus D_{n-1}$	$Q_i(t) \oplus D_i$	$Q_0(t) \oplus D_0$
AND	1 0 1	AND	$Q_{n-1}(t) \cdot D_{n-1}$	$Q_i(t) \cdot D_i$	$Q_0(t) \cdot D_0$
OC	1 1 0	One's complement	$\overline{Q}_{n-1}(t)$	$\overline{Q}_i(t)$	$\overline{Q}_0(t)$
CU	1 1 1	Count-up	$Q_{n-1}(t) \oplus (Q_{n-2}(t) \cdots Q_0(t))$	$Q_i(t) \oplus (Q_{i-1}(t) \cdot Q_{i-2}(t) \cdots Q_0(t))$	$\overline{Q}_0(t)$

The next-state equation for the most-significant flip-flop

$$\begin{aligned} Q_{n-1}(t+1) = {} & \mathrm{SR} \cdot \mathrm{SRI} + \mathrm{SL} \cdot Q_{n-2}(t) \\ & + \mathrm{PL} \cdot D_{n-1} + \mathrm{XOR} \cdot (Q_{n-1}(t) \oplus D_{n-1}) \\ & + \mathrm{AND} \cdot (Q_{n-1}(t) \cdot D_{n-1}) + \mathrm{OC} \cdot \bar{Q}_{n-1}(t) \\ & + \mathrm{CU} \cdot (Q_{n-1}(t) \oplus (Q_{n-2}(t) \cdot Q_{n-3}(t) \cdots Q_1(t) \cdot Q_0(t))) \end{aligned} \tag{2.5}$$

The next-state equation for the ith flip-flop for $i = 1, 2, \ldots, n-2$

$$\begin{aligned} Q_i(t+1) = {} & \mathrm{SR} \cdot Q_{i+1}(t) + \mathrm{SL} \cdot Q_{i-1}(t) + \mathrm{PL} \cdot D_i + \mathrm{XOR} \cdot (Q_i(t) \oplus D_i) \\ & + \mathrm{AND} \cdot (Q_i(t) \cdot D_i) + \mathrm{OC} \cdot \bar{Q}_i(t) \\ & + \mathrm{CU} \cdot (Q_i(t) \oplus (Q_{i-1}(t) \cdots Q_0(t))) \end{aligned} \tag{2.6}$$

The next-state equation for the least-significant flip-flop

$$\begin{aligned} Q_0(t+1) = {} & \mathrm{SR} \cdot Q_1(t) + \mathrm{SL} \cdot \mathrm{SLI} + \mathrm{PL} \cdot D_0 + \mathrm{XOR} \cdot (Q_0(t) \oplus D_0) \\ & + \mathrm{AND} \cdot (Q_0(t) \cdot D_0) + (\mathrm{OC} + \mathrm{CU}) \cdot \bar{Q}_0(t) \end{aligned} \tag{2.7}$$

These equations can be implemented with n 8-input multiplexers similar to those shown in Fig. 2.17 and in Fig. 2.18. Random logic can also be used to implement the excitation logic similar to that shown in Fig. 2.19. The latter approach is called *superposition*, which requires more varieties of gate functions and more IC packages, whereas the multiplexer approach requires less design effort and reduced number of IC packages.

2.8 Semiconductor RAMs and ROMs

Random Access Memory (RAM) refers to a type of solid-state memories in which the computer can read or write information from or into any memory cell with equal access time, regardless of the cell's physical location in the memory array. Both MOS and bipolar RAMs are available in IC packaged form. RAMs are utilized as high-speed scratchpad, buffer memory, or are even used as mainframe memory in a digital system. Most of the current LSI RAMs are produced with MOS-based-technology. The sizes of the faster bipolar transistor RAMs are increasing, but still at higher cost.

Figure 2.21 illustrates the basic structure of an 8-word and n-bits-per-word bipolar RAM made from standard TTL cells with triple-emitter transistor. Voltage coincidence selecting is used to address the memory word by simultaneously raising the X and Y selection lines "high." The signal current at the "on" side of the bit line will flow into the current sense amplifier, when the Read/Write Enable is held at "high" during reading. To write information into the selected cells, the R/W Enable line is held at "low" to allow external data "1" or "0" transmitted through the write amplifier to turn "on" or "off" the transistor cell. The access time, cycle time, operating power dissipation, and density of major integrated semiconductor memory technologies are given in Table 2.4.

Semiconductor memories that are used to store fixed programs or permanent data tables are called *Read-Only Memory* (ROMs). The data content of a ROM can be written only once for all. Programming ROMs is usually done through masking

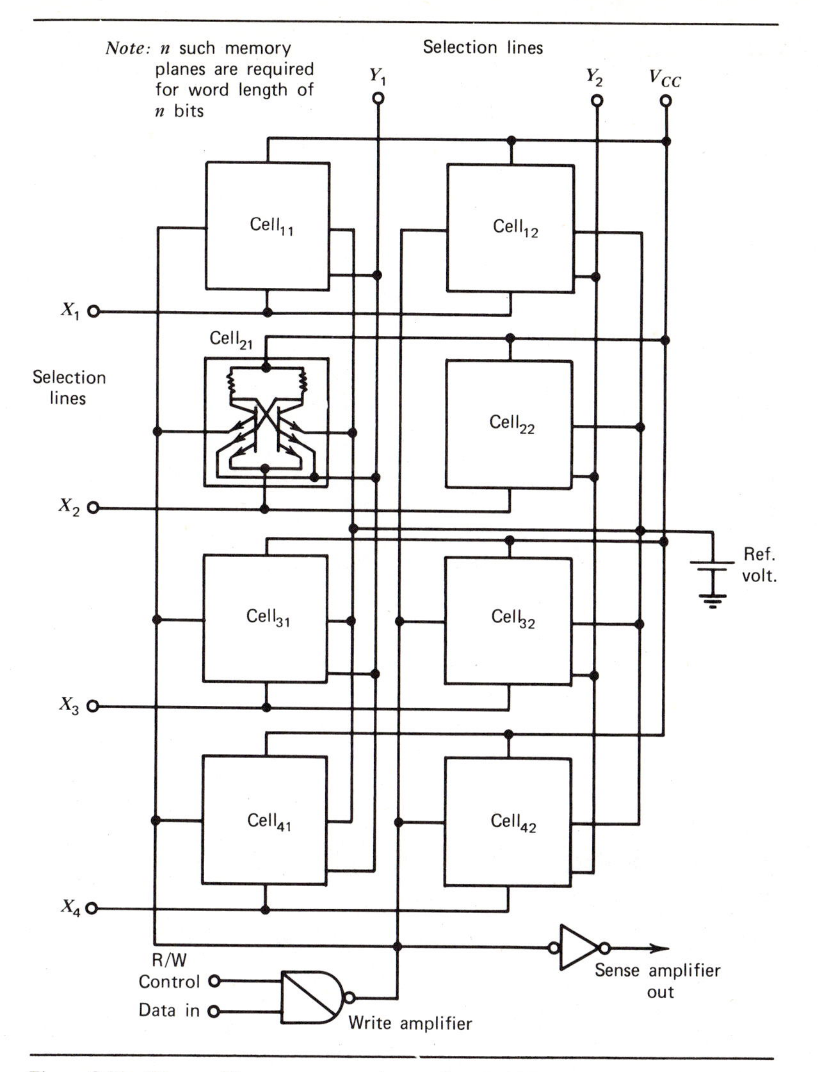

Figure 2.21 The per-bit memory array layout for a RAM system with 8 words (n bits per word) addressed by four X selection lines and two Y selection lines using coincidence technique.

during manufacture. Certain classes of ROMs that are field programmable once by the end users are called *Programmable Read-Only Memory* (PROM). ROMs can be also randomly accessed. They differ from RAMs only in that ROMs cannot be rewritten once programmed. Without this rewriting capability, ROMs are usually cheaper to manufacture. For special applications, a new class of *Electronically Alterable Read-Only Memory* (EAROM) is available. Such EAROM's can be reprogrammed at a slower write times compared with the RAMs. Because EAROMs are still used mainly for reading purposes, they are sometimes called *Read-Most*

Table 2.4 General Characteristics of Major Integrated Semiconductor Memory Technologies (1977)

	Characteristics				
Technologies	**Access Time** (nsec)	**Typical Density** (bits/chip)	**Operating Power Diss. per bit** (μW)	**Nonvolality**	**End User Cost** (¢/bit)
PMOS	300–400	16K	100	With battery	0.25
NMOS	150–250	16K	100	With battery	0.20
CMOS	50–60	4K	20	With battery	1.20
CCD	80–250	64K	10	With battery	0.08
TTL	50–60	4K	500	With battery	0.80
ECL	45–60	1K	600	With battery	2.50
I^2L	50	1K	400	With battery	4.50
MBM	3000	128K	10	Yes	0.02

Memory (RMM) or *writable ROMs*. General applications of ROMs include code conversion (encoding and decoding), alphanumeric character generation, sequential logic implementation for microprogrammed control, control-state timing, and generating elementary functions (trigonometric, logarithmic, etc.) through lookup tables.

2.9 Programmable Logic Arrays (PLAs)

A relatively new member of the memory-related semiconductor family is the *Field Programmable Logic Arrays* (FPLA), which fulfills the need for nonstandard logic functions in microprocessor and other computer systems. In practical digital design, one may often find it difficult to yield a minimal circuit with a restricted inventory of ICs. One can easily solve this problem with FPLAs. FPLA functions similarly to PROM in implementing arbitrary switching functions. The distinction of these two user-programmable devices is that only the prime implicants (necessary product terms) must be covered in a FPLA, whereas one must program all the minterms (canonical product terms) in a ROM. The presence of a fixed address decoder renders PROM addressing exhaustive, which often results in poor memory encoding efficiency. The FPLA does away with a fixed decoder to select the desired prime implicants in favor of a programmable address matrix.

The FPLA is functionally equivalent to a collection of AND gates, which may be ORed at any of its outputs. Figure 2.22 illustrates the structure of a small PLA containing 8 product terms (defined over 4 input variables) in each of the 3 sum terms. Each sum term (OR) controls an output function, which can be programmed to include up to 8 product terms by external electrical pulses. The dots in the inter-

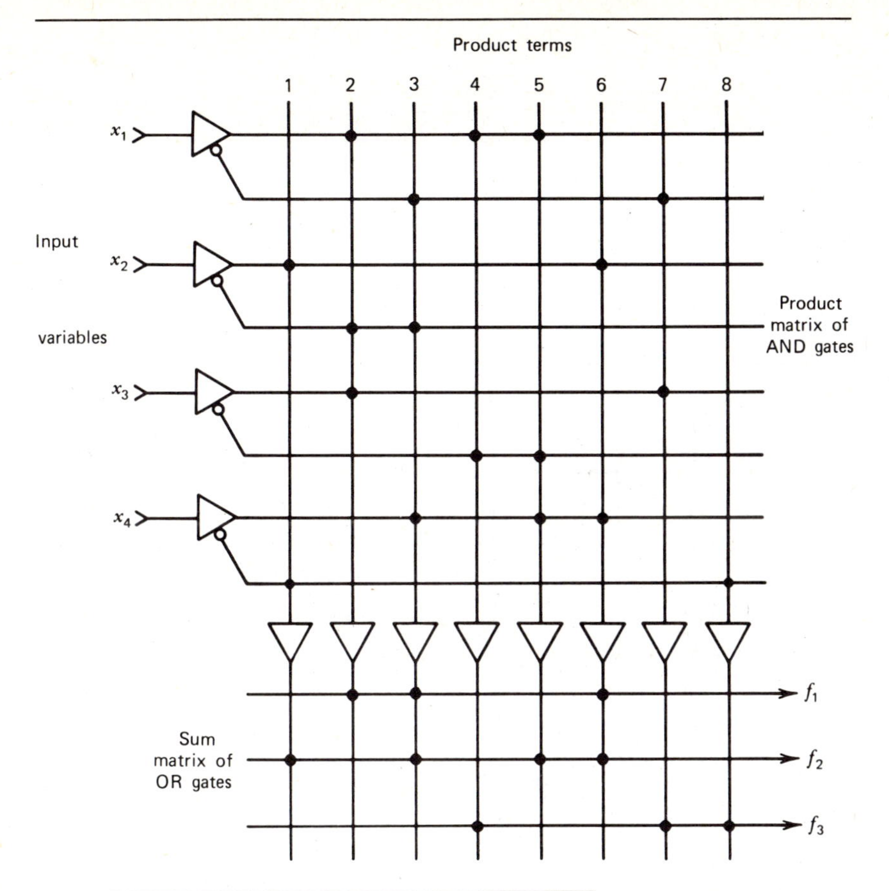

Product Term	Input Variables				Output Functions		
	x_1	x_2	x_3	x_4	f_1	f_2	f_3
1	d	1	d	0	0	1	0
2	1	0	1	d	1	0	0
3	0	0	d	1	1	1	0
4	1	d	0	d	0	0	1
5	1	d	0	1	0	1	0
6	d	1	d	1	1	1	0
7	0	d	1	d	0	0	1
8	d	d	d	0	0	0	1

d = don't care

Figure 2.22 The logic matrix and truth-table description of a PLA with 3 output functions, each containing up-to-8 product terms over 4 input variables.

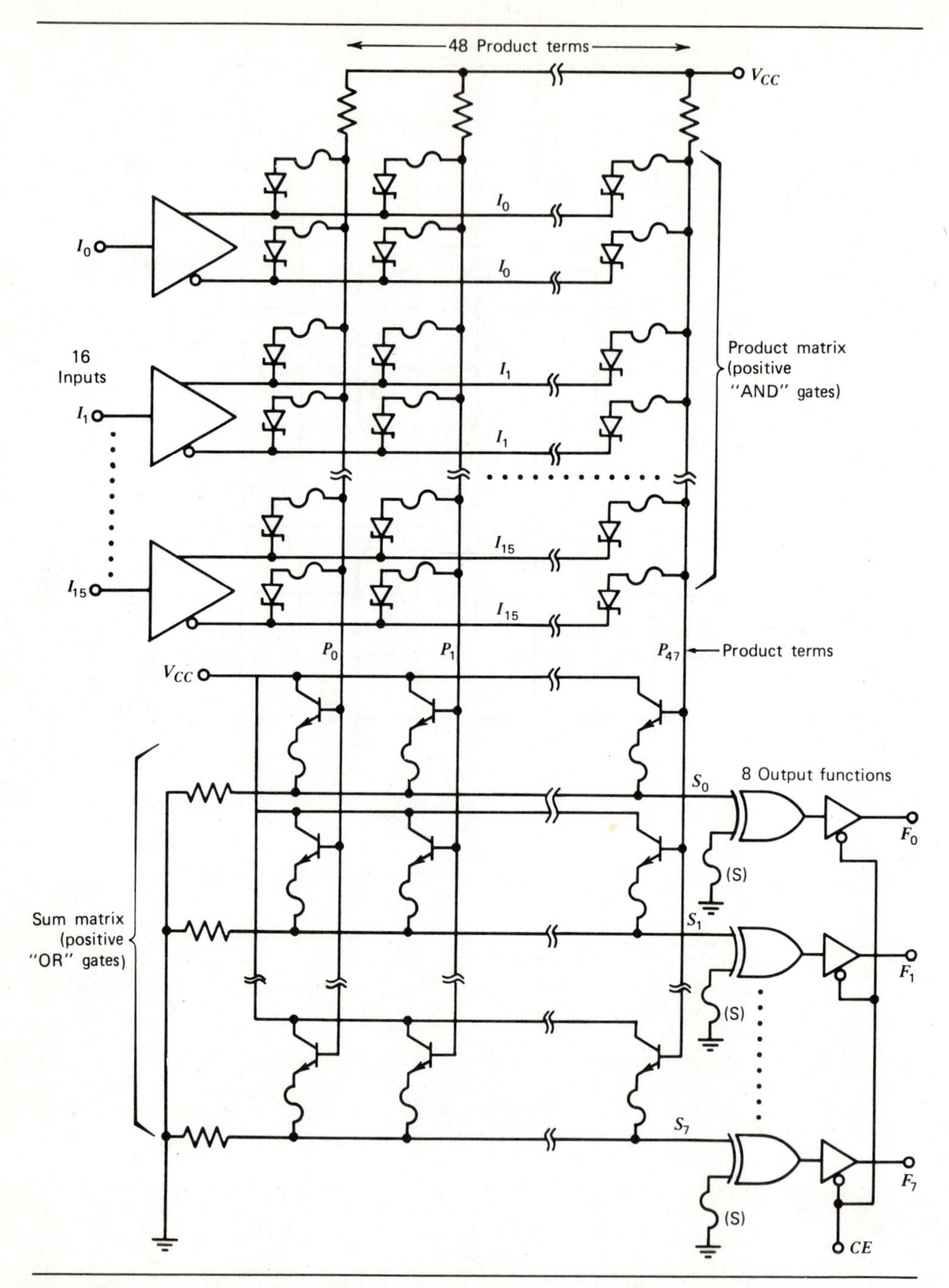

Figure 2.23 The schematic circuit diagram of the Signetics 82S101 bipolar Field Programmable Logic Array (FPLA) with 16 input variables and 8 output functions containing up-to-48 product term each. (Reprinted with permission from Signetics Corp., Sunnyvale, Calif. 94086, 1977).

sections correspond to activated AND gates in the upper matrix and correspond to OR gates in the lower matrix. Each row of input variable can be recognized as "true," "false," or "both" (don't care) by each column of the address matrix. As a result, waste storage for unused minterms is no longer required. The necessary logic output for the inactive minterms occurs by "default." Therefore, the FPLA requires less bits and less decoding logic to implement the same logic function that can be implemented with PROM's.

FPLA's containing 16 inputs and up to 48 product terms per each of the 8 output functions are now available. Figure 2.23 shows the schematic logic circuit of a Signetics 82S101 bipolar FPLA with tri-state outputs for expanding product terms. The chip enable ($\overline{CE}$) controls the expansion of input variables and inhibits the tri-state outputs. The maximum access time for this Signetics FPLA is 50 nsec with 600 mW power dissipation. Major applications of FPLAs include logic compression, memory overlays, the construction of fault detection networks, fast multibit shifters, and priority interrupt systems.

2.10 Magnetic Bubble Memories (MBMs)

Bubble memories can be viewed as solid-state integrated analogs of rotational electromechanical memories, such as discs, drums, and tape units. Both versions store information in the form of magnetized regions. The integrated bubble memories have these regions on the cylindrical domain in a thin layer of magnetic material, with magnetization opposite that of the surrounding area. The presence or absence of magnetic bubbles at specific locations corresponds to binary "1" or "0" stored at those locations. These bits are made accessible by moving the domains within the solid layers to an access device, as opposed to physically moving the storage medium, as with disc or tape units.

Storage material can be either a magnetic garnet epitaxially embedded in a nonmagnetic garnet substrate, or an amorphous metallic magnetic layer sputtered onto a substrate, such as glass. Because fewer mask levels and no stringent alignment are required in processing, the bubble memory price per bit should be lower than that of silicon integrated memories. Memory systems based on magnetic bubble technology have a number of significant applications, notably as microperipherals to replace the conventional rotating magnetic memories.

Magnetic Bubble Memories (MBMs) chips feature the basic properties of non-volatility, high density, modularity, and low-power consumption. Basic chip and system characteristics, as well as some qualitative features of MBMs, are given in Table 2.5. Because MBMs have a shift register organization and only on-bit from each minor loop is transferred to the major loop during memory accessing, block organization is probable. However, a bubble chip is bit, byte, or word addressable, because the Permalloy pattern defines discrete storage locations.

Comparing MBMs with other established memory such as RAM, CCDs, fixed-head discs cartridges, and floppy discs, Juliussen [15] presented two charts as shown in Fig. 2.24. The first chart shows the access and cost gap between RAM and

Table 2.5 General Characteristic of Magnetic Bubble Memories (MBMs) (Juliussen [15])

Chip Characteristics		**Qualitative Features**
Storage density	1M bits/in^2	Serial access memory
Chip capacity	64K to 100K bits	Shift register organization
Access time	1 to 4 ms	Block-oriented access
Transfer rate	100K bits/s	Bit addressable
Packaging	10/16-pin DIP	Nonvolatile memory
Power dissipation	< 1 W	Nondestructive read
Standby power	None	Read-modify-write cycle
Temperature range	-25 to 75°C	Stop/start operation
		Lowers effective access time
		Lowers power dissipation
System Characteristics		Permits variable transfer rate
		Minimizes buffer requirements
Transfer rate	0.1M to 1.5M bits/s	Modular storage capacities
Packaging	PC board	Give low entry price
Storage capacity	1M to 3M bits/board	Bit price almost independent of storage size
Power dissipation	10 to 20 W	Package pin-count independent of storage size
MTBF	10,000 + hours	Few manufacturing steps
Controller	Microprocessor-based	Manufacturing similar to IC production

rotating magnetic mass storage into which the CCDs, MBMs, and *Electronic Beam Addressable Memory* (EBAM) fit. The second chart shows the relationship between cost per bit and storage capacity. The chart shows that bit prices of MOS/RAM and CCD are almost independent of storage capacity; the same modularity appear in MBM. On the other hand, the cost of rotating magnetic memories depends heavily on storage capacity. MBMs and CCDs also have a packaging advantage over rotating mass memories; both can have several megabits on a printed circuit that would fit the CPU chasis.

The main advantage of MBMs over CCDs are their nonvolatility and higher bit density per chip. The main disadvantage of MBMs are the relatively low transfer rate, but this can be overcome with parallel transfer from multiple bubble chips. Bubble technology was introduced by Bell Laboratories in the late 1960s. Today there are many organizations, including Texas Instruments, IBM, Hewlett-Packard, Hitachi, and so on, that are announcing commercially available MBMs of 92K bits, 256K bits, or larger, MBM applications include mainframe peripheral, miniperipheral, microperipheral, small nonvolatile storage, and disc or tape recoder replacements.

2.11 Literature Guide and Digital IC Databooks

Readers who wish to enrich their knowledge of digital electronic devices should start with the books of Kohonen [16] and Blakeslee [2]. Kohonen has given a combined introduction of electronic circuits and of digital logic principles. Blakeslee's

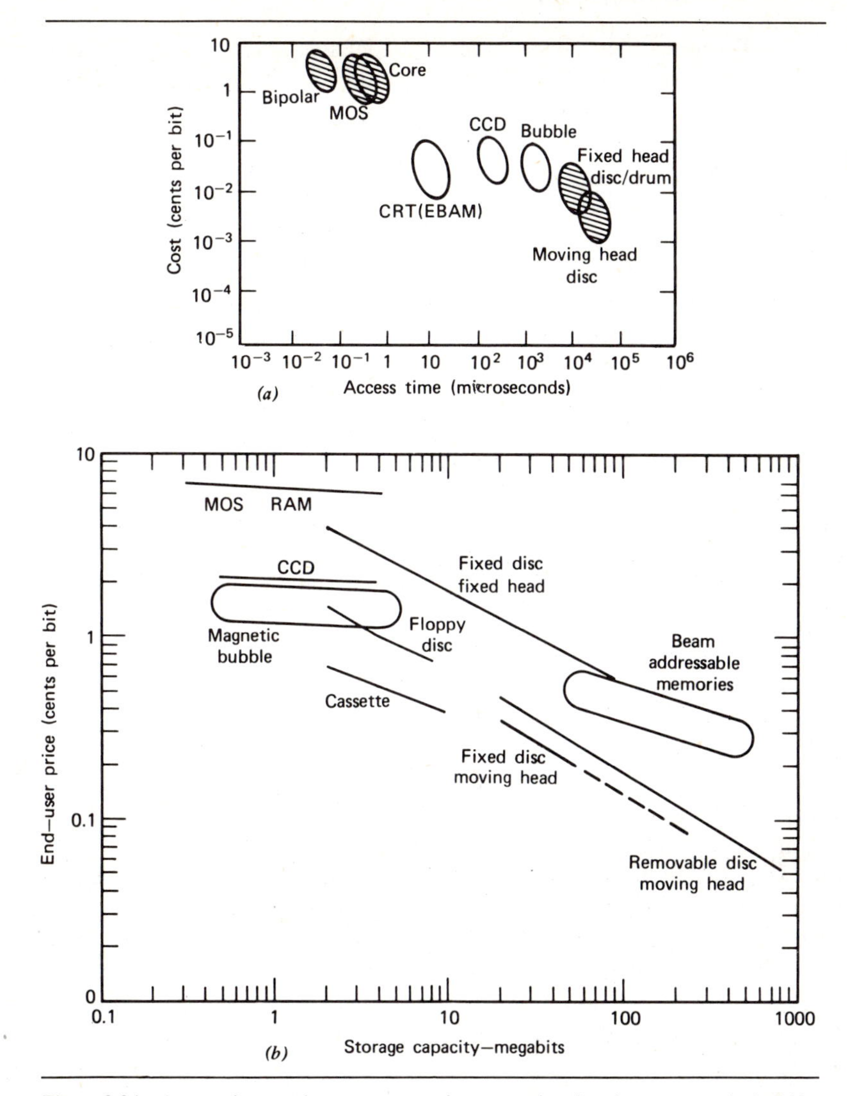

Figure 2.24 Access-time and storage capacity gaps showing how magnetic bubble memory (MBM) systems fit between conventional solid-state and rotating magnetic mass memories (*a*) The cost-per-bit versus access time. (*b*) The cost-per-bit versus storage capacity. (Reprinted with permission from *Computer Design*, 1976 [15])

book provides a summary of the commonly used standard MSI/LSI digital devices, according to their practical application potentials. Calhoun [5] has given an overview of computer hardware technology covering four generations of computers. A recent book on designing with standard IC's is by Wakerly [37]. Today there are numerous books on LSI *processing elements* (PE) and on *microprocessors* and their applications [13, 14, 29]. A false impression may exist that there is no longer any need to design arithmetic units, and one can simply use these LSI PEs or microprocessor chips. This statement is only partially correct. Because we are concerned with designing high-speed arithmetic processors with long word length, the "slow" and "short" word length PEs or microprocessors with limited computing power are not attractive, at least at the present stage. When the speed gap is closed and the basic microprocessor word length is extended in the future, however, the situation may change.

Practicing digital engineers may obtain the detailed characteristics and application parameters of IC devices from the published catalogies and data books supplied by electronic manufacturers. Among them, the most comprehensive treatment of the theory, design, and applications of modern digital IC's is the Texas Instruments Electronic Series. [18, 20], and supplemented by TI data book series [30, 31, 32, 33]. For designs with TTL or ECL logic families, readers may find these data books [4, 8, 12, 20, 23, 30] very useful. For information on new devices such as CCDs and MBMs, IEEE Press has published books [6, 19] covering theory and applications. Recently, Juliussen [15] and Tung [35] presented two interesting reports on MBMs developed separately at TI and IBM. For those who wish to have a review of digital logic design techniques, Mowle's book [22] is an appropriate choice; it offers a systematic approach starting from fundamental Boolean algebra to basic arithmetic circuit design.

There are numerous catalogies, data books, and application notes published by electronic manufacturers. The notable ones are Motorola Data Library [21], Fairchild Application Handbooks [7, 8], RCA IC's [26], National Semiconductor IC Catalogies [23, 24], and Signetics product manuals [27, 28], in addition to Texas Instruments and Intel publications mentioned above. The material on PLAs and MBMs is selected from [12, 25, 28] and from [6, 15, 33], respectively. Full adder circuit optimization is due to Liu et al. [17]. Array multipliers are from [1, 8, 11, 36].

For a review of IC digital logic families, Garrett [10] has given an excellent characterization of the logic circuit families developed up to 1970. For more recent developments and future trends of digital electronic devices and of integrated memories, readers may enjoy reading the recent articles [9, 34, 38] on computer hardware. This literature guide is by no means complete. Interested readers should check the numerous IC Data Books for detailed characteristics and application requirements of new digital IC devices.

References

[1] Advanced Micro Single-Spaced Devices, "TTL/MSI AM 2505 4-bit by 2-bit Two's Complement Multiplier," Sunnyvale, Calif., 1972.

[2] Benedek, M., "Developing Large Binary to BCD Conversion Structures," *IEEE Trans. Comp.*, Vol. C-26, No. 7, July 1977, pp. 688–689.

[3] Blakeslee, T. R., *Digital Design with Standard MSI and LSI*, Wiley-Interscience, New York, 1975.

[4] Blood, W., Jr., *MECL System Design Handbook*, Motorola Semiconductor Products, Inc., Mesa, Arizona, 1972.

[5] Calhoun, D. F., "Hardware Technologies," in *Computer Science*, (Cardenas et al. ed.), Wiley-Interscience, New York, 1972, Chap. 1.

[6] Chang, H. (ed.), *Magnetic Bubble Technology: IC Magnetics for Digital Storage and Processing*, IEEE Press, New York, 1975.

[7] Fairchild, *Easy ECL: 9500 Series High-Speed Logic*, Fairchild Semiconductor, Mountain View, Calif., 1971.

[8] Fairchild, *The TTL Application Handbook*, Fairchild Semiconductor, Mountain View, Calif., 1973.

[9] Falk, H., "Computer Hardware Software," *IEEE Spectrum*, January 1974, pp. 39–43.

[10] Garrett, L. S., "Integrated-Circuit Digital Logic Families," Parts I, II, III. *IEEE Spectrum*, October, November, December 1970.

[11] Hughes Aircraft Co., *Bipolar LSI* 8-bit Multiplier H1002MC," Newport Beach, Calif. 1974.

[12] Hwang, K., "A TTL Programmable Logic Array and its Applications," *Proc. of the IEEE*, March 1976, pp. 368–369.

[13] Intel Corp. Staff, *Microcomputer Systems Data Book*, Intel Corp., Santa Clara, Calif., 1976.

[14] Intel Corp. Staff, *Intel Data Catalog*, Santa Clara, Calif., 1976.

[15] Juliussen, J. E., "Magnetic Bubble Systems Approach Practical Use," *Computer Design*, October 1976, pp. 81–91.

[16] Kohonen, T., *Digital Circuits and Devices*, Prentice-Hall, Englewood Cliffs, N.J. 1972.

[17] Liu, T. K. et al., "Optimal One-Bit Full Adders with Different Types of Gates," *IEEE Trans. Comp.*, Vol. C-23, No. 1, January 1974, pp. 63–70.

[18] Luecke, G., et al., *Semiconductor Memory Design of Applications*, McGraw-Hill, New York, 1973.

[19] Melen, R. and Buss, D., "Charge-Coupled Devices: Technology and Applications," *IEEE Press*, New York, 1977.

[20] Morris, R. L. et al. (eds.), *Designing with TTL Integrated Circuits*, McGraw-Hill, New York, 1971.

[21] Motorola Staff, *Semiconductor Data Library*, Motorola Semiconductor Products, Inc., Phoenix, Arizona 1976.

[22] Mowle, F. J., *A Systematic Approach to Digital Logic Design*, Addison-Wesley, Reading, Mass., 1976.

[23] National Semiconductor, *Digital Integrated Circuits*, National Semiconductor Corp., Santa Clara, Calif., 1974.

[24] National Semiconductor, *CMOS Integrated Circuits*, National Semiconductor Corp., Santa Clara, Calif., 1975.

[25] Priel, V. and Holland, P. "Application of a High-Speed Programmable Logic Array," *Computer Design*, December 1973, pp. 34–96.

[26] RCA Solid-State, *RCA Integrated Circuits*, RCA Corp., Somerville, N.J., 1976.

[27] Signetics Application Book: *Digital, Linear, MOS*, Signetics Corp., Sunnyvale, Calif., 1974.

[28] Signetics Application Notes, *Bipolar Field Programmable Logic Arrays*, Signetics Corp., Sunnyvale, Calif., 1976.

[29] Soucek, B., *Microprocessor and Microcomputers*, Wiley-Interscience, New York, 1976.

[30] Texas Instruments Staff, *TTL Data Book and Supplement*, Texas Instruments, Inc., Dallas, Texas, 1973.

[31] Texas Instruments Engineering Staff, *The Integrated Circuits Catalog for Design Engineers*, Components Group, Texas Instrument, Inc., CC-401, Dallas, Texas, 1973.

[32] Texas Instruments Engineering Staff, *The Semiconductor Memory Data* Book, Texas Instruments, Inc., CC-420, Dallas, Texas, 1975.

[33] Texas Instruments Semiconductor Staff, "Magnetic Bubble Memories and System Interface Circuits," T.I., Inc., Dallas, Texas, 1977.

[34] Torrero, E. A., "Solid-State Devices," *IEEE Spectrum*, January 1977, pp. 48–54.

[35] Tung, C. et al., "Bubble Ladder for Information Processing," *IBM Research Report*, #RJ 1556, 1975.

[36] TRW Application Notes, "LSI Multipliers," Space Park, Radondo Beach, Calif., March 1977.

[37] Wakerly, J. F., *Logic Design Projects Using Standard Integrated Circuits*, Wiley, New York, 1976.

[38] Warner, R. W., Jr., "I-Squared L: A Happy Merger," *IEEE Spectrum*, May 1976, pp. 42–47.

Problems

Prob. 2.1 Implement a one-bit Controlled Add/Subtract (CAS) cell with one XOR gate and two 4-input multiplexers, one for the sum output and one for the carry output. Use the two operand bits as selection signals. Assume that both the true and complemented carry-in signals are available.

Prob. 2.2 Determine the minimal numbers of IC packages required to construct the following multiplexer trees with exclusively quad 2-input, dual 4-input, dual

8-input, and single 16-input multiplexers. Because each dual 8-input multiplexer and each single 16-input multiplexer requires a larger IC package occupying twice the PC board area, they are counted as two packages each.
(a) A 64-input multiplexer
(b) A 128-input multiplexer
(c) A 1024-input multiplexer

Prob. 2.3 Build a 16-to-65536 line decoder (a 1-out-of-2^{16} address decoder) with minimum numbers of dual 2-to-4, single 3-to-8, and 4-to-16 line decoders. Show the schematic interconnection of your design. Suppose that chip enables are available as shown in Figs. 2.4 and 2.5.

Prob. 2.4 Construct two modular code converters:
(a) Use a number of the standard 6-bit Binary-to-BCD converting modules in Fig. 2.6 to construct a 16-bit Binary-to-BCD converter.
(b) Use a number of the standard BCD-to-Binary converting in Fig. 2.8 to construct an 6-decade BCD-to-binary converter.

Prob. 2.5 Build a 48-bit magnitude comparator with a number of 4-bit comparators given in Fig. 2.12. All unused terminals should be properly specified. Show the schematic interconnection diagram of your design.

Prob. 2.6 Design an 8-by-8 additive multiply module, which will compute the following expression

$$\mathbf{P} = \mathbf{A} \times \mathbf{B} + \mathbf{C} + \mathbf{D}$$

where **A**, **B**, **C**, and **D** are all 8-bit numbers and **P** is a 16-bit number. Only full adders and 2-input AND gates are allowed in your design.

Prob. 2.7 Design an 8-bit 16-function register with D-type flip-flops and 16-input multiplexers. The function table is given below. Show the schematic diagram of your design.

Function Control WXYZ	Register Function	Function Control WXYZ	Register Function
0000	No operation	1000	Parallel Load **A**
0001	Clear to zero	1001	Parallel Load **B**
0010	Set to one	1010	**Q** OR **A** (bitwise)
0011	One's complement	1011	Two's complement
0100	Count up	1100	**Q** XOR **A** (bitwise)
0101	Count down	1101	**Q** AND **A** (bitwise)
0110	Shift right	1110	**Q** XOR **B** (bitwise)
0111	Shift left	1111	**Q** AND **B** (bitwise)

Note: **A** and **B** are 8-bit external inputs and **Q** is the current register contents

Prob. 2.8 Describe briefly the following semiconductor memory technologies:
(a) Random-Access Memory (RAM)
(b) Read-Only Memory (ROM)
(c) MOS vs. Bipolar technologies
(d) Charge-Coupled Devices (CCD)
(e) Integrated Injection Logic (I^2L)
(f) Magnetic Bubber Memory (MBM)

Compare CCD, I^2L, and MBM memory devices in terms of access time, density, power dissipation, nonvolatility, and bit cost.

Prob. 2.9 Program a Field Programmable Logic Array (FPLA) with three output lines, each containing up-to-24 product terms over 6 input variables similar to that shown in Fig. 2.22 to realize a 3-bit magnitude comparator.

Prob. 2.10 Compare Read-Only Memory (ROM) and Programmable Logic Arrays (PLA) with respect to the following technical considerations.
(a) Fabrication technologies
(b) Bit capacities
(c) Programmabilities
(d) Arithmetic design applications

Prob. 2.11 Design the following sign-complement magnitude comparators:
(a) A 4-bit one's complement comparator
(b) A 4-bit two's complement comparator
Each comparator should have three output lines indicating the result of "$>$", "$=$", and "$<$". Note that the leading bit of each input operand is reserved for the sign.

Chapter 3

Fast Two-Operand Adders/Subtractors

3.1 Introduction

High-performance adders are essential not only for addition, but also for subtraction, multiplication, and division. The speed of a digital arithmetic processor depends heavily on the speed of the adders used in the system. This chapter starts with the specification of arithmetic rules for various fixed-point binary additions and subtractions. Then we will study various carry/borrow acceleration techniques for designing fast two-operand adders. Asynchronous *carry-completion adders* are discussed first. Three classes of synchronous adders, namely, *conditional-sum*, *carry-select*, and *carry-lookahead* adders, are described subsequently. An evaluation of various two-operand adders is also provided.

Extensive literature was published in the 1950s and 1960s on high-speed adder designs. We present here only the major two-operand adder classes, but there are many other variations of these classes. We shall emphasize the design methodology associated with each class of adders. With today's LSI technology, adder construction is no longer a great burden of the designer. However, a good understanding of the carry speed-up techniques will help arithmetic design in general. For example, multiplication speed can be greatly improved when fast multiple-operand adders are used. Examples of adder applications will appear in later chapters. In most digital computers, more than one adder may be required. For example, a floating-point processor requires at least two adders, one for processing the mantissas and one for the exponents. In relative memory addressing, a dedicated fast adder may be required to compute the effective memory addresses. Only conventional two-operand adders are described in this chapter. The unconventional multiple-operand adders, signed-digit adders, and ALUs will be discussed in the next chapter.

3.2 Basic Sign-Complement and Sign-Magnitude Adders

Arithmetic algorithms for addition or subtraction in three fixed-point signed number systems are given below. The corresponding ripple-carry circuit realizations of these addition algorithms are illustrated with binary radix. It suffices to consider only the addition or subtraction of FXP integers.

Let radix-r *positive* integers be denoted by regular letters as $\mathbf{A} = a_{n-1} \cdots a_1 a_0$ with sign $a_{n-1} = 0$. The diminished-radix complement of $\mathbf{A}$ is denoted as $\overline{\mathbf{A}} = \bar{a}_{n-1} \cdots \bar{a}_1 \bar{a}_0$ where $\bar{a}_i = r - 1 - a_i$ for all i and $\overline{\mathbf{A}} = (r^n - 1) - \mathbf{A}$. The radix-complement of $\mathbf{A}$ is represented by $\overline{\mathbf{A}} + 1$ which equals the number $r^n - \mathbf{A}$. We shall use scripted letters such as $\mathscr{A} = a_{n-1} \cdots a_1 a_0$ to denote a *signed* integer, either positive or negative, in a chosen complemented form.

Radix-Complement Arithmetic

Let $\mathscr{A} = a_{n-1} \cdots a_1 a_0$ and $\mathscr{B} = b_{n-1} \cdots b_1 b_0$ be two radix-complement integers. The addition of $\mathscr{A}$ and $\mathscr{B}$ follows the following procedure.

Step 1. Add digit-by-digit directly starting from the least significant pair (a_0, b_0) to the leftmost pair of sign digits, assuming a zero initial carry input to the right end.

Step 2. Compare the carry entering the sign position with the carry out of the sign position. The result is in correct radix-complement form, if the above two carries agree (both "1" or both "0"), and the carry out of the sign position will be ignored. When the two carries differ, addition overflow has occurred and the result is invalid.

Subtraction of two radix-complement numbers is performed by first obtaining the radix-complement of the subtrahend and then adding it to the minuend following the above addition procedure. This radix-complement addition/subtraction algorithm can be proved by considering three cases.

Case 1. Both $\mathscr{A}$ and $\mathscr{B}$ are positive, in which $\mathscr{A} = \mathbf{A} = a_{n-1} \cdots a_1 a_0$ and $\mathscr{B} = \mathbf{B} = b_{n-1} \cdots b_1 b_0$ and $a_{n-1} = b_{n-1} = 0$. Then the result $\mathscr{S} = \mathscr{A} + \mathscr{B} = \mathbf{A} + \mathbf{B} = s_{n-1} \cdots s_1 s_0$. In this case, no carry can be generated out of the sign position. The resulting sum will be positive ($s_{n-1} = 0$) and in correct form if $\mathbf{A} + \mathbf{B} < r^{n-1}$, which means that no carry will enter the sign position. Otherwise ($\mathbf{A} + \mathbf{B} \geq r^{n-1}$), addition overflow will take place.

Case 2. Both $\mathscr{A}$ and $\mathscr{B}$ are negative. Let $\mathscr{A} = r^n - \mathbf{A}$ and $\mathscr{B} = r^n - \mathbf{B}$. Then

$$\mathscr{S} = \mathscr{A} + \mathscr{B} = (r^n - \mathbf{A}) + (r^n - \mathbf{B}) = 2 \cdot r^n - (\mathbf{A} + \mathbf{B}).$$

Because $a_{n-1} = b_{n-1} = 1$, there is always a carry out of the sign position. Ignoring the carry out of the sign position is equivalent to

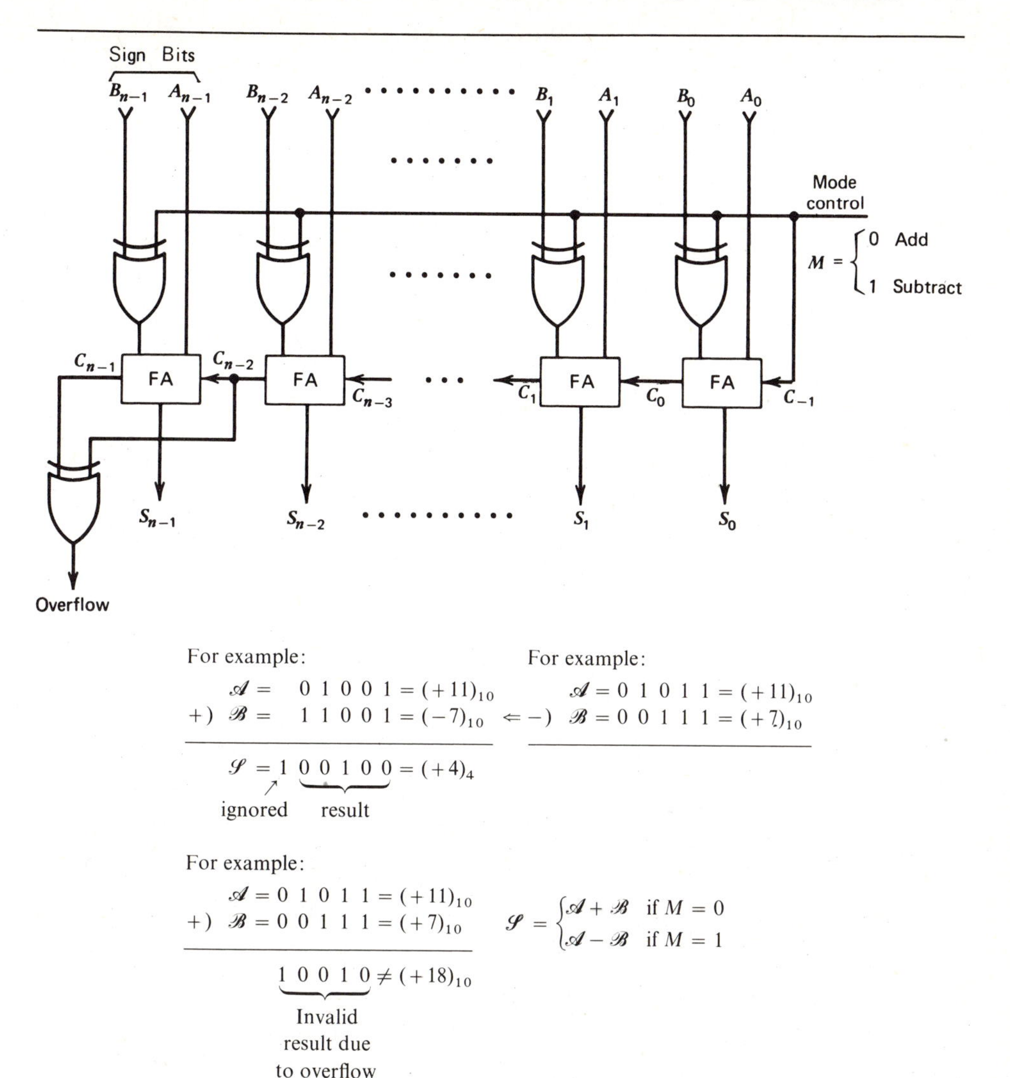

Figure 3.1 Two's complement arithmetic addition/subtraction and the corresponding ripple-carry adder design.

dividing $\mathscr{S}$ by r^n and retaining the remainder; that is,

$$\mathscr{S}/r^n = [2r^n - (\mathbf{A} + \mathbf{B})]/r^n = 1 + [r^n - (\mathbf{A} + \mathbf{B})]/r^n.$$

Therefore, the result is in correct from $r^n - (\mathbf{A} + \mathbf{B})$, if $\mathbf{A} + \mathbf{B} \leq r^{n-1}$. Otherwise ($\mathbf{A} + \mathbf{B} > r^{n-1}$), an addition overflow has occurred.

Case 3. $\mathscr{A}$ is positive and $\mathscr{B}$ is negative (or $\mathscr{A}$ negative and $\mathscr{B}$ positive, which is a similar case). In this case, no overflow will occur. Let $\mathscr{A} = \mathbf{A}$ and $\mathscr{B} = r^n - \mathbf{B}$. Then $\mathscr{S} = \mathscr{A} + \mathscr{B} = \mathbf{A} + (r^n - \mathbf{B}) = r^n + (\mathbf{A} - \mathbf{B})$.

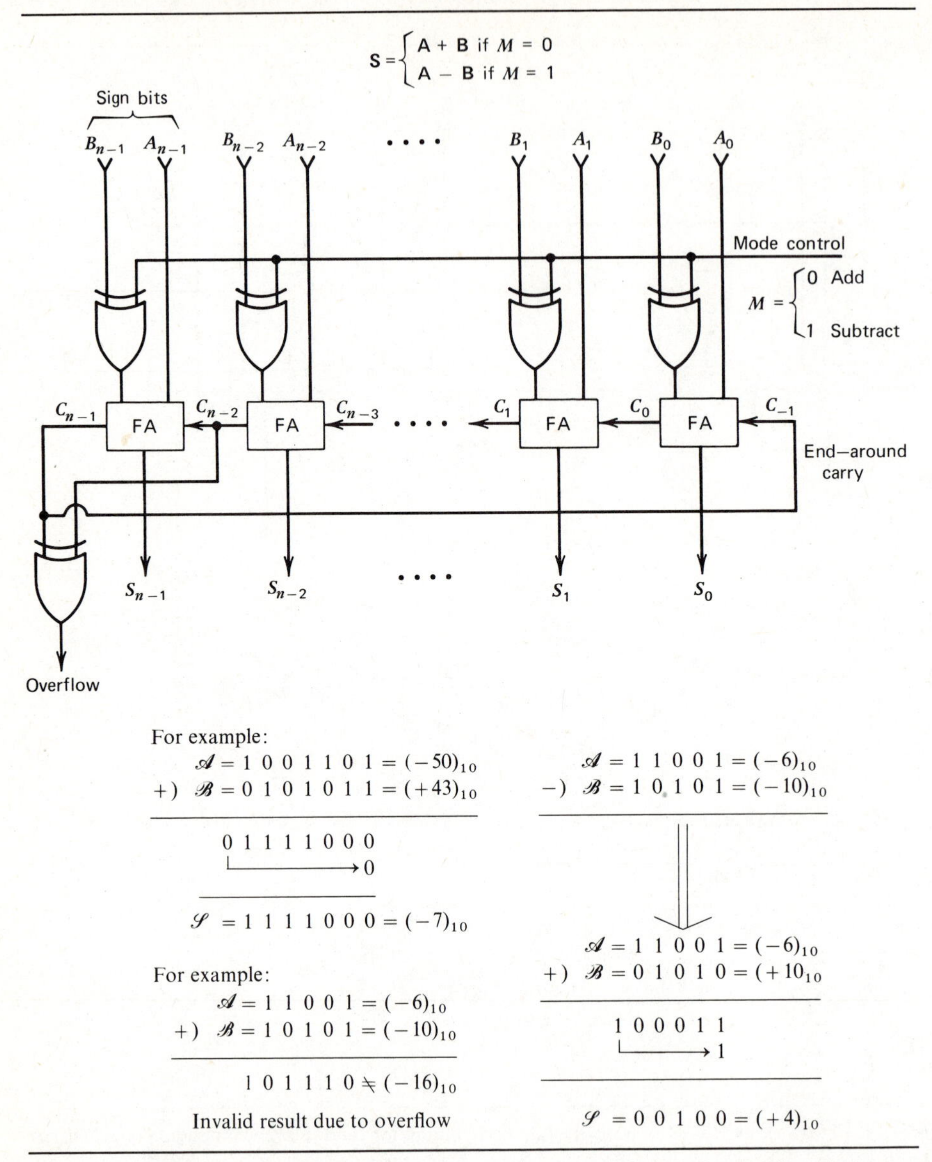

Figure 3.2 One's complement arithmetic addition/subtraction and the corresponding ripple-carry adder/subtractor.

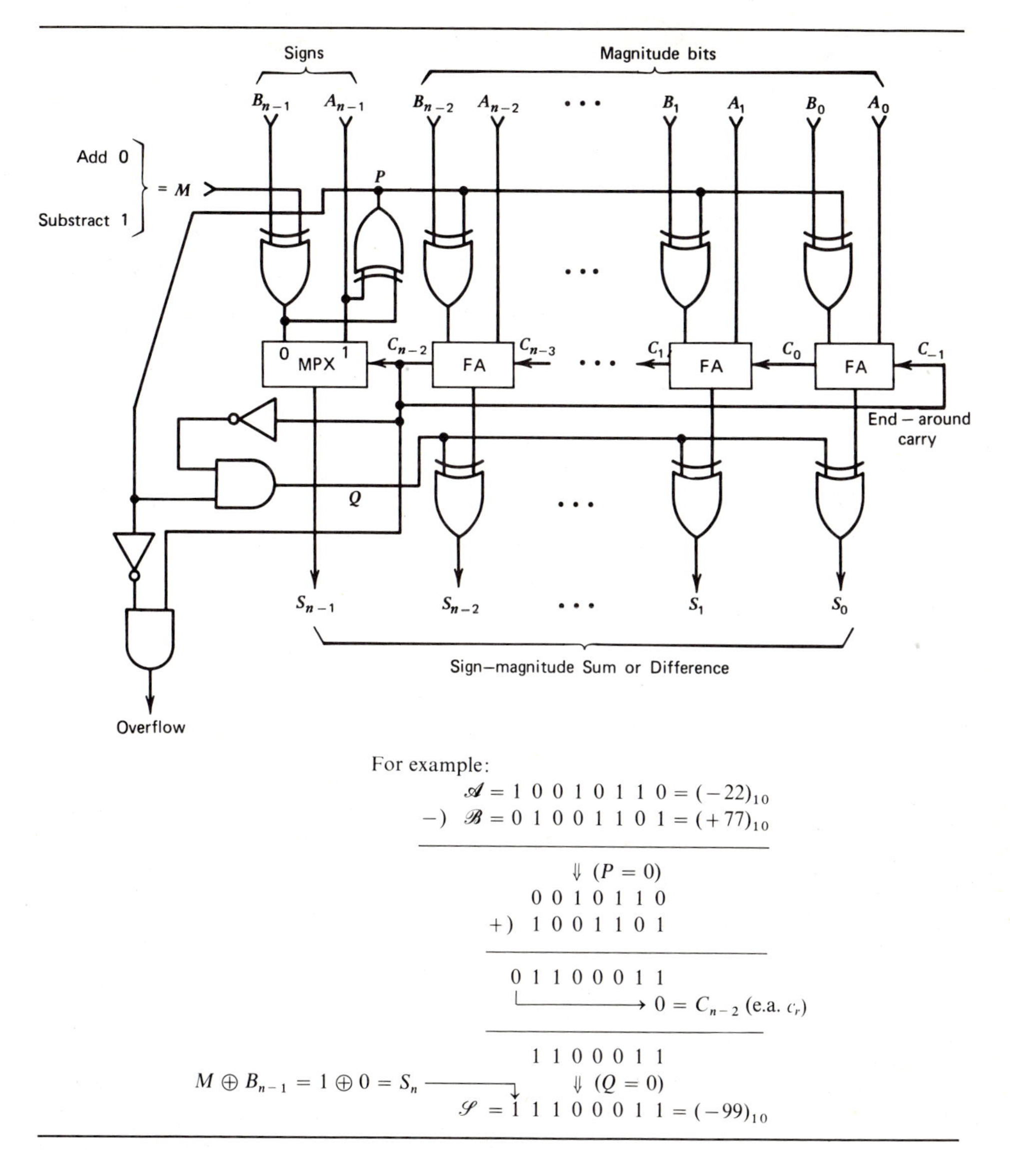

For example:

$$
\begin{array}{rl}
 & \mathscr{A} = 1\ 0\ 0\ 1\ 0\ 1\ 1\ 0 = (-22)_{10} \\
-) & \mathscr{B} = 0\ 1\ 0\ 0\ 1\ 1\ 0\ 1 = (+77)_{10} \\
\hline
 & \Downarrow\ (P = 0) \\
 & 0\ 0\ 1\ 0\ 1\ 1\ 0 \\
+) & 1\ 0\ 0\ 1\ 1\ 0\ 1 \\
\hline
 & 0\ 1\ 1\ 0\ 0\ 0\ 1\ 1 \\
 & \llcorner\!\longrightarrow 0 = C_{n-2}\ (\text{e.a. } c_r) \\
\hline
 & 1\ 1\ 0\ 0\ 0\ 1\ 1 \\
 & \Downarrow\ (Q = 0) \\
 & \mathscr{S} = 1\ 1\ 1\ 0\ 0\ 0\ 1\ 1 = (-99)_{10}
\end{array}
$$

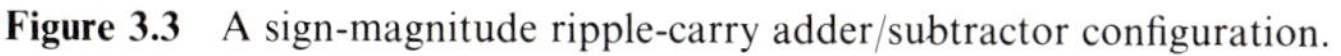

$M \oplus B_{n-1} = 1 \oplus 0 = S_n$

Figure 3.3 A sign-magnitude ripple-carry adder/subtractor configuration.

Obviously, the resulting sum will be negative and in the correct form if $\mathbf{B} > \mathbf{A}$. When $\mathbf{B} \leq \mathbf{A}$, the result will be $\mathscr{S} = r^n + (\mathbf{A} - \mathbf{B})$. By ignoring the carry out of the sign position, we divide $\mathscr{S}$ by r^n, that is, $\mathscr{S}/r^n = 1 + (\mathbf{A} - \mathbf{B})/r^n$. The remainder $\mathbf{A} - \mathbf{B} = |\mathbf{A} - \mathbf{B}|$ equals the resulting sum.

The schematic design of a binary radix-complement (two's complement) adder/subtractor is shown in Fig. 3.1. Several numerical examples are also attached to verify the operations. The initial carry input to the rightmost full adder is connected to the function mode line M, which equals "0" for addition and "1" for subtraction. Note that the radix complement of the subtrahend is obtained by forming $\overline{\mathbf{B}} + 1$.

Diminished-Radix-Complement Arithmetic

Let $\mathscr{A}$ and $\mathscr{B}$ be two diminished-radix-complement integers. The addition/subtraction rules are similar to those for radix-complement arithmetic, except that when overflow does not occur, the carry out of the sign position is added back to the least significant digit to yield the correct sum (or difference) in diminished-radix notation. This feedback carry is known as the *end-around carry*. The proof of diminished radix addition algorithm will be left as an exercise for the reader.

Numerical examples for diminished-radix addition and subtraction for the binary system $r = 2$ are given in Fig. 3.2, together with a schematic circuit diagram of an n-bit one's complement *ripple-carry adder/subtractor* with external ADD/SUBTRACT function control and overflow detection. Note that the end-around carry will never cause an oscillation in the closed-loop circuit, because any carry triggered by the end-around carry input will never propagate beyond the point where it was originated. An n-bit ripple-carry adder, either for one's complement or two's complement arithmetic, requires the following time delay

$$\Delta(n\text{-bit ripple-carry adder}) = n \cdot 2\Delta + 6\Delta = (2n + 6)\Delta \tag{3.1}$$

Sign-Magnitude Arithmetic

When the two operands are given in sign-magnitude form, one may wish to build a circuit to add or to subtract the two numbers directly, resulting in a sign-magnitude sum or difference. Both precomplement and postcomplement circuits are needed in this case. The control logic for these two complementing stages as well as the overflow detection logic are shown in Fig. 3.3. Note that the end-around carry is also needed here, because embedded within this system is the diminished-radix complement operation.

Table 3.1 summarizes the major hardware parts and operating speed for the three basic binary ripple-carry adder/subtractor configurations. It can be concluded that the sign-magnitude addition/subtraction requires the most hardware and yet results in a slower speed, compared with the sign-complement designs. Addition/subtraction using either one's complement or two's complement arithmetic requires the same

Table 3.1 Hardware Components and Execution Time of Three **FXP** Ripple-carry Adder/Subtractor Configurations

n-bit Adder/ Subtractor Types	Hardware Components			Execution Time (in terms of Δ)
	F.A.	XOR	Misc. Parts	
One's complement	n	$n + 1$	No	$(2n + 6)\Delta$
Two's complement	n	$n + 1$	No	$(2n + 6)\Delta$
Sign-magnitude	$n - 1$	$2n$	2-input MPX, 2 NOT gates, 2 AND gates	$(2n + 13)\Delta$

amount of hardware and almost equal speed. Without the feedback end around carry, however, the two's complement adder/subtractor is considered simpler. In what follows, we shall discuss only sign-complement designs, with the understanding that subtraction can be done through complemented addition.

3.3 Asynchronous Self-Timing Addition

The serial ripple carry adders just described are slow because each bit position must wait for a ripple carry to propagate through the lower positions. The ripple-carry design, based on the worst-case carry propagation, has a total time delay that is linearly proportional to the length n of the adder. Described below is a better design method resulting in a significant time saving by detecting the actual longest carry propagation in the adder, which may be much shorter than that of the worst-case propagation in an average sense.

When adding two binary numbers of n bits each, there are two types of carries that contribute to the final sum. As illustrated by the numerical example in Fig. 3.4, the carries independently generated at all columns form the *primary carry vector* **C**. The l's in vector **C** are determined by detecting ones in the pair of summand bits of the preceding column. Therefore, all the bits in vector **C** can be generated in parallel. The carries triggered by carry propagation from preceding carries form the *secondary carry vector* **D** as shown. The nonzero bits in vector **D** are generated by checking the carries triggered by primary carries in vector **C**. The resulting sum **S** is then obtained by performing bitwise mod-2 addition of the four vectors **A, B, C** and **D**.

Let us denote the four nonzero carries in vector **C** as a, b, c, and d. The left-going arrows under vector **D** depict the carry propagation lengths. The longest carry length in this example is five, instead of the worst-case $n = 15$ in a 16-bit adder. Let $\mathbf{E}_n(p)$ be the *average longest length* of carry propagation in the addition of two randomly chosen binary numbers of n bits each. In 1946, Burks, Goldstine, and Von Neumann

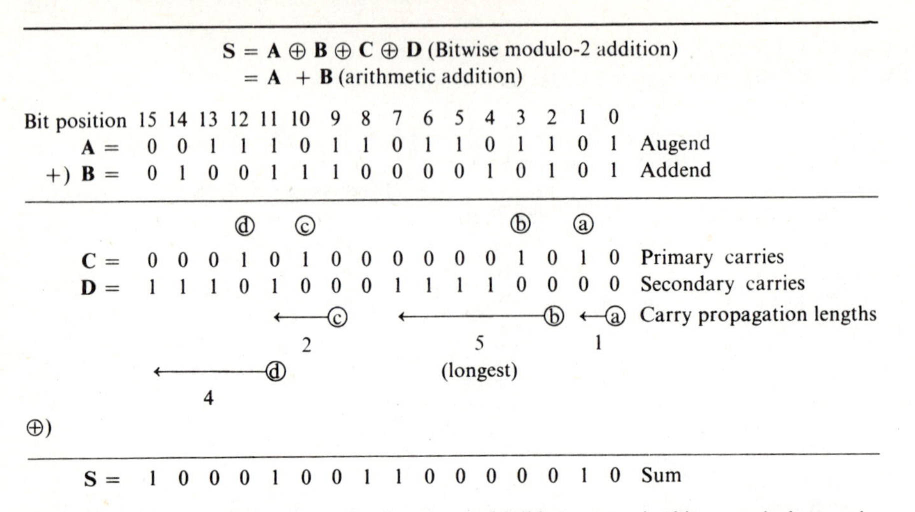

Note: The worst-case carry propagation length $n = 16$ did not occur in this numerical example.

Figure 3.4 Primary and secondary carry generations and longest carry propagation length during addition.

proved through statistical analysis over all random binary numbers that this average longest carry length is upper bounded by

$$\mathbf{E}_n(p) \leq \log_2 n \tag{3.2}$$

This gives an average longest carry length of no greater than 5 for a 32-bit sum or 6 for a 64-bit sum.

The ratio of the worst-case longest carry to that of the average value

$$\boldsymbol{\theta} = n/\log_2 n \tag{3.3}$$

becomes appreciable for large n. This implies that one can build an adder which can add $n/\log_2 n$ times faster than the ripple-carry adder in an average sense. The operating speed of an adder depends on the longest carry propagation length of the two given binary numbers. For example, it requires only 4 stages of carry propagation delay ($4 \times 2\Delta = 8\Delta$) to complete the addition of the two 16-bit numbers as shown in Fig. 3.4, about four times faster than that of a ripple-carry adder. Different pairs of binary numbers may result in different worst-case time delays, but the average time approaches $\log_2 n$.

We consider such adders *self-timing* and *asynchronous*. Self-timing refers to the fact that the adder is problem-dependent and requires variable add time to complete the operation. Asynchronous means that an unclocked control logic can be used to close up the unnecessary waiting gaps between adjacent arithmetic operations, because most additions require much less time than that of the worst-case delay. The conventional synchronous adder with a rate determined by the worst-case time delay may slow down the entire operations in an average sense. The major disadvantages of an asynchronous adder are the increased complexity of the control logic.

3.4 Carry-Completion Sensing Adders

Figure 3.5 shows a logic circuit that provides a parallel propagation of all primary carries in the vector **C** defined above. When the primary carry having the longest propagation path reaches its destination position in vector **D**, a *carry-completion* signal should be generated to indicate the completion of carry propagation. The addition can be then completed by modulo-2 addition in all the n independent full adders with just a 2Δ time delay.

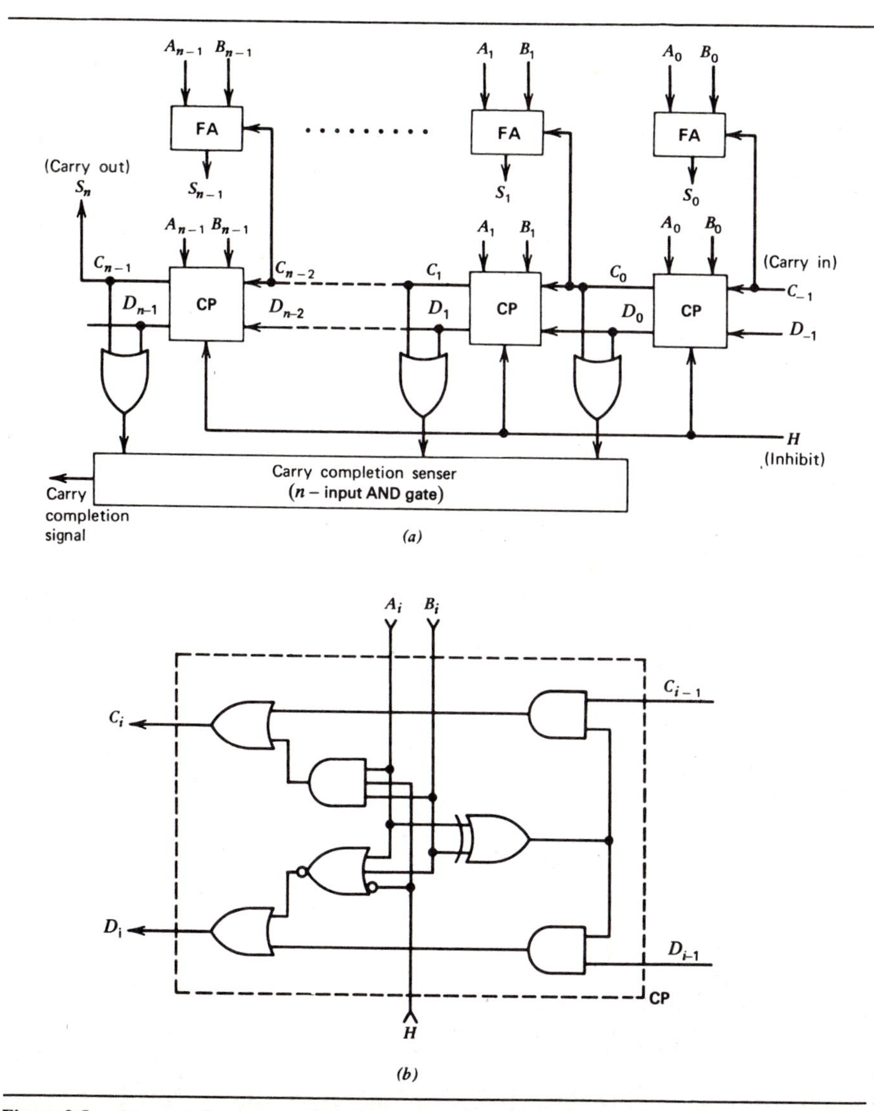

Figure 3.5 An asynchronous self-timing carry-completion sensing adder. (*a*) A carry-completion sensing adder. (*b*) The interior logic of a **CP** cell.

The *carry propagation* (**CP**) cell shown in the figure is described by the following set of Boolean equations for $i = n - 1, \ldots, 2, 1$

$$C_i = C_{i-1} \cdot (A_i \oplus B_i) + A_i \cdot B_i \cdot H \tag{3.4}$$

$$D_i = D_{i-1} \cdot (A_i \oplus B_i) + \bar{A}_i \cdot \bar{B}_i \cdot H \tag{3.5}$$

and for $i = 0$

$$C_0 = (H \cdot C_{-1}) \cdot (A_0 \oplus B_0) + A_0 \cdot B_0 \cdot H \tag{3.6}$$

$$D_0 = (H \cdot D_{-1}) \cdot (A_0 \oplus B_0) + \bar{A}_0 \cdot \bar{B}_0 \cdot H \tag{3.7}$$

where D_{-1} is the selected initial carry into the s.b. The explicit inhibit line H is used to control the operation at the beginning of an addition. Both carry chains are set "low" by setting $H = 0$. The carry sequences (on both true and complemented chains) are begun by releasing the inhibitions on all stages ($H = 1$), including the selected initial carry.

A primary carry will originate from stage i if $A_i B_i = 11$; and it will propagate to the left if the passing stage j has $A_j B_j = 01$ or 10. An n-input AND gate is added to signal the presence of a carry (1 or 0) at all activated stages when the carry propagation is completed. This carry-completion signal should not indicate completion even momentarily when an addition is still incomplete. If an input changes after an addition has been completed, the completion signal should immediately go off and remain off until the next new sum is completed. A prototype of this kind carry-completion adder was built by Gilchrist et al. [8]. Their experimental result shows that 8-to-1 time saving can be achieved for a 40-bit carry-completion adder compared with the speed of a 40-bit ripple-carry adder.

3.5 Conditional-Sum Addition Rules

Two similar types of synchronous adders are to be described. These adders overcome the carry propagation problem by generating distant carriers and using these carriers to select the true sum outputs from two simultaneously generated provisional sums under different carry input conditions. The *conditional-sum adders* are especially attractive for implementing economical arithmetic systems using low-cost gates with small fan-in and fan-out. The *carry-select adders* surpass the ripple-carry adders in significant speed gains, yet require only a moderate increase in hardware.

In what follows, *addend*, *augend*, and *true sum* bits are designated by A, B, and S respectively. *Carries* are indicated by C. Subscripts indicate the bit positions and superscripts "1" and "0" refer to the assumption that there was a carry or no carry into the lowest-order bit position of a section. The absence of a superscript indicates a true sum or a true carry. Table 3.2 describes the concept of conditional-sum addition. $S^0(k)$ and $S^1(k)$ denote two provisional sums, each consisting of multiple sections with k addend/augend columns per section. There are $\lceil n/k \rceil$ sections in $S^0(k)$ or $S^1(k)$ for an n-bit addition. Simultaneous additions are performed on all sections independently. Let $C^0(k)$ and $C^1(k)$ be the provisional carry sequences formed by the carries out of all the sections in $S^0(k)$ and $S^1(k)$, respectively.

Table 3.2 The Conditional-Sum Addition Table

	i	6	5	4	3	2	1	0	Assumed carry into each section	Step
$A = (109)_{10}$	A_i	1	1	0	1	1	0	1		
$B = (54)_{10}$	B_i	0	1	1	0	1	1	0		
	$S_i^0(1)$	1	0	1	1	0	1	1	0	1
	$C_{i+1}^0(1)$	0	1	0	0	1	0	0		
	$S_i^1(1)$	0	1	0	0	1	0		1	
	$C_{i+1}^1(1)$	1	1	1	1	1	1			
	$S_i^0(2)$	1	0	1	0	0	1	1	0	2
	$C_{i+1}^0(2)$	0	1		1		0			
	$S_i^1(2)$	0	1	0	0	1			1	
	$C_{i+1}^1(2)$	1	1		1					
	$S_i^0(4)$	0	0	1	0	0	1	1	0	3
	$C_{i+1}^0(4)$	1			1					
	$S_i^1(4)$	0	1	0					1	
	$C_{i+1}^1(4)$	1								
$S = A + B = (163)_{10}$	S	0	1	0	0	0	1	1		
	C_{out}	1								

Note: The arrows show the actual carries generated between sections. The initial carry to the rightmost section is always assumed zero.

The addition process is completed in t steps, where the integer

$$t = \lceil \log_2 n \rceil \tag{3.8}$$

At the ith step, $S^0(k)$ and $S^1(k)$ are formed with section size

$$k = 2^{i-1} \tag{3.9}$$

The grouping of the summand columns into disjoint sections starts from the lowest-order right end to the left end. When n is not an integer power of two, the leftmost section may have less than k columns. The successive section-carry outputs are used to select the true sum outputs.

In Table 3.2 we have an example of $n = 7$. Therefore, ($t = \lceil \log_2 7 \rceil = 3$) steps are required to complete the addition. At step one, each section has only a single column and the seven sections in $S^0(1)$ and $S^1(1)$ are formed by bitwise modulo-2 addition with corresponding carry sequences $C^0(1)$ and $C^1(1)$. At step two, each section contains two columns of addend/augend bits and four sections are formed in $S^0(2)$ and $S^1(2)$ each by 2-bit addition with the carry out forming the 4-bit provisional carry sequences $C^0(2)$ and $C^1(2)$. This process continues in a similar fashion except the section size doubles for each additional step. The final step reveals the true sum S and true carry out C as the desired output.

3.6 Conditional-Sum Adder Design

All the provisional sums and provisional carriers in Table 3.2 can be generated in parallel. Multilevel two-input multiplexer logic is used to select the true sums and true carries. Figure 3.6 shows the schematic diagram of a 7-bit conditional-sum adder implemented with three levels of two-input multiplexers.

Per each digital position $i = 6, \ldots, 1, 0$ a *Conditional Cell* (**CC**) is needed to generate the following provisional entities

$$S_i^0 = A_i \oplus B_i = A_i\bar{B}_i + \bar{A}_iB_i \tag{3.10}$$

$$C_{i+1}^0 = A_iB_i \tag{3.11}$$

and

$$S_i^1 = A_i \odot B_i = A_iB_i + \bar{A}_i\bar{B}_i \tag{3.12}$$

$$C_{i+1}^1 = A_i + B_i \tag{3.13}$$

Seven **CC**'s are shown on the top row of Fig. 3.6. The detailed logic of the **CC** is also shown. Two of these **CC**'s can be built into a 14-pin IC package with only 2Δ time delay using wired logic. Dual, triple, and quintuple 2-input multiplexers are used respectively at levels 1, 2, and 3 in Fig. 3.6, corresponding to steps 1, 2, and 3 in Table 3.2, respectively. These multiplexers select the true sum bits S_0 at level 1, S_2S_1 at level 2, and $S_6S_5S_4S_3$ at level 3. The scheme can be extended to adders with longer word length by increasing the number of **CC**'s used and multiplexer levels accordingly.

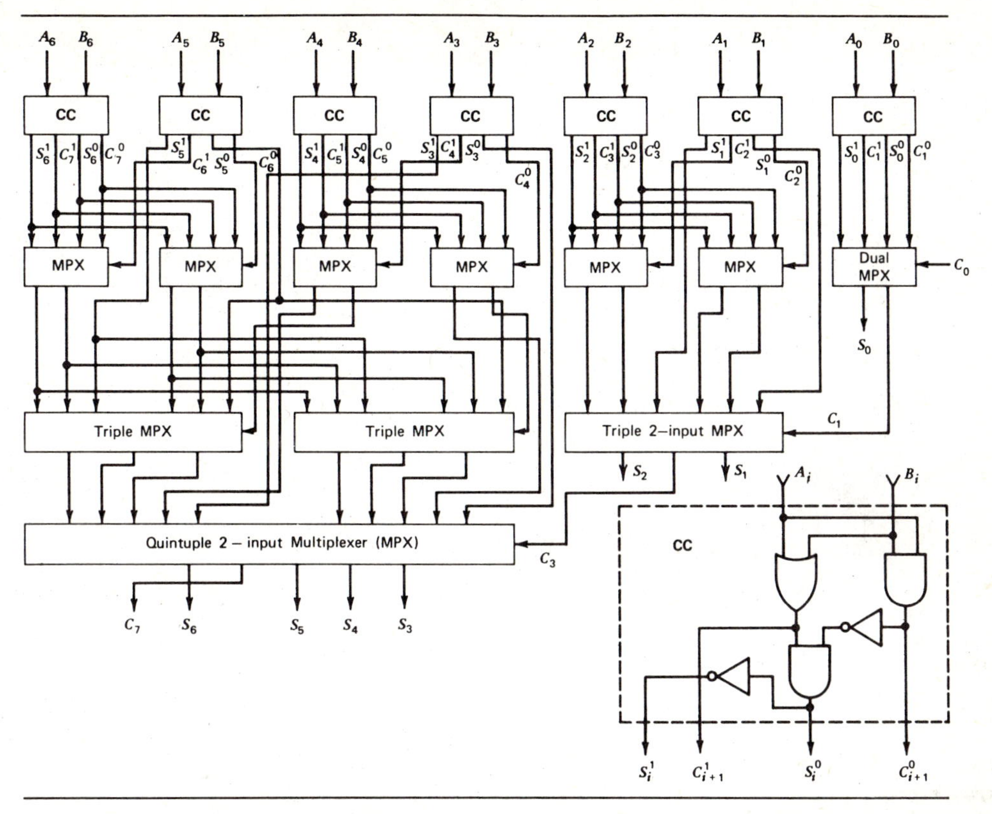

Figure 3.6 The schematic logic diagram of a 7-bit conditional-sum adder implemented with MSI multiplexers.

In general, an n-bit conditional-sum adder requires

$$\Delta(\text{Conditional-Sum Add Time}) = (\lceil \log_2 n + 1 \rceil + 1) \cdot 2\Delta \tag{3.14}$$

to complete the addition of two n-bit numbers. Each level of multiplexing contributes 2Δ delay. Using current MSI circuits, $\lceil n/2 \rceil$ packages of conditional cells and $\lceil m/4 \rceil$ quad two-input multiplexer packages are required to implement an n-bit conditional-sum adder where the total number of two-input multiplexers, m, is calculated by

$$m = 2 \cdot n + 3 \cdot \left\lceil \frac{n-1}{2} \right\rceil + 5 \cdot \left\lceil \frac{\left\lceil \frac{n-1}{2} \right\rceil - 1}{2} \right\rceil + \cdots + (\lceil \log_2 n \rceil + 2) \cdot 1 \tag{3.15}$$

For the design in Fig. 3.6, 11 IC packages are needed.

3.7 Carry-Select Adders

Instead of starting with generating bitwise provisional sums and carriers, one can partition a long adder into fixed-size adder sections and proceed with simultaneous section additions with appropriate carry input to select the true sum output. We choose

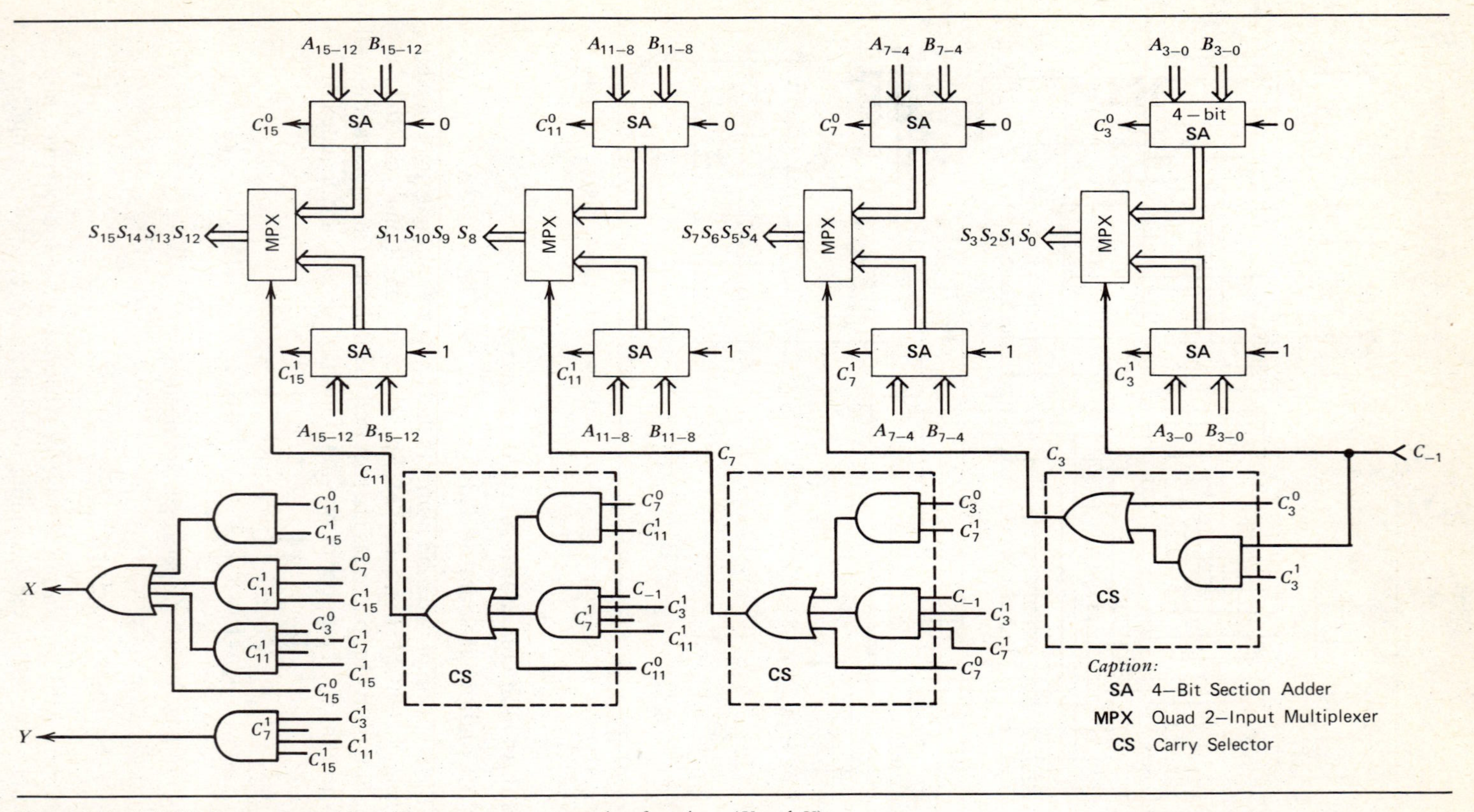

Figure 3.7 A 16-bit carry-select adder with group-carry generation functions (X and Y).

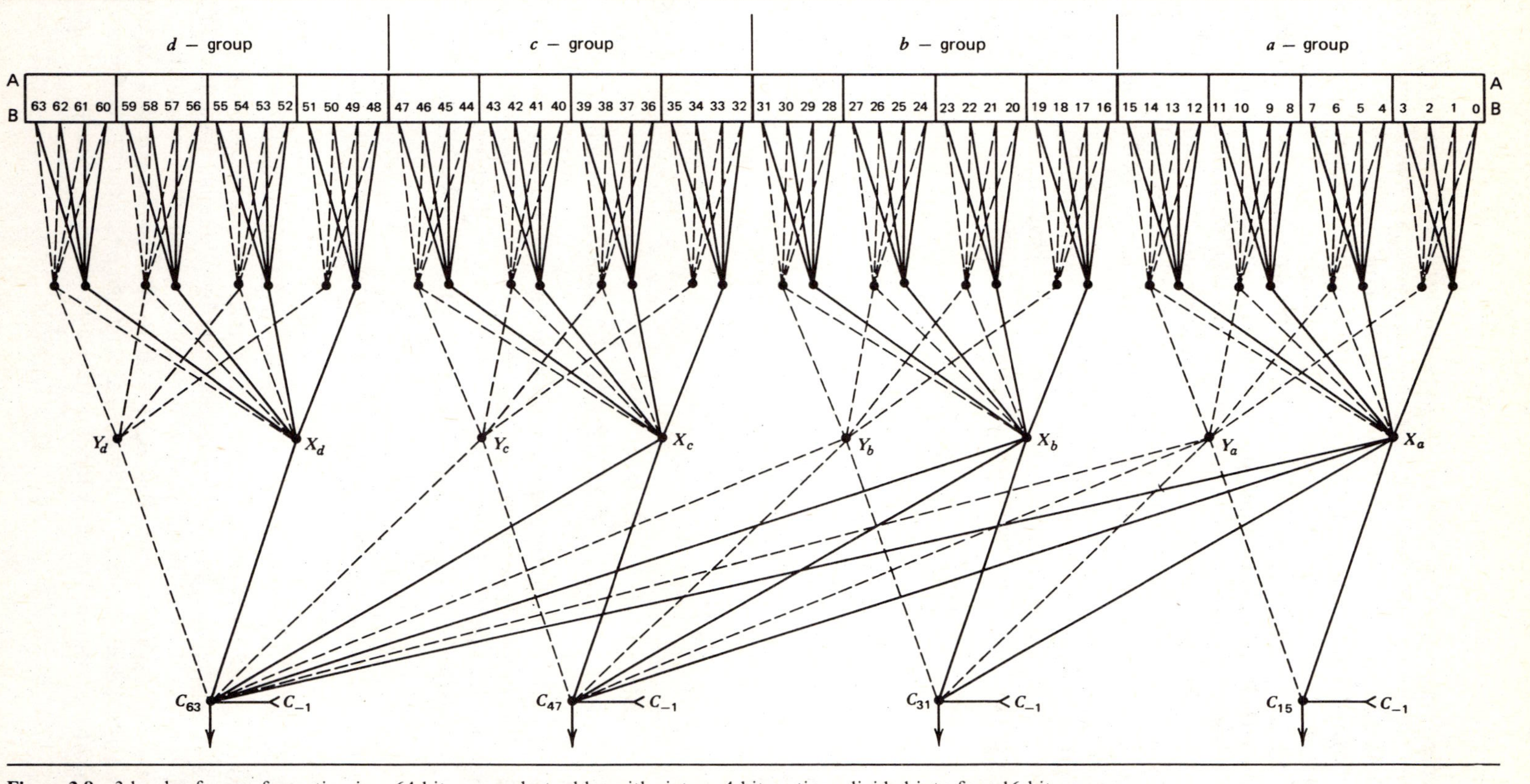

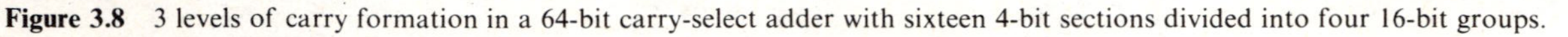

Figure 3.8 3 levels of carry formation in a 64-bit carry-select adder with sixteen 4-bit sections divided into four 16-bit groups.

a 16-bit adder design to illustrate the carry-select addition (Fig. 3.7). The 16-bit adder is divided into four 4-bit adder sections. Each adder section is in duplicate, one with a carry and one without a carry into the lowest-order bit in the section. We use a forced "1" or "0" to distinguish the two section adders configurations. Ripple-carry propagation is assumed within each section subadder for clarity.

The selection carry input to each section is generated sequentially by the cascaded *carry selection* (**CS**) boxes. The complexity of the **CS** circuits increases rapidly when more high-order adder sections are added to increase the total adder length. However, one can extend the high-order radix carry. (The radix equals $2^4 = 16$ in Fig. 3.7) to even higher order ones so that multilevel carry selection will be possible. For example, Fig. 3.8 demonstrates the use of sixteen 4-bit adders to form a 64-bit adder with multilevel carry sections.

The carry formation paths are shown in Fig. 3.8 by solid or dashed lines corresponding to the assumption that a carry "0" or "1" was brought into the section. The group carries X and Y are defined by the following equations

$$X = C_3^0 C_7^1 C_{11}^1 C_{15}^1 + C_7^0 C_{11}^1 C_{15}^1 + C_{11}^0 C_{15}^1 + C_{15}^0 \qquad (3.16)$$

$$Y = C_3^1 C_7^1 C_{11}^1 C_{15}^1 \qquad (3.17)$$

where carries C_i^0 and C_i^1 are generated in each 4-bit adder section. The indices a, b, c, and d identify the 16-bit groups. The carries from each 16-bit group are combined to generate the carry C_{15}, C_{31}, C_{41}, and C_{63} according to the following equations

$$\begin{aligned} C_{15} &= C_{-1} Y_a + X_a \\ C_{31} &= C_{-1} Y_a Y_b + X_a Y_b + X_b \\ C_{47} &= C_{-1} Y_a Y_b Y_c + X_a Y_b Y_c + X_b Y_c + X_c \\ C_{63} &= C_{-1} Y_a Y_b Y_c Y_d + X_a Y_b Y_c Y_d + X_b Y_c Y_d + X_c Y_d + X_d \end{aligned} \qquad (3.18)$$

The above carries are returned to the appropriate groups of sections, where they are used to simultaneously select the true sum outputs in the four 16-bit groups. Comparing this 64-bit multilevel carry-select adder with its ripple-carry counterpart, one may obtain a speedup of about 15 with only approximately twice the hardware.

3.8 Carry Generate, Propagate, and Lookahead Functions

We describe below a *carry lookahead* (*CLA*) technique to speed up the carry (or borrow) propagation in an adder/subtractor complex. The carries entering all the bit positions of a "parallel" adder are generated simultaneously by additional logic circuitry. This results in a constant addition time independent of the length of the adder.

Let $\mathbf{A} = A_{n-1} \cdots A_1 A_0$ and $\mathbf{B} = B_{n-1} \cdots B_1 B_0$ be the *augend* and *addend* inputs to an n-bit adder, where A_{n-1} and B_{n-1} are the sign bits. Let C_{i-1} be the carry input to the ith bit position. The carry input to the least significant position is denoted as C_{-1}. Let S_i and C_i be the sum and carry outputs of the ith stage. We define two auxililary functions as follows:

$$G_i = A_i \cdot B_i \tag{3.19}$$

$$P_i = A_i \oplus B_i \tag{3.20}$$

The *carry generate* function G_i reflects the condition that a carry is originated at the ith stage. The function P_i, called *carry propagate*, is true when the ith stage will

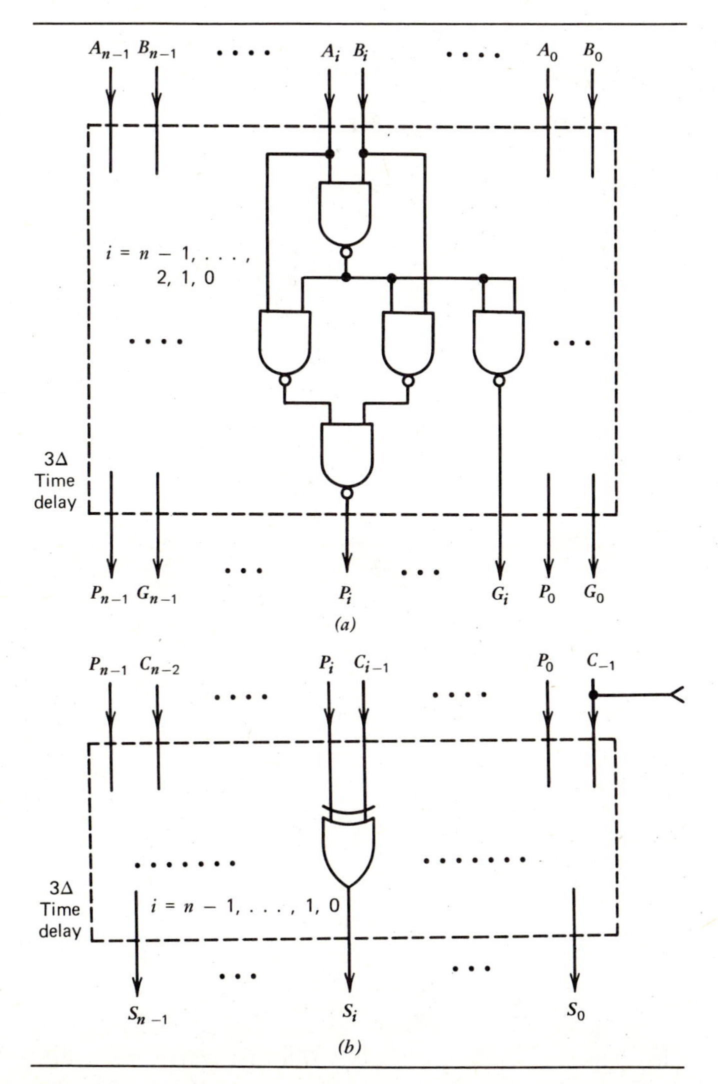

Figure 3.9 Logic circuits for realizing the carry-generate, carry-propagate, and sum functions. (*a*) Carry generate/propagate unit. (*b*) Summation unit.

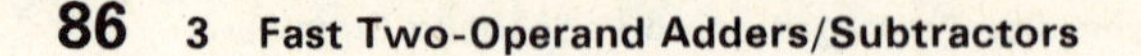

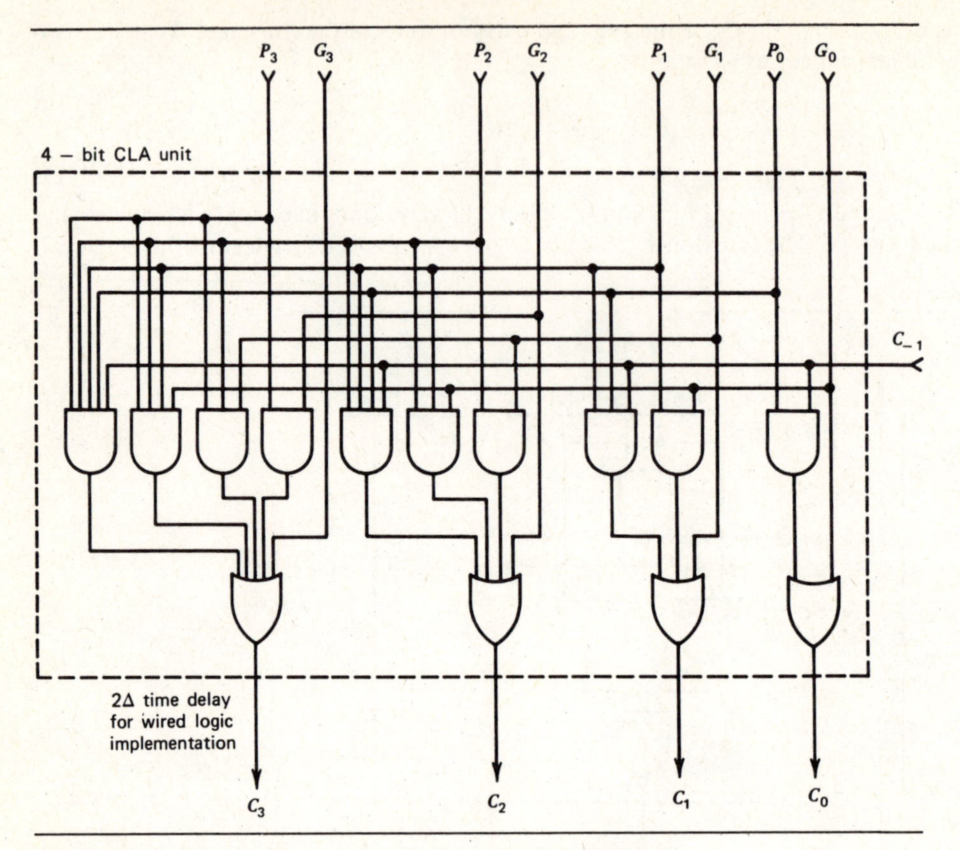

Figure 3.10 The schematic logic circuit of a 4-bit carry lookahead unit.

pass the incoming carry C_{i-1} to the next higher stage. Substituting P_i and G_i into Eq. 2.2, we obtain for $i = n - 1, \ldots, 1, 0$

$$\begin{aligned} S_i &= (A_i \oplus B_i) \oplus C_{i-1} \\ &= P_i \oplus C_{i-1} \end{aligned} \tag{3.21}$$

$$\begin{aligned} C_i &= A_i \cdot B_i + (A_i \oplus B_i) \cdot C_{i-1} \\ &= G_i + P_i \cdot C_{i-1}. \end{aligned} \tag{3.22}$$

These equations reveal the fact that all P_i and G_i for $i = n - 1, \ldots, 1, 0$ can be generated simultaneously from the external inputs **A** and **B** as illustrated in Fig. 3.9. Equation 3.21 implies that all the sum bit S_i for $i = n - 1, \ldots, 1, 0$ can be generated in parallel as shown, if all the carry inputs $C_{n-2}, \ldots, C_1, C_0, C_{-1}$ are available simultaneously. We shall next show the circuit that anticipates these carries.

One can repeatedly apply the recursive formula in Eq. 3.22 to obtain the following set of carry equations in terms of the variables P_i's; G_i's, and the initial carry C_{-1}.

$$
\begin{aligned}
C_0 &= G_0 + C_{-1}P_0 \\
C_1 &= G_1 + C_0 P_1 \\
&= G_1 + G_0 P_1 + C_{-1}P_0 P_1 \\
C_2 &= G_2 + C_1 \cdot P_2 \\
&= G_2 + G_1 P_2 + G_0 P_1 P_2 + C_{-1}P_0 P_1 P_2 \\
&\vdots \\
C_k &= G_k + G_{k-1} \cdot P_k + G_{k-2}P_{k-1}P_k + \cdots \\
&\quad + G_0 P_1 P_2 \cdots P_k + C_{-1}P_0 P_1 \cdots P_k \\
&\vdots \\
C_{n-1} &= G_{n-1} + G_{n-2}P_{n-1} + \cdots + C_{-1}P_0 P_1 \cdots P_{n-1}
\end{aligned}
\tag{3.23}
$$

This set of equations can be realized with a combinational logic circuit, known as an n-bit *carry lookahead* unit, as illustrated in Fig. 3.10 for the case of $n = 4$. Two additional terminal functions, called the *block carry generate* G^* and the *block carry propagate* P^*, can be added to form an alternate design as shown in Fig. 3.11 with

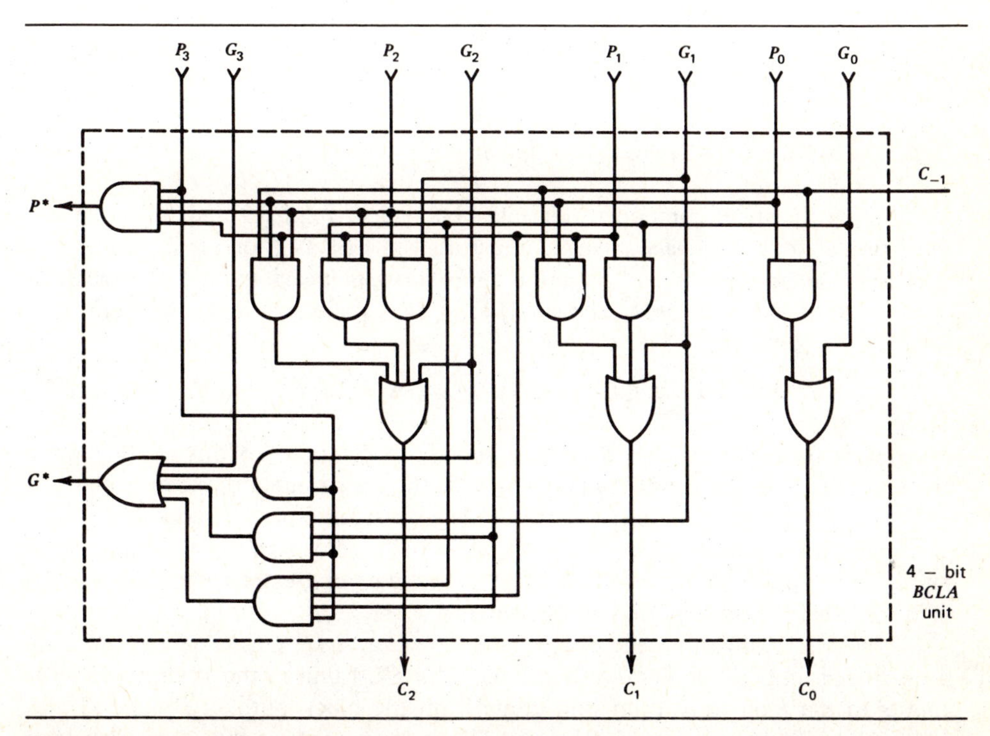

Figure 3.11 The schematic logic circuit of a 4-bit block carry lookahead unit with block carry generate/propagation functions.

only three carry output terminals C_0, C_1, and C_2 and output terminals

$$P^* = P_0 P_1 P_2 P_3 \tag{3.24}$$

$$G^* = G_3 + G_2 P_3 + G_1 P_2 P_3 + G_0 P_1 P_2 P_3 \tag{3.25}$$

The variable P^* is true if a carry into the block (4-bit slice) would result in a carry out of the block. The variable G^* corresponds to the condition that the carry generated out of the most significant position of the block was originated within the block itself. Therefore, the carry out of the block can be written as

$$C_{\text{out}} = G^* + P^* \cdot C_{\text{in}}, \tag{3.26}$$

where C_{in} is the carry into the block. We call this alternate design a *block carry lookahead* (*BCLA*) unit.

At present, monolithic 4-bit or 8-bit *CLA* units and *BCLA* units are made with bipolar transistor gates. The main reason that the circuit is prevented from growing too large, say $n > 10$, is because of the fan-in and fan-out limitations of current TTL, ECL, or I^2L gates, if one insists on a two-level (2Δ delay) implementation of the *CLA* unit. This situation may change in the future.

3.9 Carry Lookahead Adders

Combining the three component circuits obtained in Figs. 3.9 and 3.10, one can construct a *carry lookahead adder* as illustrated in Fig. 3.12 for the case of $n = 8$. The *carry generate-propagate unit* and the *summation unit* each require 3Δ time delay because XOR gates are used. The single-level *CLA* adder requires a constant time delay

$$\Delta(\text{One-Level } CLA \text{ Add Time}) = 3\Delta + 2\Delta + 3\Delta = 8\Delta \tag{3.27}$$

Theoretically, one should be able to build full *CLA* adders of any word length if the *CLA* unit can be freely expanded. Due to the constraints described in the preceding section, single-level *CLA* applies only to designing parallel adders of length $5 \leq n \leq 16$ at present. *CLA* does not pay off for adders of length $n \leq 4$, because short ripple-carry adders would be equally fast. For adders of greater length, say $n > 16$, *multilevel* carry lookahead may be needed.

Figure 3.13 illustrates the design of a 32-bit *two-level CLA* adder using eight 4-bit *BCLA* units in the first level and one 8-bit *CLA* unit in the second level. The inputs to the 8-bit lookahead unit come from the block outputs P_j^* and G_j^* for $j = 0, 1, \ldots, 7$, of the eight 4-bit blocks. The carry outputs at the second level are connected to the block carry inputs in the first level. Such a dual-level design requires

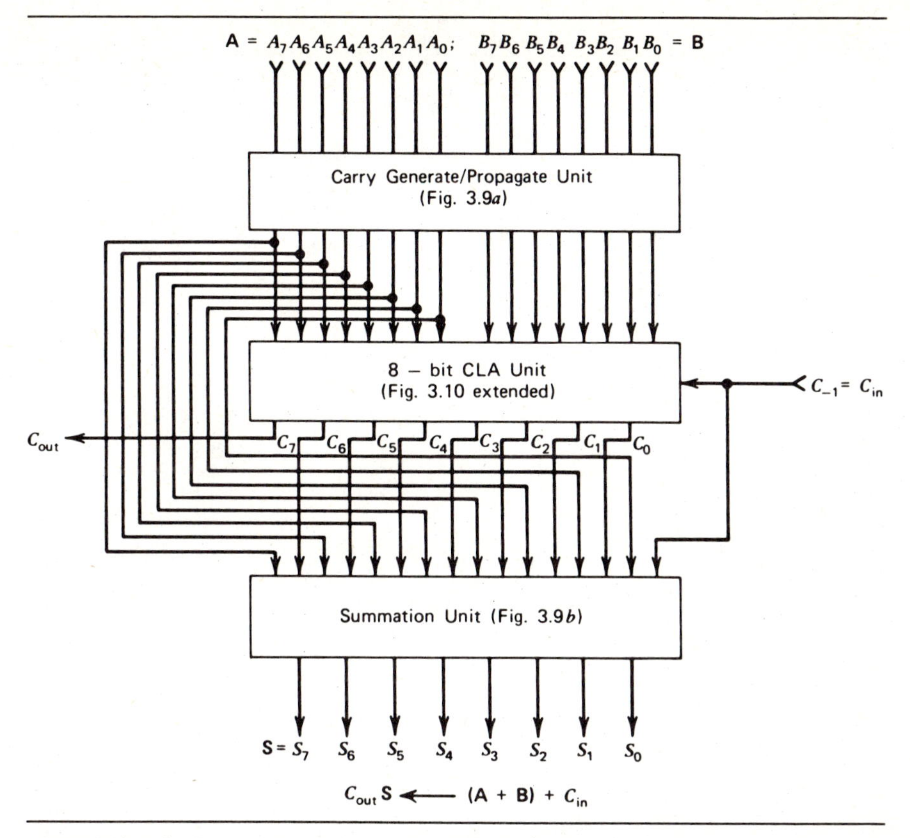

Figure 3.12 The functional block diagram of a full carry lookahead adder (8-bit CLA adder shown).

6Δ gate delays to generate all the required carries. In total, we have the constant time delay

$$\Delta(\text{Two-Level } CLA \text{ Add Time}) = 3\Delta + 6\Delta + 3\Delta = 12\Delta \qquad \textbf{(3.28)}$$

The same idea can be extended to designing long adders with more than two levels. Each additional level of carry lookahead will contribute a 4Δ additional time delay. In practical situations, we rarely use adders beyond 64 bits. It is sufficient to consider only single-level or two-level CLA adders. Table 3.3 summarizes the design and application parameters associated with various combinations of n-bit carry lookahead adder design. The notation **CLA** $(m \times k)$ is used to denote the two-level configuration, in which m units of k-bit *BCLA* units are used in the first level and one m-bit *CLA* unit is employed in the second level, such that $m \times k = n$. The design shown in Fig. 3.13 corresponds to a parallel adder configuration of $32 = 8 \times 4$. Three additional gate delays 3Δ should be added to each of the total add time figures in Table 3.3, if pre-complement and overflow-detection circuits are included to facilitate subtraction as well.

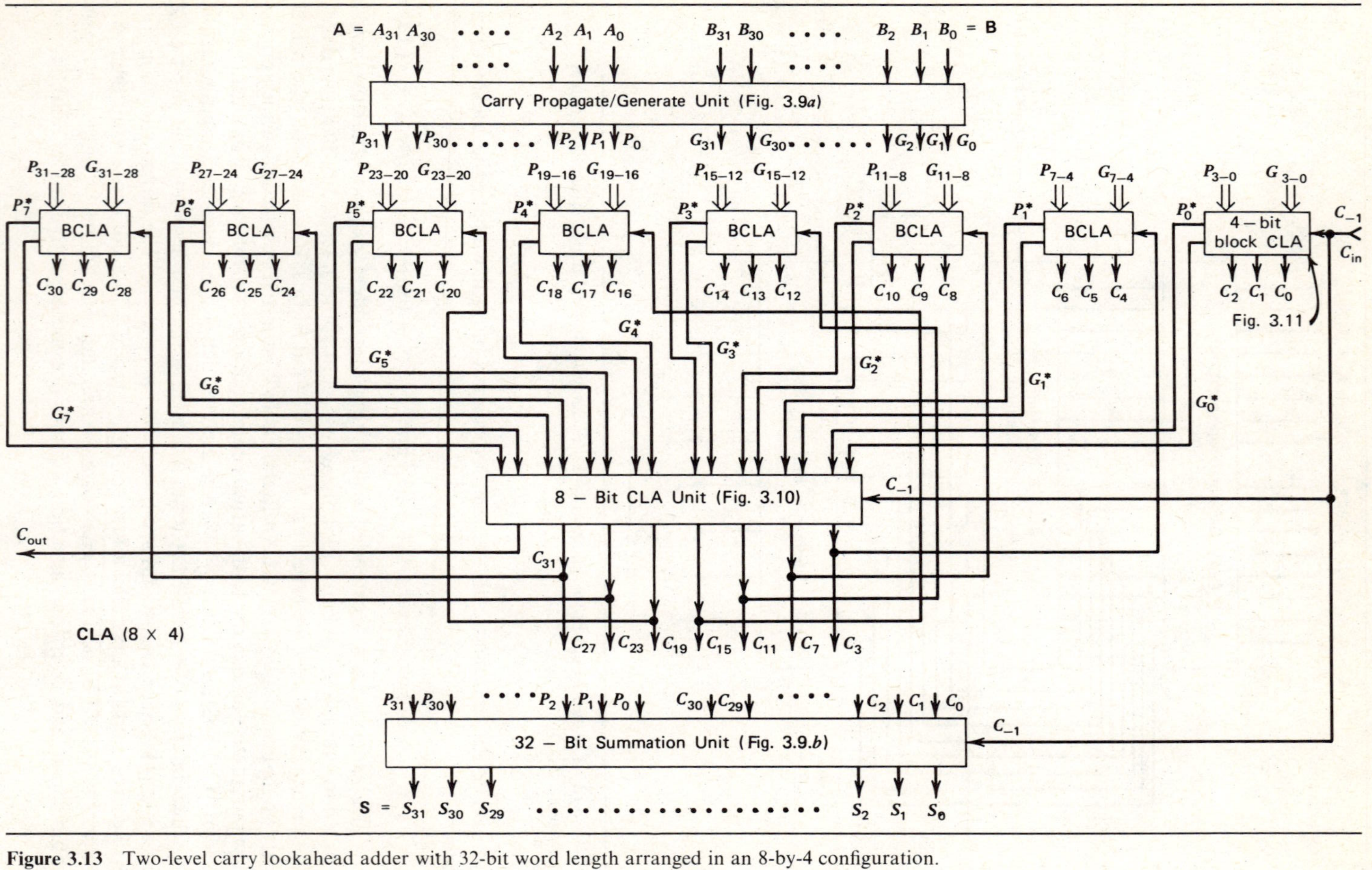

Figure 3.13 Two-level carry lookahead adder with 32-bit word length arranged in an 8-by-4 configuration.

Table 3.3 Carry Lookahead Adder Configurations, Design and Application Parameters

CLA Adder Configurations	CLA Unit Required	BCLA Units Required	Operating Speed[a]
CLA(8)	CLA(8)	None[b]	8Δ
CLA(16)	CLA(16)	None[b]	8Δ
CLA(4 × 4)	CLA(4)	Four BCLA(4)	12Δ
CLA(4 × 6)	CLA(4)	Four BCLA(6)	12Δ
CLA(6 × 4)	CLA(6)	Six BCLA(4)	12Δ
CLA(8 × 4)	CLA(8)	Eight BCLA(4)	12Δ
CLA(8 × 6)	CLA(8)	Eight BCLA(6)	12Δ
CLA(8 × 8)	CLA(8)	Eight BCLA(8)	12Δ
CLA(16 × 4)	CLA(16)	Sixteen BCLA(4)	12Δ
CLA(16 × 8)	CLA(16)	Sixteen BCLA(8)	12Δ

[a] 3Δ additional delay may be needed if delays due to complemented subtraction and overflow detection are included.
[b] Full-length *CLA* was assumed.

3.10 An Evaluation of Various Two-Operand Adders

The computational *efficiency* η of an adder is related to two factors: the *investment* **I** and the *computing power* **W** of the adder, by the following ratio

$$\eta = \frac{\mathbf{W}}{\mathbf{I}} \tag{3.29}$$

The investment **I** is usually reflected by the product of *hardware complexity* ϕ and the total *addition time* ρ which determines the operating speed of the adder. The computing power **W** can be easily measured by the width n of the adder, which reflects the operating range of the adder. There is no unique or general formula to measure the hardware complexity ϕ of the adder designs with different structures based on different technology devices. Sklansky [17] has proposed to define ϕ by the logarithm of the *gate count* **G** in an adder construction.

$$\phi = \log_2 \mathbf{G} \tag{3.30}$$

Base two is assumed in the logarithm because only two-input logic gates, AND, OR or Inverters, were assumed in Sklansky's analysis. Although the above equation may not apply directly to designs with modern multiple-input gate functions, the analytic method will aid in the development of more general approaches to evaluate computer arithmetic systems. Table 3.4 provides the formulas for **G** and ρ for the four types of adders we have studied in the preceding sections. Both **G** and ρ are expressed in terms of the addend/augend length n under the two-input gate assumption. The above discussions lead to the following adder efficiency measure

$$\eta = \frac{n}{\rho \cdot \log_2 \mathbf{G}} \tag{3.31}$$

Table 3.4 Sklansky's Formula for Evaluating the Gate Counts and Delay Time of Four Two-operand Adders (Sklansky [17])

Adder type	Number of 2-input Gates Required[a]	Total Delay Time in Terms of Δ[b]
Ripple-carry adder	$7n$	$2n\Delta$
Carry-completion adder	$17n - 1$	$(n + 4)\Delta$ (lower bound)
Conditional-sum adder	$3n[2 + \log_2(n + 1)]$	$[2 + 2\log_2(n + 1)]\Delta$
Carry-lookahead adder $\mathbf{CLA}\left(\frac{n}{p} \times p\right)$	$6n + 1 + \frac{p+1}{p-1}\cdot q + \frac{p^2 + 2p - 1}{p} \cdot \left[k \cdot \left(n + \frac{1}{p-1}\right) - \frac{p \cdot q}{(p-1)^2}\right]$, where $k \triangleq \log_p[1 + n(p - 1)] - 1$, $q \triangleq 1 + (n - 1)p - n$.	$[4 + k(p + 1)]\Delta$ where $k = \log_p[1 + n(p - 1)] - 1$

[a] n is the adder word length. p is the carry span in CLA adder.

[b] Delays due to complementer or overflow detector are excluded. The fact that only 2-input **AND**, **OR**, or **NOT** gates are allowed affects the delay time, especially in the CLA cases.

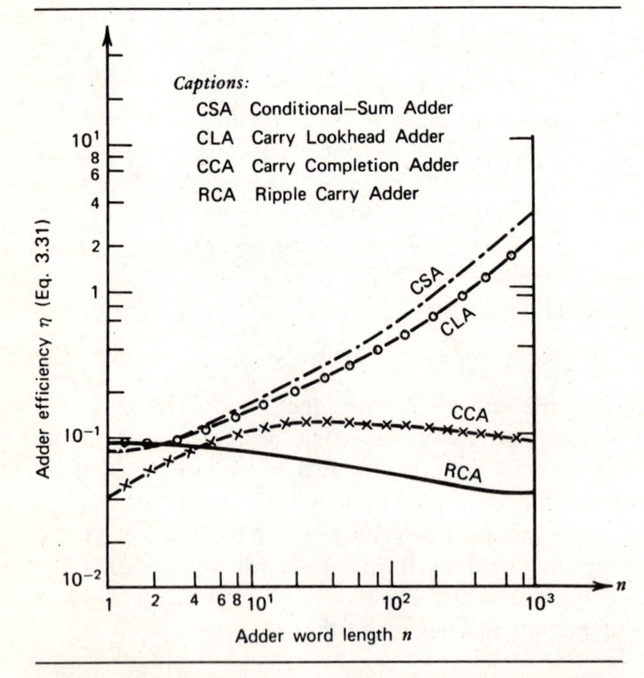

Figure 3.14 Comparison of the efficiency of four types of two-operand adder designs on the basis of Eq. 3.31 (Sklansky [17]).

This efficiency equation is plotted in Fig. 3.14 as a function of the adder length n for all four adder configurations. The average longest carry length ($\log_2 n$) was used to estimate the expected addition time of a carry-completion adder. The carry span in the carry-lookahead adders was assumed to be $p = 5$.

Efficient adder designs involve the matches of device technology and many other system factors such as device fan-in and fan-out capabilities, the consideration of overflow, end-around carry, various complementing schemes, and tie-in with other arithmetic operations. Synchronous adders require simpler clocked controls and have fixed add time, whereas the asynchronous adders offer variable add time at the expense of more complicated local timing control logic. Other factors such as ease of design and fabrication and error detection capability should be also taken into account in the final choice of adder design.

3.11 Bibliographic Notes

Most early research efforts on machine arithmetic were devoted to faster and economical adders designs. Practical implementations of adders with modern devices can be found in Fairchild's application notes [6]. Asynchronous adders were studied by Gilchrist et al. [8]. Studies on carry propagation length were reported in Reitwiesner [15] and Briley [5]. The conditional-sum addition was proposed by Sklansky [16] and further studies by Kruy [9]. Carry-select adders were introduced by Bedrij [4].

Among many other researchers, Aleksander [1], Avizienis [2], Ferrari [7], Lehman [10, 11], MacSorley [13], Sklansky and Lehman [19], and Weller [21] have investigated the carry propagation speed-up techniques and their possible implementations. It is fair to say that the popular class of carry lookahead adders is the end product of these researchers on parallel carry generation. In particular, the work of Lehman and Sklansky [19] is comprehensive. The evaluation of various two-operand adders is due to Sklansky [17].

Bartee and Chapman's work [3] on general-purpose accumulators affects the design of modern digital adders. Majerski [14] analyzed the carry skip distributions in adder design. Recently, Shedletsky [18] commented on the end-around carry adder designs. In the theoretical aspects, Winograd [23] and Spira [21] have studied the ultimate limits of the time required for machine addition.

References

[1] Aleksander, I., "Array Networks for a Parallel Adder and Its Control," *IEEE Trans. on Electr. Computers*, Vol. EC-16, No. 2, April 1967.

[2] Avizienis, A., "Logic Nets for Carry and Borrow Propagation," *Class Notes*, Dept. of Engineering, University of California, Los Angeles, 1968.

[3] Bartee, T. C., Chapman, D. J., "Design of an Accumulator for a General Purpose Computer," *IEEE Trans.* EC-14, No. 4, August 1965, pp. 570–574.

[4] Bedrij, O. J., "Carry-Select Adders," *IRE Trans.*, EC-11, No. 3, June 1962, pp. 340–346.

[5] Briley, B. E., "Some New Results on Average Worst-Case Carry," *IEEE Trans. Compt.*, Vol. C-22, No. 5, May 1973, pp. 459–463.

[6] Fairchild Semiconductor Staff, *The TTL Applications Handbook*, Mountain View, Calif., August 1973.

[7] Ferrari, D., "Fast Carry-Propagation Iterative Networks," *IEEE Trans. Compt.*, Vol. C-17, No. 2, February 1968, pp. 132–145.

[8] Gilchrist, B. et al., "Fast Carry Logic for Digital Computers," *IRE Trans.* EC-4, December 1955, pp. 133–136.

[9] Kruy, J. F., "A Fast Conditional Sum Adder Using Carry Bypass Logic," *AFIPS Conf. Proceedings*, Vol. 27, FJCC 1965, pp. 695–703.

[10] Lehman, M. and Burla, N., "Skip Techniques for High-Speed Carry-Propagation in Binary Arithmetic Units," *IRE Trans.* EC-10, No. 4, December 1961, pp. 691–698.

[11] Lehman, M., "A Comparative Study of Propagation Speed-Up Circuits in Binary Arithmetic Units," *Inform. Processing*, 1962, Elsevier-North Holland, Amsterdam, 1963, pp. 671–677.

[12] Ling, H., "High-Speed Binary Parallel Adder," *IEEE Trans. Comput.*, EC-15, No. 5, October 1966, pp. 799–802.

[13] MacSorley, O. L., "High-Speed Arithmetic in Binary Computers," *Proc. IRE*, Vol. 49, No. 1, January 1961, pp. 67–91.

[14] Majerski, S., "On Determination of Optimal Distributions of Carry Skips in Adders," *IEEE Trans.* EC-16, No. 1, February 1967, pp. 45–58.

[15] Reitwiesner, G. W., "The Determination of Carry Propagation Length for Binary Addition," *IRE Trans.* EC-9, No. 1, March 1960, pp. 35–38.

[16] Sklansky, J., "Conditional-Sum Addition Logic," *IRE Trans.* EC-9, No. 2, June 1960, pp. 226–231.

[17] Sklansky, J., "An Evaluation of Several Two-Summand Binary Adders," *IRE Trans.* EC-9, No. 2, June 1960, pp. 213–226.

[18] Shedletsky, J. J., "Comment on the Sequential and Indeterminate Behavior of and End-Around-Carry Adder," *IEEE Tran. Comput.*, Vol. C-26, No. 3, March 1977, pp. 271–272.

[19] Sklansky, J. and Lehman, M., "Ultimate-Speed Adders," *IRE Trans.* EC-12, No. 2, April 1963, pp. 142–148.

[20] Spira, P. M., "Computation Times of Arithmetic and Boolean Functions in (d, r) Circuits," *IEEE Trans.* Comput. Vol. C-22, No. 6, June 1973, pp. 552–555.

[21] Weller, C. W., "A High-Speed Carry Circuit for Binary Adders," *IEEE Trans. Comput.* Vol. C-18, No. 8, August 1969, pp. 728–732.

[22] Winograd, S., "On the Time Required to Perform Addition," *Journal of ACM*, Vol. 12, No. 2, April 1965, pp. 277–285.

Problems

Prob. 3.1 Prove that overflow will not occur if the carry entering the sign position is equal to the carry out of the sign position in either one's complement or two's complement addition/subtraction. The proof should cover all possible cases of input combinations.

Prob. 3.2 Prove that the average longest carry propagation length $\mathbf{E}_n(p)$ is upper bounded by $\log_2 n$, where n is the wordlength of the adder (Ref. [15]).

Prob. 3.3 Show the complete schematic diagram of a 16-bit asynchronous self-timing carry completion sensing adder. What is the projected speed ratio of this sensing adder over the conventional 16-bit ripple-carry adder?

Prob. 3.4 Design a 15-bit conditional-sum adder with the conditional cells (CC) and 2-input multiplexers (MPX). Show the schematic logic diagram and estimate the total number of IC packages that may be required, assuming that triple CCs and quad 2-input MPXs are each available in one IC package.

Prob. 3.5 Draw the complete schematic logic diagram of the 64-bit carry-select adder with the carry formation tree as shown in Fig. 3.8. Compare the speed and hardware requirements of this carry-select adder with those of a 64-bit, two-level, 8-by-8 carry lookahead adder.

Prob. 3.6 Suppose that only 4-bit block carry lookahead (*BCLA*) with (Fig. 3.11) and 4-bit carry lookahead (*CLA*) units (Fig. 3.10) are available. Construct a three-level, 4-by-4-by-4, two's complement carry lookahead adder/subtractor. Show the schematic interconnections similar to Fig. 3.13. Estimate the total time delay of this three-level **CLA** $(4 \times 4 \times 4)$ adder, including the delays of the complementer and of the overflow detection logic.

Prob. 3.7 A (d, r)-circuit is a d-valued logic circuit, which is composed of logic modules with maximum fan-in of r inputs each (note that binary logic has $d = 2$). Each module computes an r-argument d-valued logic function, $g: \{0, 1, \ldots, d-1\}^r \rightarrow \{0, 1, \ldots, d-1\}$, in unit time. The total computation time of a (d, r)-circuit is determined by the number of module levels in the longest path through the (d, r)-circuit (Fig. 3.15).
(a) Prove that any n-variable d-valued logic function $f: \{0, 1, \ldots, d-1\}^n \rightarrow \{0, 1, \ldots, d-1\}$ can be computed by a (d, r) circuit with at least $\lceil \log_r n \rceil$ computation time. You can assume $n \gg r$ (Spira [20]).
(b) Prove the Ofman bound: Two n-bit binary numbers can be added by a $(3, 2)$ circuit in time $O(\log n) + 1$.
(c) Prove the Winograd bound [22]: Two n-digit d-ary numbers can be added by a (d, r) circuit in time $c_1 + \lceil c_2 \log_r n \rceil$, where $c_1 = 1 + \log_{\lfloor (r+1)/2 \rfloor} \lfloor r/2 \rfloor$ and $c_2 = \log_{\lfloor (r+1/2) \rfloor} r$ for $r \geq 3$. Modulo-d^n radix complement arithmetic may be assumed in the proof.

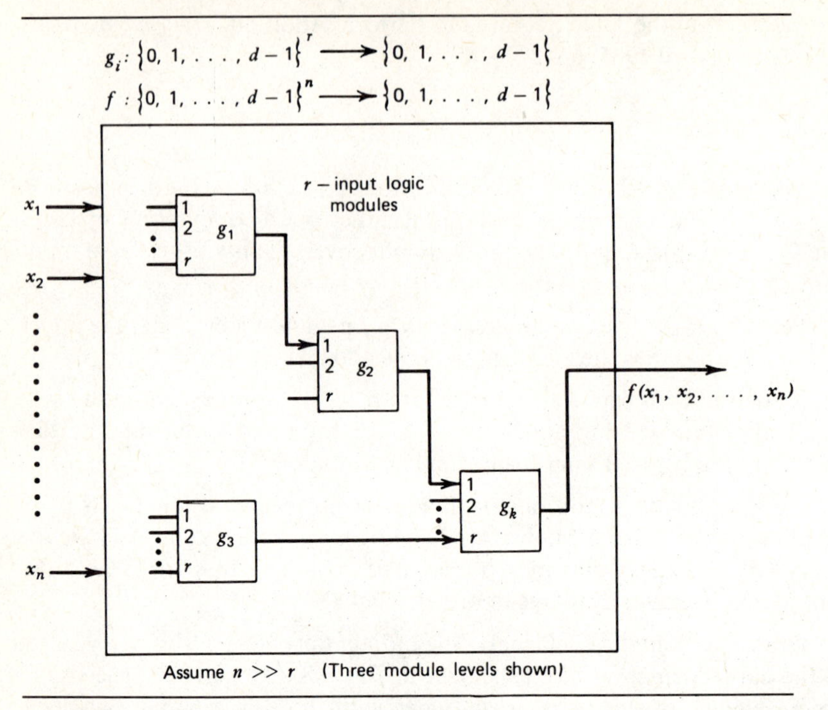

Figure 3.15 A sample (d, r)-circuit composed of logic modules with at most r inputs each.

Prob. 3.8 Prove that in a straight sense, a multilevel carry lookahead adder of long word length n should have a total add time on the order of $\mathbf{0}(2 \log_r n)$, where r is the maximum fan-in limitation of the logic gates used.

Prob. 3.9 Prove that the direct *diminished-radix complement* addition/subtraction algorithm works for all possible cases of the input operands. Justify the necessity of using the end-around carry.

Prob. 3.10 Repeat Prob. 3.9 for the cases of *sign magnitude* addition/subtraction algorithm as illustrated in Fig. 3.3.

Chapter

4

Multioperand Adders, Signed-Digit Arithmetic, and **ALU's**

4.1 Introduction

This chapter extends the design methodologies associated with fast two-operand adders in three directions. The reader should have a thorough understanding of the two-operand adder designs discussed in the preceding chapter before challenging the advanced approaches presented in this chapter. The conventional two-number addition schemes can be extended as follows:

A. Multiple-Operand Addition

A much more powerful adder can be designed by modifying the two-operand adders to a form which can add a large number of numbers simultaneously, instead of just two at a time. Such multioperand addition plays a central role in the implementation of high-speed hardware MULTIPLY or DIVIDE units to be described in later chapters. We shall describe the *carry-save adders* in the next two sections and then the bit-slice *partitioned adders* in sections 4.4 and 4.5. Both types of adders were suggested to handle multiple-operand addition.

B. Signed-Digit Addition/Subtraction

The redundant *signed-digit* (**SD**) representations described in section 1.5 are used to design *totally parallel adder/subtractors.* Redundancy in **SD** number representation allows a method of fast addition/subtraction, in which each sum/difference digit depends only on the digits in two adjacent digital positions of the operands. In other words, the carry propagation chains are eliminated. We shall specify the **SD** addition/subtraction rules in section 4.6 and their physical implementations in section 4.7.

C. Arithmetic Logic Units

We are interested in converting a parallel adder into a more powerful device, known as *Arithmetic Logic Unit* (**ALU**). The **ALU**s play a vital role in constructing practical arithmetic processors. Our presentation will focus on describing the functional features and system applications of **ALU**s. A case study is also presented for the design and applications of a 32-function **ALU** (the SN 74181).

4.2 Multioperand Carry-Save Adders

The traditional adders are designed to add two numbers at a time. There are arithmetic operations that require the addition of a large number of binary numbers. We present below a type of multiple-operand adder, which can be used to perform fast multiple additions, multiplication, and special-function generation with limited amount of hardware components. When multiple additions are performed, it is possible to save the carry propagation until all the additions are completed and then take a final cycle (or a number of final cycles) to complete the carry propagation for all the additions. Adders designed with such carry-saving capability are called *carry-save adders* (**CSA**).

An n-bit **CSA** is described in Fig. 4.1. The adder consists of n full adders. Instead of connecting the n stages in cascade as we did in ripple-carry adders, the carry outputs of all full adders of a **CSA** go to an intermediate register, called a *carry save register* (CSR). The carry-in terminals of the n full adders form a third input vector to the adder. The three sets of input terminals, designated as A, B, and C, can be used interchangeably without affecting the outputs. The two output vectors are designated as **S** and **R** for the *sum* and *carry* outputs, respectively. Again, subscripts i for $0 \leq i \leq n-1$ indicate the bit positions of all numbers. Superscripts are used to identify the input numbers.

The addition of k n-bit numbers, $\mathbf{N}^1 + \mathbf{N}^2 + \cdots + \mathbf{N}^k$ (Modulo-2^n sum), by a CSA is described by the following procedures:

Step 1. Load the first three input numbers $\mathbf{N}^1$, $\mathbf{N}^2$, and $\mathbf{N}^3$ into **A**, **B**, and **C** input lines.

Step 2. Enter the outputs **S** and **R** obtained from step 1 back into **A** and **B**, and the fourth input number $\mathbf{N}^4$ into **C**. In this operation, sum bit S_i goes into input line A_i, and carry bit R_i goes into input line B_{i+1} for all bit lines $0 \leq i \leq n-1$, except a zero carry is entered into line B_0, and the leftmost carry R_{n-1} is ignored.

Step 3. Repeat step 2 for the remaining $k-4$ input numbers, one during each add cycle, until all the numbers have entered the adder. The above steps need $k-2$ add cycles to complete.

Step 4. Repeat step 3, except enter zeros into the **C** lines until the **R** outputs of all stages become zero. Then the sum of the k numbers is available at the **S** outputs. This final step may take up to $n-1$ add cycles to complete the final carry propagation

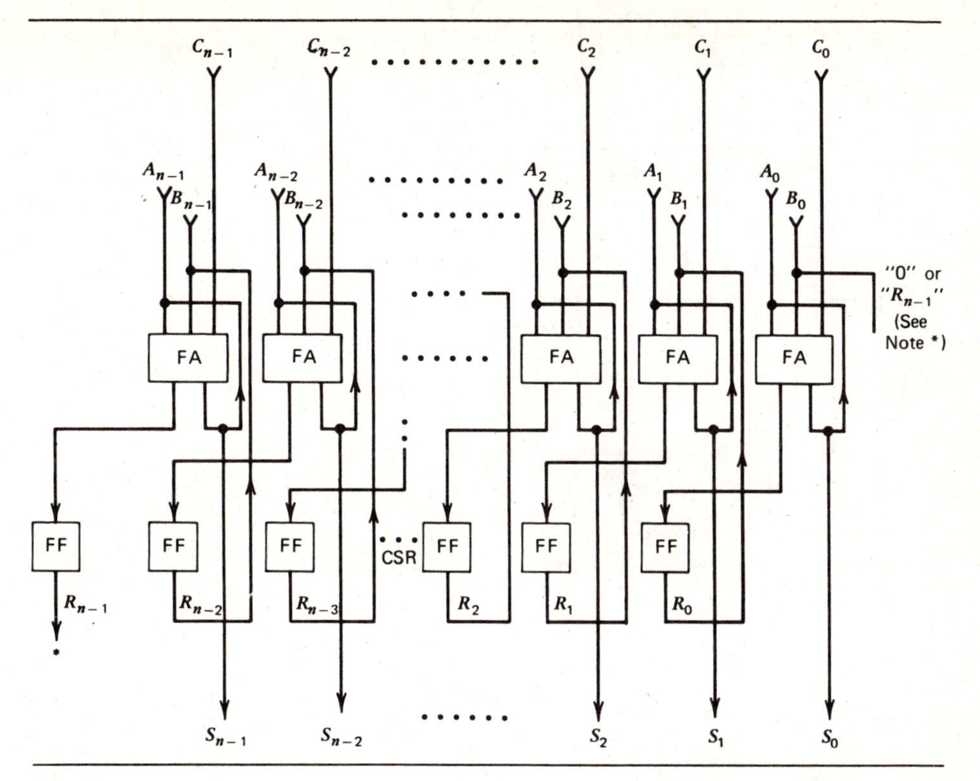

Figure 4.1 The schematic logic diagram of an n-bit carry-save adder (CSA) with a carry-save register (CSR). (* This carry terminal R_{n-1} is ignored for two's complement (Modulo-2^n sum) addition and is connected to B_0 input for one's complement (Modulo-($2^n - 1$) sum) addition. B_0 is grounded ("0") for 2's complement operation.)

Note that the leftmost carry R_{n-1} should be connected to the right-most input B_0, when a Modulo-($2^n - 1$) sum is desired, such as the end-around carry in one's complement arithmetic. The Modulo-2^n addition corresponds to the two's complement addition operation. The total add time of the above procedure is **p** cycles such that

$$k - 2 \leq \mathbf{p} \leq n + k - 3 \tag{4.1}$$

The basic **CSA** is attractive for its low hardware requirement. However, the final stages of carry propagation in step 4 of the above procedure is conducted in a ripple-carry fashion, which may render the whole addition process intolerably slow, especially when the word length n is long. An improvement in speed can be achieved by using a separate two-input adder at step 4, to merge (add) the *sum vector* and the *carry vector* of the **CSA** into one "final" *sum.* This conventional two-input adder is called a *carry propagate adder* (**CPA**). The interconnection of the **CSA** with a **CPA** is shown in Fig. 4.2. The **CPA** provides the final sum of k numbers at the end of the process. With this combined **CSA/CPA** adding unit, only $k - 1$ add cycles are needed to complete

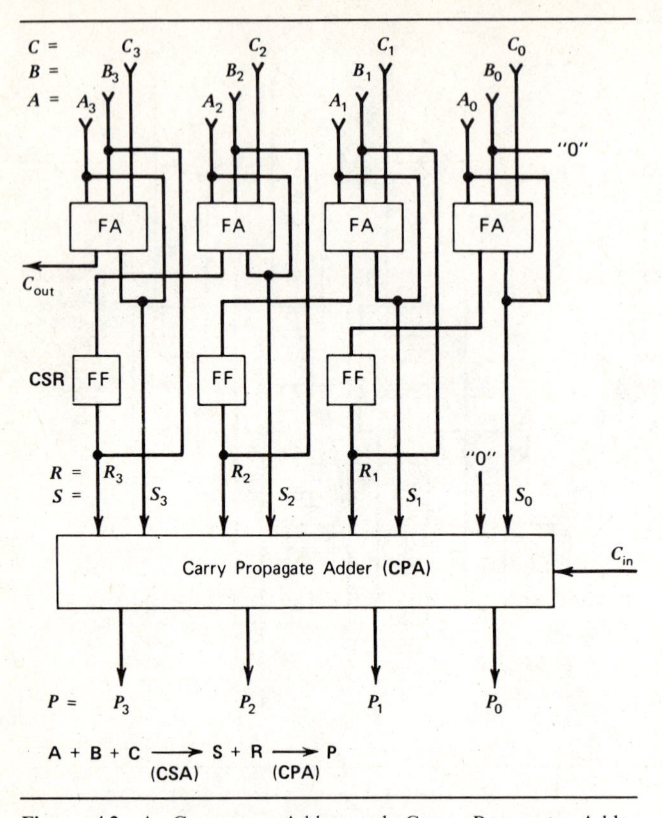

Figure 4.2 A Carry-save Adder and Carry Propagate Adder (CSA/CPA) complex ($n = 4$ bits shown) in two's complement arithmetic (modulo-2^n sum) for multiple operand addition.

the multiple additions of the k input numbers. The **CPA** may assume any of the two-input adder configurations described in Chapter 3.

The final carry propagation can be speeded up by the carry lookahead logic in the **CPA**. The basic cycle period in the combined **CSA/CPA** design may be a little longer than that of the **CSA** alone, if one uses the same clock to control both the **CSA** and the **CPA**. However, this **CSA/CPA** scheme is still much faster and more desirable if all the cycles are counted.

4.3 Multilevel Carry-Save Adders

The designs shown in Figs. 4.1 and 4.2 still input one additional number per cycle. Therefore, in a strict sense, its operation is still sequential. We can use a number of CSAs interconnected as a multilevel adder tree to add more than one number per cycle.

Figure 4.3 shows the **CSA** tree structures for the cases of adding $k = 4, 5, 6,$ and 7 numbers simultaneously. Two feedback inputs in each case are needed to accumulate

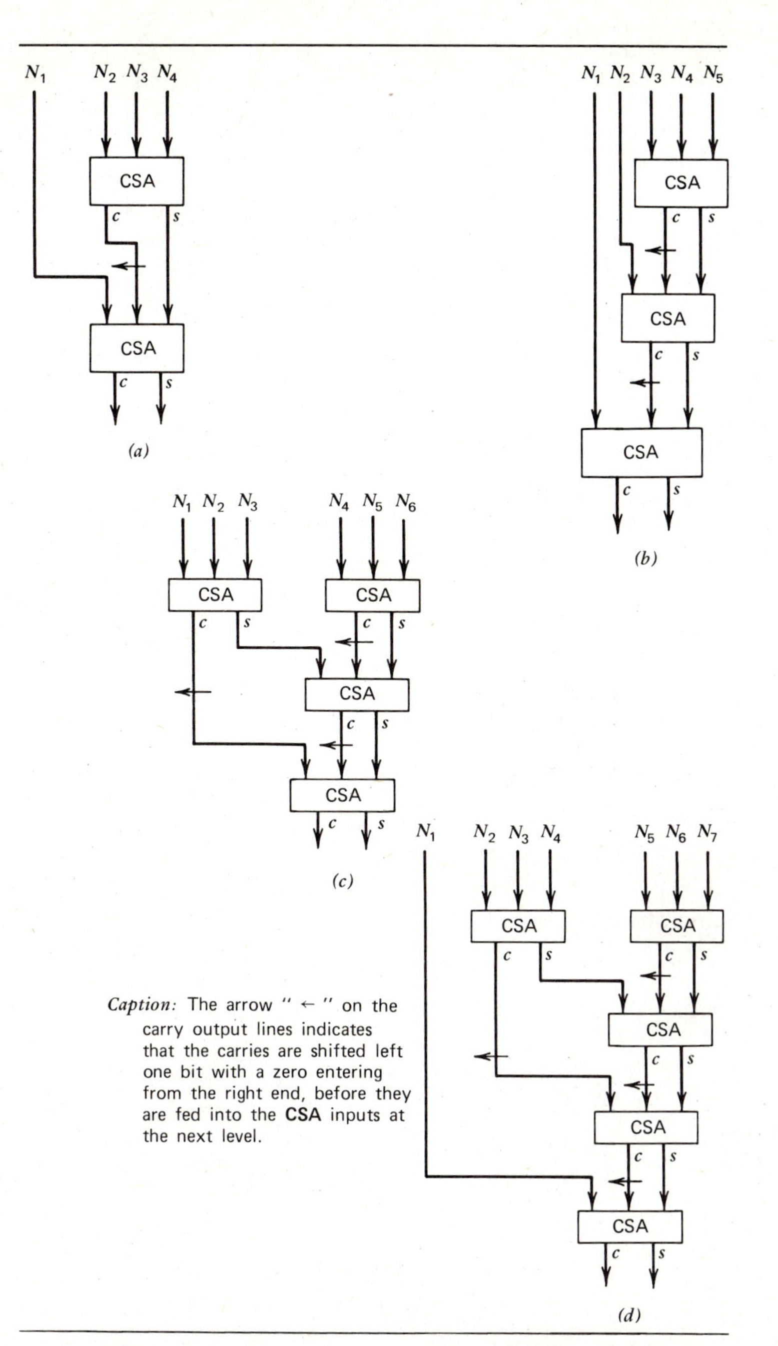

Figure 4.3 Carry-save Adder (CSA) trees of various numbers of inputs. (*a*) A 4-input **CSA** tree. (*b*) A 5-input **CSA** tree. (*c*) A 6-input **CSA** tree. (*d*) A 7-input **CSA** tree.

the *partial sum* and the one-bit left-shifted *partial carry*. The number of levels of the **CSA** tree determines the basic cycle time of the addition process. Each additional input operand increases the number of **CSA**s by one. To reduce k operands to a two-vector result, $k - 2$ **CSA**s are required. Each level of **CSA** contributes a 2Δ time delay equal to that of one stage of full adder. An optimal design should require a minimum number of levels on **CSA** tree. Let $\mathbf{v}$ be number of levels on a **CSA** tree. Define the maximum number of operands that can be processed with a $\mathbf{v}$-level **CSA** tree as,

$$\theta(\mathbf{v}) \tag{4.2}$$

where $\theta(\mathbf{v})$ also includes the two feedback inputs. Avizienis [2] has derived a recursive formula to evaluate this number

$$\theta(\mathbf{v}) = \left\lfloor \frac{\theta(\mathbf{v}-1)}{2} \right\rfloor \times 3 + (\theta(\mathbf{v}-1))\text{Mod } 2 \tag{4.3}$$

for $\mathbf{v} = 2, 3, \ldots,$ and initially

$$\theta(1) = 3 \tag{4.4}$$

Table 4.1 summarizes the value of $\theta(\mathbf{v})$ versus the values of $\mathbf{v}$ for practical situations. For example, adding 28 numbers requires a seven-level CSA tree, which means 14Δ gate delays.

In later chapters these **CSA** adder trees will be used in designing high-speed multiply and divide processors. It should be noted that appropriate columnwise

Table 4.1 The Maximum Number of Operands $\theta(\mathbf{v})$ Processable by a **CSA** Tree of $\mathbf{v}$ Levels (Avizienis [2])

$\mathbf{v}$ No. of Levels on a CSA Tree	$\theta(\mathbf{v})$ Max. No. of Input Operands
1	3
2	4
3	6
4	9
5	13
6	19
7	28
8	42
9	63
10	94

Note: Loop inputs (feedback) are included. The add time for a $\mathbf{v}$-level **CSA** tree is equal to $2\mathbf{v} \cdot \Delta$

alignment of the inputs and of the feedback inputs is essential to the correct performance of the **CSA** tree. The initial contents of the carry-save register must be zero to ensure the correct operations. Carry outputs must be shifted one bit to the left before feeding to the next stage.

4.4 Bit-Partitioned Multiple Addition

Contrary to the rowwise parallel addition used in carry-save adders, the *bit-partitioned adders* perform parallel additions of k input numbers in a bit-slice columnwise fashion. The bit columns of all numbers are divided into partitioned blocks of adjacent columns. Within each block, the numbers are added column by column. All the partitioned subadders operate simultaneously. The sum outputs of all the partitioned subadders are combined to form the final sum of the k input numbers. This bit-partitioning scheme is attractive in at least three aspects. *First*, it is suitable for LSI

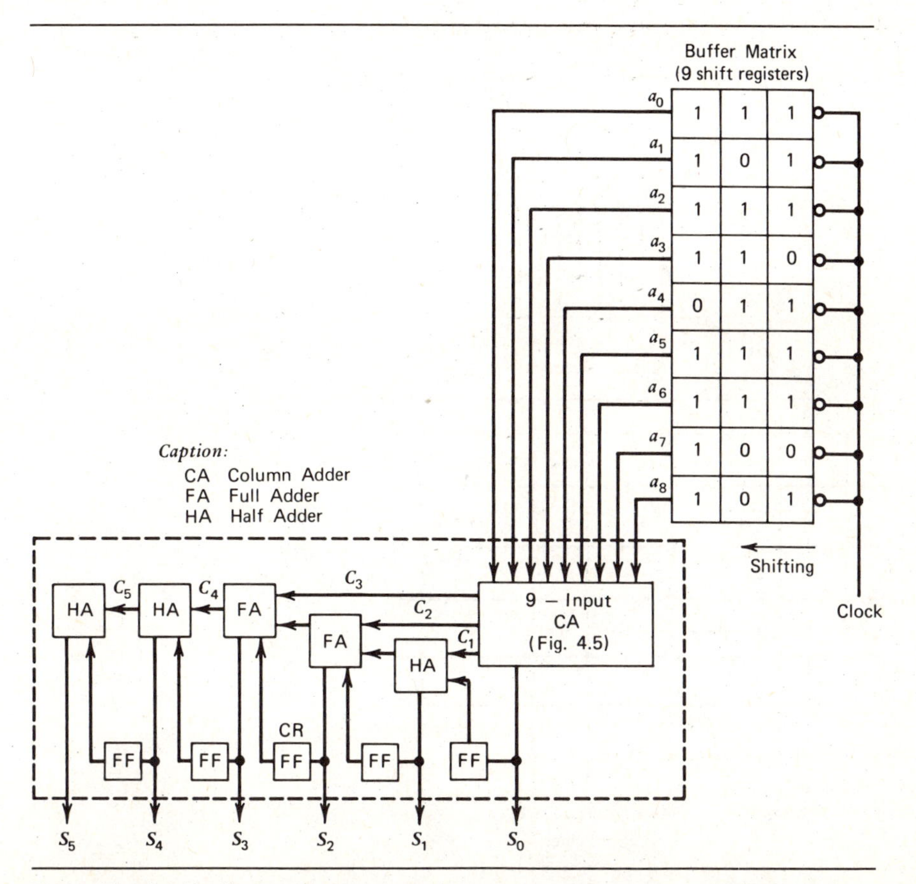

Figure 4.4 A 9-bit-slice partitioned adder realized with 9 shift registers, one column adder, and several full adders and half adders.

by using ROMs or PLAs. *Second*, it is much faster than the rowwise carry-save approach. *Third*, it matches the high-speed pipelined arithmetic to be discussed in Chapter 11.

The basic idea involved in the *partitioned subadder* design is illustrated by the example shown in Fig. 4.4. This design considers the addition of nine 3-bit numbers (partition size $m = 3$). The nine numbers are available from a sequential-access buffer storage (a set of nine 3-bit shift registers). During each cycle, the column information stored in the buffer matrix shifts one column position to the left. The nine bits $a_0, a_1, a_2, \ldots, a_7, a_8$ in the leftmost column of the buffer are fed into a nine-input *column adder* (CA), which produces three carriers C_3, C_2, C_1, and one sum bit S_0, simultaneously.

This column adder can be implemented with random logic as well as with monolithic ROM or PLA chips. Figure 4.5 shows the partial column addition table as programmed into a 2048-bit ROM (512-by-4). The nine bits $a_0, a_1, a_2, \ldots, a_7, a_8$ form the address input to the ROM which selects one word out of the 512 words as the sum outputs C_3, C_2, C_1, S_0 of the CA. Figure 4.4 shows a 5-bit carry-save adder formed with three half adders, two full adders, and one 5-bit carry register. This

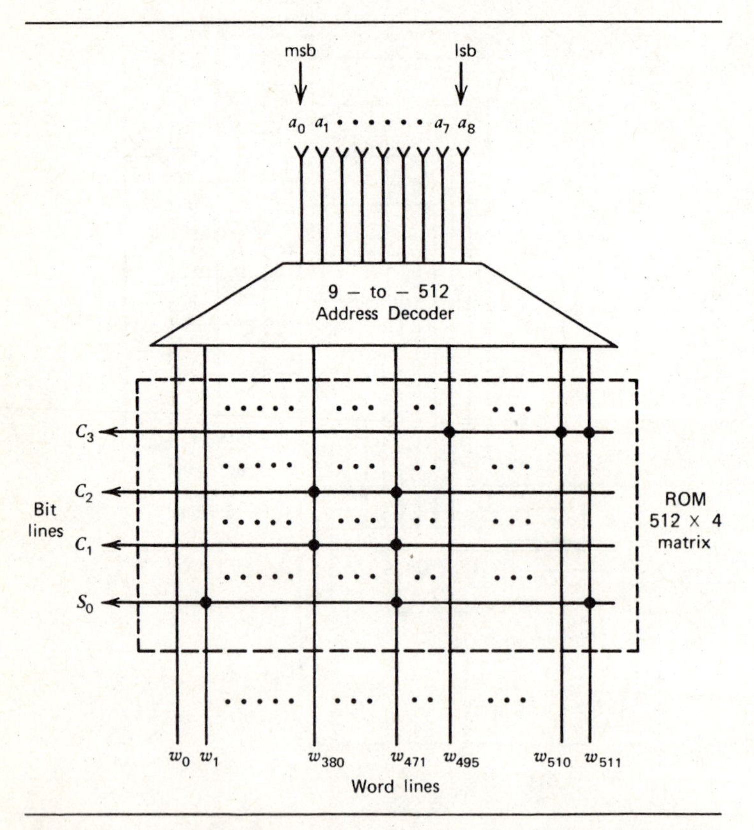

Figure 4.5 A ROM-implementation of the 9-bit-slice Column Adder (**CA**) used in Figure 4.4.

circuit is used to propagate the five carries $C_5, \ldots, C_1$ to the high-order positions during each column addition. The subscripts correspond to bit positions of the final sum outputs. The carry register is cleared to zero before starting the columnwise additions.

4.5 Bit-Partitioned Multioperand Adders

The detailed circuit operations of the bitwise multiple addition is explained by the following arithmetic procedures against the set of numerical data given in the buffer matrix registers.

First Add Cycle Shift the leftmost column into the column adder to produce the carries and sum bit C_3, C_2, C_1, S_0 as 1000, because there are eight 1's among the nine first-column bits. The first 6-bit partial sum appears as terminals $S_5, S_4, S_3, S_2, S_1, S_0$ as 001000. The 5-bit carry register is now loaded as 01000.

Second Add Cycle Shift the current leftmost column (the second column in the buffer matrix) into the column adder to produce C_3, C_2, C_1, S_0 as 0110. The three carries C_3, C_2, $C_1 = 011$ are added with current content of the carry register to yield the leftmost 5 bits of the second partial sum as 01011 (00011 + 01000). The complete second partial sum appears as 010110.

Third Add Cycle Shift the remaining column from the buffer into the column adder. Repeat the same procedure in the next cycle to yield the final sum $S_5 S_4 S_3 S_2 S_1 S_0 =$ 110011 of the nine 3-bit numbers.

The above procedures enable the addition of k numbers in n cycles, where n is the word length of the operands. The partitioned subadder designed in Fig. 4.4 has a partition size $m = 3$. We proceed now with an example design of a full length ($n = 32$ bits) nine-number adder using 11 of the 3-bit partitioned subadders. As shown in Fig. 4.6, the lower 3-bit sum outputs of all the 11 subadders are stored in the 33-bit register **A** and all the upper 3-bit sums are stored in the 33-bit register **B**. Three add cycles as just described are required to obtain the register sum in **A** and **B** from the 11 partitioned subadders. In the fourth cycle, the contents of registers **A** and **B** are fed into a 36-bit carry lookahead adder following the column lineups shown below.

$$\begin{array}{rl} & 0\ \ 0\ \ 0\ \ A_{32}A_{31}\cdots A_4A_3A_2A_1A_0 \\ +) & B_{32}B_{31}B_{30}B_{29}B_{28}\cdots B_1B_0\ \ 0\ \ 0\ \ 0 \\ \hline & S_{35}S_{34}S_{33}S_{32}S_{31}\cdots S_4S_3S_2S_1S_0 \end{array}$$

Altogether, four cycles are needed to produce the 36-bit sum of nine 32-bit input numbers.

Singh and Waxman [11] have proved that it is possible to add k numbers in $m + 1$ cycles, where m is the partition width lower bounded by

$$m \geq \lceil \log_2(k - 1) \rceil \qquad \textbf{(4.5)}$$

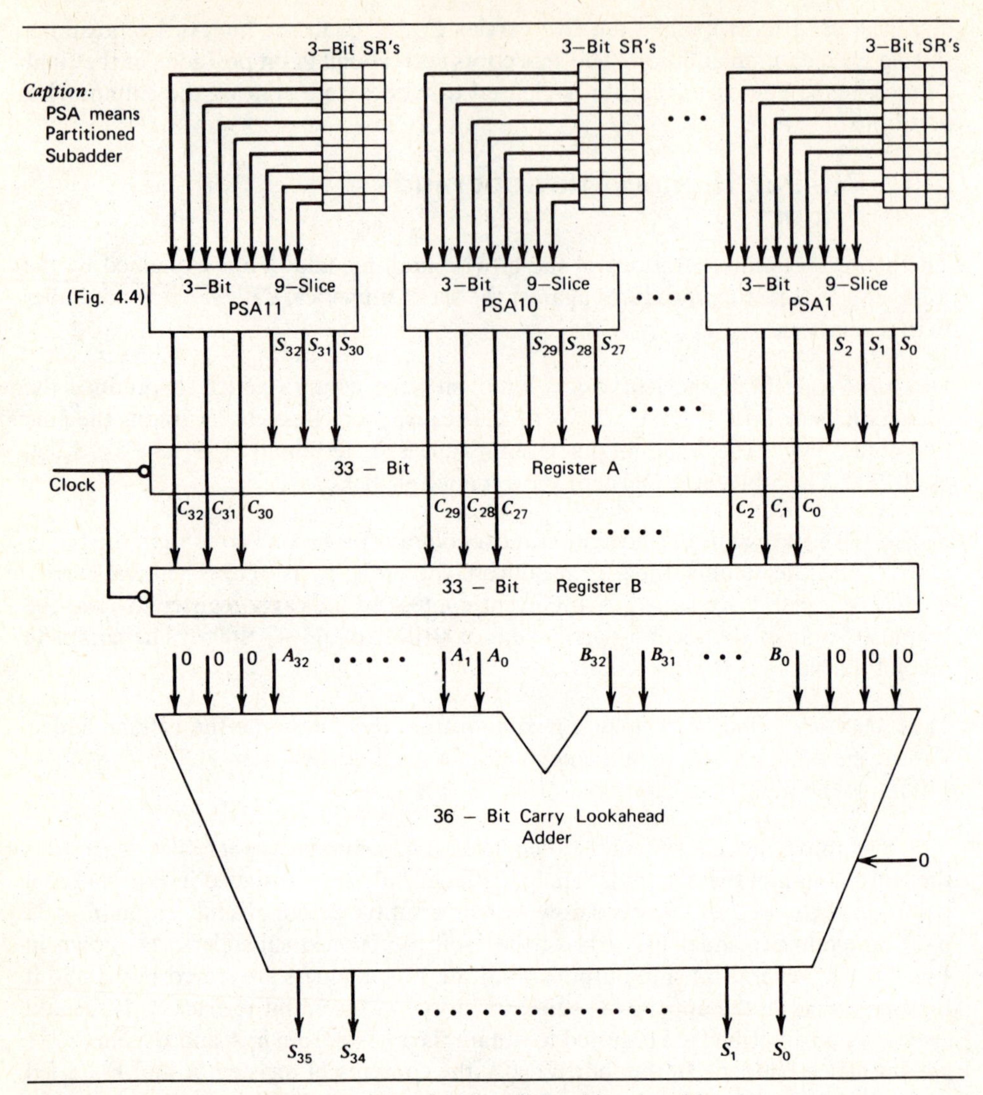

Figure 4.6 A 32-bit 9-number bit-partitioned adder with partition size of 3.

The multiple-number adder shown in Fig. 4.6 requires only $m + 1 = 3 + 1 = 4$ cycles to add the $k = 9$ numbers with 32-bit word length. This offers a significant improvement over the n cycles required with a single subadder of full size $m = n$. Table 4.2 provides the minimum partition widths m for various size multiple-operand adder designs. For example, the addition of 33 full-length numbers (from 16 bits up to 64 bits) with 5-bit column partitions ($m = 5$) will require only six ($m + 1 = 6$) machine cycles. Note that the total add time is proportional to $m + 1$ but independent of the word length n of the adder.

Table 4.2 The Minimum Partition Widths for Bit-Partitioned Addition of Various Numbers of Operands.

No. of Operands to Be Added	Minimal Partition Width
5	2
9	3
17	4
33	5
65	8
k	$m = \lceil \log_2 (k - 1) \rceil$

4.6 Signed-Digit Addition/ Subtraction

We describe below the representational properties of signed-digit arithmetic for "parallel" addition or subtraction of numbers using a restricted class of **SD** redundant codes. The method limits carry-propagation to one position to the left during the operations of ADD and SUBTRACT in digital computers. The addition time for such **SD** numbers of any length is, therefore, equal to the time required to add only two adjacent digits. In order to achieve this interdigit independency for fast parallel arithmetic, the following four requirements on the **SD** number representations are necessary.

1. The radix r is a positive integer greater than 2.
2. The algebraic value of zero has a unique **SD** representation.
3. There exist transformations between conventional sign-magnitude m-digit representation and **SD** m-digit representation for every algebraic value within the machine range.
4. *Totally parallel* addition/subtraction is possible for all digits in corresponding positions of the two **SD** operands.

The addition or subtraction of two **SD** numbers **Z** and **Y** is considered *totally parallel*, if the following two conditions are satisfied for all digital positions $n - 1 \geq i \geq -k$ as shown in Eq. 1.22. $(n + k)$-digit **SD** numbers are assumed here,

$$\mathbf{Z} = (z_{n-1} \cdots z_1 z_0 . z_{-1} \cdots z_{-k})_r \text{ and } \mathbf{Y} = (y_{n-1} \cdots y_0 . y_{-1} \cdots y_{-k})_r .$$

1. Let s_i be the ith *sum digit* of the resulting sum $\mathbf{S} = (s_{n-1} \cdots s_1 s_0 . s_{-1} \cdots s_{-k})_r = \mathbf{Z} + \mathbf{Y}$ and t_i be the *transfer digit* from the $(i - 1)$th digital position. Then s_i is a function only of three variables

$$s_i = f(z_i, y_i, t_i) \tag{4.6}$$

where z_i, y_i are the i-th digits of the augend **Z** and addend **Y** respectively.

2. The transfer digit t_{i+1} to the $(i + 1)$th digital position on the left is a function only of the augend digit z_i and the addend digit y_i.

$$t_{i+1} = g(z_i, y_i) \tag{4.7}$$

Totally-parallel subtraction of the subtrahend digit y_i from the minuend digit z_i is performed as the totally-parallel addition of z_i to $\bar{y}_i$, the *additive inverse* of y_i.

$$z_i - y_i = z_i + (\bar{y}_i) \tag{4.8}$$

It should be noted that the transfer digit t_i may assume both positive and negative values in either **SD** addition or **SD** subtraction, unlike the "carry" or "borrow" of conventional addition or subtraction, which can only assume nonnegative values. The transfer digit is never propagated past the first adder position on the left. The above definitions imply a two-step **SD** addition process as described below, given digits z_i, y_i and t_i.

Step 1. The outgoing transfer digit t_{i+1} and an *interim sum* digit ω_i are generated by the addition of z_i to y_i

$$r \cdot t_{i+1} + \omega_i = z_i + y_i \tag{4.9}$$

Step 2. The sum digit s_i is obtained by adding ω_i to t_i, the transfer digit from digital position $i - 1$.

$$s_i = \omega_i + t_i \tag{4.10}$$

To satisfy the conditions given in Eq. 4.6 and Eq. 4.7, the range of values which s_i may assume in Eq. 4.10 should not exceed those for the digits z_i and y_i in Eq. 4.9. The **SD** subtraction specified in Eq. 4.8 is possible, only if for every allowed nonzero value of digit y_i, there exists an additive inverse $\bar{y}_i$ in the set of all allowed values of y_i; that is, for every $y_i = a$, there exists $\bar{y}_i = -a$ such that

$$y_i + \bar{y}_i = a + (-a) = 0 \tag{4.11}$$

The second requirement for unique zero representation is satisfied by the condition that the magnitude of allowed digit values may not exceed $r - 1$, that is

$$|z_i| \leq r - 1 \tag{4.12}$$

where z_i is the ith digit of any **SD** number $\mathbf{Z} = (z_{n-1} \cdots z_0 . z_{-1} \cdots z_{-k})_r$.

The third requirement of convertibility is satisfied if the procedure of totally parallel addition, when applied to convert any conventional radix digit $x_i \in \{0, 1, \cdots r - 1\}$, will yield an allowed value of base-r **SD** digit z_i

$$x_i = r \cdot t_{i+1} + \omega_i \tag{4.13}$$

$$z_i = \omega_i + t_i \tag{4.14}$$

The above procedure is an extension of those specified in Eqs. 1.26, 1.27, and 1.28 by allowing the transfer digit t_i to be negative if necessary. The above conditions as specified in Eqs. 4.9 through 4.12 establish the following set of values for each transfer digit

$$t_i = -1, 0, 1 \quad \textbf{(4.15)}$$

The following condition

$$|\omega_i| \leq r - 2 \quad \textbf{(4.16)}$$

is obtained as the upper limit for the magnitude of the interim sum ω_i, if t_i is restricted to $-1, 0, 1$. An immediate result of Eq. 4.16 is the restriction

$$r \geq 3 \quad \textbf{(4.17)}$$

because, for $r = 2$, the only allowed value of $\omega_i = 0$ does not satisfy Eq. 4.13 for the value $x_i = 1$. Requirements in Eqs. 4.13, 4.15, and 4.16 establish a set of all allowed values for each interim digit ω_i.

$$\omega_i \in \{-(r-2), \ldots, -1, 0, 1, \ldots, r-2\} \quad \textbf{(4.18)}$$

The above derivations lead to two *enlarged* digit sets for the operand digits in a totally parallel **SD** add/subtraction system.

For an odd radix $r_0 \geq 3$, we may choose

$$\Sigma'_{r_0} = \left\{\frac{-(r_o+1)}{2}, \ldots, -1, 0, 1, \ldots, \frac{r_o+1}{2}\right\} \quad \textbf{(4.19)}$$

For an even radix $r_e \geq 4$, we have

$$\Sigma'_{r_e} = \left\{-\left(\frac{r_e}{2}+1\right), \ldots, -1, 0, 1, \ldots, \left(\frac{r_e}{2}+1\right)\right\} \quad \textbf{(4.20)}$$

These two sets contain more redundancy than those given in Eq. 1.18 and Eq. 1.21. Otherwise, the required total parallelism cannot be achieved. For instance, given $r = 4$ the digit set $\Sigma'_4 = \{-3, -2, -1, 0, 1, 2, 3\}$ is used, instead of

$$\Sigma_4 = \{-2, -1, 0, 1, 2\}$$

implied by Eq. 1.21. It should also be noted that for $r > 4$, there may exist more than one set of allowed digit values. The redundancy of an **SD** number is considered ***minimal,*** **when each digit assumes the smallest value** $\alpha = (r_o + 1)/2$ or $\alpha = r_e/2 + 1$ as specified above. The redundancy is considered maximal when $\alpha = r - 1$ is chosen, as shown in Eq. 1.19. The multiple number of allowed digit sets for **SD** systems with radix $r > 4$ results from various degrees of increased redundancy, upper bounded by $r - 1$ as shown in Eq. 4.12.

4.7 Totally Parallel SD Adders/Subtractors

The totally parallel **SD** addition and subtraction methods described in the preceding section are realized in this section. The two-step addition process can be implemented with the adder configuration shown in Fig. 4.7. The Roman numerals I and II are used to identify two types of adding cells realizing Eq. 4.9 and Eq. 4.10, respectively.

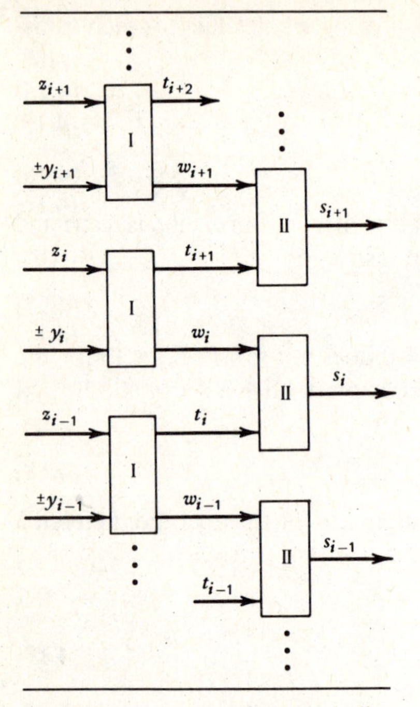

Figure 4.7 A section of the block diagram of a totally parallel adder/subtractor in signed-digit arithmetic (3 adjacent digits shown).

Only a section of the totally parallel adder corresponding to three adjacent digits is shown in the block diagram.

The detailed execution steps of an **SD** addition are to be demonstrated by a numerical example. In a general **SD** arithmetic unit, addition, subtraction, left shifting, and right shifting are the elementary operations, whereas multiplication and division are executed as sequences of additions or subtractions and shifts. To perform **SD** subtraction, the sign of the subtrahend is changed by inverting the sign of all nonzero digit in **SD** vector **Y**, then totally parallel **SD** addition of corresponding digits is performed. Given the allowed values of ω_i as the sequence

$$\{\omega_{\min}, \ldots, -1, 0, 1, \ldots, \omega_{\max}\} \tag{4.21}$$

the rules for forming ω_i, t_{i+1} and s_i are rewritten from Eq. 4.9 as follows in decreasing order of the index $i = n - 1, n - 2, \ldots, -k$.

$$\omega_i = (z_i + y_i) - r \cdot t_{i+1} \tag{4.22}$$

where

$$t_{i+1} = \begin{cases} 0, & \text{if } \omega_{\min} \le z_i + y_i \le \omega_{\max} \\ 1, & \text{if } z_i + y_i > \omega_{\max} \\ -1, & \text{if } z_i + y_i < \omega_{\min} \end{cases} \tag{4.23}$$

The sum digit s_i is still obtained by Eq. 4.10. It is interesting to note that the addition or subtraction of **SD** numbers may be initiated from either end of the number. The following example demonstrates the addition of two radix-10 **SD** numbers under the assumption that we start from the most significant digit (left end) to the right end.

Example. Given $r = 10$, and the following allowed digit sets for a radix-10 **SD** system

$$\omega_i \in \{\bar{5}, \bar{4}, \bar{3}, \bar{2}, \bar{1}, 0, 1, 2, 3, 4, 5\} \tag{4.24}$$

where $\omega_{min} = \bar{5}$ and $\omega_{max} = 5$;

$$t_i \in \{\bar{1}, 0, 1\}; \tag{4.25}$$

and

$$y_i, z_i, s_i \in \{\bar{6}, \bar{5}, \bar{4}, \bar{3}, \bar{2}, \bar{1}, 0, 1, 2, 3, 4, 5, 6\} \tag{4.26}$$

Consider the **SD** addition of the following pair of **SD** numbers with $n = 1$ and $-k = -5$.

$$\begin{aligned} \text{Augend } \mathbf{Z} &= 1.\bar{3}65\bar{1}\bar{4} \\ \text{Addend } \mathbf{Y} &= 0.\bar{4}053\bar{1} \end{aligned} \tag{4.27}$$

with algebraic values

$$\begin{aligned} \tilde{\mathbf{Z}} &= (0.76486)_{10} \\ \tilde{\mathbf{Y}} &= (-0.39471)_{10} \end{aligned} \tag{4.28}$$

The detailed steps of **SD** addition of **Z** and **Y** is described in Table 4.3. The resulting sum is in **SD** form as shown in the bottom row of the table.

$$\mathbf{S} = \mathbf{Z} + \mathbf{Y} = 0.4\bar{3}02\bar{5} \tag{4.29}$$

Table 4.3 The Addition of Two Radix-10 **SD** Numbers **S** = **Z** + **Y**.

Digital Position i	**1**	**0**	**−1**	**−2**	**−3**	**−4**	**−5**
Augend **Z** z_i		1	$\bar{3}$	6	5	$\bar{1}$	$\bar{4}$
Addend **Y** y_i		0	$\bar{4}$	0	5	3	$\bar{1}$
Direct sum $z_i + y_i$		1	$\bar{7}$	6	10	2	$\bar{5}$
Transit Digit t_i	0	$\bar{1}$	1	1	0	0	
Interim Sum ω_i		$(1 - 0) = 1$	$(\bar{7} + 10) = 3$	$(6 - 10) = \bar{4}$	$(10 - 10) = 0$	$(2 - 0) = 2$	$(\bar{5} - 0) = \bar{5}$
Sum S s_i		$(\bar{1} + 1) = 0$	$(1 + 3) = 4$	$(1 + \bar{4}) = \bar{3}$	$(0 + 0) = 0$	$(0 + 2) = 2$	$(0 + \bar{5}) = \bar{5}$

with an algebraic value

$$\tilde{S} = \tilde{Z} + \tilde{Y} = (0.37015)_{10} \tag{4.30}$$

Shifting (left or right) of **SD** numbers is performed as follows: If the digits z_i of a number pass through the adder logic during the shifting operation, transfer digits t_{i+1} will be generated when $z_i > \omega_{max}$ or $z_i < \omega_{min}$ and added to the interim sum ω_{i+1} at the left according to Eq. 4.22 and Eq. 4.23. During the right shift of one position, $z_0 = 0$ is inserted as the most significant digit of the shifted number. If this shift includes transfer generation, the least significant digit z_{-k} in the shift register will retain the property

$$|z_{-k}| \leq |z_i|_{max} - 1 \tag{4.31}$$

which is useful in a multiple precision operation. Before the left shift of one position, the inspection of digits z_1 and z_2 will predict overflow after the shift.

A totally parallel adder for **SD** numbers of m digits each ($m = n + k$ as shown in Eq. 1.18) consists of m identical *digit-adders.* Each digit adder consists of two adding cells I and II, as shown in Fig. 4.7. These m digit adders are linked to their immediate neighbors by transfer digit lines, an output to the left and an input to the right. All transfer digits t_i and interim sum digits ω_i are generated simultaneously; thus, the **SD** addition time is independent of the word length m. It is determined by the add time of only one stage of the digit adder.

The minimal-redundancy **SD** representations are preferable for saving storage space. Furthermore, less complicated digit-adder logic may be expected when the least possible number of digit values is employed.

The choice of the radix r depends on the desired balance between the increase in storage requirements and the corresponding logical complexity of one digit adder. For example, radix-4 **SD** arithmetic requires seven values (-3 to 3) for the digits of a subtrahend at the input to the digit adder, and at least six of these values (-3 or 3 may be avoided) for the sum digits s_i plus variant requirements for the quotient and multiplier digits. Avizienis [3] has concluded that *three* binary storage elements per digit are required for all radices $3 \leq r \leq 6$ and *four* elements for radices $7 \leq r \leq 14$. For example, the radix 10 (with 13 values) requires four storage elements per digit, the same as for conventional BCD adders. A radix-4 **SD** digit adder was actually designed with a seven-value digit set (-3, -2, -1, 0, 1, 2, 3) requiring three binary digits weighted by -4, 2, and 1, respectively. A radix-4 **SD** adder requires the equivalent logic circuitry of approximately 12 half adders of the conventional radix-4 adder. This is about three times more than that required by one position of a conventional radix-4 adder with ripple carry propagation.

4.8 Multifunction Arithmetic Logic Unit

By imposing additional combinational logic into the parallel two-operand adders developed in previous chapters, one can easily obtain an *arithmetic logic unit* (**ALU**) that performs not only a number of arithmetic operations, but also various logic operations on two input numbers. Additional external function-select lines are needed to determine these operations.

At present, most **ALU**s are available in MSI 4-bit slices. Monolithic 8-bit or up-to-16-bit **ALU**s are possible with current LSI circuitry. Because full carry lookahead is usually embedded within the 4-bit slices and provision is made for multilevel carry lookahead interconnections, **ALU**s of longer word lengths (unless restricted by special space and power requirements) are obtained by interconnecting several 4-bit slices. The number of functions performed by commercially available **ALU**s ranges from 4 to 32, depending on the package sizes.

The design of a basic 4-function, 4-bit **ALU** is presented as an example. The unit can perform ADD, SUBTRACT, and several logic functions on two 4-bit input words $\mathbf{A} = A_3A_2A_1A_0$ and $\mathbf{B} = B_3B_2B_1B_0$ to yield a 4-bit output word $\mathbf{S} = S_3S_2S_1S_0$. Two function-select lines S and T are used to determine these operations. As shown in Fig. 4.8, the device can operate on either *active-high* or *active-low* data buses. Both one's complement and two's complement arithmetic operations can be realized with this **ALU**.

Internally, carry lookahead logic is used for its high speed. Provision is made for second-level carry lookahead from preceding **ALU** slices without the need for additional external logic. The internal circuit operations of this **ALU** are described with assumed active-low data inputs and outputs. The internal logic of the **ALU** can be divided into several functional sections as shown in Fig. 4.9. The top section includes a set of gates, which produces the AND function Λ_i and the OR functions V_i, for $i = 3, 2, 1$, and 0, under the control of S, T lines

$$\Lambda_i = \begin{cases} A_i\bar{B}_i, & \text{if } S = 0, \\ A_iB_i, & \text{if } S = 1, \end{cases}$$

$$V_i = \begin{cases} A_i + \bar{B}_i, & \text{if } S = 0, \\ A_i + B_i, & \text{if } ST = 10, \\ 1, & \text{if } ST = 11 \end{cases} \tag{4.32}$$

The various forms of gate functions Λ_i and V_i are used to vary the functions performed by the circuit.

The left section of the circuit includes a set of gates for the generation of a lookahead carry input

$$C_{\text{in}} = G^*_{-1} + P^*_{-1}G^*_{-2} + P^*_{-1}P^*_{-2}G^*_{-3} \tag{4.33}$$

where the block carry generate G_i^* and the block carry propagate P_i^* were defined in Eq. 3.20, except the carry propagate functions $P_i = A_i \oplus B_i$ for all i are now replaced by $V_i = A_i + B_i$. The negative subscripts -1, -2, and -3, refer to block-carry-generate or propagate signals generated from the immediate preceding, the second preceding, and the third preceding **ALU** slices, respectively. The bottom section of the circuit contains the adding logic, which adds (Exclusive-OR) the

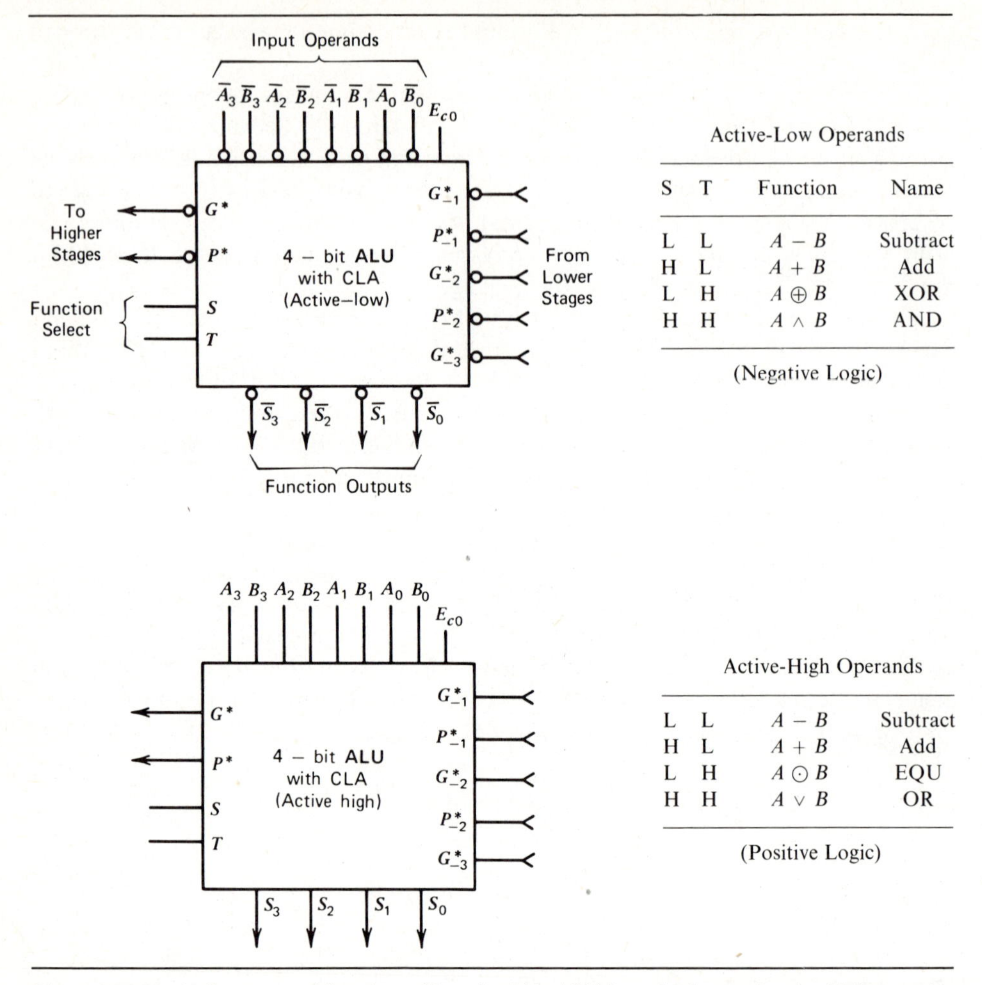

Active-Low Operands

S	T	Function	Name
L	L	$A - B$	Subtract
H	L	$A + B$	Add
L	H	$A \oplus B$	XOR
H	H	$A \wedge B$	AND

(Negative Logic)

Active-High Operands

S	T	Function	Name
L	L	$A - B$	Subtract
H	L	$A + B$	Add
L	H	$A \odot B$	EQU
H	H	$A \vee B$	OR

(Positive Logic)

Figure 4.8 Block diagrams and function tables of a 4-bit ALU in active-low and active-high operating modes.

appropriate Λ, V signals, and the carry signal to each bit position. The internal carries into the four-bit positions are listed below.

$$\begin{aligned} C_0 &= C_{in} + T \\ C_1 &= (\Lambda_0 + C_{in} V_0) + T \\ C_2 &= (\Lambda_1 + \Lambda_0 V_1 + C_{in} V_0 V_1) + T \\ C_3 &= (\Lambda_2 + \Lambda_1 V_2 + \Lambda_0 V_1 V_2 + C_{in} V_0 V_1 V_2) + T \end{aligned} \quad \textbf{(4.34)}$$

The ith output bit S_i of the **ALU** can be written as

$$S_i = (\bar{\Lambda}_i \cdot V_i) \oplus C_i \quad \textbf{(4.35)}$$

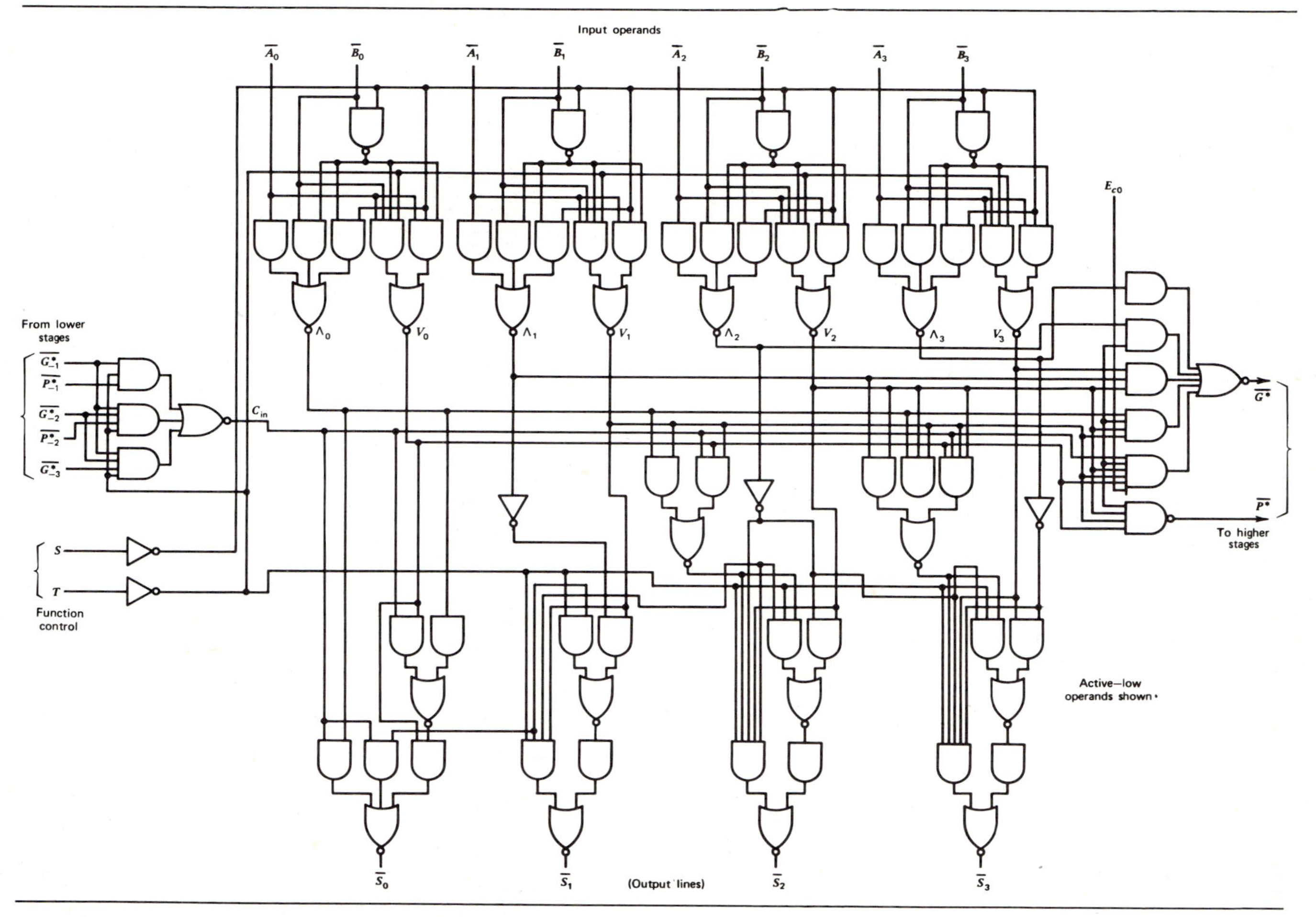

Figure 4.9 The schematic logic diagram of a 4-bit ALU with carry lookahead as specified in Figure 4.8.

Note that for logic operations with T being high, a carry $C_i = 1$ is forced into each bit position, which makes

$$S_i = (\bar{\Lambda}_i \cdot V_i) \oplus 1 = \overline{\bar{\Lambda}_i \cdot V_i} \tag{4.36}$$

Each output digit S_i is now independent of carry propagations as expected. For completeness, the right side of the circuit generates two block carry auxiliary functions

$$\begin{aligned} G^* &= \Lambda_3 + \Lambda_2 V_3 + \Lambda_1 V_2 V_3 + \Lambda_0 V_1 V_2 V_3 + C_{\text{in}} V_0 V_1 V_2 V_3 E_{\text{co}} \\ P^* &= V_0 V_1 V_2 V_3 \end{aligned} \tag{4.37}$$

A carry out enable E_{co} is provided such that the carry out of the **ALU** slice can be written as

$$C_{\text{out}} = G^* + P^* \cdot C_{\text{in}} \cdot E_{\text{co}} \tag{4.38}$$

The circuit operations corresponding to active-high data will be left as an exercise for the readers. A more sophisticated **ALU** design and its applications will be presented as a case study in a later section.

4.9 System Applications of ALU's

Methods of interconnecting a number of basic 4-bit **ALU** slices to form a full-length **ALU** are described in this section. Then we shall use an **ALU** to compare the magnitude of binary numbers in various FXP notations. System-matching considerations in tying **ALU**s with other arithmetic operations are also discussed.

The **ALU** slice design described in the preceding section is used as a building block in constructing full-size **ALU**s of word lengths equal to multiples of 4. A 16-bit **ALU** is constructed in Fig. 4.10 using four 4-bit **ALU**s. Full carry lookahead is achieved in this 16-bit **ALU** via external block carry connections as shown. Additional **ALU** slices can be connected to form ripple block addition and still maintain rather high-speed performance over large word lengths. Only two additional gate delays (2Δ) are incurred for each 12-bit increment. For example, one can connect several 12-bit blocks, similar to that shown in the dotted area of the diagram, to the left end of the 16-bit **ALU** circuit to form a 28-bit **ALU** or a 40-bit **ALU**, and so on. Typical add times of various **ALU**s constructed with cascading the above **ALU** slices are shown in Table 4.4. Subtraction can be done in only one more gate delay than the add time.

The **ALU**s constructed above can work in either one's or two's complement arithmetic and in either active-high or active-low conventions. The external circuit connections using the above **ALU**s in one's complement or two's complement arithmetic operations are shown in Fig. 4.11. The end-around carry in one's complement arithmetic may cause an extra time delay. A carry is forced into the least significant stage of a two's complement **ALU** when subtraction is desired. The auxiliary carry in terminal G^*_{-1} can be used for inputting the initial carry.

Magnitude comparisons of two binary numbers are frequently required in arithmetic processes such as division, square rooting, and so on. The comparison is

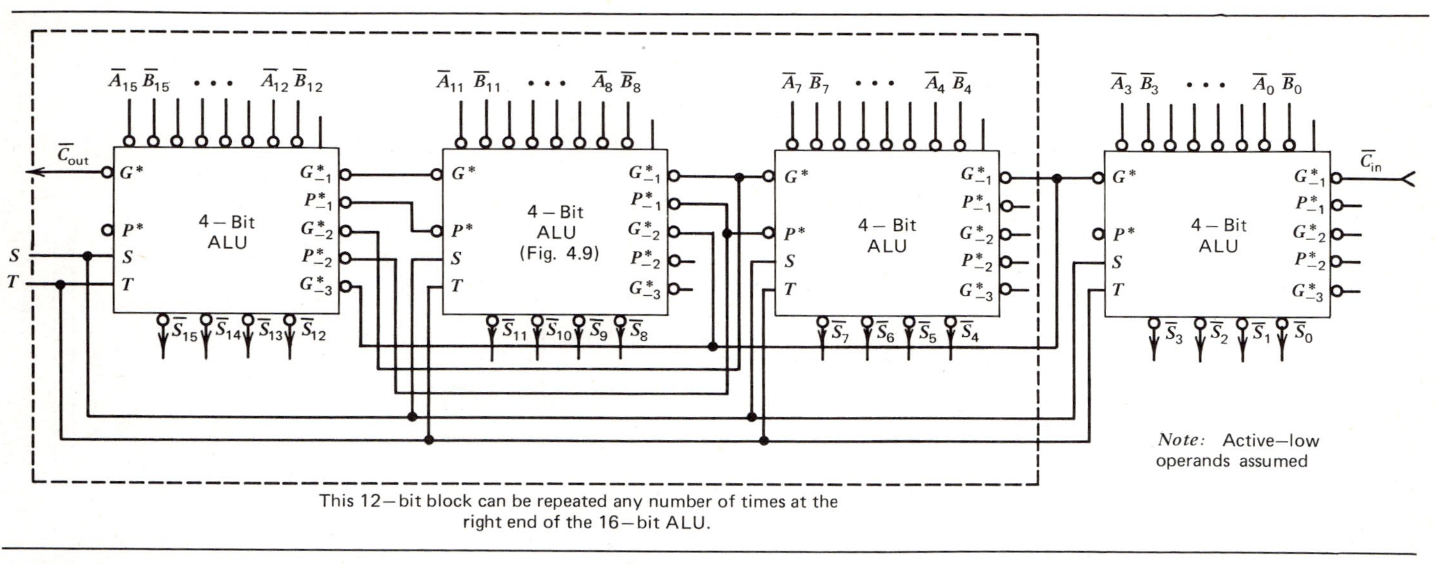

Figure 4.10 A 16-bit ALU obtained by cascading the 4-bit ALUs with embedded CLA.

Table 4.4 Typical Add Time and Subtract Time of **ALU**s Constructed as in Fig. 4.10 using the 4-bit Slices as Shown in Figs. 4.8 and 4.9

Word Length (bits)	Add (nsec)[a]	Subtract (nsec)[a]
1–4	24	27
5–16	38	41
17–28	53	56
19–40	68	71
41–52	83	86
53–54	98	101

[a] Data from Fairchild Semiconductor [5].

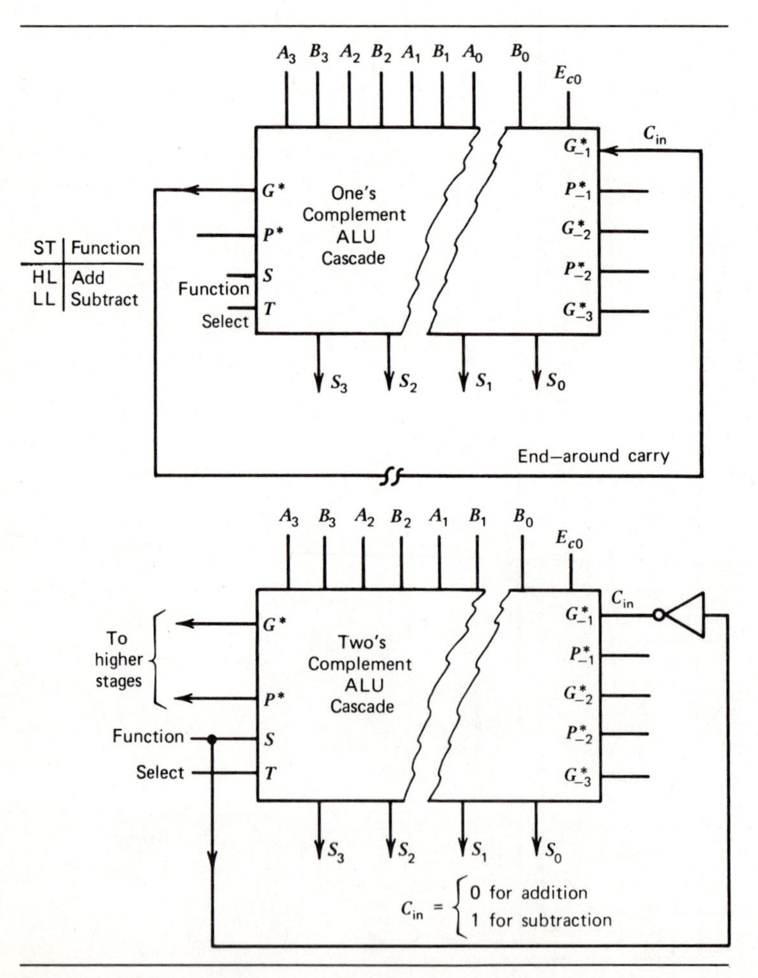

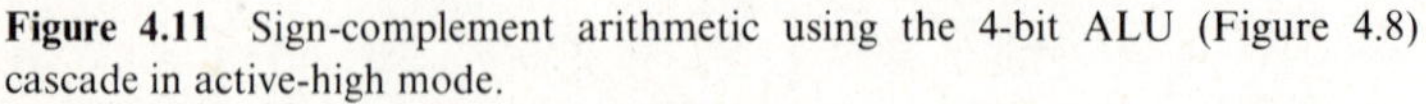
Figure 4.11 Sign-complement arithmetic using the 4-bit ALU (Figure 4.8) cascade in active-high mode.

performed with the device in the subtract mode and by observing the carry-out signal. In a two's complement subtraction (with carry in active), the function performed is **A** − **B**. If the carry-out signal from the sign is active, then **A** ≥ **B**. If the carry input is removed, the function to be performed is **A** − **B** − 1. If the carry out is active for this operation, then **A** > **B**. When only the first case is true, obviously **A** = **B**.

Equivalence between **A** and **B** may be detected directly by using the carry-generate and carry-propagate outputs of the **ALU**. If **A** = **B**, then in the subtract mode or in the Exclusive-Or mode, all the internal V_i signals will be active and none of the internal Λ_i signals will be active. This is detected at the outputs by the following condition

$$\mathbf{A} = \mathbf{B}, \quad \text{if } P^* \cdot \bar{G}^* = 1 \tag{4.39}$$

Equivalence over the entire word is detected by forming the above function for each unit and ANDing all the signals.

The selection or the design of **ALU**s has to be tied in with the rest of a computing system. First, the hardware features embedded within the **ALU** chips should be maximally utilized. They should be geared with the various microoperations associated with either fixed-point or floating-point arithmetic functions. Not only addition and subtraction, but multiplication, division, square rooting, and other useful operations should also be tied in.

The **ALU** should be matched to the logic convention or the data buses and to the control circuitry. Synchronous versus asynchronous controls make a big difference. Moreover, the **ALU** should match the main memory or the high-speed scratch pad buffer organizations. RAMs, sequential-access memories, or stack memories require different intermediate data formats. Fixed-word length and variable-word length memory words also require different data acquisition schemes. If all these factors are taken into account, we can then achieve a meaningful design of arithmetic processors.

4.10 Case Study I: Design and Applications of The SN 74181 ALU

We describe below a real 4-bit **ALU** that can perform 16 arithmetic operations, including add, subtract, increment, double, invert, pass, and so on, and all the 16 possible logic operations on two 4-bit parallel words. This device has been commercially labeled as the 74181 **ALU**. To select among the 32 operations requires at least five control lines. The logic block diagram corresponding to the two logic conventions of the 74181 **ALU** is shown in Fig. 4.12 together with the function table. The mode-select line M distinguishes logic operations ($M = 1$) from arithmetic operations ($M = 0$). Logic operations are performed on a bitwise basis. Depending on whether or not there is a carry in, the arithmetic operations are described by two different columns in the function table.

Figure 4.13 illustrates the internal logic circuit schematic of the 74181 **ALU**. The auxiliary AND/OR functions are generated on the top of the diagram under the control of four function-select lines $S_3 S_2 S_1 S_0$. These are then used to generate the

Mode Select Inputs				Active Low Inputs and Outputs		Active High Inputs and Outputs	
S_3	S_2	S_1	S_0	Logic (M = H)	Arithmetic (M = L)(C_n = L)	Logic (M = H)	Arithmetic[b] (M = L)(C_n = H)
L	L	L	L	$\bar{A}$	A minus 1	$\bar{A}$	A
L	L	L	H	$\overline{AB}$	AB minus 1	$\overline{A + B}$	$A + B$
L	L	H	L	$\overline{A + B}$	$A\bar{B}$ minus 1	$\bar{A}B$	$A + \bar{B}$
L	L	H	H	Logical 1	minus 1	Logical 0	minus 1
L	H	L	L	$\overline{A + B}$	A plus $(A + \bar{B})$	$\overline{AB}$	A plus $A\bar{B}$
L	H	L	H	$\bar{B}$	AB plus $(A + \bar{B})$	$\bar{B}$	$(A + B)$ plus AB
L	H	H	L	$\overline{A \oplus B}$	A minus B minus 1	$A \oplus B$	A minus B minus 1
L	H	H	H	$A + \bar{B}$	$A + \bar{B}$	$A\bar{B}$	$A\bar{B}$ minus 1
H	L	L	L	$\bar{A}B$	A plus $(A + B)$	$\bar{A} + B$	A plus AB
H	L	L	H	$A \oplus B$	A plus B	$\overline{A \oplus B}$	A plus B
H	L	H	L	B	$A\bar{B}$ plus $(A + B)$	B	$(A + \bar{B})$ plus AB
H	L	H	H	$A + B$	$A + B$	AB	AB minus 1
H	H	L	L	Logical 0	A plus A[a]	Logical 1	A plus A[a]
H	H	L	H	$A\bar{B}$	AB plus A	$A + \bar{B}$	$(A + B)$ plus A
H	H	H	L	AB	$A\bar{B}$ plus A	$A + B$	$(A + \bar{B})$ plus A
H	H	H	H	A	A	A	A minus 1

H = High voltage level

L = Low voltage level

[a] Each bit is shifted to the next more significant position, i.e., $A^* = 2A$

[b] Arithmetic operations expressed in 2's complement notation

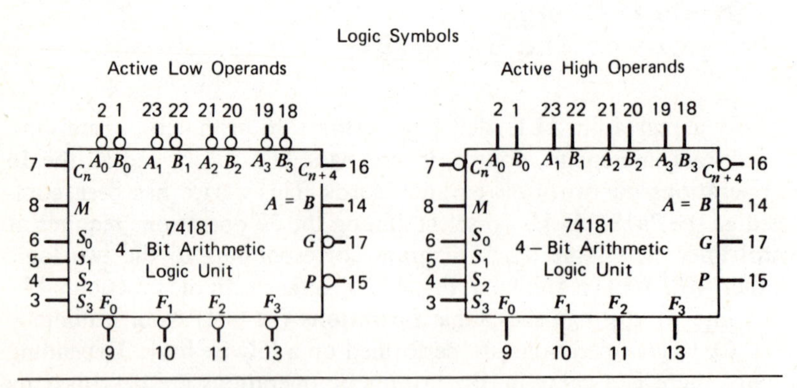

Figure 4.12 Block diagrams and function table of the 74181 ALU operated on either active-high or active-low operands.

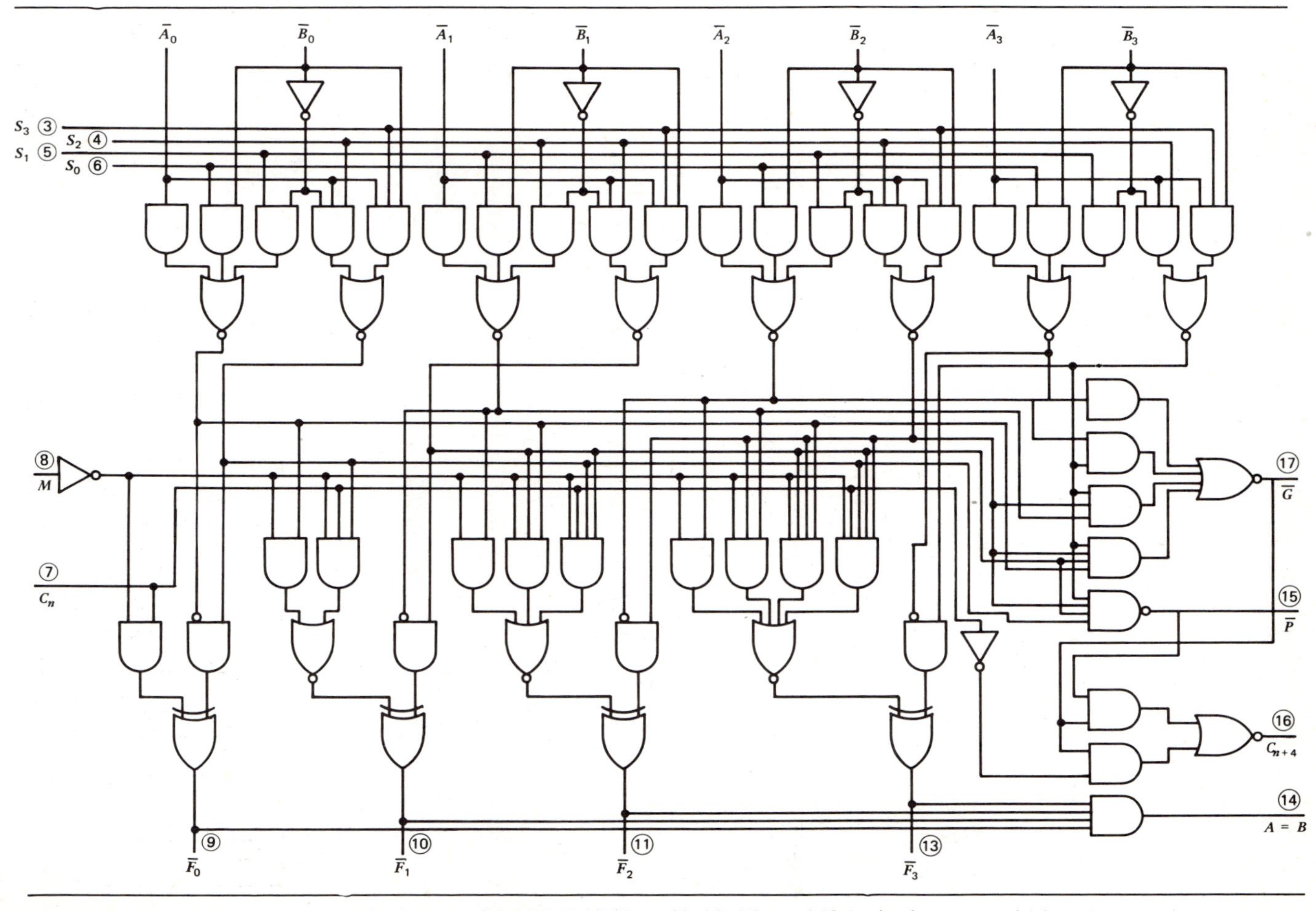

Figure 4.13 The schematic logic circuit diagram of the 74181 ALU specified in Figure 4.12 (active-low operands shown).

sum and carry terms at the lower portion of the circuit. Carries are inhibited from propagation when M is high.

Internal carry lookahead is used, but no provision is made to anticipate carries from preceding **ALU** slices. One carry-in and one carry-out terminal are available for ripple-carry cascading of similar units. The device has two auxiliary carry functions $\bar{G}$ and $\bar{P}$, defined similarly to G^* and P^* in Chapter 3. The open-collector **A** = **B** output can be AND-tied to the **A** = **B** outputs of other **ALU**s to detect the all-one condition for several units. Subtraction is performed by one's complement addition, where the one's complement subtrahend is generated internally. The resulting output is **A** − **B** − 1 which requires an end-around carry or a forced carry to provide **A** − **B**.

Comparing the two operating modes of the 74181 with respect to active-high and active-low data, we observed that the functions available with the device form a *closed* set such that inversion of the logic inputs produces a function which is still in the set. Therefore, the device performs the same set of arithmetic and logic operations in the active-high data convention as it does in the active-low data convention. Mixed conventions are also allowed in the use of this **ALU** without losing a majority of the useful functions. Two mixed operating modes, one with active-low A and F and active-high B and the other with active-high A and F and active-low B, are left as exercises.

Next we show how to use a number of 74181 **ALU** slices together with external CLA units (74182) to form full-length **ALU**s. A 32-bit **ALU** with ripple-carry lookahead is shown in Fig. 4.14. Eight 74181 **ALU**s and two 4-bit 74182 **CLA** units are used in this circuit. Typical add times and the necessary IC package count of various-length **ALU**s designed with 74181 and 74182 are summarized in Table 4.5.

The comparison operations performed by the 74181 **ALU** are described next. The **A** = **B** output is internally decoded from the function output $S_3S_2S_1S_0$ so that it will assume a "high" to indicate equality, when two words of equal magnitude are

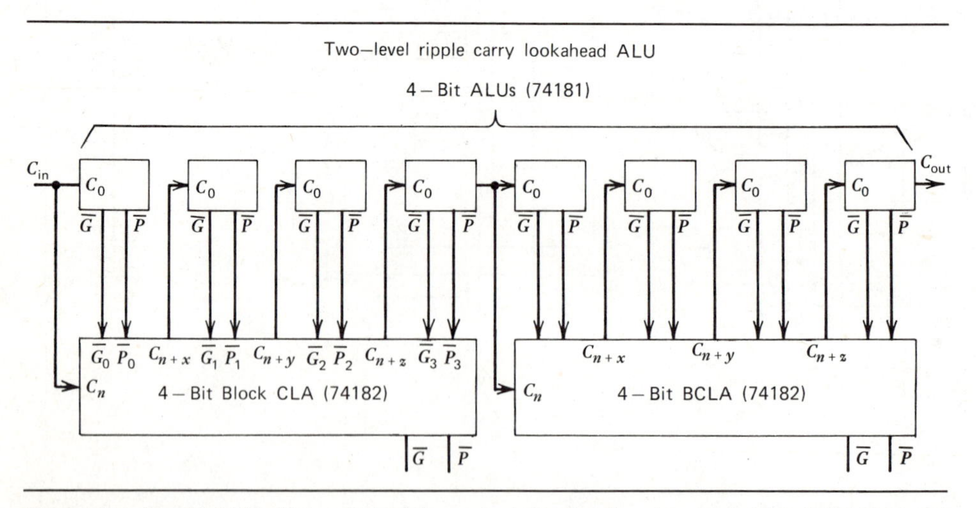

Figure 4.14 A 32-bit ALU with two 16-bit full carry lookahead sections in cascade (eight 74181 ALUs and two 74182 BCLA units are used).

Table 4.5 Add Times and Package Counts of **ALU** Designs with Various Word Length and Configurations

Word Length (bits)	ALU Configuration	Total Add Time[a] (ns.)	IC Package Count	
			74181	74182
4	1-level CLA	21	1	0
8	Ripple CLA	36	2	0
12	Ripple CLA	48	3	0
12	2-level CLA	36	3	1
16	Ripple CLA	60	4	0
16	2-level CLA	36	4	1
32	Ripple CLA	110	8	0
32	2-level Ripple CLA	62	8	2
48	Ripple CLA	160	12	0
48	2-level Ripple CLA	136	12	3
48	3-level CLA	88	12	4
64	Ripple CLA	210	16	0
64	2-level Ripple CLA	101	16	4
64	3-level CLA	64	16	5

[a] Data from Fairchild Semiconductor [5].

applied at the **A** and **B** inputs. There are several functions of the 74181 **ALU** that can be used to reveal the following relationships of two input numbers:

$$=, >; \geq; <; \leq, \neq \tag{4.40}$$

These useful functions are shown in Table 4.6. By examining the values on output terminals $\mathbf{A} = \mathbf{B}$ and the carry-out C_{out}, one can determine the relative magnitudes of **A** and **B** as shown in the right columns of Table 4.6. For example, the equivalence operation $\overline{\mathbf{A} \oplus \mathbf{B}}$ can be used to indicate the complementary relationship $\mathbf{A} = \overline{\mathbf{B}}$. Because subtraction is performed by one's complement addition, the existence of an input carry ($C_{\text{in}} = 0$ or 1) will make a difference of "one" during subtraction, which is why the input carry should also be specified when using subtraction to compare two numbers. For unsigned numbers with the most significant bit positive (a "0"), the carry out of the **ALU** indicates their relative magnitude.

Table 4.6 Comparison Operations Using the 74181 **ALU**

Output	State	Operation	Active LOW Logic	Active HIGH Logic
$A = B$	H	A minus B	$A = B$	$A =$ (B minus 1)
	H	$\overline{A \oplus B}$	$A \neq B$	$A = B$
	H	$A \oplus B$	$A = B$	$A \neq B$
Carry Out	H	A minus B	$A \geq B$	$A < B$
($\bar{C}_4$ for active High operands)	L	A minus B	$A < B$	$A \geq B$
(C_4 for active Low operands)	H	A minus B minus 1	$A > B$	$A \leq B$
	L	A minus B minus 1	$A \leq B$	$A > B$

4.11 Bibliographic Notes

The carry-save adders presented in this chapter are based on the work of MacSorley [7], Robertson [9], Rohatsch [10], and Wallace [14]. Typical applications of a carry-save adder tree to implement high-speed multiply/divide units can be found in the reports by Wallace [14] and by IBM technical staff members [1, 4]. The bit-partitioned scheme for multiple-operand addition was proposed by Singh and Waxman [11]. They also described how to use the bit-partitioned multioperand adders for fast multiplication. Another system for simultaneous addition of several binary numbers is proposed by Kouvaras et al. [6], using an arithmetic procedure based on a set of mathematical recurrence formulas.

The signed-digit arithmetic rules were originally formulated by Avizienis [2, 3]. Only **SD** two-operand addition and subtraction are presented in this chapter. Other arithmetic operations, such as **SD** multiplication, **SD** division, multiple-operand **SD** addition, overflow detection, multiple-precision, and round-off operations in **SD** arithmetic, were also studied by Avizienis and by Robertson [9] of the University of Illinois. Metze and Robertson [8] also presented an interesting report on the elimination of carry-propagation in digital computers.

For more design and application information of **ALU**s, readers may check the TTL application handbook by the Fairchild Semiconductor staff [5] and the TTL Data Books by TI [12]. Texas Instruments recently announced the monolithic I^2L 4-bit parallel binary processor element [13]. This device contains not only a 16-operation **ALU** implemented with full carry lookahead, but also a number of working registers and control logic within the chip. In this sense, the monolithic microprocessor chips are obviously extended from the basic **ALU**s. The recent announcement of bipolar bit-slice microprocessors with high-speed performance may achieve a new level of performance/cost efficiency in the design and manufacture of digital arithmetic processors.

References

[1] Anderson, S. F. et al., "The IBM System/360 Model 91: Floating-Point Execution Unit," *IBM Journal of R & D*, Vol. 11, No. 1, January 1967, pp. 34–53.

[2] Avizienis, A., "A Study of Redundant Number Representations for Parallel Digital Computers," Ph.D. Thesis, University of Illinois, Urbana, Illinois, May 1960.

[3] Avizienis, A., "Signed-Digit Number Representations for Fast Parallel Arithmetic," *IEEE Trans. Elec. Comp.*, Vol. EC-10. September 1961, pp. 389–400.

[4] Bratun, J. M. et al., "Multiply/Divide Unit for A High-Performance Digital Computer," *IBM Tech. Disc. Bulletin*, Vol. 14, No. 6, November 1971, pp. 1813–1316.

[5] Fairchild Semiconductor Staff, *The TTL Application Handbook*, Fairchild Semiconductor, Mountain View, CA 1973.

[6] Kouvaras, N. D. et al., "Digital System of Simultaneous Addition of Several Binary Numbers," *IEEE Trans. Comput.*, Vol. C-17, No. 10, October 1968, pp. 992–997.

[7] MacSorley, O. L. "High-Speed Arithmetic in Binary Computers," *Proc. of IRE*, Vol. 49, No. 1, January 1961, pp. 67–91.

[8] Metze, G. and Robertson, J. E., "Elimination of Carry Propagation in Digital Computers," *Proc. International Conf. on Inf. Processing*, Paris, France, June 1959, pp. 389–396.

[9] Robertson, J. E., "A Deterministic Procedure for the Design of Carry-Save Adders and Borrow-Save Subtractors," University of Illinois, Dept. of Computer Science, *Report No.* 235, July 1967.

[10] Rohatsch, F. A., "A Study of Transformations Applicable to the Development of Limited Carry-Borrow Propagation Adders," Ph.D. Thesis, University of Illinois, Urbana, Illinois, June 1967.

[11] Singh, S. and Waxman, R., "Multiple Operand Addition and Multiplication," *IEEE Trans. Comput.*, Vol. C-22, No. 2, February 1973, pp. 113–119.

[12] Texas Instruments Staff, *The TTL Data Book and Supplement*, Dallas, Texas, 1973, pp. 38–395 and pp. S312–S313.

[13] Texas Instruments Application Note, *I^2L SBP-0400 4-bit Parallel Binary Processing Element*, Texas Instrument Inc., Dallas, Texas 75222, 1976.

[14] Wallace, C. S., "A Suggestion for a Fast Multiplier," *IEEE Trans. Elec. Comp.*, Vol. EC-13, February 1964, pp. 14–17.

Problems

Prob. 4.1 Construct a 16-bit, two's complement, multiple-operand adder with one-level **CSA** and one full carry lookahead **CPA**. Estimate the total add time required to yield the sum of 100 16-bit two's complement numbers using this **CSA/CPA** adding unit.

Prob. 4.2 Modify the one-level **CSA/CPA** design in Prob. 4.1 into a multilevel design, in which the **CSA** loop can receive 10 input numbers per add cycle. Show the schematic logic circuit diagram and estimate the total time to add one hundred 16-bit numbers in two's complement notation using this multilevel **CSA/CPA** adding unit.

Prob. 4.3 Prove that Avizienis' formula (Eq. 4.3) for determining the maximum number of operands $\boldsymbol{\theta}(\mathbf{v})$ processable by a **CSA** tree of $\mathbf{v}$ levels works for all possible $\mathbf{v}$-level **CSA** tree configurations.

Prob. 4.4 Design a 5-bit, 33-slice, *Bit-Partitioned SubAdder* (PSA) with shift register array, ROM, full adders, and half adders, similar to that shown in Fig. 4.4.

Then construct a 55-bit, 33-number adder with 11 of the PSA's, two 55-bit registers, and one 60-bit CLA adder. Explain how 33 numbers (55 bit word length) can be added in six machine cycles using this multiple-number adding unit.

Prob. 4.5 Prove the Singh and Waxman lower bound (Eq. 4.5) on the number of cycles required to add k numbers using the bit-partitioned multiple addition method described in the text. Compare your proof with the original proof given in Ref. [11].

Prob. 4.6 Show the schematic logic design of a totally parallel adder, which can add radix-4 signed-digit (**SD**) numbers, and verify your design with a table similar to Table 4.3. Note that the seven possible values of radix-4 **SD** $(-3, -2, -1, 0, 1, 2, 3)$ are represented by three binary digits weighted -4, 2, and 1.

Prob. 4.7 Show the schematic interconnections of a 28-bit **ALU** using seven of the 4-bit **ALU** slices specified in Section 4.8. Assume two's complement arithmetic in your design.

Prob. 4.8 Establish the functional equivalence of the active-high and the active-low operating modes of the 74181 **ALU** by listing the one-to-one correspondence of all the 32 functions in either mode.

Prob. 4.9 Figure out the function table of the 74181 **ALU** with mixed operating modes with active-low A and F and active-high B as specified below.

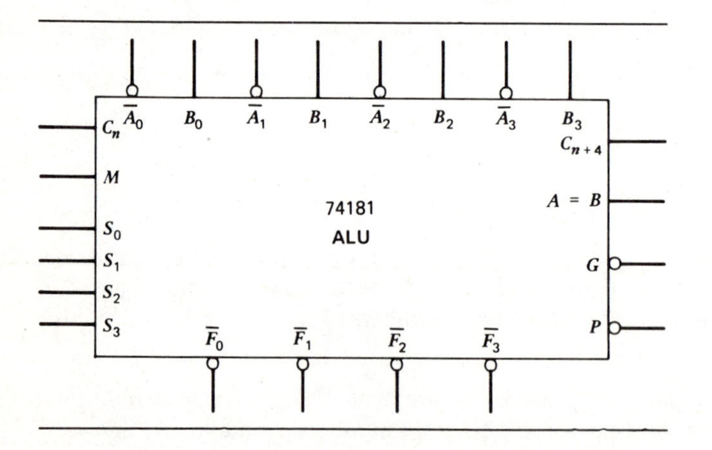

Prob. 4.10 Show the schematic block diagrams for all of the three 64-bit **ALU** configurations: Ripple CLA, 2-level Ripple CLA, and 3-level CLA with the number of IC packages (74181's & 74182's) specified in the last three rows of Table 4.5.

Prob. 4.11 Given two 4-bit unsigned numbers **A** = 1011 and **B** = 1010, verify the entries of Table 4.6 by feeding these two binary numbers into the 74181 **ALU**, applying the functions suggested in the table, and computing the outputs at terminals **A** = **B** or carry out.

Prob. 4.12 Construct a multilevel **CSA** tree which can receive 30 external input numbers simultaneously (excluding the feedback loop inputs). Try to minimize the number of carry-save adders (**CSA**'s) used and to reduce the number of levels on the **CSA** tree.

Chapter

5

Standard and Recoded Multipliers

5.1 Introduction

A hardware multiply/divide unit has become a standard feature in modern digital computers built with IC technology. Arithmetic processors for high-speed multiplication based on add-shift methods under various number notations are to be described. The detailed design of a universal multiplication processor is given. Fast multiplication with multiple-bit scanning and multiplier recoding techniques are explained. The use of carry-save adders to implement high-speed multipliers is illustrated. Booth multiplier [4] shall be described from the viewpoint of multiplier recoding.

In what follows, we use $\leftarrow$ as a *replacement* operator, $\cdot$ as a *concatenation* or *cascading* operator and (**R**) to denote the contents of a register **R** or of a vector variable **R**. Logic OR and AND operations are denoted by $\vee$ and $\wedge$ respectively. On the other hand, arithmetic operations of addition, subtraction, multiplication, and division, are still denoted as $+$, $-$, $\times$, $/$, respectively. The multiplication of two n-bit unsigned integers A and B will create a $2n$-bit *product*

$$P = A \times B \tag{5.1}$$

where A is called the *multiplicand* and B is the *multiplier*. When A and B are signed integers (in any of the three FXP notations), the product is only $(2n - 1)$ bits long. In order to facilitate a universal design, which handles both unsigned and signed numbers upon the user's choice, we assume a $2n$-bit product with a dummy bit between the sign and the most significant bit, when signed numbers are processed. The basic hardware host machine and four multiplication algorithms corresponding to *unsigned*, *sign-magnitude*, *one's complement*, and *two's complement* arithmetic systems are described first.

5.2 Indirect Multiplication Algorithms and Hardware

A typical indirect multiply unit is composed of several functional devices as shown in Fig. 5.1. Three n-bit parallel-load registers are needed, namely, *Accumulator* (**AC**), *Multiplier Register* (**MR**), and *Auxiliary register* (**AX**). The multiplicand A and the multiplier B are initially loaded into the registers **AX** and **MR**, respectively. The initial contents of the **AC** should be zero corresponding to a zero initial partial product. Two separate flip flops, labeled as A_s and B_s, are used to hold the signs of A and B, respectively (zeros are loaded into A_s and B_s for unsigned operation). An n-bit parallel adder is required, which may be selected from those fast adder configurations given in Chapter 3.

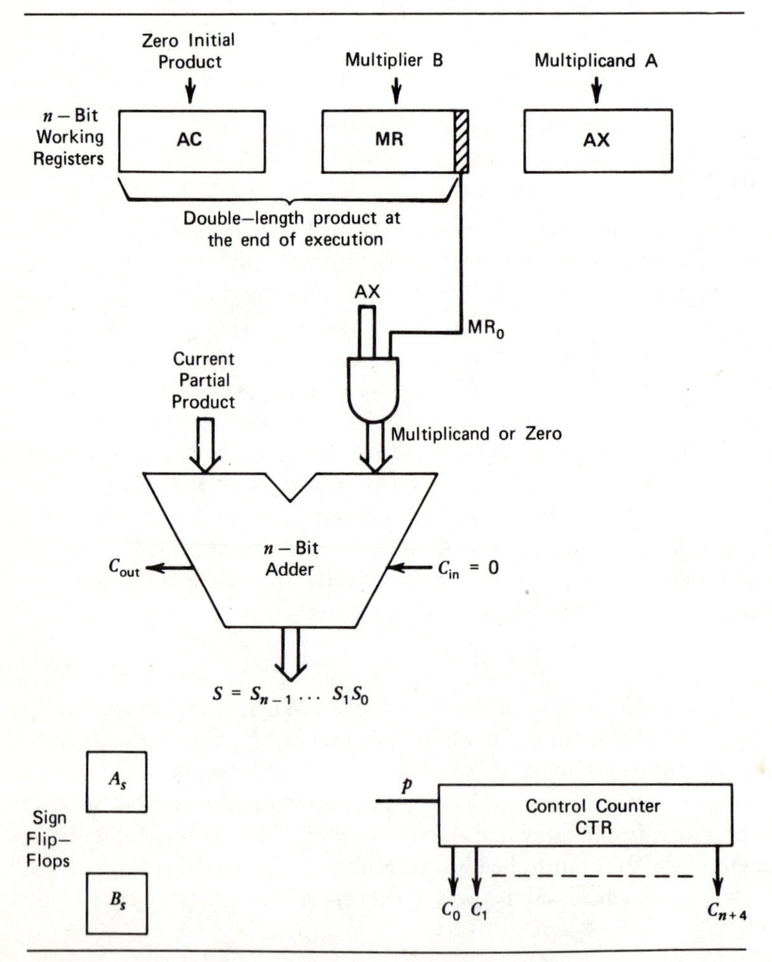

Figure 5.1 The basic hardware host for fixed-point standard add-shift multiplication with uniform single right shift (**AC** · **MR**) per machine cycle.

The resulting $2n$-bit product P should appear in the *cascaded register* $\mathbf{AC} \cdot \mathbf{MR}$, formed by concatenating the **MR** to the right of **AC**. The leading bit AC_{n-1} in **AC** carries the sign of the resulting product, when signed multiplication is performed. A *Control Counter* (**CTR**) is provided to keep track of the number of add-shifts and to recognize the completion of the multiplication. Detailed interconnections of these functional blocks and associated control logic will be shown in subsequent sections.

The following operations take place during each counter cycle. Let

$$\mathbf{S} = S_{n-1} \cdots S_1 S_0$$

be the sum outputs of the adder. C_{in} and C_{out} stand for the initial carry-in and carry-out of the adder. We have to define an auxiliary vector $\mathbf{AX} \wedge \mathrm{MR}_0$ obtained by ANDing each of the n outputs of the **AX** register with the current least significant bit MR_0 of the **MR**.

$$\mathbf{AX} \wedge \mathrm{MR}_0 = \begin{cases} \mathbf{AX}, & \text{if } \mathrm{MR}_0 = 1 \\ \mathbf{0}, & \text{if } \mathrm{MR}_0 = 0 \end{cases} \tag{5.2}$$

The boldface zero refers to a *zero vector* of n bits. The adder performs the following addition with a zero initial carry-in ($C_{in} = 0$) at each counter cycle.

$$C_{out} \cdot \mathbf{S} \leftarrow (\mathbf{AC}) + (\mathbf{AX} \wedge \mathrm{MR}_0) \tag{5.3}$$

A possible C_{out} may be generated in this addition. The n-bit adder could be a full carry lookahead adder with a constant time delay. The following right shift operation is performed immediately after the sum S is formed at the outputs of the adder.

$$\mathbf{AC} \cdot \mathbf{MR} \leftarrow C_{out} \cdot S_{n-1} \cdots S_1 \cdot S_0 \cdot \mathrm{MR}_{n-1} \cdots \mathrm{MR}_1 \tag{5.4}$$

This means that the right-shifted version of the sum outputs (including a C_{out} bit into AC_{n-1}) are loaded into **AC**, S_0 into MR_{n-1}, and the MR_0 has been pushed off the right end of the **MR**. This *stored shifting* can be implemented with multiplexer logic, which selects the appropriate inputs to the working registers when a shifted version is required.

The above procedure illustrates that at each counter cycle, either a copy of the multiplicand A or a zero vector is added to the upper n bits of the partial product depending on whether the multiplier bit MR_0 being examined, is a "1" or "0". The shifting of the successive partial product to the right is equivalent to the left-shift performed in paper-pencil method, in which the multiplicand bits are entered from fixed column positions. This process may be repeated n times, until all the multiplier bits are examined, one at a time. At the end of the computation, the n-bit multiplier B will be pushed off the right end of the **MR** and replaced by the lower half of the resulting product. The **AC** holds the upper half of the product.

Unsigned and Sign-Magnitude Multiplications

The multiplication of the two n-bit unsigned number is described in detail by the flow chart shown in Fig. 5.2. Only $n + 1$ cycles are needed to complete this operation. The load of data into appropriate registers and flip-flops occurs at cycle C_0 as indicated.

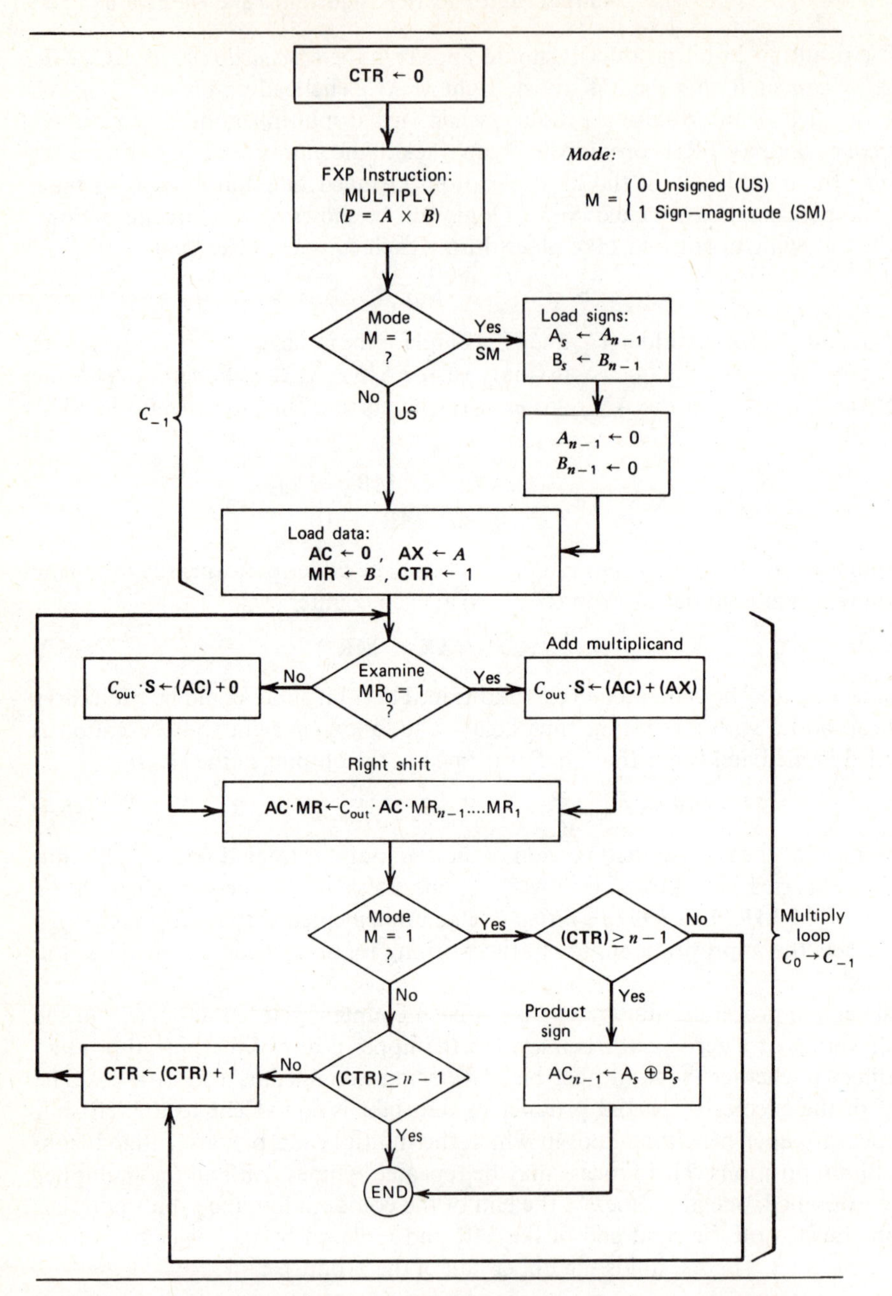

Figure 5.2 A basic unsigned and sign-magnitude add-shift multiplication algorithm.

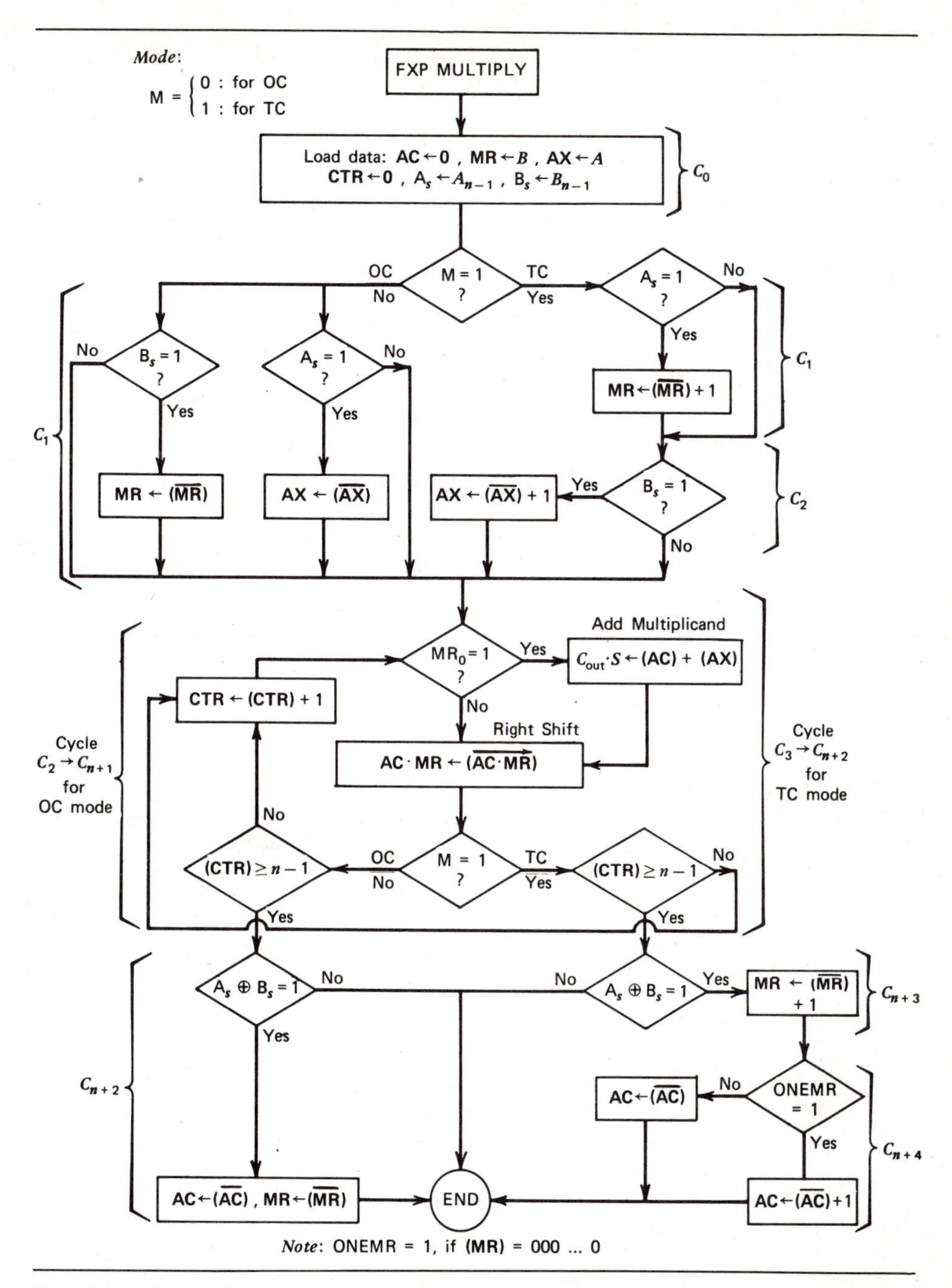

Figure 5.3 A basic indirect one's complement and two's complement add-shift multiplication algorithm.

The multiply loop consists of n cycles $C_1, \ldots, C_n$ as shown. To multiply two n-bit sign-magnitude numbers requires only $n - 1$ cycles, because there are only $n - 1$ magnitude bits. The sign bits a_{n-1} of A and b_{n-1} of B are loaded into the flip-flops A and B_s, respectively, during the initial cycle C_0. The leading bits of the registers AX and MR are initially loaded with zeros. At the end of the multiplication process, the expected sign of the product $A_s \oplus B_s$ should be loaded into the sign position AC_{n-1} in the accumulator. The resulting double-length product will be held in the cascaded register **AC** · **MR**. The contents of a register, say **(AC)** for **AC** as shown in Fig. 5.2 refer to the value stored in the register.

Indirect Sign-Complement Multiplications

Indirect algorithms for multiplying sign-complement numbers, either in one's complement or two's complement notation, require two complementing stages: the input operands if negative should be converted to sign-magnitude form before entering the multiply loop and the resulting product should be converted to the desired form if it is negative. As illustrated in Fig. 5.3, both one's complement and two's complement multiplications require the use of a *precomplementer* and a *postcomplementer* to carry out these operations.

The one's complement multiplication requires two cycles (C_1 and C_{n+2}) for the bitwise pre- and postcomplementing operations. Four cycles (C_1, C_2, C_{n+3}, and

Table 5.1 Summary of Operations of Four Standard Add-Shift Indirect Multiplication Schemes

	Multiplier Cycle			
Operation Performed	**US**	**SM**	**OC**	**TC**
Load the operands from memoy to registers and signs to flags	C_0	C_0	C_0	C_0
Bitwise complement negative operands for OC or TC mode. Load 0 into MR_{n-1} and into AX_{n-1} for SM mode	/	/	C_1	C_1
Increment the complemented negative operand by 1 for TC mode	/	/	/	C_2
Multiply loop consisting of add-shift operation per each loop cycle	C_1 ↓ C_n	C_1 ↓ C_n	C_2 ↓ C_{n+1}	C_3 ↓ C_{n+2}
Bitwise complement the product in **AC** · **MR** for OC or TC mode if the product sign is negative. Load sign of product into AC_{n-1} for SM mode	/	/	C_{n+2}	C_{n+3}
Increment the complemented double-length product in **AC** · **MR** by 1 for TC mode, if it is negative	/	/	/	C_{n+4}

Note: n is the word length of the operands.

C_{n+4}) are needed for the two's complement operations. The two's complement can be obtained by first producing the one's complement of the given number and then adding a 1 to the l.s.b. of the result. This addition can be achieved by feeding a 1 into the C_{in} terminal of an adder; loading the number being incremented into one input of the adder and a zero vector into the other adder input, and then performing the addition through the adder. One can also use registers with count-up feature to achieve this addition of 1.

When a combinational two's complementer (Fig. 2.11b) is available, only one cycle is needed to generate the two's complement of a given number. In other words, both the precomplementing cycles C_1 and C_2 and the postcomplementing cycles C_{n+3} and C_{n+4} can be combined into one cycle. The reduction of two cycles into one may not always pay off, because the cycle time may be prolonged to accommodate the long carry delay in the two's complementer circuit. A summary of the above four indirect multiplication methods is given in Table 5.1. The UnSigned (US), Sign-Magnitude (SM), One's Complement (OC), and Two's Complement (TC) multiplications require $n + 1, n + 1, n + 3$, and $n + 5$ machine cycles, respectively.

5.3 An Indirect Universal Multiplier Design

The sign-magnitude and indirect sign-complement multiplication methods described in Figs. 5.2 and 5.3 can be combined to yield a universal design. Two external mode control lines X and Y are required to determine the operating modes as shown in Fig. 5.4. The abbreviations US, SM, OC, and TC are used to represent the decoded mode control signals *U*n*S*igned, *S*ign-*M*agnitude, *O*ne's *C*omplement, and *T*wo's *C*omplement multiplications, respectively. This universal multiplication method is described by the attached function table. In case of two's complement multiply, $n + 5$ counter cycles are needed. The major operations taking place within each cycle for all four operating modes are summarized in the table. The processor may be idling in some of the cycles, except for the two's complement operation.

The hardware host described in Fig. 5.1 must be expanded to yield the universal design. The specification of all register functions and the control line assignment are shown in Fig. 5.5. The function mode control is implemented with a 2-to-4 line decoder as shown. Each of the three working registers **AC**, **MR**, and **AX** can assume four of the following functions: bitwise complement, increment by 1, parallel load external inputs, shift right one bit with an SRI serial input, or clear the register to zero. For example, when the EAC line ① turns high and the two selection lines ② and ③ are low, the right shifted version of the adder outputs will be loaded into the **AC**. All the possible register functions are shown in the schematic diagram. During the final postcomplement cycle (C_{n+4}) of a two's complement multiplication, a 1 should be added to the bitwise complemented product in **AC** · **MR**, if the resulting product is negative. This addition takes two steps as described below:

First, we complement both registers $\mathbf{AC} \cdot \mathbf{MR} \leftarrow (\overline{\mathbf{AC}}) \cdot (\overline{\mathbf{MR}})$ at cycle C_{n+3}.

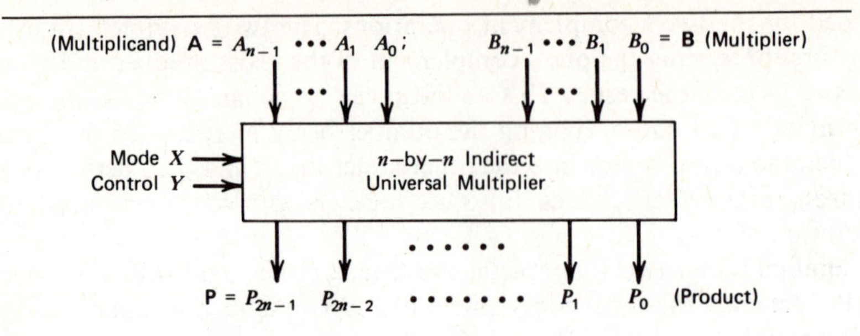

	Mode XY			
Cycles	00 US	01 SM	10 OC	11 TC
C_0	Load data	Load data and signs	Load data and signs	Load data and signs
C_1		Complement negative signs	Bitwise complement negative operands	Bitwise complement negative operands
C_2				Increment the complemented operand by 1
C_3 ⋮ C_{n+2}	n-cycle Multiply Loop with uniform one right shift per cycle			
C_{n+3}		Establish sign of resulting product	Bitwise complement negative product	Bitwise complement the negative product in **AC** · **MR**
C_{n+4}				Increment MR by 1 and increment **AC** if Eq. 5.6 is true

Figure 5.4 The specification of an n-by-n indirect universal multiplier.

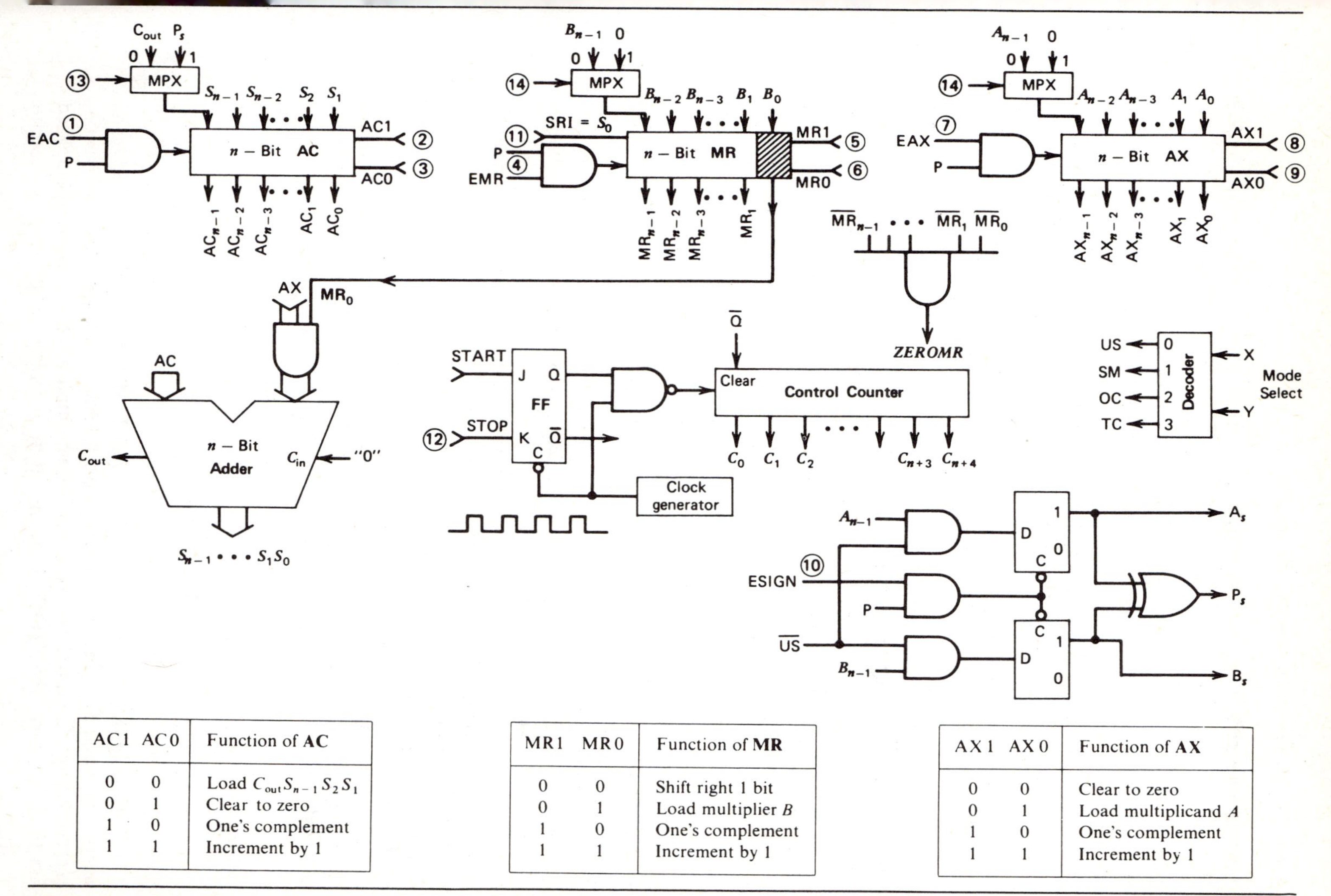

AC1	AC0	Function of **AC**
0	0	Load $C_{out} S_{n-1} S_2 S_1$
0	1	Clear to zero
1	0	One's complement
1	1	Increment by 1

MR1	MR0	Function of **MR**
0	0	Shift right 1 bit
0	1	Load multiplier *B*
1	0	One's complement
1	1	Increment by 1

AX1	AX0	Function of **AX**
0	0	Clear to zero
0	1	Load multiplicand *A*
1	0	One's complement
1	1	Increment by 1

Figure 5.5 The schematic logic circuit diagram of an *n*-by-*n* indirect universal multiplier.

Then the register **MR** is incremented by one at cycle C_{n+4}. The net operation on **MR** is

$$\mathbf{MR} \leftarrow (\overline{\mathbf{MR}}) + 1 \qquad (5.5)$$

If the resulting contents of **MR** is zero, a carry-out must flow into the **AC**. This condition is represented by the Boolean variable *ZEROMR* which is generated by an n-input NOR-logic. When the following condition is met

$$C_{n+4} \wedge \mathrm{TC} \wedge ZEROMR \wedge P_s = 1 \qquad (5.6)$$

the following operation is demanded at cycle C_{n+4} so that the negative product can be correctly represented in two's complement form in the cascaded register **AC** · **MR**.

$$\mathbf{AC} \leftarrow (\mathbf{AC}) + 1 \qquad (5.7)$$

The sign of the resulting product is

$$P_s = \begin{cases} A_s \oplus B_s, & \text{if SM} \vee \text{OC} \vee \text{TC} = 1; \\ 0, & \text{if US} = 1 \end{cases} \qquad (5.8)$$

The control counter is essentially a modulo-$(n + 5)$ autonomous clock ($\lceil \log_2(n + 5) \rceil$ bits) with decoded outputs $C_0, C_1, \ldots, C_{n+4}$. The rate of the counter is identical to that of the master clock of the processor. A MULTIPLY flip-flop is used to control the start or ending of a multiplication process.

Described below are the procedures to determine the values assigned to the control lines (circled labels) for each processor cycle. The intrinsic operations, implied by various combinations of control-line values, are called *microoperations.* A matrix with the control lines as the column headings and the counter cycles as the row headings is shown in Table 5.2. Such a matrix is called *microoperation table* or *control matrix.* The row entries of the table correspond to the desired control line values per each of the operating cycles. These values can be determined by checking against the flow chart. Most entries are simply 0's or 1's. Some entries are entered with logic conditions. For example, the condition shown in Eq. 5.6 determines if at cycle C_{n+4} the **AC** Enable line ① should be a "1". This testing condition is then entered at the intersection of cycle C_{n+4} and control line ① on the table.

One can immediately derive the logic equation for each of the control lines by examining the microoperation table. For example, the AX Enable line ⑦ should be specified as

$$⑦ = C_0 \vee C_1 \wedge A_s \vee C_2 \wedge A_s \wedge \mathrm{TC} \qquad (5.9)$$

The **AC** Enable line EAC is specified by

$$① = \mathrm{TC} \wedge ZEROMR \wedge P_s \wedge C_{n+4} \vee C_0 \vee C_3 \vee \cdots$$
$$\vee\, C_{n+2} \vee C_{n+3} \wedge P_s \wedge X \qquad (5.10)$$

The START signal to the **J** terminal of the MULTIPLY flip-flop is generated by the instruction decoder in the central processor. Whenever a MULTIPLY instruction is executed, this flip-flop should be set until the multiplication operation is completed. On the other hand, the STOP signal should be generated at the very last counter

Table 5.2 The Microoperation Table for the Universal Multiplier in Fig. 5.5

Machine Cycles	Control Lines													
	(1) EAC	(2) AC1	(3) AC0	(4) EMR	(5) MR1	(6) MR0	(7) EAX	(8) AX1	(9) AX0	(10) ESG	(11) SR1	(12) STP	(13) SPD	(14) SMZ
C_0	1	0	1	1	0	1	1	0	1	1	d	0	0	0
C_1	0	d	d	B_s	1	0	A_s	1	0	0	d	0	0	SM
C_2	0	d	d	$B_s \wedge$ TC	1	1	$A_s \wedge$ TC	1	1	0	d	0	0	0
C_3	1	0	0	1	0	0	0	d	d	0	S_0	0	0	0
$\vdots$	1	0	0	1	0	0	0	d	d	0	S_0	0	0	0
C_{n+2}	1	0	0	1	0	0	0	d	d	0	S_0	0	0	0
C_{n+3}	$P_s \wedge X$	1	0	$P_s \wedge X$	1	0	0	d	d	0	S_0	0	SM	0
C_{n+4}	$ZEROMR \wedge P_s \wedge$ TC	1	1	$P_s \wedge$ TC	1	1	0	d	d	0	d	1	0	0

None: "d" means "don't care" condition. X = OC $\vee$ TC

cycle. Whenever the MULTIPLY flip-flop is cleared, no event pulses can be generated to enable the multiply unit.

The excitation logic described by the microoperation table can be implemented in two different ways. The first method is to use hardwired random combinational logic circuitry to implement control line equations such as Eq. 5.9 or Eq. 5.10 derived above. An arithmetic processor implemented this way is called *hardwire-controlled.* The second method employs the control memory (ROMs) to store the encoded version of the microoperation table. The format of the microinstruction is the subject matter of microprogramming. An arithmetic processor governed by firmware control is called *microprogrammed.* The clock rate should be determined by the longest propagation delay along all the possible signal paths in the circuit. This may include the delays contributed by the excitation logic, the multiplexer, the register, the adder, and so on. With the current 7400 series TTL gates and flip-flops, and a carry lookahead adder, the worst-case delay in this multiplication processor can be less than 300 nsec corresponding to a clock rate of more than three megacycles per second.

5.4 Multiple Shifts in Multiplication

The execution of a multiply instruction in a digital computer can be speeded up, if more than one multiplier bit is examined per cycle. Such multiple-bit scanning requires multiple shifts after each addition. For example, the total number of add-shift cycles can be reduced by *half,* if two multiplier bits are examined at a time. The successive scanning blocks can be *disjoint* or *overlapped.* In this section, we study the multiplication method with nonoverlapped multiple-bit scanning. Both approaches are readily implementable with **CSA** trees. The gain in speed is obvious at the expense of extra hardware. With the **CSA** loops, however, the increase in hardware is rather limited. For illustrative purpose, we present the method with unsigned integer multiplication. Minor modifications can be made to accommodate other arithmetic notations.

The initial register assignments are same as before, that is multiplicand $A = (\mathbf{AX})$; multiplier $B = (\mathbf{MR})$; and $\mathbf{0} = (\mathbf{AC})$. Let us describe the principle with a specific example of examining two multiplier bits at a time. Without loss of generality, we assume even word length $n = 2k$ (otherwise, an artificial zero can be always inserted into the high-order end to make it even in length). Operation starts at the lower-order end of the multiplier, which means shifting right two positions at a time.

The four possible actions, after scanning the least significant pair of multiplier bits (MR_1, MR_0) are described in Table 5.3. A *multiple* of the multiplicand A, say $m \times A$ for $m = (\mathrm{MR}_1, \mathrm{MR}_0)_2$, is added to the current partial product to produce the next partial product. Note that the resulting sum may have more than n bits with multiple shifts.

$$C_{\text{out}} \cdot \mathrm{S} \leftarrow (\mathbf{AC}) + m \times A \qquad \textbf{(5.11)}$$

Each multiple of the multiplicand is obtained by a linear sum of several shifted versions of A. Shifting a binary number to the left i position with zero entering from the

Table 5.3 Multiples of the Multiplicand to Be Added to the Partial Product After Scanning a Pair of Multiplier Bits

MR_1	MR_0	Multiples to Be Added
0	0	0
0	1	A
1	0	$2A$
1	1	$A, 2A$

right end is equivalent to multiplying the number by a factor of 2^i. Therefore, we have

$$m \times A = \sum_{i=0}^{d} b_i \times 2^i \times A \tag{5.12}$$

if $m = (b_d \cdots b_1 b_0)_2$. The entry of each row in Table 5.3 is obtained by using Eq. 5.11 with $m = (b_1 b_0)_2 = (\mathrm{MR}_1, \mathrm{MR}_0)_2$.

Following the addition, the cascaded register **AC · MR** is shifted right two positions as

$$\mathbf{AC} \cdot \mathbf{MR} \leftarrow \overbrace{C_{out} S_n S_{n-1} \cdots S_2}^{n \text{ bits}} \cdot \overbrace{S_1 S_0 \mathrm{MR}_{n-1} \cdots \mathrm{MR}_3 \mathrm{MR}_2}^{n \text{ bits}} \tag{5.13}$$

Note that the next higher-order pair of multiplier bits has been shifted into the least significant positions and is ready to be examined at the next cycle. This 2-bit scanning process repeats itself for $k = \lceil n/2 \rceil$ cycles, until all the multiplier bits are exhausted. Again, the final product will be available in cascaded register **AC · MR**. The above scheme can be extended to any scanning block sizes. In general, the largest multiplicand to be added in a d-bit scanning system can be written as

$$(2^d - 1) \times A = 2^{d-1} \times A + \cdots + 4A + 2A + A \tag{5.14}$$

This nonoverlapped multiple-bit-scanning multiplication can be realized with *carry-save adders* (**CSA**) and *carry-propagate adder* (**CPA**) described in Section 4.2. The design is demonstrated by scanning two multiplier bits per cycle ($d = 2$). The addition of the multiplicand multiples is implemented with one **CSA** with 3 inputs and 2 outputs and one **CPA** with 2 inputs and one output interconnected as in Fig. 5.6. Two inputs of the **CSA** are reserved for the multiples $2A$ and A. These multiples are enabled by ANDing with the multiplier bits in MR_1 and MR_0, respectively. The remaining input to the **CSA** is from the current partial product residing in **AC**. The two outputs of the **CSA** are then summed up by the $(n + 2)$-bit **CPA** to form the new partial product, and eventually the final product.

After each iteration, the new partial product is shifted right two positions according to the transfer pattern shown in Eq. 5.13. The multiply loop consists of

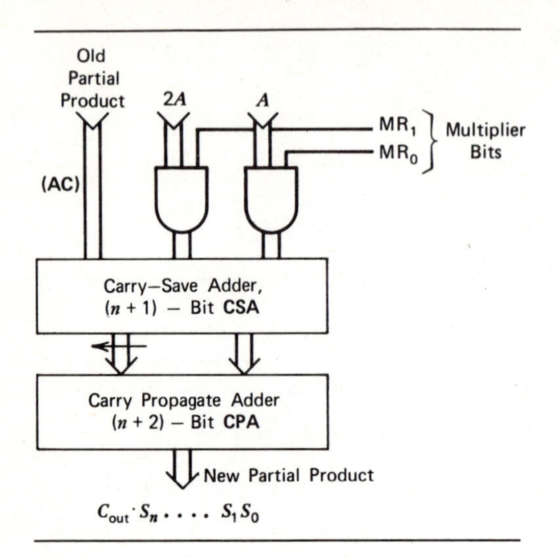

Figure 5.6 Carry-save adder/carry-propagate adder complex for adding the multiplicand multiples in a 2-bit multiplier scanning system.

$k = \lceil n/2 \rceil$ iterations. With one initial cycle for loading data, the total multiply time will be t cycles, where

$$t = \left\lceil \frac{n}{m} \right\rceil + 1 \tag{5.15}$$

for the general case of scanning m bits per cycle. For the hardware multiply unit in Fig. 5.6, $t = \lceil n/2 \rceil + 1 = \lceil 2k/2 \rceil + 1 = k + 1$ cycles are needed to complete the multiplication of two $2k$-bit numbers.

In order to increase the scanning block size, a multilevel **CSA** tree may be required. Each additional level of **CSA** contributes 2Δ time delay. The **CPA** with full CLA has a constant time delay, say 12Δ for 2-level CLA. The delay of the working registers, the multiplexers, and the multiple selection gates are lumped together as 8Δ. Therefore, the basic multiply cycle (clock period of one iteration) of an m-bit scanning multiplier using a v-level **CSA** tree can be estimated as

$$p = 12\Delta + 8\Delta + 2v\Delta = (20 + 2v)\Delta \tag{5.16}$$

The total multiply time is obtained by multiplying p by t

$$\text{Multiply Time} = p \times t = (20 + 2v) \times \left(\left\lceil \frac{n}{m} \right\rceil + 1 \right) \Delta \tag{5.17}$$

The number of **CSA**'s required for an m-bit scanning multiplier is equal to

$$\#(\textbf{CSA}) = m - 1 \tag{5.18}$$

In the next section, we shall show how to reduce this number with overlapped scanning. It should be noted that the "m-bit scanning" can be visualized as a type of high-radix, $r = 2^m$, multiplication, to be discussed in later sections.

5.5 Overlapped Multiple-Bit Scanning

In the nonoverlapped bit scanning method each multiplier bit generates one multiple of the multiplicand to be added to the partial product. When the scanning block size m becomes large, this means a large number of multiplicand multiples to be added. The execution time of a MULTIPLY instruction is determined mainly by the number of additions performed. Therefore it is desirable to reduce the multiple numbers. An overlapped multiple-bit scanning method is presented below which will reduce the number of multiplicand multiples by *half* per each add cycle. This means that only half the **CSA**'s used in nonoverlapped design are actually needed in an overlapped design.

The idea is based on the fact that the execution time can be reduced by shifting across a string of zeros in the multiplier. The greater the number of *zeros* in the multiplier, the faster the operation will be. Consider a string of k consecutive 1's in the multiplier as shown below:

$$\begin{array}{lllll} \text{Column Position} & \ldots, i+k, & i+k-1, & i+k-2, \ldots, & i, i-1, \ldots \\ \text{Bit Content} & \ldots, \quad 0, & \underbrace{1, \quad\quad 1, \ldots,} & & 1, 0, \ldots \\ & & k \text{ consecutive 1's} & & \end{array} \tag{5.19}$$

By the following **string property**

$$2^{i+k} - 2^i = 2^{i+k-1} + 2^{i+k-2} + \cdots + 2^{i+1} + 2^i \tag{5.20}$$

we can replace the k consecutive 1's by the following string

$$\begin{array}{llllll} \text{Column Position} & \ldots, i+k+1, & i+k, & i+k-1, \ldots, i+1, & i, & i-1, \ldots \\ \text{Bit Content} & \ldots, \quad 0, & 1, & \underbrace{0, \ldots, 0}_{k-1 \text{ consecutive 0's}}, & \bar{1}, & 0, \ldots \\ & & \uparrow & & \uparrow & \\ & & \text{One Addition} & & \text{One Subtraction} & \end{array} \tag{5.21}$$

The low-order $\bar{1}$ is overbarred indicating a -1 corresponding to a subtraction to be performed. With this multiplier recoding we can replace k consecutive additions by only one addition at the beginning and one subtraction at the ending of the string. This obviously saves a significant amount of add time, especially when the value of k is large. We are using this string property to explain why overlapped bit scanning is advantageous.

The multiplier bits are still divided into two-bit groups (pairs). But three bits (triplets) are scanned at a time, two bits from the present pair and the third bit from

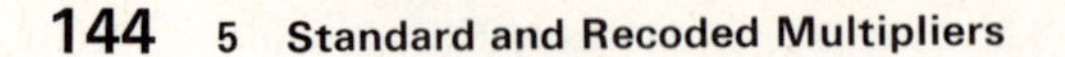

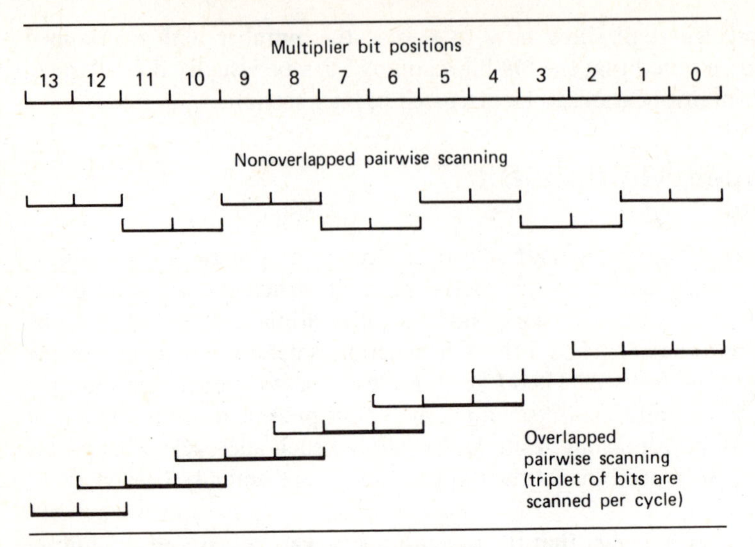

Figure 5.7 Nonoverlapped versus overlapped 2-bit scanning patterns.

the low-order bit of the next high-order pair. In effect, the low-order bit of each pair is examined twice. This scanning can start from either end of the multiplier. In order to be consistent with the previous illustrations, we assume the convention of scanning the multiplier from the right end (least significant pair) to the left. The scanning patterns for both overlapped and nonoverlapped examinations are shown in Fig. 5.7 for a multiplier of 14 bits long.

The actions after each triplet is examined are described in Table 5.4. The table shows that either a pure *shift*, or an *addition* or a *subtraction* is performed per each

Table 5.4 Multiples of the Multiplicand to Be Added After Scanning a Triplet of Multiplier Bits in an Overlapped Pairwise Scanning System

Multiplier Bits				
The Low-Bit of the Next Higher Pair X_{i+2}	The Present Pair X_{i+1}	X_i	Multiplicand Multiples to Be Added	Reasoning By String Property in Eq. 5.20
0	0	0	0	No string of 1's
0	0	1	$+2A$	End of string
0	1	0	$+2A$	Isolated 1
0	1	1	$+4A$	End of string
1	0	0	$-4A$	Beginning of string
1	0	1	$-2A$	Beginning and ending of strings
1	1	0	$-2A$	Beginning of string
1	1	1	0	Center of string

machine cycle. Only the multiple $2A$ or $4A$ may be needed. When the low-order bit x_{i+2} of the next pair is a 0, the leftmost 1 among the three always indicates the *left end* (ending) of a string of ones. Addition should be performed with this nonzero multiplier bit as implied by the string property described in Eq. 5.20. On the other hand, when $x_{i+2} = 1$, the *right end* (beginning) or the *center* of a string of ones is implied, which requires a subtraction by the string property. The partial product is shifted right two positions per each add cycle. This makes the partial product smaller than it should be by four times the multiplicand ($-4A$). This may be corrected by adding the difference between four and the desired multiplicand multiple in a later scanning step. The entry point of the multiples $2A$ or $4A$ to the adder is important. Thus, if a pair ends in a *zero*, the resulting partial product is correct and the next operation is an addition. If a pair ends in a *one*, the resulting partial product is too large and the next operation will be a subtraction.

It should be noted that a complemented partial product may result in some of the added cycles. This means that a shifter must be designed to handle the shifting of complemented numbers also. On the very first cycle, there is a zero initial partial product. Therefore, the partial product input lines should be zero. If the least significant bit of the multiplier is a *one*, enter the complement of A into the adder by way of the partial product inputs, which were receiving zeros otherwise. At the same time, the multiple of the multiplicand selected by the decoding of the first pair of multiplier bits is entered at the other **CSA** inputs. The above scheme can be extended to steps of the three bits (four bits, if including the overlapped bit from the next triplet). Readers are encouraged to explore the rules associated with higher-order overlapped bit scanning. Following is the construction of a hardware multiply unit based on the overlapped two-bit scanning method described above.

5.6 Multiplier Design with Overlapped Scanning

The technique described above ensures the detection of the *beginning* and the *ending* of a string of *ones*. We describe below an iterative hardware processor that will decode six overlapped triplets of multiplier pairs (13 bits) at a time. Only six multiplicand multiples of magnitude 0, $\pm 2A$, or $\pm 4A$ are needed per iteration. In constrast, using the equivalent nonoverlapped design requires 12 multiples per iteration. Figure 5.8 shows the overlapped scanning patterns of the six generated multiples per each iteration for a multiplier design which has been actually built in IBM **360/91** computer. Five iterations are sufficient to retire all the 61 multiplier bits, 13 bits at a time in an overlapped fashion. Note that in IBM **360/91**, 56-bit fractions or 14-digit hexadecimal mantissas are considered for multiplication. Four dummy zeros were attached to the right end making up the 15 hexadecimal digits as shown in Fig. 5.8. Altogether 61 bits are contained in the operand including the sign bit, even though only 56 bits are effective magnitude bits in the fraction.

Two different trees of carry-save adders are shown in Fig. 5.9. Both **CSA** trees will merge the six generated multiples into two vectors the *partial sum* and the *partial carry*. Each **CSA** on the tree has a length equal to that of the multiplicand. The partial

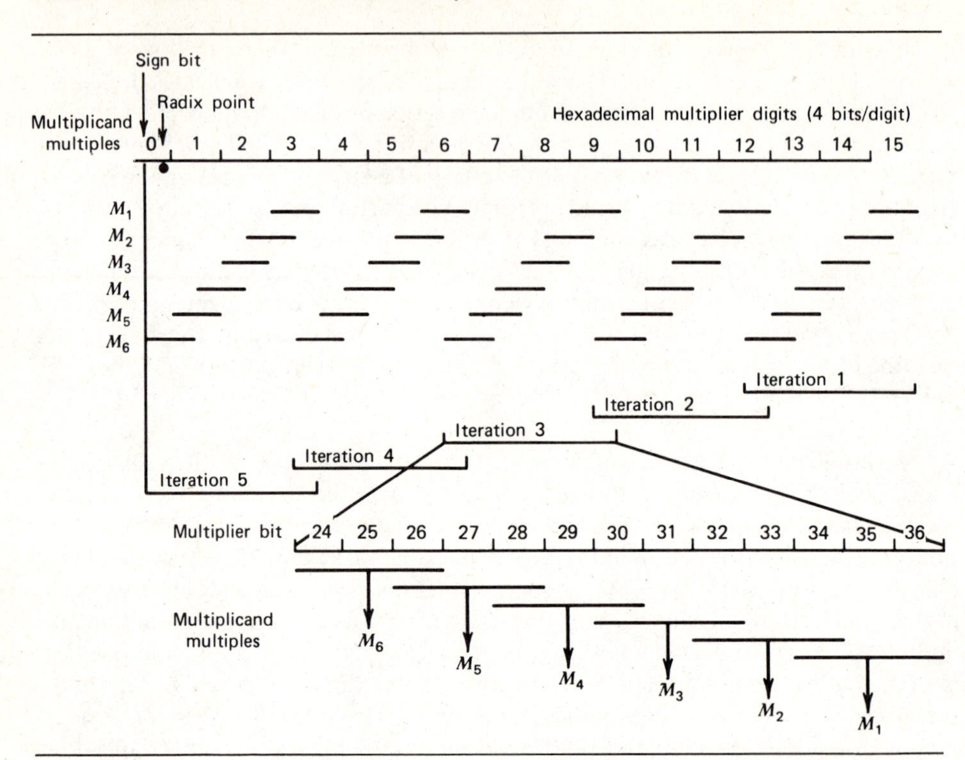

Figure 5.8 Iterations and multiple generation for the multiply unit in IBM System/**360** Model **91**. (Anderson, et al.[1]).

sum and partial carry are shifted right 12 positions and looped back to become the feedback inputs. After the multiplier has been assimilated, these two vectors from the **CSA** tree are added through a carry propagate adder to form the final product. This **CPA** should have a length that can accommodate the double-length final product, say 112 bits for the 56-by-56 multiplier.

The two **CSA** trees differ only in the way the feedback loop is formed. The **CSA** tree on the left of Fig. 5.9 has a feedback loop from the output, whereas the right-hand **CSA** tree has an accumulating loop at the output. The major difference is the iteration period associated with each tree. The left-hand tree has an iteration period equal to the time required to make a complete pass through the four levels of the tree. The arrangement in the right-hand design allows the iteration period to approach the delay through only the last two levels of **CSAs**. A schematic block diagram of the 12-bit-scanning multiply unit built in IBM **360/91** is shown in Fig. 5.10. The data flow is illustrated by directed arrows in the diagram. More details of this hardware multiply unit will be given as a case study in Chapter 9.

The multiplier recoding is implemented by examining the last 12 bits of the **MR**. The six multiplicand multiples are generated by gating the left-shifted multiplicand with six sets of AND gates. A bitwise one's complementer is needed to provide the true and complemented multiplies. The **CSA** tree may assume either of the designs

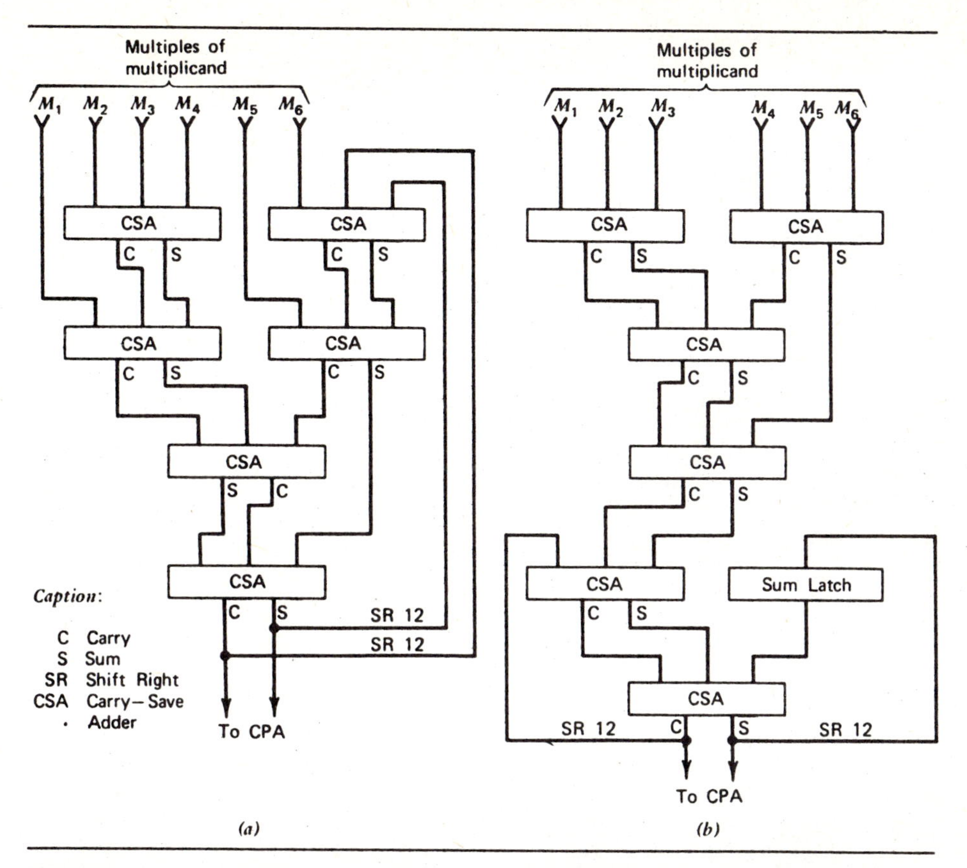

Figure 5.9 Two possible **CSA** trees for adding the six multiplicand multiples (Fig. 5.8) to the accumulated partial product (Anderson, et al.[1]). (*a*) Feedback loop from output (*b*) Accumulating loop at output.

in Fig. 5.9 depending on the required speed and hardware limit. The right shifter should be able to shift the sign-complement number to the right with the sign extended to fill the vacated leftmost 12 positions, because the partial product sometimes may be negative in complemented form.

The alignment of the multiples and feedback lines to the appropriate input lines of the **CSA** tree is very important. Initially, the two shifters should be cleared so that initial partial product feedback would be zero. When the least significant bit of the multiplier is a *one*, the complemented multiplicand should be loaded through these feedback lines in the very first cycle so that appropriate subtraction can be performed. If two's complement arithmetic is used, there should be an initial carry entering the least significant position of the **CPA** during the summation of the final product.

With an m-bit-scanning multiplier where m is an even number, one can complete the multiplication of two n-bit numbers using the above multiplier in k iterations, where $k = \lceil (n + 4)/m \rceil$, where 4 accounts the four dummy low-order zero appended. With one initial cycle for loading data into the appropriate registers and one final

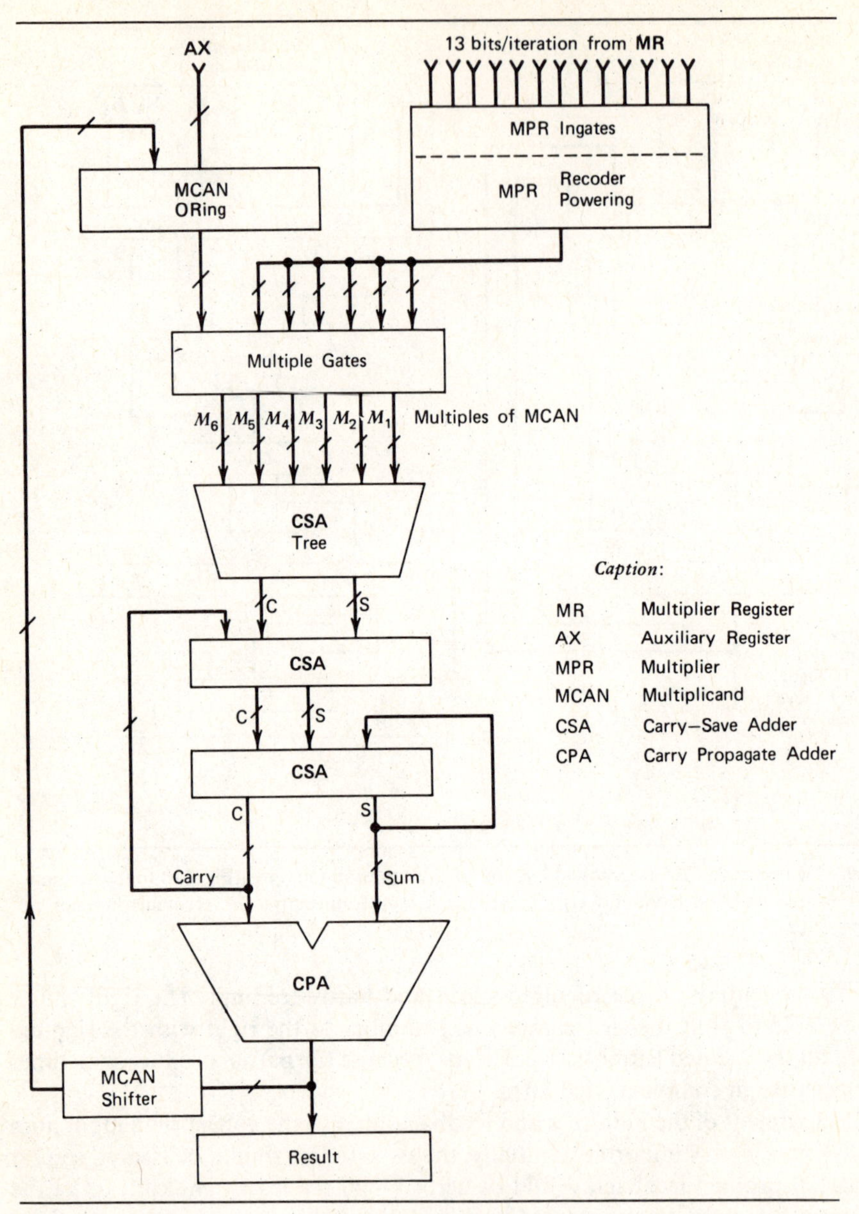

Figure 5.10 The 56-by-56 recoded multiplier with overlapped 12-bit scanning built in IBM System/**360** Model **91** computer (Anderson, et al.[1]).

cycle for summing the partial sum and partial carry vectors, the above process needs **C** cycles to complete, where

$$\mathbf{C} = \left\lceil \frac{n + 4}{m} \right\rceil + 2 \tag{5.22}$$

Because the **CPA** requires a constant time delay, which is usually lower than that of the **CSA** tree, the cycle period is determined primarily by the delays through the path leading to the output level of the **CSA** tree. These may include the delays in the multiplier recoder, the multiple complementer and gate control, the first three levels of **CSAs**, and the last two levels of feedback loop plus the shifter delay. With today's IC devices, these delays can be added as low as 30 nsec. Therefore, one can use the above hardware multiply unit to complete the multiplication of two 56-bit fractions in less than $\{\lceil(56 + 4)/(12 - 1)\rceil + 2\} \times 30 \simeq 225$ nsec, which is quite satisfactory to most of today's computers with memory cycle time in the neighborhood of one μsec.

5.7 Canonical Multiplier Recoding

One of the more recent advances in machine arithmetic is the use of redundancy in **SD** code to replace the conventional multiplier digits, so that the add-type operations can be reduced in a multiplication with the increase of average shift length across the zeros in the multiplier.

Canonical Signed-Digit Code

In an **SD** number with radix 2, the allowed digit set is $\{\bar{1}, 0, 1\}$. Among all the possible redundant representations of an n-digit number with a prespecified value α, we are particularly interested in the minimal **SD** vector, which has minimal weight as described in section 1.5. There may be more than one n-digit **SD** vectors that have equal minimal weight for a given value α. For example, given $n = 6$ and $\alpha = 3$, there are eight distinct radix-2 **SD** vectors that have value 3. The weights of these vectors are listed below. Among the eight, the top two vectors are both minimal with weight *two*.

Signed-Digit Vectors	Value	Weight
$(0\ 0\ 0\ 0\ 1\ 1)_2$	$2 + 1 = 3$	2
$(0\ 0\ 0\ 1\ 0\ \bar{1})_2$	$4 - 1 = 3$	2
$(0\ 0\ 1\ \bar{1}\ 0\ \bar{1})_2$	$8 - 4 - 1 = 3$	3
$(0\ 1\ \bar{1}\ \bar{1}\ 0\ \bar{1})_2$	$16 - 8 - 4 - 1 = 3$	4
$(1\ \bar{1}\ \bar{1}\ \bar{1}\ 0\ \bar{1})_2$	$32 - 16 - 8 - 4 - 1 = 3$	5
$(0\ 0\ 1\ \bar{1}\ \bar{1}\ 1)_2$	$8 - 4 - 2 + 1 = 3$	4
$(0\ 1\ \bar{1}\ \bar{1}\ \bar{1}\ 1)_2$	$16 - 8 - 4 - 2 + 1 = 3$	5
$(1\ \bar{1}\ \bar{1}\ \bar{1}\ \bar{1}\ 1)_2$	$32 - 16 - 8 - 4 - 2 + 1 = 3$	6

A minimal **SD** vector $\mathbf{D} = D_{n-1} \cdots D_1 D_0$ that contains no adjacent nonzero digits is called a *canonical signed-digit vector*. This means that all nonzero digits in a canonical **SD** vector are separated by zeros as formally defined by

$$D_i \times D_{i-1} = 0 \quad \text{for} \quad 1 \leq i \leq n-1 \tag{5.23}$$

Reitwiesner [17] has proved that there exists a "unique" canonical **SD** form **D** for any digital number with a fixed value α and a fixed vector length n, provided the product of the two leftmost digits in **D** does not equal one, that is

$$D_{n-1} \times D_{n-2} \neq 1 \tag{5.24}$$

for $\mathbf{D} = D_{n-1} D_{n-2} \cdots D_0$. This property can be always satisfied by imposing an additional digit $D_n = 0$ to the left end of the vector **D**. Without loss of generality, we shall consider $(n + 1)$-digit **SD** vectors with a leading digit *zero* in the following discussions.

A procedure to transform a conventional binary vector to a canonical **SD** vector is described below.

Canonical Recoding Algorithm

Given an $(n + 1)$-digit binary vector $\mathbf{B} = B_n B_{n-1} \cdots B_1 B_0$ with $B_n = 0$ and $B_i \in \{0, 1\}$ for $0 \leq i \leq n - 1$. We wish to obtain the $(n + 1)$-digit canonical **SD** vector $\mathbf{D} = D_n D_{n-1} \ldots D_1 D_0$ with $D_i = \{\bar{1}, 0, 1\}$ and $D_n = 0$ such that both vectors **D** and **B** represent the same value

$$\alpha = \sum_{i=0}^{n} B_i \times 2^i = \sum_{i=0}^{n} D_i \times 2^i \tag{5.25}$$

Step 1. Start with the low-order end of **B** by setting the index $i = 0$ and initial carry $C_0 = 0$.

Step 2. Examine two adjacent bits B_{i+1} and B_i of vector **B** conditioned by the carry-in C_i and generate the next carry C_{i+1} according to the same rule of conventional binary arithmetic, that is, $C_{i+1} = 1$ if and only if there are two or three 1's among the three inputs B_{i+1}, B_i, and C_i.

Step 3. Generate the ith digit D_i of vector **D** by the following arithmetic equation

$$D_i = B_i + C_i - 2C_{i+1} \tag{5.26}$$

Step 4. Increment the index i by one and check if $i = n$. Go to Step 2 if no, and halt otherwise.

The above procedure is summarized in the Table 5.5. Let us use a numerical example to straighten out any ambiguity. Given the following 9-bit binary vector

$$\mathbf{B} = (0\ 0\ 1\ 0\ 1\ 0\ 1\ 1\ 1)_2 \tag{5.27}$$

Table 5.5 The Canonical Multiplier Recoding Algorithm With Radix $r = 2$.

Conventional Multiplier Bits B_{i+1}	B_i	Assumed Carry-in C_i	Recoded Bit[a] D_i	Carry-out[b] C_{i+1}
0	0	0	0	0
0	1	0	1	0
1	0	0	0	0
1	1	0	$\bar{1}$	1
0	0	1	1	0
0	1	1	0	1
1	0	1	$\bar{1}$	1
1	1	1	0	1

[a] Use Eq. 5.26 to obtain D_i.
[b] $C_{i+1} = B_{i+1}B_i + B_iC_i + B_{i+1}C_i$.

The corresponding canonical **SD** vector is obtained below using the above procedure.

$$\mathbf{D} = (0\ 1\ 0\ \bar{1}\ 0\ \bar{1}\ 0\ 0\ \bar{1})_{SD} \tag{5.28}$$

D has a value $128 - 32 - 8 - 1 = 87$ equal to that of **B**. Note that all the nonzero digits (1's or $\bar{1}$'s) in **D** are separated by zeros. Furthermore, vector **D** has a weight of 4 and **B** has a weight 5. In fact, the conventional binary representation of a digital number can be considered as a special case of the **SD** numbers.

The canonical recoding can be used in multiplier design. Only the additions of the multiples, A or $-$A, are required per each cycle. All the zeros, which separate the nonzero digits, in the recoded multiplier causes more "shifts" to be performed. The method can be extended to generating two or more signed-digits at a time. For instance, two digits D_i, D_{i+1} of **D** and the next carry C_{i+2} can be generated in one step upon the examination of three adjacent digits B_{i+2}, B_{i+1}, B_i of **B** and the incoming carry C_i. Because **D** must be canonical, adjacent digits D_{i+1} and D_i can not be both nonzero. This renders the following five possible choices of the pair, D_{i+1}, D_i, excluding 11, $\bar{1}\bar{1}$, $\bar{1}1$, or $1\bar{1}$.

$$\{\bar{1}0,\ 0\bar{1},\ 00,\ 01,\ 10\} \tag{5.29}$$

This process of generating two signed digits at a time can be visualized as a radix-4 **SD** vector with the following digit set

$$\{\bar{2},\ \bar{1},\ 0,\ 1,\ 2\} \tag{5.30}$$

corresponding to the five pairs in Eq. 5.29, respectively.

The radix-4 canonical vector can be used to design fast recoded multipliers with two-digit shifting per cycle, provided the multiples 0, $\pm$A, $\pm$2A are generated in the arithmetic processor. The use of redundancy and higher radix methods have been applied to the design of arithmetic units in the series of **ILLIAC** computers.

5.8 String Recoding and Booth Multiplier

Another interesting multiplier recoding scheme is based on the string property described in section 5.5. We shall term this method *string recoding*. In fact, the famous Booth multiplier is based on string recoding. Table 5.6 specifies the rules associated

Table 5.6 The Radix-2 String Multiplier Recoding Algorithm

Inputs B_i	B_{i-1}	Outputs D_i	Remark on String Property
0	0	0	No string
0	1	1	End of string
1	0	$\bar{1}$	Beginning of string
1	1	0	Center of string

with string recoding. The output digit D_i for $0 \leq i \leq n$ is obtained by the following discrimination rule

$$D_i = \begin{cases} 0, & \text{if} \quad B_i = B_{i-1} \\ 1, & \text{if} \quad B_i < B_{i-1} \\ \bar{1}, & \text{if} \quad B_i > B_{i-1} \end{cases} \tag{5.31}$$

The remarks given in Table 5.6 correspond to the detection of the *beginning*, the *interior*, and the *ending* of a string of ones. The procedure requires the attachment of two dummy digits $B_{-1} = B_n = 0$ to both ends of an n-digit binary vector

$$\mathbf{B} = B_{n-1} \cdots B_1 B_0.$$

Because there are no carries to be generated in the procedure, each transformed digit D_i is only a function of two adjacent bits B_i and B_{i-1} of the input string **B**. Therefore, all the digits in **D** can be generated in parallel by examining all the adjacent pairs B_i, B_{i-1} in vector **B** simultaneously.

This string recoding is attractive in converting binary vectors containing long subsequences of *ones*. However, it is less efficient in converting binary vectors having many isolated *ones*. In other words, nonzero digits may appear adjacent to each other after this string recoding, which may result in an **SD** vector with even higher weight than before. This phenomenon can be seen by the following example:

$$\text{Binary Vector } \mathbf{B} = (0\ 0\ 1\ 0\ 1\ 0\ 1\ 0\ 1\ 0)_2 \tag{5.32}$$

$$\text{Transformed String Vector } \mathbf{D} = (0\ 1\ \bar{1}\ 1\ \bar{1}\ 1\ \bar{1}\ 1\ \bar{1}\ 0)_{\mathbf{SD}} \tag{5.33}$$

Note that after the string recoding adjacent nonzero digits must have opposite signs, that is $D_i \times D_{i+1} \neq 1$ for all i.

To extend the above method to higher radices, one can partition the recoded string into sections of pairs, or of triplets. Let us illustrate the *radix-4 string recoding* by partitioning the recoded **SD** vector into disjoint pairs. Each pair, (D_{i+1}, D_i), in **D** can be considered as a radix-4 digit F_j such that the following correspondence is established

$$\text{Radix-2 } (D_{i+1}, D_i)\text{: } \bar{1}0,\ 0\bar{1},\ 00,\ 01,\ 10 \quad \text{or } \bar{1}1 \ (\text{for } 0\bar{1}), \quad \text{or } 1\bar{1} \ (\text{for } 01) \tag{5.34}$$

$$\text{Radix-4 } (F_j)\text{: } \bar{2},\ \bar{1},\ 0,\ 1,\ 2 \tag{5.35}$$

With this transformation, one can design a radix-4 string recoded multiplier with uniform two shifts per step. It should be noted that the overlapped scanning method for multiplication described in Table 5.4 is indeed a radix-4 string recoding scheme, in which three input digits B_{i+1}, B_i, B_{i-1} are scanned and two output signed digits D_{i+1}, D_i are generated per each step.

A hardware multiply unit based on the string recoding theory is illustrated below. The unit can multiply two 2's complement numbers directly without concern about the signs of the two numbers. In other words, there is no need for precomplement or postcomplement. This method is known as Booth multiplication.

After a pair of multiplier bits B_i and B_{i-1} is examined, the multiply unit performs the following operations.

$$\begin{aligned}
&\text{Shifting right partial product, if } B_i = B_{i-1}.\\
&\text{Adding } A \text{ to partial product and then shifting right, if } B_i < B_{i-1}.\\
&\text{Subtracting } A \text{ from partial product, and then shifting right, if } B_i > B_{i-1}.
\end{aligned} \tag{5.36}$$

The shifting can be thought as a dummy addition of zero vector to the partial product. The subtraction is actually performed by two's complement addition. All pairs of multiplier bits can be simultaneously compared. The design is illustrated in Fig. 5.11. Three overlapped pairs of multiplier bits (4 bits) are scanned per each iteration. Successive iterations also overlapped in one bit. The scanning pattern of a 32-bit Booth multiplier using,this method is shown in Fig. 5.12. The scanning of three multiplier pairs at once means the simultaneous addition of three multiplicand multiples to the partial product. Therefore, a five-input **CSA** tree is used to assume the task of multiple operand addition.

Initially a zero partial product is loaded into the **AC**, the multiplicand A into **AX**, and the n-bit multiplier B with a appended zero into the $(n + 1)$-bit **MR**. Each add cycle, the four rightmost positions MR_2, MR_1, MR_0, MR_{-1} of the register **MR** are examined to determine the necessary operations. Gate logic is shown to select the appropriate inputs to the **CSA** tree. The multiply unit can complete the multiplication of two n-bit numbers in $\lceil (n + 1)/3 \rceil$ iterations. Uniform 3-bit right shift is assumed in each iteration. With the example of $n = 32$ only $\lceil (n + 1)/3 \rceil = 11$ machine cycles are needed. There exists a tradeoff between the *operating speed* and *hardware complexity* .The simplest design is to scan just one pair of multiplier bits at a time, which requires n cycles to complete the multiplication.

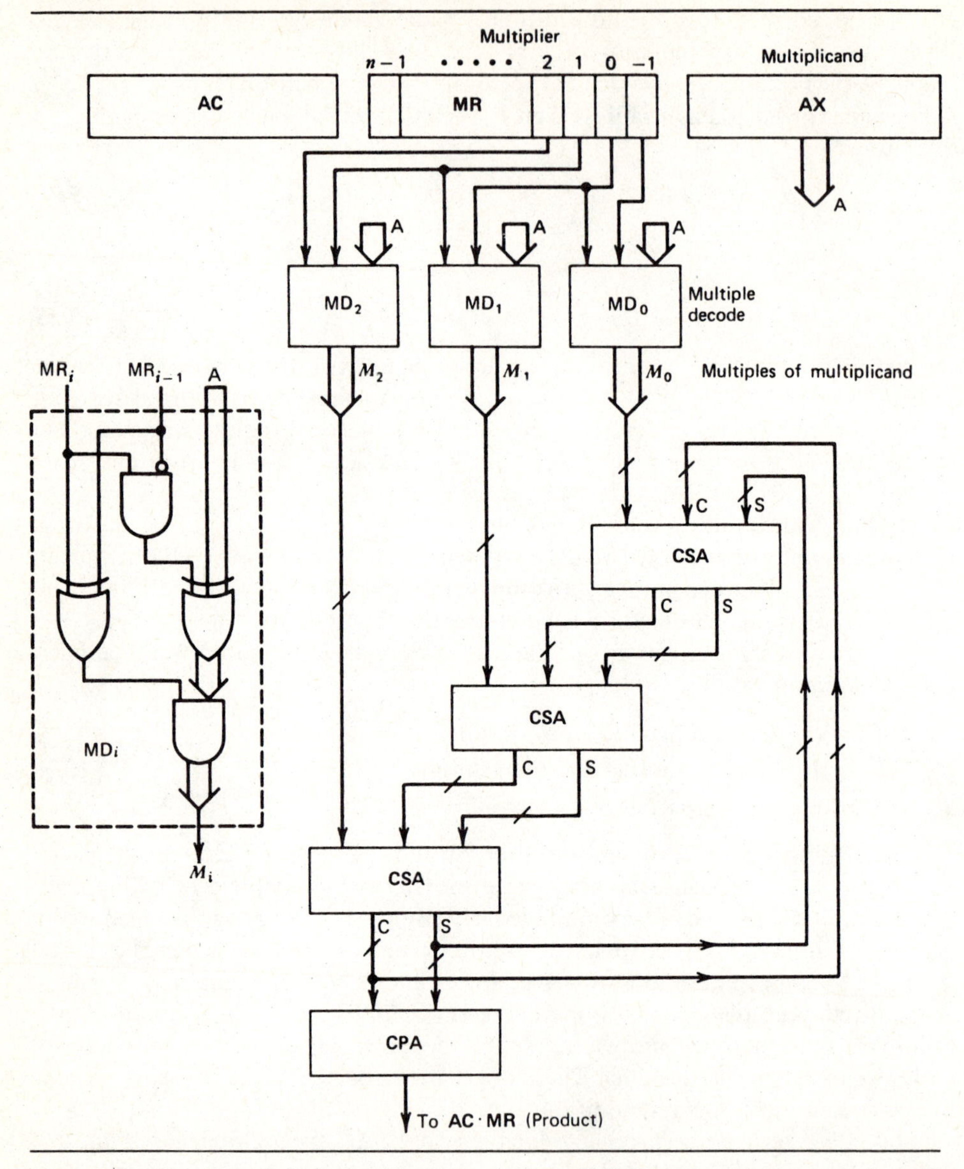

Figure 5.11 The schematic logic showing the necessary multiplier decode, multiple generation, and addition using **CSA/CPA** for a 4-bit scanning generalized Booth's multipler.

The above multiply unit can also start from the left end provided left shifting is provided. In fact, all the multiplier pairs can be simultaneously examined to produce the desired **SD** string code. The Booth algorithm can be also implemented with fast iterative logic arrays of combinational add-subtract-shift cells to be studied in Chapter 6.

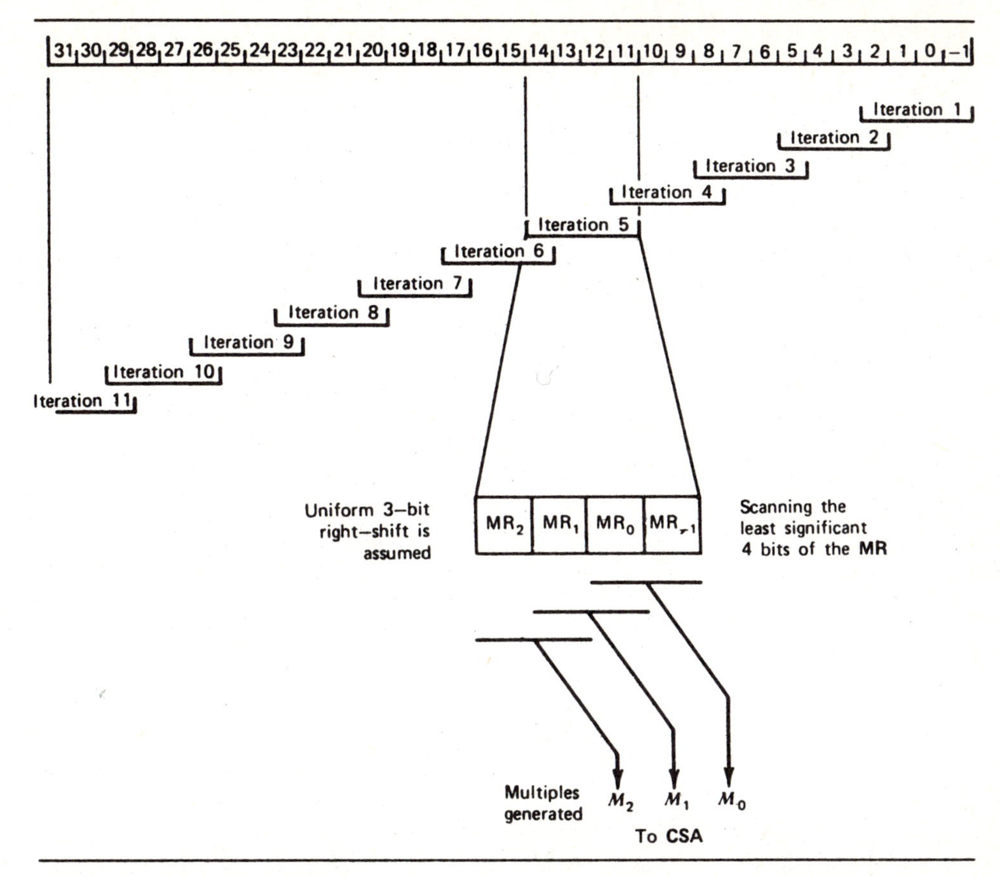

Figure 5.12 The 4-bit scanning pattern of a 32-bit generalized Booth's multiplier implemented with carry-save adders as shown in Fig. 5.11.

5.9 An Evaluation of Various Add-Shift Multipliers

A good measurement of the efficiency of arithmetic processors is always demanded by machine designers. The effectiveness of an arithmetic method should be determined by averaging over all possible input operands. Two commonly accepted measures of multiplier efficiency are

1. *The average number of add-type operations required in a multiplication.*
2. *The average shift length between successive additions in a multiplication.*

Before we compare the performance of various multiplier designs with respect to these two measures, let us summarize those multiplier design alternatives into five categories.

(1) Adder Bypassing

This refers to shifting across strings of zeros in the multiplier. Because nonuniform shifts may be required, this method is used mainly in asynchronous computers. The disadvantage is the increased control complexity.

(2) Reducing Addition Time

Either parallel carry generation using CLA technique or carry-save methods can be used to reduce the basic addition time. With SD arithmetic, totally parallel adders are possible.

(3) Multiple Shifts

Provisions must be made to shift more than one digit per iteration. Multiple-bit decoding results in increased hardware. The optimal choice is centered around the tradeoffs between speed and practical economic factors.

(4) Multiplier Recoding

The introduction of redundancy into multiplier through recoding may not necessarily increase the circuit complexity due to the simplicity in generating the multiplicand multiples $-A$ or $+A$. Recoding tends to result in longer average shift length.

(5) High-Radix Multiplication

The number of iterations required in a multiplication can be reduced by a factor of $k = \log_2 r$, where $r = 2^k$ is the number radix. The rare use of high-radix in commercial machines is mainly due to the much increased hardware complexity.

The *average number of additions* and the *average shift length* associated with various multiplication schemes are compared in Table 5.7. The simplest design requires n additions with a unity average shift length. Averaging all possible shifts in multipliers, the mean value of shift lengths is *two* instead of *one*. With multiplier recoding, one can achieve an average shift length of *three*, because more zeros are created and all the nonzero digits are sparse, With increased shift lengths, the number of add-type operations, on the average, can be reduced by one-half or even one-third as shown in the table. For example, the recoded multiplier using radix-8 with an average shift length of 3, may require less than $n/3$ additions to complete the multiplication of two n-bit numbers.

The multiply unit requires roughly 10 percent of the total semiconductor components in a modern digital computer, but costs less than 10 percent of the total cost of computer. It is desirable, therefore, to have other operations such as squaring, division, and so on, sharing the hardware associated with multiplication. In general, the equipment cost rises as the square of the word length, whereas the time increases as the logarithm of the word length. At present, fixed-point machine multiplication takes about four times as long as machine addition. However, this ratio is closer to one for floating-point operations. About one-third of all arithmetic operations in scientific computers are multiplications. The use of a high-speed multiplier unit would approximately double the speed of computation of most machines.

Table 5.7 Comparison of Standard and Recoded Multiplier Configurations (Garner [9])

Category	Multiplier Configuration	Average No. of Additions	Average Shift Length
Standard multipliers	Basic add-shift	n	1
	Skip zero add	$n/2$	1
	Skip zero add with single/double shift	$n/2$	$1\frac{1}{2}$
	Skip zero add with 1, 2, and 3 shifts	$n/2$	$1\frac{3}{4}$
	Skip zero add with all possible shifts	$n/2$	2
Recoded multipliers	Skip zero add with 1 and 2 shifts	$n/3$	$1\frac{3}{4}$
	Skip zero add with all possible shifts	$n/3$	3
	Double shift only (radix-4)	$n/3$	2
	Triple shift only (radix-8)	$<n/3$	3

5.10 Bibliographic Notes

The conventional computers used the standard add-shift techniques for multiplication. Mowle [15] has given a detailed description of these basic multiplication algorithms and their logic design problems. The design of a universal multiplier demonstrates a systematic approach to typical arithmetic design, starting from algorithmic flow charts to microoperation tables and then to actual realizations. Direct two's complement multiplication schemes were reported by Booth [4], Robertson [18], and Kamal et al. [10]. One's complement arithmetic was specially studied by Metze [14]. Multiple shifts in multiplication were considered by many authors, among which MacSorley's treatment [13] is most instructive. He demonstrated the design merits of using either uniform shifts (2 or 3 shifts per cycle) or variable shifts in realizing multipliers with **CSA** loops. Freeman [8] also gave an analysis of the mean shift length for a computer using uniform shifts of 2, 3, or 4 bits at a time.

The use of redundancy to improve multiplier efficiency was promoted by Robertson [20], and was later implemented in the **ILLIAC III** computer as reported by Atkins [2]. Ling [12] proposed a multiple-bit decoding scheme for fast multiplication. His method decomposes a multiplication into a handful of additions and subtractions using a simple set of mathematical identities. Chen [6] proposed a binary multiplication method based on squaring. The overlapped multiple-bit scanning methods were studied in references [1], [5], [7], [11], [13], [16], and [20]. Garner [9] has proposed a ring model for the study of sign-complement multiplication methods. Winograd [23] studied the theoretical limit of the time required for multiplication.

The origin of multiplier recoding methods is unclear. The canonical recoding was studied by Reitwiesner [17], who proved that no other method in an average sense is more efficient than the canonical **SD** recoding. This does not rule out the existence of equally efficient methods. Pennhollow [16] discovered a class of recodings having an average shift length of 3, of which the canonical is the simplest. Multiplier recoding was suggested by Wallace [22] in **CSA** realization of multipliers. Avizienis [3] and Robertson [20] have given comprehensive treatment of various multiplier recoding schemes. The string recoding was applied by Booth [4]. Robertson [19] has established the correspondence between methods of digital division and multiplier recoding procedures. Recently, Trivedi and Ercegovac [21] published some on-line algorithms for fast division and multiplication.

References

[1] Anderson, S. F. et al. "The IBM System/360 Model 91: Floating-Point Execution Unit," *IBM Journal of Research and Development* Jan. 1967, pp. 34–53.

[2] Atkins, D. E., "Design of the Arithmetic Units of Illiac III: Use of Redundancy and Higher Radix Methods," *IEEE Trans. Comput.*, Vol. C-19, No. 8, Aug. 1970, pp. 720–733.

[3] Avizienis, A., "Recoding of the Multiplier," *Class Notes*, Engr. 225A, UCLA, Los Angeles, CA., 1971.

[4] Booth, A. D., "A Signed Binary Multiplication Technique," *Quart. Journ. Mech. and Appl. Math.*, Vol. 4, Part 2, 1951, pp. 236–240.

[5] Burstev, V. S., "Accelerating Multiplication and Division Operations in High-Speed Digital Computers," *Tech. Report,* Institute of Exact Mechanics and Computing Technique, The Academy of Sciences of the USSR, Moscow, 1958.

[6] Chen, T. C., "A Binary Multiplication Scheme Based on Squaring," *IEEE Trans. Comput.*, Vol. C-20, No. 6, June 1971, pp. 678–680.

[7] Fenwick, P. M., "Binary Multiplication with Overlapped Addition Cycles," *IEEE Trans. Comput.*, Vol. C-18, No. 1, Jan. 1969, pp. 71–74.

[8] Freeman, H., "Calculation of Mean Shift for a Binary Multiplier Using 2, 3, or 4-bit at a Time," *IEEE Trans.*, Vol. EC-16, No. 6, Dec. 1967, pp. 864–866.

[9] Garner, H. L., "A Ring Model for the Study of Multiplication for Complement Codes," *IRE Trans.*, Vol. EC-8, No. 1, March 1959, pp. 25–30.

[10] Kamal, A. A., and Ghanam, M., "High-Speed Multiplication Systems," *IEEE Trans., Comput.*, Vol. C-21, No. 9, Sept. 1972, pp. 1017–1021.

[11] Lehman, M., "Short-Cut Multiplication and Division in Automatic Binary Digital Computers," *Proc. IEEE*, Vol. 10, Sept. 1958, pp. 496–504.

[12] Ling, H., "High-Speed Computer Multiplication Using a Multiple-Bit Decoding Algorithm," *IEEE Trans. Comput.*, Vol. C-19, No. 8, Aug. 1970, pp. 706–709.

[13] MacSorley, O. L., "High-Speed Arithmetic in Binary Computers," *Proc. of IRE*, Vol. 49, Jan. 1961, pp. 91–103.

[14] Metze, G., "A Study of Parallel One's Complement Arithmetic Units with Separate Carry or Borrow Storage," *Ph.D. Thesis*, Univ. of Illinois, Urbana, Ill., 1958.

[15] Mowle, F. J., "Simplified Logic Deisgn Using Digital Circuit Elements," Vol. 3, *Class Notes*, Dept. of Elec. Eng., Purdue University, Sept. 1974, Chap. 14.

[16] Pennhollow, J. O., "Study of Arithmetic Recoding with Applications in Multiplication and Division," *Ph.D. Thesis*, Univ. of Illinois, Urbana, Ill., Sept. 1962.

[17] Reitwiesner, G. W., "Binary Arithmetic," in *Advances in Computers*, Vol. 1, Academic Press, N.Y. 1960, pp. 261–265.

[18] Robertson, J. E., "Two's Complement Multiplication in Binary Parallel Digital Computers," *IRE Trans.*, Vol. EC-4, No. 3, Sept. 1955, pp. 118–119.

[19] Robertson, J. E., "The Correspondence Between Methods of Digital Division and Multiplier Recoding Procedures," *IEEE Trans. Comput.*, Vol. C-19, No. 8, Aug. 1970, pp. 692–701.

[20] Robertson, J. E., "Increasing the Efficiency of Digital Computer Arithmetic Through Use of Redundance," *Class Notes*, Univ. of Illinois, Urbana, Illinois.

[21] Trivedi, K. S. and Ercegovac, M. D., "On-Line Algorithms for Division and Multiplication," *IEEE Trans. on Computers*, Vol. C-26, No. 7, July 1977, pp. 681–687.

[22] Wallace, C. S., "A Suggestion for a Fast Multiplier," *IEEE Trans. on Electronic Computers*, Vol. EC-14, No. 1, Feb. 1964, pp. 14–17.

[23] Winograd, S., "On the Time Required to Perform Multiplication," *Journal of ACM*, Vol. 14, No. 4, Oct. 1967, pp. 793–802.

Problems

Prob. 5.1 Consider the universal n-by-n multiplier in Fig. 5.5 operating in four different modes with word length $n = 16$. Figure out the successive register contents (**AC**), (**MR**), (**AX**), and the successive adder outputs $S_{n-1} \cdots S_1 S_0$ for all machine cycles with respect to each of the four operation modes US, OC, TC, and SM, separately.

The two operands fetched from main memory have the following binary form:

$$\text{Multiplicand} = (0101101101111010)_2$$
$$\text{Multiplier} = (1010111100000101)_2$$

Note that these binary patterns may represent different operand values in different fixed-point notations.

Prob. 5.2 Based on the microoperations specified in Table 5.2, figure out all 14 control line equations of the universal multiplier and implement them with minimal 3-level NAND circuits. Estimate the number of gate levels of the longest delay path in the circuit of Fig. 5.5, assuming 8Δ delays for registers, 12Δ delays for the adder, and the delays of gate functions follow Table 2.1. Let the average value of unit gate delay Δ be 4 nsec. Determine the maximum clock rate (machine cycle time) that can be applied to drive the universal multipler.

Prob. 5.3 Show the schematic circuit diagram of a fast 40-by-40 multiplier with uniform shifts of 5 bits per each machine cycle, assuming a nonoverlapped 5-bit scanning system. Multilevel carry-save adders may be used. Estimate the machine cycle time of this multiple scanning multiplier, assuming that each level of carry-save adder contributes 2Δ delay with $\Delta = 4$ nsec.

Prob. 5.4 Draw a multiplicand multiple table similar to Table 5.4 for an overlapped 4-bit-scanning multiplication system, in which triplet of present multiplier bits $X_{i+2}X_{i+1}X_i$, plus the low-order bit X_{i+3} of the next high-order triplet are scanned per each machine cycle. Comment on the 16 possible 4-bit multiplier patterns scanned with respect to string property.

Prob. 5.5 Prove that there exists a unique canonical signed-digit form for a number with a given value α and given word length n, provided Eq. 5.24 is satisfied (Retweisner [17]).

Prob. 5.6 Given a 16-bit binary number $\mathbf{B} = (0011010111110101)_2$, find the corresponding canonical signed-digit vector $\mathbf{D}$, which represents the same value as $\mathbf{B}$. Show step-by-step how each recoded signed digit in $\mathbf{D}$ is generated.

Prob. 5.7 Extend Table 5.5 to generate two canonical signed digits (radix 4) at a time. The column headings of this extended canonical recoding table should be B_{i+2}, B_{i+1}, B_i for the conventional binary digits; C_i for the assumed carry-in; D_{i+1}, D_i for the recoded signed digits, and C_{i+2} for carry out. Note that, with four input bits, the extended radix-4 canonical recoding table has 16 row combinations.

Prob. 5.8 Figure the entries of a radix-8 string recoding table (similar to Table 5.6) using the string recoding technique described in section 5.8. Compare the resulting table with one obtained in Prob. 5.4. The two tables should be equivalent.

Prob. 5.9 The Booth's string recoded multiplier design shown in Fig. 5.11 is based on a 4-bit scanning system. Modify the design to scan seven overlapped pairs of multiplier bits (8 bits) per iteration. Show the 8-bit scanning pattern of the generalized string recoded multiplier (similar to Fig. 5.12). Provide a schematic logic diagram of the modified design using multilevel carry-save adders.

Chapter

6

Iterative Cellular Array Multipliers

6.1 Array Multiplication Basics

Conventional designs of hardware multipliers using the serial-shift and parallel-add techniques described in the preceding chapter are less expensive to implement, especially with microprogrammed controls. However, the add-shift approach is still considered too slow to satisfy today's scientific and engineering demands for faster multiplication. Ever since the advent of LSI circuits, high-speed cellular array multipliers have been a logical and feasible improvement over the serial-parallel designs. In the past decade or so, various iterative array multiplication schemes have been proposed. As expected, their gains in speed are obtained at the expense of extra hardware investment.

Both unsigned and signed array multipliers and their modular network realizations are introduced in this chapter. Several methods for direct multiplication of two's complement numbers are presented. A class of universal multiplication arrays and networks is shown. Other array multiplication schemes using controlled add/subtract cells, ROM-adder networks, and logarithmic transformations are also included in this chapter.

Consider two unsigned binary integers $\mathbf{A} = a_{m-1} \cdots a_1 a_0$ and $\mathbf{B} = b_{n-1} \cdots b_1 b_0$ with values A_v and B_v, respectively

$$A_v = \sum_{i=0}^{m-1} a_i 2^i$$

$$B_v = \sum_{i=0}^{n-1} b_i 2^i \qquad \textbf{(6.1)}$$

In a binary multiplication, the $(m + n)$-bit *product* $\mathbf{P} = P_{m+n-1} \cdots P_1 P_0$ is formed by multiplying the *multiplicand* **A** by the *multiplier* **B**. The *product* **P** has a value

$$P_v = A_v B_v = \left(\sum_{i=0}^{m-1} a_i 2^i\right)\left(\sum_{j=0}^{n-1} b_j 2^j\right)$$
$$= \sum_{i=0}^{m-1} \sum_{j=0}^{n-1} (a_i b_j) 2^{i+j}$$
$$= \sum_{k=0}^{m+n-1} P_k 2^k \qquad \textbf{(6.2)}$$

The operations required to carry out this multiplying process are illustrated in Fig. 6.1. Each of the partitial product terms $a_i b_j$ is called a *summand*. The $m \times n$ summands $\{a_i b_j | 0 \le i \le m - 1 \text{ and } 0 \le j \le n - 1\}$ are generated in parallel by $m \times n$ AND gates (Fig. 6.2). Hence the basic problem in designing a high-speed parallel multiplier is to reduce the time to add the 1's in each column of the summand matrix. Three different schemes are given below to realize the summand-summation process.

We choose the design of a 5-by-5 unsigned array multiplier ($m = n = 5$) to demonstrate the parallel array multiplication. The arrangement of all the summand terms follows the same matrix layout in Fig. 6.1. The schematic circuit diagram of a 5-by-5 Braun's array multiplier [5] is shown in Fig. 6.3. Such a multiplier of size n-by-n needs $n(n - 1)$ full adders and n^2 AND gates to implement it. The total multiply time of this multiply unit is estimated as follows:

Let Δ_a, Δ_x, and Δ_f be the propagation delay times for AND gate, XOR gate, and Full Adder (FA), respectively. From Table 2.1, we have $\Delta_x = 3\Delta$ and $\Delta_a = \Delta_f = 2\Delta$, if 2-level NAND logic or AND-OR-INVERTER wired logic is assumed in implementing these basic functions. Tracing the worst-case delay path along the right-most diagonal and the lowest row of the array, we obtain the total multiply time for n-by-n *UnSigned* (US) array multiplier.

$$\Delta_{US} = \Delta_a + [(n - 1) + (n - 1)] \times \Delta_f = 2\Delta + (2n - 2) \times 2\Delta = (4n - 2)\Delta \qquad \textbf{(6.3)}$$

				a_{m-1}	a_{m-2}		$\cdots$		a_1	a_0 = **A**
			×)				b_{n-1}	$\cdots$	b_1	b_0 = **B**
				$a_{m-1}b_0$	$a_{m-2}b_0$		$\cdots$		a_1b_0	a_0b_0
			$a_{m-1}b_1$	$a_{m-2}b_1$		$\cdots$		a_1b_1	a_0b_1	
		$\cdot\cdot\cdot$					$\cdot\cdot\cdot$			
+)	$a_{m-1}b_{n-1}$	$a_{m-2}b_{n-1}$		$\cdots$		a_1b_{n-1}	a_0b_{n-1}			
P_{m+n-1}	P_{m+n-2}	P_{m+n-3}		$\cdots$			P_{n-1}	$\cdots$	P_1	P_0 = **P**

P = **A** × **B**

Figure 6.1 Summand matrix describing the add-shift operations in an m-by-n unsigned integer array multiplication.

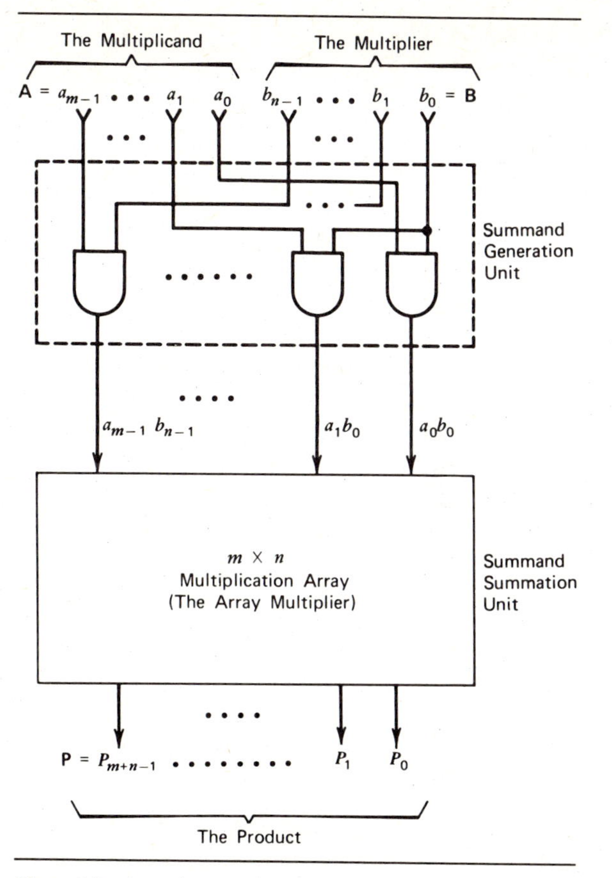

Figure 6.2 An m-by-n unsigned array multiplier.

The last row of the array, surrounded by dash lines, forms a ripple-carry adder with a time delay of $(n - 1)2\Delta$. One can replace this row by a carry lookahead adder having a constant time delay, say 8Δ if full carry lookahead is assumed. With this improvement, the total multiply time can be reduced to

$$\Delta_{US}(\text{CLA}) = \Delta_a + (n - 1)\Delta_f + 8\Delta = 2\Delta + (n - 1)2\Delta + 8\Delta = (2n + 8)\Delta \qquad \textbf{(6.4)}$$

The structures of signed array multipliers differ by the specific number representation used. We shall describe how to incorporate the complementers described in Section 2.4 into the unsigned array multiplier to perform sign-magnitude, one's complement, and two's complement multiplications. As illustrated in Fig. 6.4, the precomplementer is used to convert the two operands to positive integers before they can be multiplied by the core of the unsigned multiplication array. The postcomplementer will convert the result to the signed number representation if the two input operands do not agree in their signs.

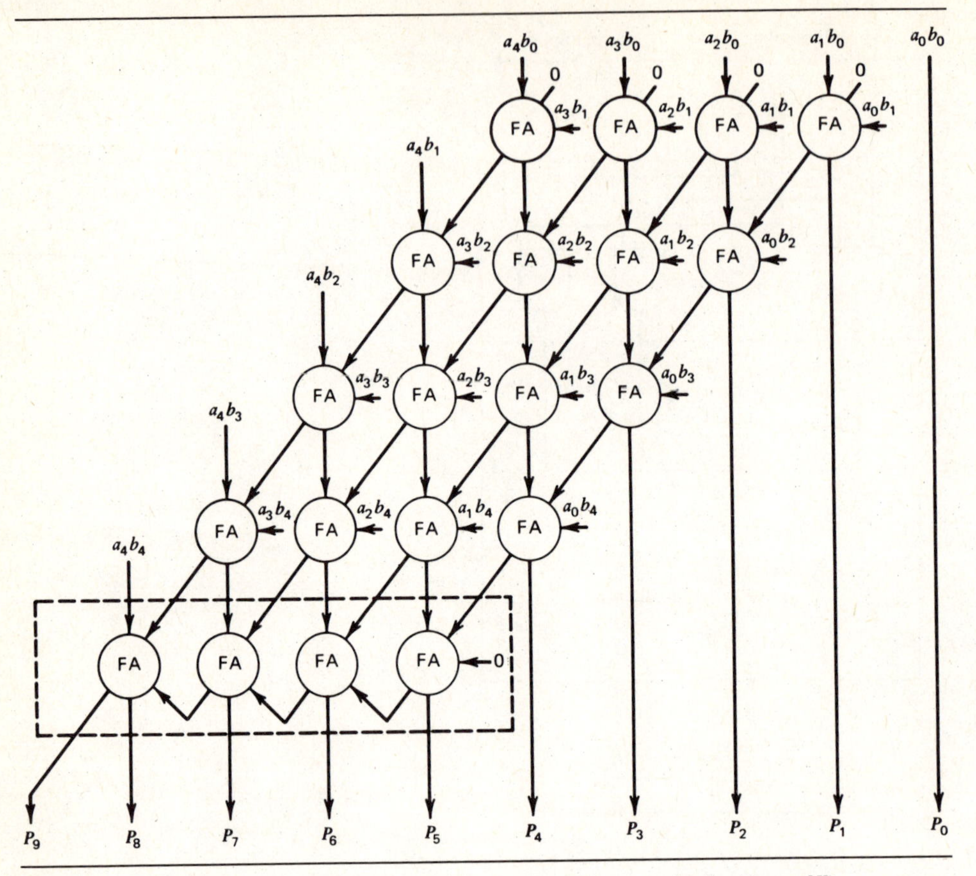

Figure 6.3 The schematic circuit diagram of a 5-by-5 unsigned array multiplier (Braun [5]).

Multipliers containing these complementing stages are called *sign complemented array multipliers.* Let $\mathbf{A} = a_n a_{n-1} \cdots a_1 a_0$ and $\mathbf{B} = b_n b_{n-1} \cdots b_i b_0$ be $(n + 1)$-bit signed integers in any FXP representation. After the necessary complementing operation, the true magnitudes of **A** and **B** are fed into an n-by-n unsigned array multiplier to produce the $2n$-bit true product $\mathbf{A} \times \mathbf{B} = \mathbf{P} = p_{2n-1} \cdots p_1 p_0$ with a sign $p_{2n} = a_n \oplus b_n$. In sign-magnitude multiplication, the pre- and postcomplementers are not required, because true magnitudes are immediately available. In one's complement multiplication, the complementers are simply made with XOR gates (Fig. 2.11*a*), each with a time delay of 3Δ. In two's complement multiplication, each of the three complementers assumes the circuit in Fig. 2.11*b*, with a time delay

$$\Delta(n\text{-bit two's complementer}) = (n - 1) \times 2\Delta + 5\Delta = (2n + 3)\Delta \qquad \textbf{(6.5)}$$

Let Δ_{SM}, Δ_{OC}, and Δ_{TC} be the total time delays of complemented array multipliers

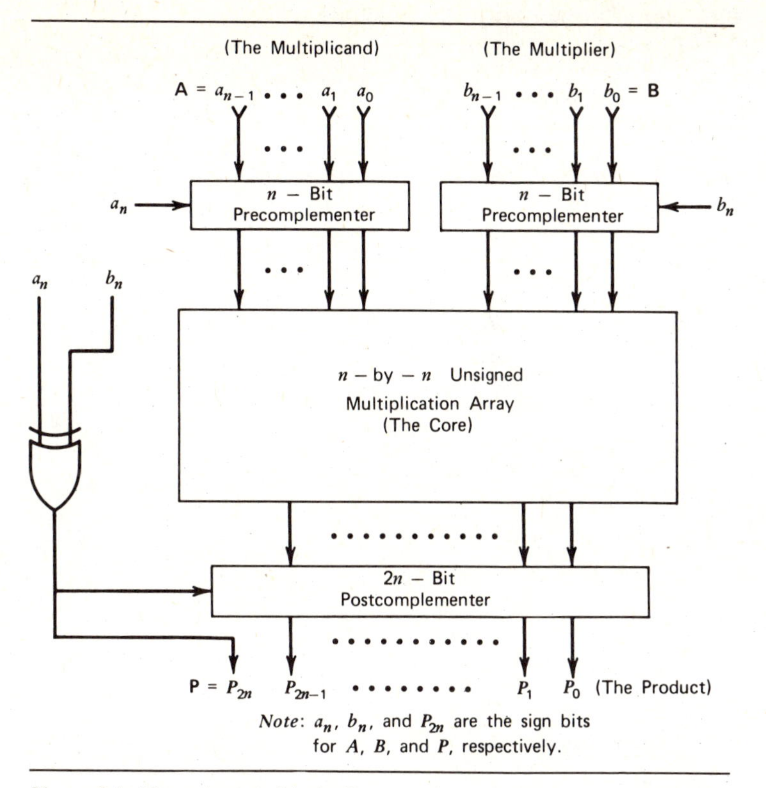

Figure 6.4 The schematic block diagram of an $(n + 1)$-by-$(n + 1)$ complemented array multiplier.

(with dimension as specified in Fig. 6.4) in sign-magnitude, one's complement and two's complement conventions, respectively. We have

$$\Delta_{SM} = (4n - 2)\Delta \tag{6.6}$$

$$\Delta_{OC} = 3\Delta + (4n - 2)\Delta + 3\Delta = (4n + 4)\Delta \tag{6.7}$$

$$\Delta_{TC} = 2(2n + 3)\Delta + (4n - 2)\Delta = (8n + 4)\Delta \tag{6.8}$$

Hardware requirements and time delays in $(n + 1)$-by-$(n + 1)$ unsigned, sign-magnitude, one's complement, and two's complement array multipliers are summarized in Table 6.1. One can conclude from this table that both sign-magnitude and one's complement notations are suitable for high-speed complemented array multiplication, whereas the indirect two's complement array multiplication requires too much extra hardware and takes about twice as long to complete the necessary complementing and multiplication operations. Remedies to overcome these difficulties will be given in later sections.

Table 6.1 Hardware Components and Time Delays of Basic Multiplication Arrays

Hardware and Speed	Notation			
	Sign-Magnitude	**One's Complement**	**Two's Complement**	**Unsigned**
XOR gates	1	$4n + 1$	$4n + 1$	0
AND gates	n^2	n^2	$n^2 + 4n$	$(n + 1)^2$
OR gates	0	0	$4n - 3$	0
Full adders	$n(n - 1)$	$n(n - 1)$	$n(n - 1)$	$n(n + 1)$
Total time delay	$(4n - 2)\Delta$	$(4n + 4)\Delta$	$(8n + 4)\Delta$	$(4n + 2)\Delta$

Note: Each operand is assumed to have $n + 1$ bits.

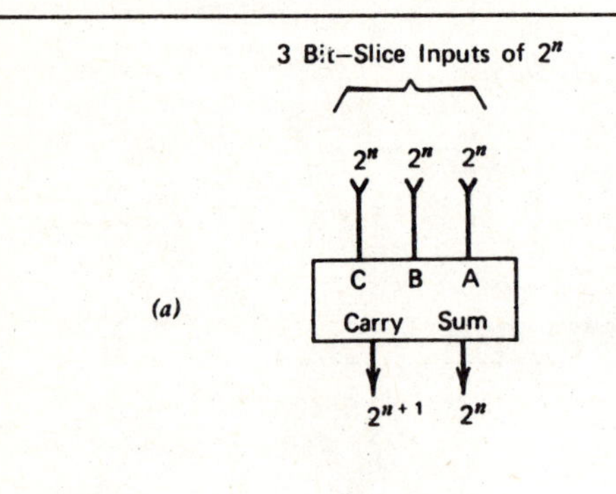

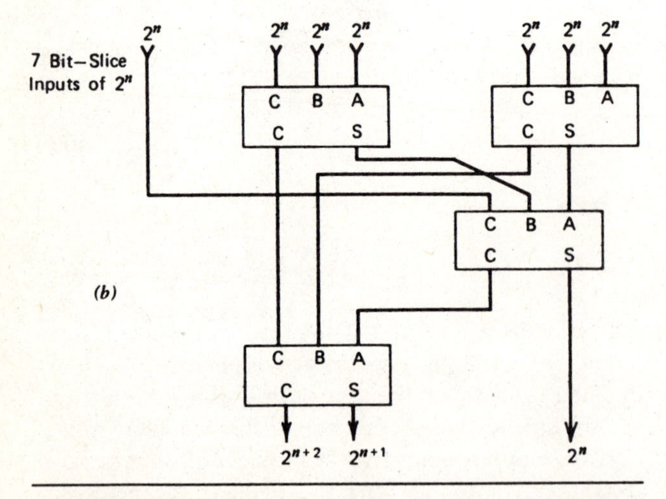

Figure 6.5 Bit-slice Wallace trees built with carry-save full adders. (*a*) A 3-bit-slice Wallace Tree W_3 (*b*) A 7-bit-slice Wallace tree W_7

6.2 Modular Multiplication and Wallace Trees

This section demonstrates the modular network realization of large-scale multiplication arrays using the basic multiply modules introduced in section 2.5. A bit-sliced version of the carry-save adders, similar to the column adders used in section 4.4, are presented first. These bit-slice adders, known as *Wallace trees*, will be used to sum up the subproducts independently generated by the *Nonadditive Multiply Modules* (NMMs).

A k-input Wallace tree is a bit-slice summing circuit, which produces the sum of k bit-slice inputs. Figure 6.5 presents two Wallace trees, one with 3 inputs and the other with 7 inputs. The 3-input Wallace tree is nothing but a 3-to-2 carry-save full adder with two outputs representing the binary sum of the three input bits. The 7-input Wallace tree adds 7 bit-slice inputs to yield a 3-bit sum output. Texas Instruments has produced a 7-bit-slice Wallace tree with TTL/LSI circuit having a typical time delay of 45 nsec. Five 3-to-2 carry-save full adders are used in the construction of the 7-input Wallace tree as shown. For larger Wallace trees, one can use more levels of bit-slice CSAs.

The 4-by-4 NMMs are used to produce the local partial products, called *subproducts* of 8-bits each. In order to generate all the subproducts simultaneously, the input operands (multiplicand and multiplier) must be decomposed into 4-bit slices. Let us consider the construction of an 8-by-8 array multiplier using four 4-by-4 NMMs. The 16-bit product can be written as

$$\begin{aligned} \mathbf{P} = \mathbf{A} \times \mathbf{B} &= (A_H \cdot A_L) \times (B_H \cdot B_L) \\ &= A_H \times B_H + A_H \times B_L + A_L \times B_H + A_L \times B_L \\ &= P_{HH} + P_{HL} + P_{LH} + P_{LL} \end{aligned} \tag{6.9}$$

where

$$\begin{aligned} \mathbf{A} &= A_H \cdot A_L \\ \mathbf{B} &= B_H \cdot B_L \end{aligned} \tag{6.10}$$

and the subscripts identify the "high" and "low" 4-bit slices in each 8-bit number. The dot "·" refers to concatenation. The resulting product is the sum of four 8-bit subproducts shifted 4-bits apart.

The generation of these four subproducts is shown in the upper portion of Fig. 6.6. The summation of these subproducts is carried out by the 3-input Wallace trees used in the middle portion of the network. The column alignments are important to ensure the correctness of the result in the form of two vectors, the *sum vector* and the *carry vector*. The low portion of the network is a 12-bit *Carry Propagate Adder* (CPA) which merges the two vectors into the final product. Note that the least significant slices, corresponding to subproduct $P_{LL} = A_L \times B_L$, need not go through the CPA.

In Fig. 6.7, we demonstrate the modular arrangement for various array multiplication networks from size 4-by-4 to size 32-by-32. Each rectangle represents an 8-bit

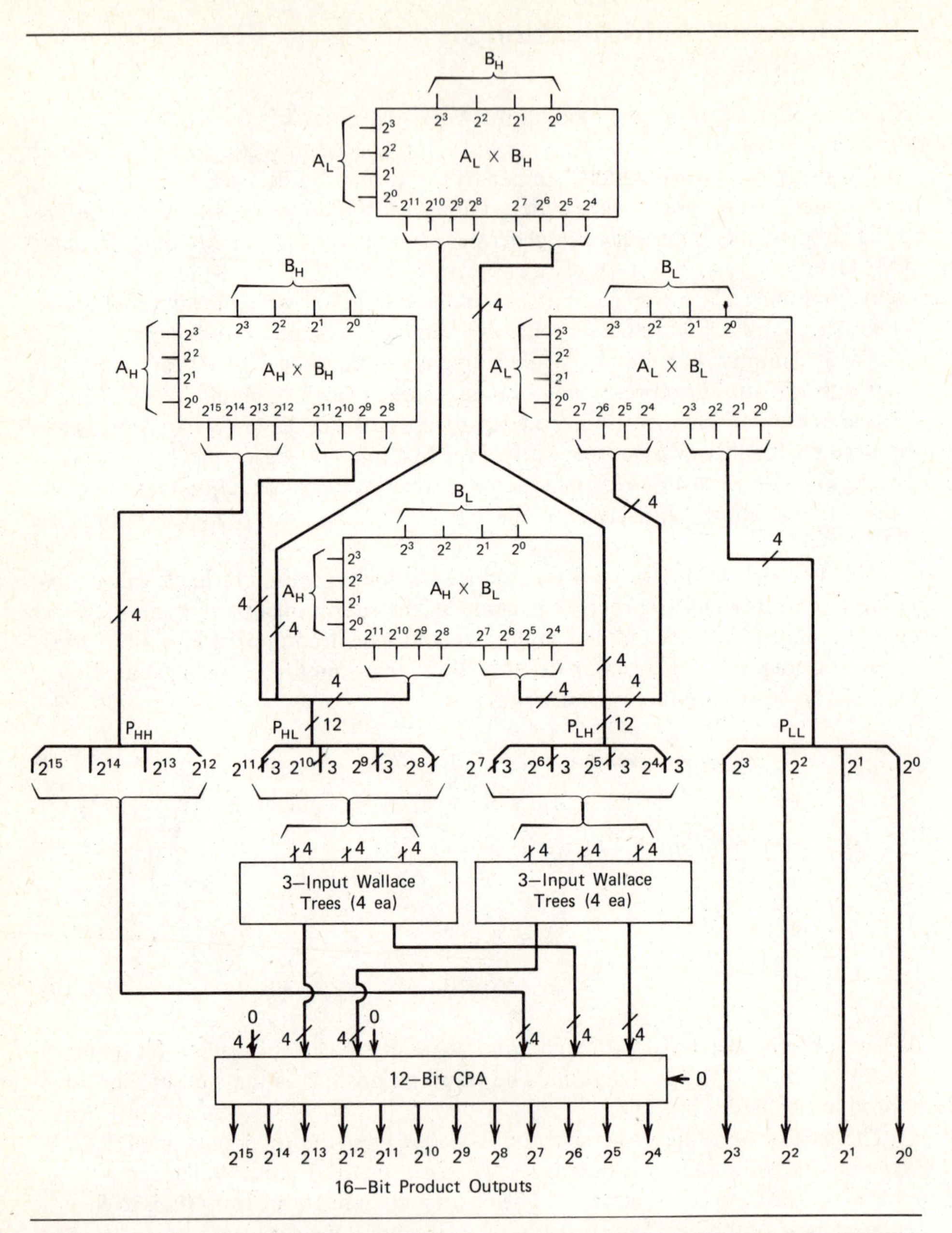

Figure 6.6 An 8-by-8 array multiplier built with 4-by-4 Nonadditive Multiply Modulus (NMMs), Wallace trees, and Carry Propagate Adder (CPA).

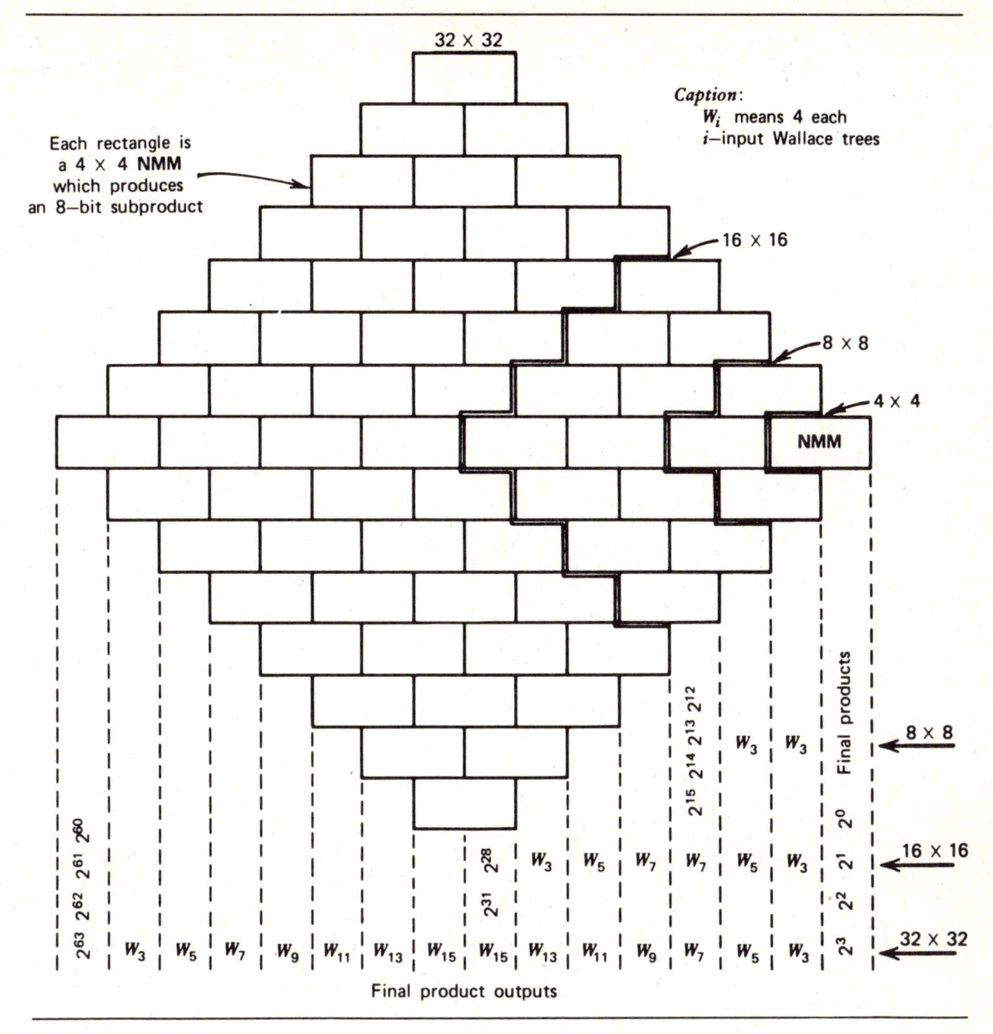

Figure 6.7 Array arrangement for various multipliers from size 4-by-4 to 32-by-32.

subproduct divided into "high" and "low" 4-bit slices. All the slices are added in a columnwise fashion by Wallace trees of odd-numbered inputs. For an 8-by-8 multiplication network, only 3-input trees are required. For 16-by-16 network, Wallace trees of sizes 3, 5, 7, 7, 5, 3 are needed, and for 32-by-32 multiplication, tree sizes up to 15 inputs are required. Note that the two 4-bit slices at both ends do not require Wallace trees. A 15-input Wallace tree can be constructed from three 7-input trees. All the unused inputs of Wallace trees must be grounded to ensure that "0" inputs. The TI NMMs (SN74S274) and Wallace trees (SN74S275) can yield a 16-bit product in 75 nsec and a 32-bit product in 116 nsec using the above network realization method.

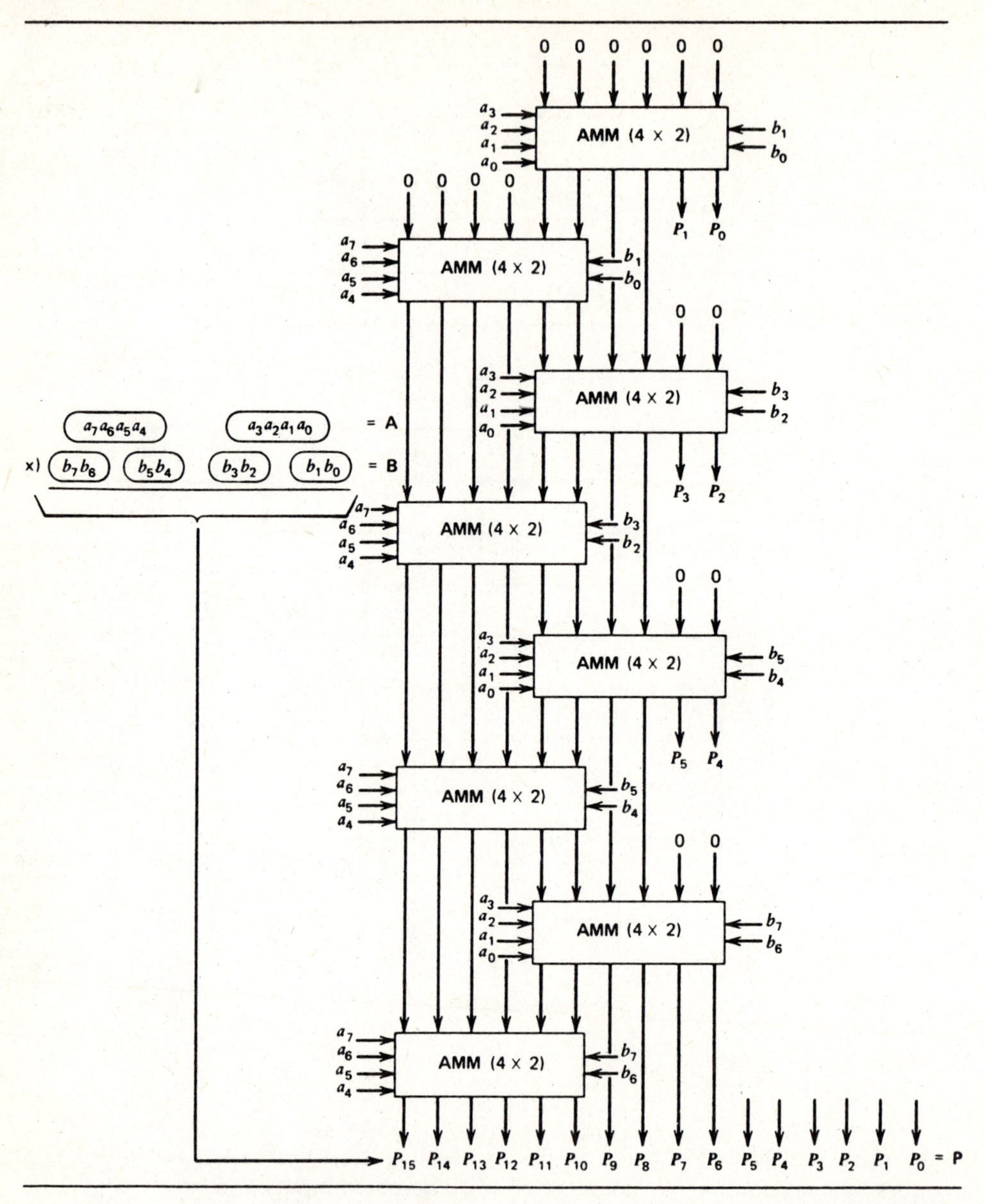

Figure 6.8 The schematic interconnection diagram of an 8-by-8 multiplication network made from 8 4-by-2 **AMMs** (see Fig. 2.15 for details of **AMM**).

Described below are the methods of interconnecting the *Additive Multiply Modules* (AMMs), introduced in Chapter 2, which require no bit-slice summing trees. An 8-by-8 array multiplier is shown in Fig. 6.8 using eight 4-by-2 AMMS. In general, a $4m$-by-$4m$ multiplication network can be constructed with $2m^2$ 4-by-2 AMMs. These $2m^2$ modules are arranged in the form of a skewed matrix of $2m$ 45°-diagonals and m interleaved columns as shown in Fig. 6.8. AMMs in the same column have identical 4-bit-slice multiplicands, and AMMs on the same diagonal have identical 2-bit-slice multipliers. The total time delay of such a $4m$-by-$4m$ multiplication network of 4-by-2 AMMs is estimated as

$$\Delta_N(4m \times 4m) = (3m - 1) \times 10\Delta + 2\Delta = (30m - 8)\Delta \qquad \textbf{(6.11)}$$

where 10Δ accounts the time delay in each 4-by-2 module and 2Δ due to the AND gate delay for generating the summand terms. With a high-speed TTL gate delay $\Delta = 3$ nsec, the AMM network multiplication requires less than 400 nsec for a product of 32 bits or less. This approach can be extended to using larger AMMs such as 4-by-4 or 8-by-8 as the building blocks. Table 6.2 summarizes the number of NMM, AMM, Wallace tree types and time delays for array multiplier of various sizes.

Table 6.2 Module Counts, Wallace-Tree Types, and Speed of Modular Multiplication Networks of Three Practical Sizes

	Nonadditive Network			Additive Network	
Network Size	No. of 4 × 4 NMMs	Wallace-Tree Types	Multiply Time	No. of 4 × 2 AMMs	Multiply Time
16-by-16	16	W_3, W_5, W_7	65Δ	32	112Δ
32-by-32	64	W_3, W_5, W_7 W_9, W_{11}, W_{13}, W_{15}	90Δ	128	232Δ
64-by-64	256	W_3, W_5, W_7 ..., W_{25}, W_{27}, W_{29}, W_{31}	250Δ	512	472Δ

6.3 Direct Two's Complement Multiplication

Array multipliers, which can perform "direct" multiplication of two's complement numbers without requiring the complementing stages, are to be described. This direct approach significantly speeds up the multiplication process by eliminating the slow two's complementing operations. The mathematical properties associated with such direct two's complement multiplication are described first. After characterizing several generalized full adder types, we describe several direct two's complement array multiplication algorithms and their possible implementations in the next two sections.

Thus far, we have treated two's complement numbers as positional numbers with an unweighted sign and positively weighted coefficients. A better approach is to evaluate the values of two's complement numbers as positional numbers with a negatively weighted sign and positively weighted coefficients. Let us start with a brief review of the conventional FXP representation of two's complement integers. Consider a two's complement number $\mathbf{N} = (a_{n-1}a_{n-2}\cdots a_1a_0)_2$ where a_{n-1} is the designated sign. The value N_v of the number **N** can be represented as follows depending on the sign of **N**.

$$N_v = \begin{cases} +\sum_{i=0}^{n-2} a_i 2^i & \text{if } a_{n-1} = 0 \quad (\mathbf{N} \text{ positive}) \\ -\left[1 + \sum_{i=0}^{n-2} (1 - a_i)2^i\right] & \text{if } a_{n-1} = 1 \quad (\mathbf{N} \text{ negative}) \end{cases} \qquad \mathbf{(6.12)}$$

We used to represent the two's complement number $-\mathbf{N}$ as

$$-\mathbf{N} = \bar{a}_{n-1}\bar{a}_{n-2}\cdots\bar{a}_1\bar{a}_0 + 1 \qquad \mathbf{(6.13)}$$

where $\bar{a}_i = 1 - a_i$ for $0 \le i \le n-1$. If we impose a negative weighting factor -2^{n-1} to the sign bit a_{n-1}, we can combine the two positional representations in Eq. 6.12 into the following form

$$N_v = -a_{n-1}2^{n-1} + \sum_{i=0}^{n-2} a_i 2^i \qquad \mathbf{(6.14)}$$

It suffices to prove that the number $-\mathbf{N}$ represented in Eq. 6.13 can be evaluated by

$$-N_v = -(1 - a_{n-1})2^{n-1} + \left[\sum_{i=0}^{n-2} (1 - a_i)2^i\right] + 1 \qquad \mathbf{(6.15)}$$

Multiplying both sides of Eq. 6.14 by -1, we obtain

$$\begin{aligned} -N_v &= a_{n-1}2^{n-1} - \sum_{i=0}^{n-2} a_i 2^i \\ &= a_{n-1}2^{n-1} + (2^{n-1} - 2^{n-1}) - \sum_{i=0}^{n-2} a_i 2^i \\ &= (a_{n-1} - 1)2^{n-1} + 2^{n-1} - \sum_{i=0}^{n-2} a_i 2^i \\ &= -(1 - a_{n-1})2^{n-1} + \left(1 + \sum_{i=0}^{n-2} 2^i\right) - \sum_{i=0}^{n-2} a_i 2^i \\ &= -(1 - a_{n-1})2^{n-1} + \left[\sum_{i=0}^{n-2} (1 - a_i)2^i\right] + 1 \end{aligned} \qquad \mathbf{(6.16)}$$

Therefore, we conclude from Eq. 6.16 that expressions given in Eq. 6.14 and Eq. 6.15 are equivalent, and both are legitimate representations for two's complement numbers. The above derivations can be best illustrated by a numerical example. Consider the case of $n = 5$, $\mathbf{N} = (+13)_{10} = (01101)_2$ and $-\mathbf{N} = (-13)_{10} = (10011)_2$. It can be easily verified that $\mathbf{N} = (01101)_2$ has a value $N_v = -0 \times 2^4 + 1 \times 2^3 + 1 \times 2^2 + 1 \times 2^0 = (+13)_{10}$ and $-\mathbf{N} = (10011)_2$ has a value $-N_v = -1 \times 2^4 + 0 \times 2^3 + 0 \times 2^2 + 1 \times 2^1 + 1 \times 2^0 = -16 + 3 = (-13)_{10}$.

The conventional 1-bit full adder assumes positive weights to all of its 3 inputs and 2 outputs. Such adders can be generalized to four types of adding cells by imposing positive and negative weights to the input/output terminals. Table 6.3 lists

Table 6.3 Names and Logic Symbols of Four Types of Generalized Full Adders

Type	Logic Symbol	Operation
Type 0 Full Adder	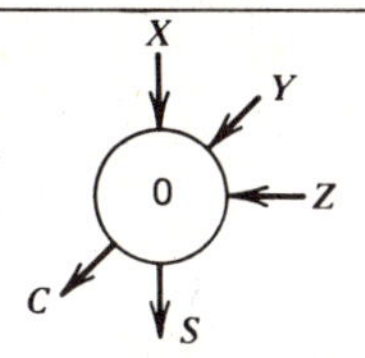 0 (inputs X, Y, Z; outputs C, S)	X Y +) Z CS
Type 1 Full Adder	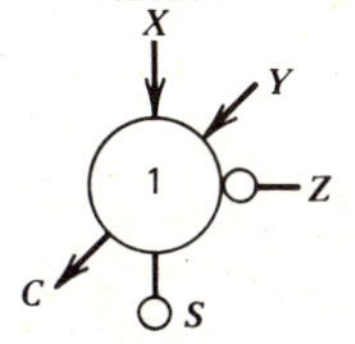 1 (inputs X, Y, Z; outputs C, S)	X Y +) $-Z$ $C(-S)$
Type 2 Full Adder	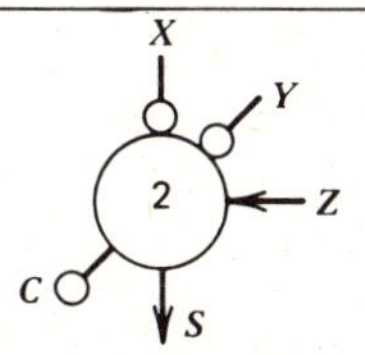 2 (inputs X, Y, Z; outputs C, S)	$-X$ $-Y$ +) Z $(-C)\ S$
Type 3 Full Adder	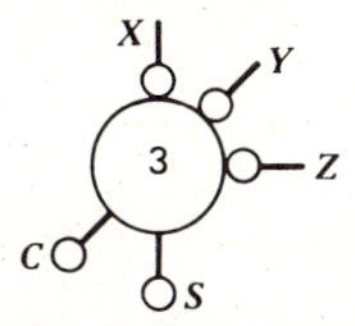 3 (inputs X, Y, Z; outputs C, S)	$-X$ $-Y$ +) $-Z$ $(-C)\ (-S)$

the names and logic symbols of four types of generalized full adders. Each type of full adder is named by the number of negatively weighted inputs contained in it. Therefore, Type **0** full adder has no negative input and Type **2** full adder has two negative inputs and one positive input, and so on.

Listed below are four arithmetic equations that describe the input/output relationships of the four types of generalized full adders.

Type 0. $C2^1 + S2^0 = X2^0 + Y2^0 + Z2^0$ **(6.17a)**

Type 1. $C2^1 + (-S)2^0 = X2^0 + Y2^0 + (-Z)2^0$ **(6.17b)**

Type 2. $(-C)2^1 + S2^0 = (-X)2^0 + (-Y)2^0 + Z2^0$ **(6.17c)**

Type 3. $(-C)2^1 + (-S)2^0 = (-X)2^0 + (-Y)2^0 + (-Z)2^0$ **(6.17d)**

These four arithmetic equations lead to the truth-table descriptions of the four generalized full adders given in Table 6.4. Note that Type **0** and Type **3** adders are

Table 6.4 Truth Table Describing the Four Types of Generalized Full Adders

Full Adder	Weighted Inputs			Weighted Outputs	
Type **0**	$X \cdot 2^0$	$Y \cdot 2^0$	$Z \cdot 2^0$	$C \cdot 2^1$	$S \cdot 2^0$
Type **3**	$-X \cdot 2^0$	$-Y \cdot 2^0$	$-Z \cdot 2^0$	$-C \cdot 2^1$	$-S \cdot 2^0$
	0	0	0	0	0
	0	0	1	0	1
	0	1	0	0	1
Truth	0	1	1	1	0
Table	1	0	0	0	1
	1	0	1	1	0
	1	1	0	1	0
	1	1	1	1	1
Type **1**	$X \cdot 2^0$	$Y \cdot 2^0$	$-Z \cdot 2^0$	$C \cdot 2^1$	$-S \cdot 2^0$
Type **2**	$-X \cdot 2^0$	$-Y \cdot 2^0$	$Z \cdot 2^0$	$-C \cdot 2^1$	$S \cdot 2^0$
	0	0	0	0	0
	0	0	1	0	1
	0	1	0	1	1
Truth	0	1	1	0	0
Table	1	0	0	1	1
	1	0	1	0	0
	1	1	0	1	0
	1	1	1	1	1

describable by the same truth table with different weighted column headings. Type **1** full adder and Type **2** full adder also share the same truth table but with different column headings. One can easily derive the Boolean equations governing the four types of full adders from the table entries.

$$\begin{array}{l} \textbf{Type 0} \\ \text{or} \\ \textbf{Type 3} \end{array} \begin{cases} S = \overline{X}\overline{Y}Z + \overline{X}Y\overline{Z} + X\overline{Y}\overline{Z} + XYZ \\ C = XY + YZ + ZX \end{cases} \qquad \textbf{(6.18a)}$$

$$\begin{array}{l} \textbf{Type 1} \\ \text{or} \\ \textbf{Type 2} \end{array} \begin{cases} S = \overline{X}\overline{Y}Z + \overline{X}Y\overline{Z} + X\overline{Y}\overline{Z} + XYZ \\ C = XY + X\overline{Z} + Y\overline{Z} \end{cases} \qquad \textbf{(6.18b)}$$

Note that Type **0** and Type **3** full adders are characterized by the same pair of logic equations, identical to that of the conventional 1-bit full adder (Type **0**). This is because Type **3** full adder can be obtained from Type **0** full adder by simply negating all its input and output values and vice versa. Similar relationships can be established between Type **1** and Type **2** full adders. The logic in Eq. 6.18 in 2-level sum-of-product form can be implemented with wired logic (AND-OR-INVERTER TTL gates) with a delay of 2Δ. In the following sections, we use different logic symbols to represent Type **0** and Type **3** full adders, even though they have identical internal structure; the same situation applies to Type **1** and Type **2** full adders. This distinction will make the schematic drawings of the multiplication arrays more comprehensible.

6.4 Pezaris Array Multiplier and Modifications

A direct two's complement array multiplier using mixture types of full adders is described below. For clarity, we show the structure of a 5-by-5 two's complement array multiplier to illustrate Pezaris' original design. Then we describe two possible modifications of his design. Consider two 5-bit two's complement numbers, the multiplicand $\mathbf{A} = (a_4)a_3a_2a_1a_0$ and the multiplier $\mathbf{B} = (b_4)b_3b_2b_1b_0$, with their negatively weighted signs a_4 and b_4 marked by parentheses. In what follows, we shall use parentheses to mark a negative summand term as (a_ib_j). The operations included in the process of multiplying **A** and **B** are illustrated by the summand matrix and numerical example in Fig. 6.9. The schematic logic circuit diagram of the Pezaris multiplier is shown in Fig. 6.10. In the general case of an n-by-n Pezaris multiplier, $(n-2)^2$ Type **0**, $n-2$ Type **1**, $2n-3$ Type **2**, and one Type **3** full adders are required, making a total of $n(n-1)$ full adders. The total multiply time required in Pezaris multiplier would be

$$\Delta_P = \Delta_a + 2(n-1)\Delta_f = 2\Delta + (2n-2)2\Delta = (4n-2)\Delta \qquad \textbf{(6.19a)}$$

If full carry lookahead is used in the last row, the total delay can be reduced to

$$\Delta_P(\text{CLA}) = \Delta_a + (n-1)\Delta_f + 8\Delta = 2\Delta + (n-1)2\Delta + 8\Delta = (2n+8)\Delta \qquad \textbf{(6.19b)}$$

By replacing those Type **2** full adders in the upper left diagonal of Fig. 6.10 by Type **1** full adders, we can modify the Pezaris design into an array as shown in Fig. 6.11. This modification scheme is called a *Tri-Section Array Multiplier*. Each section of the

					(a_4)	a_3	a_2	a_1	a_0	$= \mathbf{A}$
				×)	(b_4)	b_3	b_2	b_1	b_0	$= \mathbf{B}$
					(a_4b_0)	a_3b_0	a_2b_0	a_1b_0	a_0b_0	
				(a_4b_1)	a_3b_1	a_2b_1	a_1b_1	a_0b_1		
			(a_4b_2)	a_3b_2	a_2b_2	a_1b_2	a_0b_2			
		(a_4b_3)	a_3b_3	a_2b_3	a_1b_3	a_0b_3				
+)	a_4b_4	(a_3b_4)	(a_2b_4)	(a_1b_4)	(a_0b_4)					
P_9	P_8	P_7	P_6	P_5	P_4	P_3	P_2	P_1	P_0	$= \mathbf{P}$

```
                       (0)  1  1  0  1   = +13
                    ×) (1)  1  0  1  1   = - 5
                   ---------------------
                       (0)  1  1  0  1
                   (0)  1   1  0  1
               (0)  0   0   0  0
          (0)  1    1   0   1
  +)  0   (1)  (1)  (0) (1)
 ----------------------------------------
      0   (1)  0    1   1   1  1  1  1
     (1)   1   0    1   1   1  1  1  1   = -65
 Extended  ↑
   Sign   Sign
```

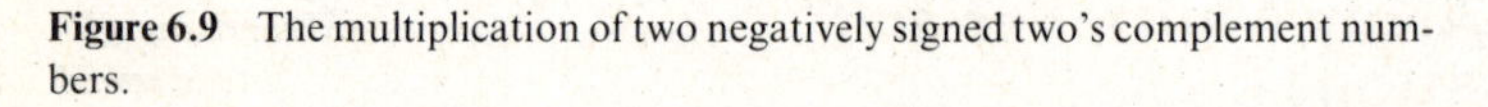

Figure 6.9 The multiplication of two negatively signed two's complement numbers.

array, as separated by dash lines, is made from only one type of full adders. Only Type **0** full adders are used in the upper right triangle, Type **1** full adders in the upper left triangle, and Type **2** full adders in the last two rows of the array. One can rearrange the summand matrix in Fig. 6.9 into a more uniformly structured array without alterning the columnwise summation relationship. The method is demonstrated by the rearranged matrix shown in Fig. 6.12. With this rearranged matrix, we can segregate the positive summands from the negative summands. This segregation enables the use of only two types of full adders, as illustrated in the *Bi-Section Array Multiplier* in Fig. 6.13. The upper section, consisting of $(n-1) \times (n-2)$ Type **0** full adders, functions exactly the same as an $(n-1)$-by-$(n-1)$ integer multiplier without any negative summands in it. The lower section, consisting of two rows of $2(n-1)$ Type **2** full adders, adds all the negative summands to the intermediate sum of all the positive summands from the upper section.

The total multiply time of this bi-section array multiplier is slightly slower than that of the two previous arrays. The longest carry propagation path along the rightmost diagonal and the last row has a total time delay of

$$\Delta_{\mathrm{BS}} = \Delta_a + (n-1)\Delta_f + (n-1+1)\Delta_f = 2\Delta + (2n-1)2\Delta = 4n\Delta \qquad \textbf{(6.20)}$$

With full carry lookahead in the last row, we can achieve the speed

$$\Delta_{\mathrm{BS}}(\mathrm{CLA}) = \Delta_a + n\Delta_f + 8\Delta = (2n+10)\Delta.$$

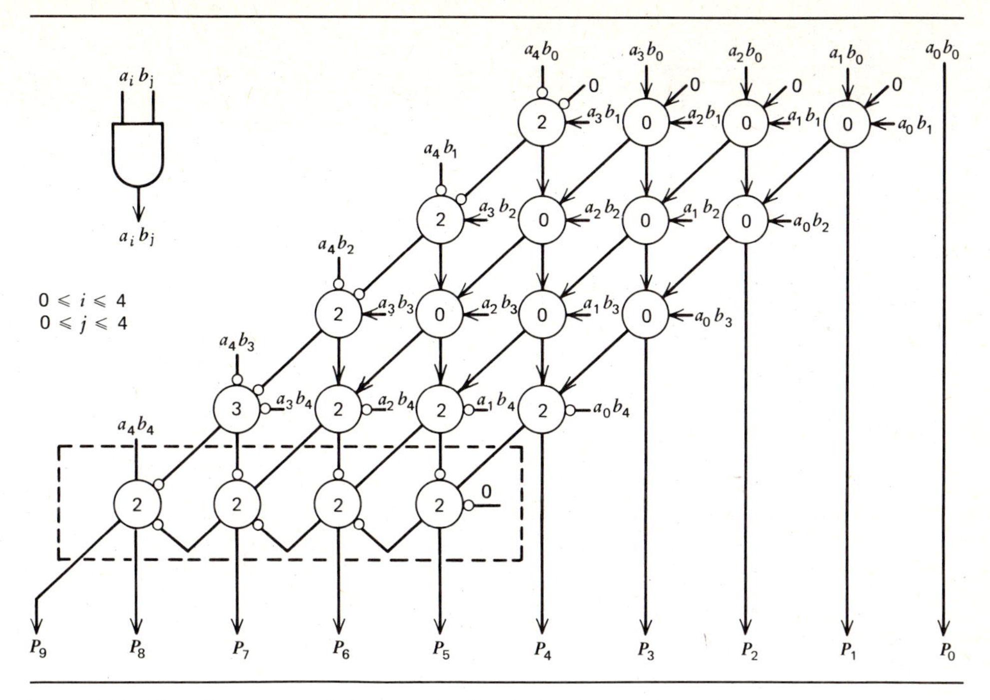

Figure 6.10 The schematic circuit diagram of a 5-by-5 Pezaris array multiplier.

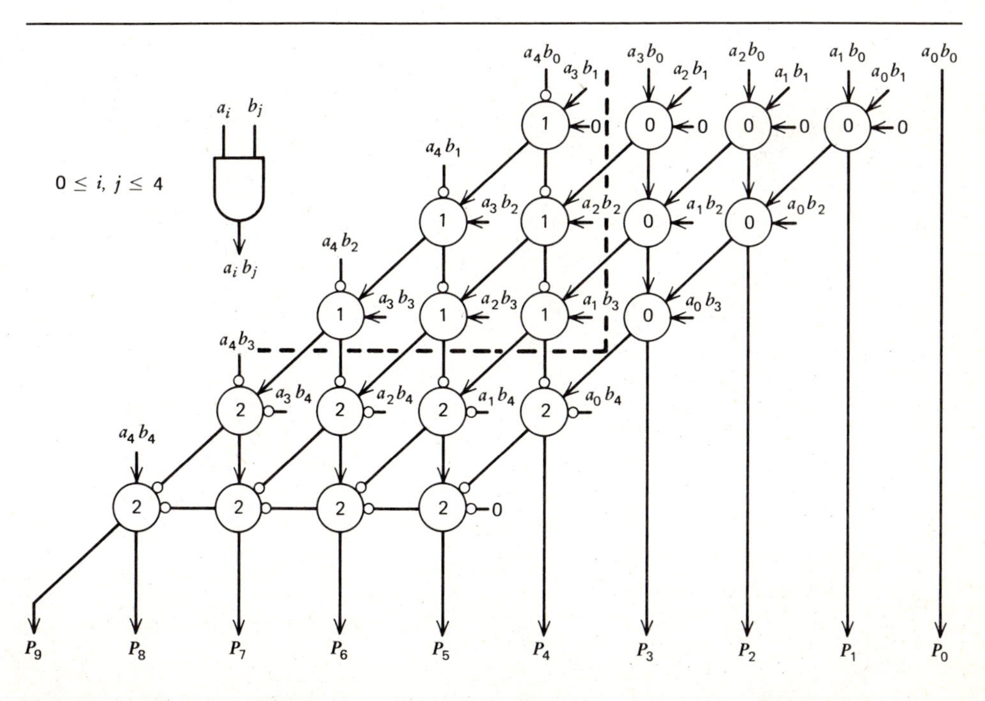

Figure 6.11 The schematic logic circuit diagram of a 5-by-5 tri-section array multiplier obtaned from modifying the Pezaris multiplier.

						(a_4)	a_3	a_2	a_1	a_0	$= \mathbf{A}$
					$\times)$	(b_4)	b_3	b_2	b_1	b_0	$= \mathbf{B}$
Positive section							a_3b_0	a_2b_0	a_1b_0	a_0b_0	
						a_3b_1	a_2b_1	a_1b_1	a_0b_1		
					a_3b_2	a_2b_2	a_1b_2	a_0b_2			
		a_4b_4	0	a_3b_3	a_2b_3	a_1b_3	a_0b_3				
Negative section			(a_4b_3)	(a_4b_2)	(a_4b_1)	(a_4b_0)					
			(a_3b_4)	(a_2b_4)	(a_1b_4)	(a_0b_4)					
	(P_9)	P_8	P_7	P_6	P_5	P_4	P_3	P_2	P_1	P_0	$= \mathbf{P}$

$$\mathbf{P} = \mathbf{A} \times \mathbf{B}$$

A, B, and P are all negatively signed two's complement numbers.

Figure 6.12 The segregation of positive and negative summands in a 5-by-5 bisection array multiplier for two's complement numbers.

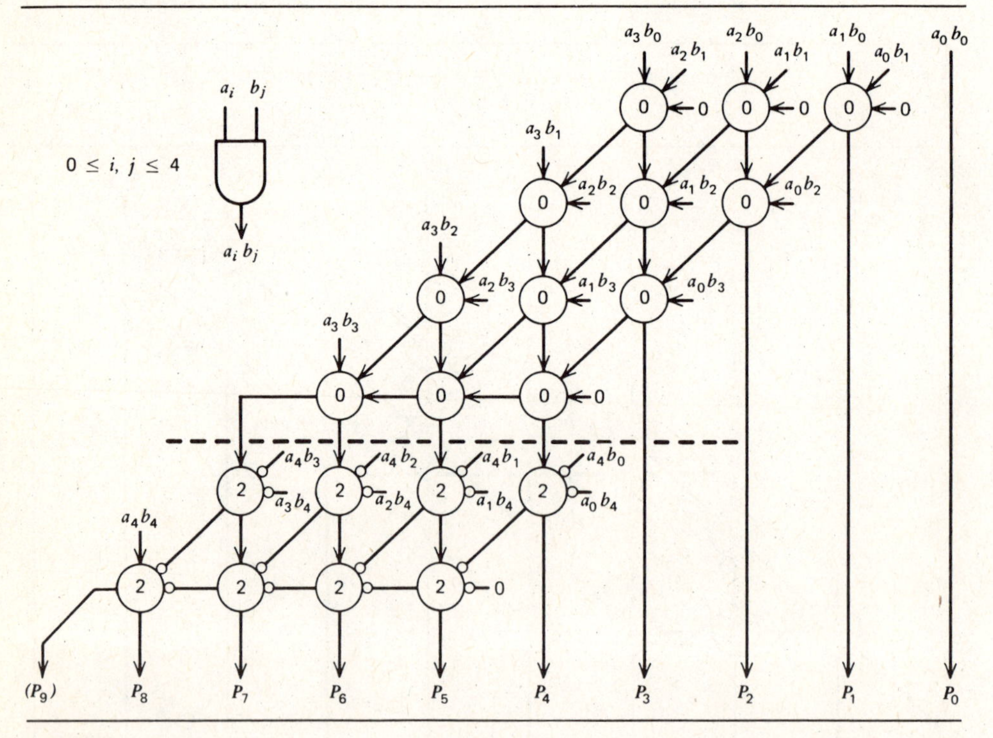

Figure 6.13 The 5-by-5 two's complement bi-section array multiplier based on the matrix in Fig. 6.12.

6.5 Baugh–Wooley Two's Complement Multipliers

Baugh and Wooley [3] have proposed an algorithm for direct two's complement array multiplication. The principal advantage of their algorithm is that the signs of all the summands are positive, thus allowing the array to be constructed entirely with the conventional Type **0** full adders. This uniform structure is very attractive for LSI.

Consider two 2's complement integers, an m-bit multiplicand

$$\mathbf{A} = (a_{m-1}a_{m-2}\cdots a_1a_0)_2$$

and an n-bit multiplier $\mathbf{B} = (b_{n-1}b_{n-2}\cdots b_1b_0)_2$. The values of **A** and **B** denoted as A_v and B_v can be written by Eq. 6.14 as

$$A_v = -a_{m-1}2^{m-1} + \sum_{i=0}^{m-2} a_i 2^i$$

$$B_v = -b_{n-1}2^{n-1} + \sum_{i=0}^{n-2} b_i 2^i \tag{6.21}$$

The value P_v of the product $\mathbf{P} = A \cdot B = (p_{m+n-1}p_{m+n-2}\cdots p_1p_0)_2$ in two's complement notation can be written as follows in terms of the coefficient product of a_i's and b_j's with appropriate weighting factors attached.

$$\begin{aligned} P_v &= -p_{m+n-1}2^{m+n-1} + \sum_{i=0}^{m+n-2} p_i 2^i = A_v B_v \\ &= \left(-a_{m-1}2^{m-1} + \sum_{i=0}^{m-2} a_i 2^i\right)\left(-b_{n-1}2^{n-1} + \sum_{i=0}^{n-2} b_i 2^i\right) \\ &= a_{m-1}b_{n-1}2^{m+n-2} + \sum_{i=0}^{m-2}\sum_{j=0}^{n-2} a_i b_j 2^{i+j} \\ &\quad - \sum_{i=0}^{m-2} a_i b_{n-1} 2^{n-1+i} - \sum_{i=0}^{n-2} a_{m-1} b_i 2^{m-1+i} \end{aligned} \tag{6.22}$$

In Eq. 6.22, the signs of the summands $a_i b_{n-1}$ for $i = 0, \ldots, m-2$, and of $a_{m-1}b_j$ for $j = 0, \ldots, n-2$ are all negative. By placing all the negative summands in the last two rows as shown in Fig. 6.14, the product can be formed by adding the first $n-2$ summand rows and subtracting the last two rows as we did in the bi-section array multiplier. Instead of subtracting the negative summands, the negation of the summands are added. Using Eq. 6.15, we can replace the subtraction of the third term in Eq. 6.22

$$\sum_{i=0}^{m-2} a_i b_{n-1} 2^{n-1+i} = 2^{n-1}\left(-0 \cdot 2^m + 0 \cdot 2^{m-1} + \sum_{i=0}^{m-2} a_i b_{n-1} 2^i\right) \tag{6.23a}$$

by the addition of

$$2^{n-1}\left(-1 \cdot 2^m + 1 \cdot 2^{m-1} + 1 + \sum_{i=0}^{m-2} \overline{a_i b_{n-1}} 2^i\right) \tag{6.23b}$$

$$
\begin{array}{ccccccccccccccl}
 & & & & & a_{m-1} & a_{m-2} & \cdots & & a_3 & a_2 & a_1 & a_0 & = \mathbf{A} & \\
 & & & & \times) & & & b_{n-1} & \cdots & & b_2 & b_1 & b_0 & = \mathbf{B} & \\
\hline
 & & & & & & a_{m-2}b_0 & & \cdots & a_3b_0 & a_2b_0 & a_1b_0 & a_0b_0 & & \\
 & & & & & a_{m-2}b_1 & \cdots & & a_3b_1 & a_2b_1 & a_1b_1 & a_0b_1 & & & \\
\mathbf{P} = \mathbf{A} \times \mathbf{B} & & & & a_{m-2}b_2 & \cdots & & a_3b_2 & a_2b_2 & a_1b_2 & a_0b_2 & & & & \text{Positive summands} \\
 & & & & \cdot^{\cdot^{\cdot}} & & & & & \cdot^{\cdot^{\cdot}} & & & & & \\
 & a_{m-1}b_{n-1} & 0 & a_{m-2}b_{n-2} & \cdots & & a_2b_{n-2} & a_1b_{n-2} & a_0b_{n-2} & & & & & & \\
\hdashline
0 & 0 & a_{m-2}b_{n-1} & a_{m-3}b_{n-1} & \cdots & & a_1b_{n-1} & a_0b_{n-1} & & & & & & & \text{Negative summands} \\
0 & 0 & a_{m-1}b_{n-2} & a_{m-1}b_{n-3} & \cdots & a_{m-1}b_0 & & & & & & & & & \\
\hline
P_{m+n-1} & P_{m+n-2} & P_{m+n-3} & P_{m+n-4} & \cdots & P_{m-1} & \cdots & P_{n-1} & \cdots & & P_2 & P_1 & P_0 & = \mathbf{P} &
\end{array}
$$

Note: The above matrix shows the case that $m > n$

Figure 6.14 The segregation of positive and negative summands in an m-by-n two's complement multiplication.

Note that Eq. 6.23b has the value

$$\begin{cases} 0 & \text{for } b_{n-1} = 0 \\ 2^{n-1}\left(-2^m + 2^{m-1} + 1 + \sum_{i=0}^{m-2} \bar{a}_i 2^i\right) & \text{for } b_{n-1} = 1 \end{cases} \tag{6.23c}$$

From Eq. 6.23c, Eq. 6.23b can be rewritten as

$$2^{n-1}\left(-2^m + 2^{m-1} + \bar{b}_{n-1}2^{m-1} + b_{n-1} + \sum_{i=0}^{m-2} \bar{a}_i b_{n-1} 2^i\right) \tag{6.23d}$$

This implies that the second to the last row vector in Fig. 6.14

$$0 \quad 0 \quad a_{m-2}b_{n-1} \quad a_{m-3}b_{n-1} \quad \cdots \quad a_0 b_{n-1} \tag{6.24a}$$

can be replaced by the sum of the following two row vectors

$$\begin{array}{cccccc} 0 & \bar{b}_{n-1} & \bar{a}_{m-2}b_{n-1} & \bar{a}_{m-3}b_{n-1} & \cdots & \bar{a}_0 b_{n-1} \\ 1 & 1 & 0 & 0 & \cdots & b_{n-1} \end{array} \tag{6.24b}$$

Similarly, we can replace the subtraction of the fourth term $\sum_{i=0}^{m-2} a_{m-1}b_i \times 2^{m-1+i}$ in Eq. 6.22 by the addition of

$$2^{m-1}\left(-2^n + 2^{n-1} + \bar{a}_{m-1}2^{n-1} + a_{m-1} + \sum_{i=0}^{n-2} a_{m-1}\bar{b}_i 2^i\right) \tag{6.25}$$

This implies that the last row vector in Fig. 6.14

$$0 \quad 0 \quad a_{m-1}b_{n-2} \quad a_{m-1}b_{n-3} \quad \cdots \quad a_{m-1}b_0 \tag{6.26a}$$

can be replaced by the sum of the following two row vectors

$$\begin{array}{cccccc} 0 & \bar{a}_{m-1} & a_{m-1}\bar{b}_{n-2} & a_{m-1}\bar{b}_{n-3} & \cdots & a_{m-1}\bar{b}_0 \\ 1 & 1 & 0 & 0 & \cdots & a_{m-1} \end{array} \tag{6.26b}$$

Furthermore, one can combine the second row vector in Eq. 6.24b and the second row vector in Eq. 6.26b into a single row vector as shown below, ignoring the carry out of the sign column P_{m+n-1} due to the nature of two's complement addition.

$$\begin{array}{lccccccccccccc} & 1 & 0 & \cdots & 0 & a_{m-1} & 0 & \cdots & 0 & b_{n-1} & 0 & \cdots & 0 \\ & \uparrow & & & & \uparrow & & & & \uparrow & & & \uparrow \\ \text{Column} & p_{m+n-1} & & & & p_{m-1} & & & & p_{n-1} & & & p_0 \end{array} \tag{6.27}$$

After the substitution of the last two rows in Fig. 6.14 using formulas (6.24a), (6.26b), and (6.27), we obtain the new summand matrix as shown in Fig. 6.15. The

$$
\begin{array}{ccccccccccccccccc}
 & & & & & & & a_{m-1} & a_{m-2} & \cdots & & & a_3 & a_2 & a_1 & a_0 & = \mathbf{A} \\
 & & & & & \times) & & & & & b_{n-1} & \cdots & & b_2 & b_1 & b_0 & = \mathbf{B} \\
\hline
 & & & & & & & & a_{m-2}b_0 & & \cdots & & a_3b_0 & a_2b_0 & a_1b_0 & a_0b_0 & \\
 & \mathbf{P} = \mathbf{A} \times \mathbf{B} & & & & & & a_{m-2}b_1 & & & \cdots & & a_2b_1 & a_1b_1 & a_0b_1 & & \\
 & & & & & a_{m-2}b_2 & & & & & \cdots & & a_1b_2 & a_0b_2 & & & \\
 & & & & \iddots & & & & & & & \iddots & & & & & \\
 & a_{m-1}b_{n-1} & 0 & a_{m-2}b_{n-2} & & & \cdots & & & a_2b_{n-2} & a_1b_{n-2} & a_0b_{n-2} & & & & & \\
 & \bar{a}_{m-1} & \bar{a}_{m-2}b_{n-1} & \bar{a}_{m-3}b_{n-1} & & & \cdots & & \bar{a}_2b_{n-1} & \bar{a}_1b_{n-1} & \bar{a}_0b_{n-1} & & & & & & \\
 & \bar{b}_{n-1} & a_{m-1}\bar{b}_{n-2} & a_{m-1}\bar{b}_{n-3} & \cdots & a_{m-1}\bar{b}_1 & & a_{m-1}\bar{b}_0 & & & & & & & & & \\
1 & & & & & & & a_{m-1} & & & b_{n-1} & & & & & & \\
\hline
P_{m+n-1} & P_{m+n-2} & P_{m+n-3} & P_{m+n-4} & \cdots & & & P_{m-1} & & \cdots & P_{n-1} & & P_3 & P_2 & P_1 & P_0 & = \mathbf{P}
\end{array}
$$

Figure 6.15 The Baugh–Wooley two's complement multiplication algorithm with all positive summands obtained by using the two's complement of the negative summands in Fig. 6.14 ($m > n$ is again assumed here).

principal characteristic of this new matrix is its uniformity in consisting of only positive summands. Therefore, the product can be obtained by performing addition using only Type **0** full adders. The schematic logic circuit diagram for a 6 × 4 two's complement multiplier based on Baugh-Wooley's algorithm is shown in Fig. 6.16. Note that if the complemented inputs are not directly available from the data bus, inverters have to be used. In general, an m-by-n Baugh-Wooley multiplier requires $m(n - 1) + 3$ full adders for its implementation. When $m = n$, the circuit design is slightly different, as revealed by the schematic circuit for a 5-by-5 Baugh-Wooley multiplier in Fig. 6.17. In this n-by-n multiplier $n^2 - n + 3$ full adders are required. In Table 6.5, we have provided a comparison of the four different designs for direct two's complement multiplication. The speeds of these two's complement array multipliers are almost the same.

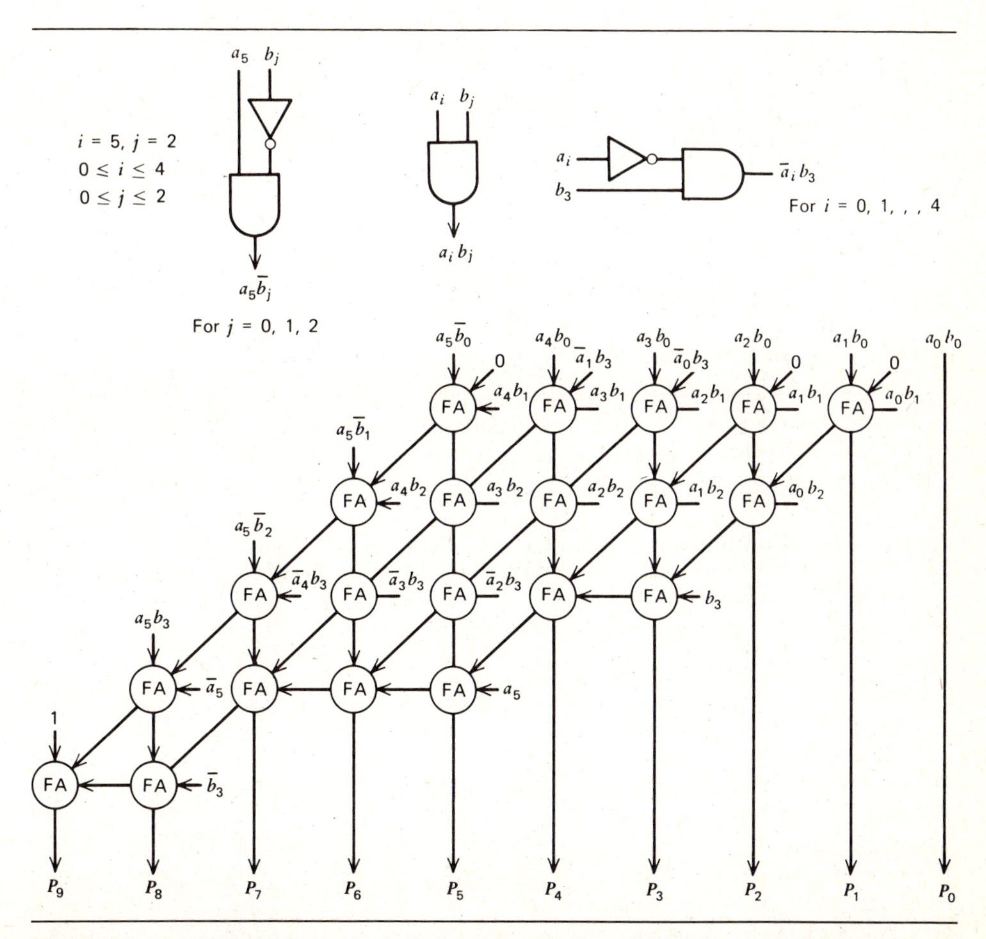

Figure 6.16 The schematic logic circuit diagram of a 6-by-4 Baugh–Wooley two's complement array multiplier.

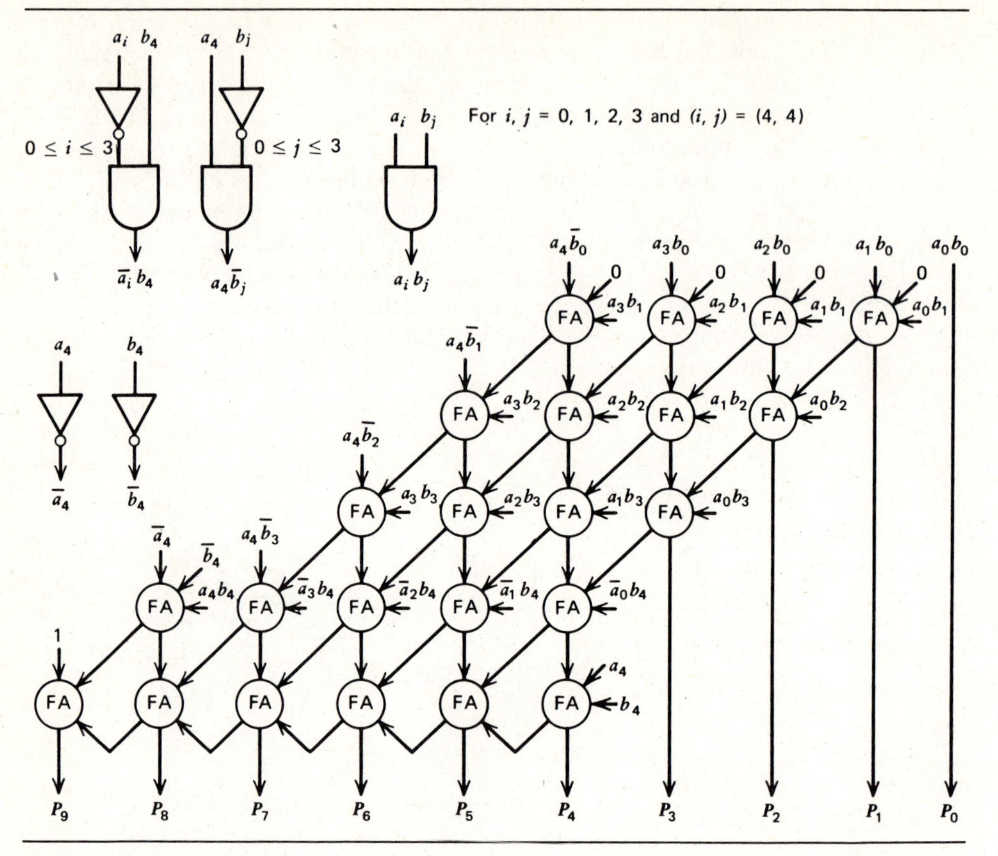

Figure 6.17 The schematic logic circuit diagram of a 5-by-5 Baugh–Wooley two's complement array multiplier (Hwang [18]).

Table 6.5 Comparisons among the Pezaris Multiplier, the Tri-Section Multiplier, the Bi-Section Multiplier and the Baugh-Wooley Multiplier for Two's Complement Numbers

		$m \times n$ **Two's Complement Array Multiplier**			
		Pezaris' (Fig. 6.10)	**Tri-Section (Fig. 6.11)**	**Bi-Section (Fig. 6.13)**	**Baugh-Wooley (Fig. 6.16)**
Full adder used	Type **0**	$(m - 2)(n - 2)$	$(m - 1)(n - 2)/2$	$(m - 2)(n - 1)$	$m(n - 1) + 3$
	Type **1**	$m - 2$	$(m - 1)(n - 2)/2$	0	0
	Type **2**	$m + n - 3$	$2(m - 1)$	$2(m - 1)$	0
	Type **3**	1	0	0	0
Total time delay (Multiply time)		$2(m + n)\Delta - 2\Delta$	$2(m + n)\Delta - 2\Delta$	$2(m + n)\Delta$	$2(m + n)\Delta$

6.6 Universal Multiplication Arrays

Most of the parallel array multiplication schemes we have discussed thus far treat unsigned and signed numbers separately. In this section, we study a generalized universal approach to designing high-speed array multipliers. A multiplier is said to be *universal* if it can multiply two binary numbers and yield the product in any prespecified number notation. The implied notation may be any of the following four: unsigned, sign-magnitude, one's complement, and two's complement. The input-output relationship of an n-by-n *Universal Multiplication Array* (UMA) is sketched in Fig. 6.18. The two inputs are the *multiplicand*, $\boldsymbol{\alpha} = \alpha_{n-1} \cdots \alpha_1 \alpha_0$, and the *multiplier*, $\boldsymbol{\beta} = \beta_{n-1} \cdots \beta_1 \beta_0$, and the output is the *product* of $\boldsymbol{\alpha}$ and $\boldsymbol{\beta}$, denoted as $\boldsymbol{\Pi} = \pi_{2n-1} \cdots \pi_1 \pi_0$. Both $\boldsymbol{\alpha}$, $\boldsymbol{\beta}$, and $\boldsymbol{\Pi}$ should assume the same number representation.

The summation operations involved in such a UMA are illustrated by the flow chart in Fig. 6.19. The precomplement and postcomplement stages are only required to convert one's complement numbers. These two stages are bypassed when dealing with unsigned, sign-magnitude, or two's complement numbers.

The parallel multiplication algorithms are given corresponding to four distinct number notations. Let $\mathbf{A} = a_{n-1} \cdots a_1 a_0$ and $\mathbf{B} = b_{n-1} \cdots b_1 b_0$ be obtained from $\boldsymbol{\alpha}$

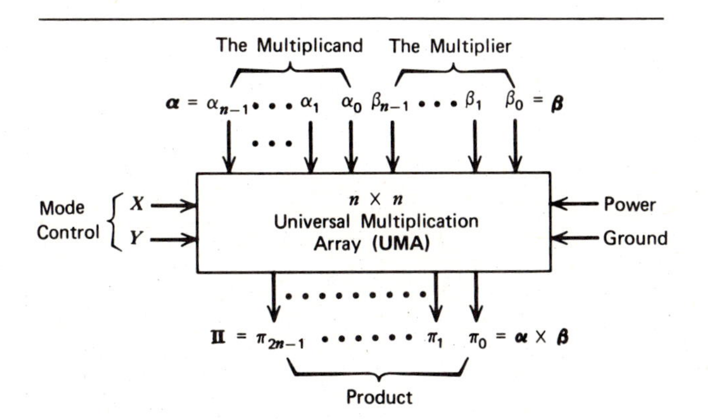

Mode Control X Y	Multiplicative Operation Performed
0 0	Unsigned multiply
0 1	Sign-magnitude multiply
1 0	One's complement multiply
1 1	Two's complement multiply

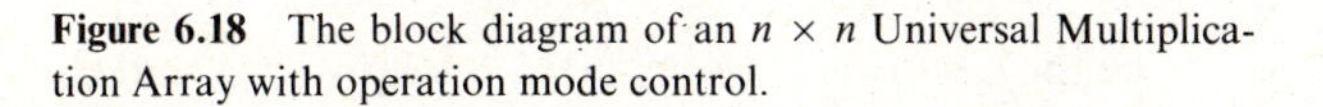

Figure 6.18 The block diagram of an $n \times n$ Universal Multiplication Array with operation mode control.

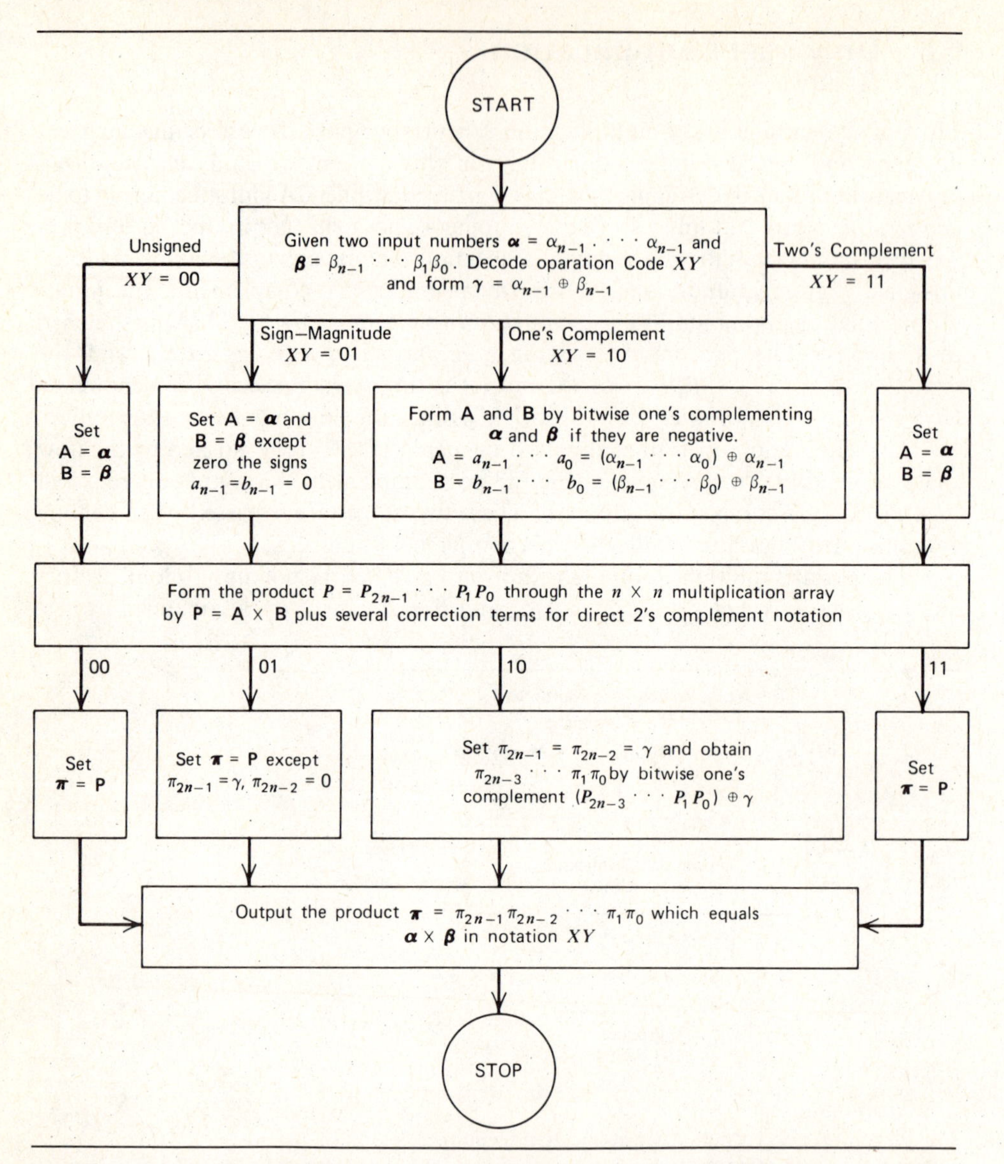

Figure 6.19 Flow chart description of the operations involved in a Universal Multiplication Array (UMA).

and $\boldsymbol{\beta}$ by exclusively ORing the bits α_i and β_i with the sign bit α_{n-1} and β_{n-1}, for $0 \le i \le n-1$ for one's complement mode ($XY = 10$)

$$a_i = \alpha_i \oplus (X\overline{Y}\alpha_{n-1})$$
$$b_i = \beta_i \oplus (X\overline{Y}\beta_{n-1}) \tag{6.28}$$

Let $\mathbf{P} = P_{2n-1}P_{2n-2} \cdots P_1P_0$ be the resulting product after multiplying **A** and **B**, with appropriate correction terms added only for the two's complement multiplication. Define a Boolean variable $\gamma = \alpha_{n-1} \oplus \beta_{n-1}$. $\mathbf{P}$ can be related to the final output $\mathbf{\Pi}$ as follows:

$$\begin{aligned} \Pi_{2n-1} &= P_{2n-1}(\overline{X \oplus Y}) + \gamma(X \oplus Y) \\ \Pi_{2n-2} &= P_{2n-2}(\overline{X \oplus Y}) + X\bar{Y}\gamma \end{aligned} \quad \textbf{(6.29)}$$

Note that only one's complement multiplication requires the complementing operations specified above. For multiplications in unsigned, sign-magnitude and two's complement mode, we have the simple relationships $\mathbf{A} = \boldsymbol{\alpha}$, $\mathbf{B} = \boldsymbol{\beta}$ and $\Pi_k = P_k$ for all $0 \leq k \leq 2n - 2$; except in sign-magnitude multiply, we set the signs $a_{n-1} = b_{n-1} = 0$.

In an *unsigned multiplication* with $(X, Y) = (0, 0)$, as illustrated in Fig. 6.1, the product $\mathbf{P} = \mathbf{A} \times \mathbf{B}$ is obtained by direct multiplication using Eq. 6.2 for $m = n$.

For *sign-magnitude multiplication* with $(X, Y) = (0, 1)$, the sign bit of the product $P_{2n-1} = \gamma$ and the next bit $P_{2n-2} = 0$. The sign bits should be zeroed $a_{n-1} = b_{n-1} = 0$ before applying Eq. 6.2 to obtain the remaining product bits $P_{2n-3} \cdots P_1 P_0$.

For *one's complement multiplication* with $(X, Y) = (1, 0)$, the leading two bits (the extended sign) are obtained as $P_{2n-1} = P_{2n-2} = \gamma$. After the precomplement operation, both $\mathbf{A}$ and $\mathbf{B}$ are positive sign-magnitude numbers with $a_{n-1} = b_{n-1} = 0$. This implies that the remaining bits $P_{2n-3} \cdots P_1 P_0$ can be also obtained using Eq. 6.2.

We recommend the use of Baugh-Wooley's algorithms described in section 6.5 for *two's complement multiplication* with $(X, Y) = (1, 1)$. Adding the entries in the summand matrix of Fig. 6.19 produces the following two's complement number for a general $n \times n$ Baugh-Wooley multiplier.

$$\begin{aligned} \mathbf{P} = {} & 1 \times 2^{2n-1} + (\bar{a}_{n-1} + \bar{b}_{n-1} + a_{n-1}b_{n-1}) \times 2^{2n-2} \\ & + \left(\sum_{i=0}^{n-2} a_i \times 2^i\right) \times \left(\sum_{j=0}^{n-2} b_j \times 2^j\right) + (a_{n-1} \times 2^{n-1}) \times \left(\sum_{j=0}^{n-2} \bar{b}_j \times 2^j\right) \\ & + \left(\sum_{i=0}^{n-2} \bar{a}_i \times 2^i\right) \times (b_{n-1} \times 2^{n-1}) + (a_{n-1} + b_{n-1}) \times 2^{n-1} \end{aligned} \quad \textbf{(6.30)}$$

Among the four operation modes, the two's complement multiplication involves the largest number of summand terms. The sign-magnitude and one's complement multiplications involve the least number of summand terms. This is also reflected by the hardware requirements for adding these terms, as described in previous sections.

Logic circuits to realize the universal parallel multiplication will now be presented. The schematic diagram of a 5-by-5 UMA is shown in Fig. 6.20. The precomplement and postcomplement stages are implemented with two sets of exclusive OR gates under the control of sign bits and mode lines. The two leading bits π_{2n-1} and π_{2n-2} are retrieved by a dual 4-to-1 multiplexer to cover all the four cases. The core of the UMA is the iterative array of gated full adders in the middle section of the figure. The core carries out the actual summation of the shifted summands.

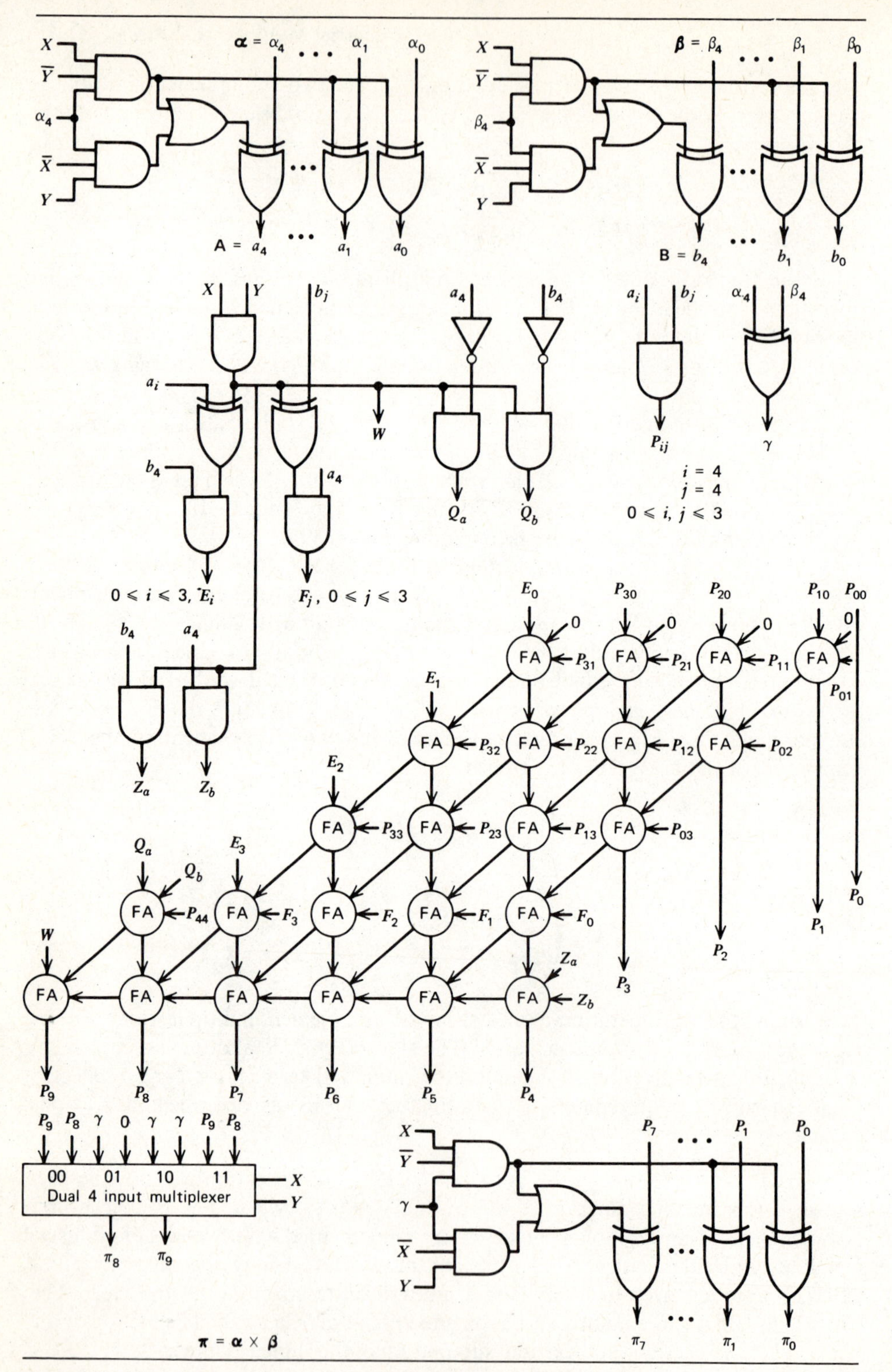

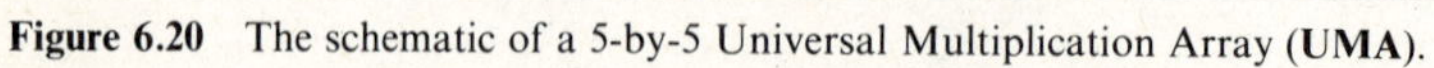

Figure 6.20 The schematic of a 5-by-5 Universal Multiplication Array (**UMA**).

The input terminals of most of these full adders are connected to some excitation logic circuits. The logic equations for these logic circuits are listed below.

$$P_{ij} = a_i b_j \quad \text{for} \begin{matrix} 0 \le i \le n-2, \\ 0 \le j \le n-2 \end{matrix}$$

$$P_{n-1,n-1} = a_{n-1} b_{n-1}$$

$$\begin{aligned} E_i &= (XY \oplus a_i) b_{n-1} \quad \text{for } 0 \le i \le n-2 \\ F_j &= a_{n-1}(XY \oplus b_j) \quad \text{for } 0 \le j \le n-2 \\ Q_a &= XY\bar{a}_{n-1}; \quad Q_b = XY\bar{b}_{n-1} \\ Z_a &= XYa_{n-1}; \quad Z_b = XYb_{n-1} \\ W &= XY \end{aligned} \tag{6.31}$$

The design parameters of UMAs are summarized below: A general UMA denoted as UMA ($n \times n$), requires $n^2 - n + 3$ full adders, $6n - 3$ XOR gates, $n^2 + 11$ AND gates, 3 OR gates and a dual 4-to-1 multiplexer for its implementation. In total $4n + 4$ external connection terminals are required in UMA ($n \times n$). The time delay of an $n \times n$ UMA will be analyzed in a later section.

6.7 Programmable Additive Multiply Modules

The hardware parts and the number of connection terminals of a UMA grow rather quickly with its size. This indicates the fundamental limitation in fabricating the whole n-by-n UMA on single IC chip for very large n. Small multiply modules must be used to build a large universal multiplication network. These modules must have not only the *additive capability* but also *external programmability*.

The block diagram of a *Programmable Additive Multiply* (PAM) module is illustrated in Fig. 6.21. We use the symbol $M(k \times k)$ to denote a PAM module. The design of an $M(4 \times 4)$ module has been chosen as an illustrative example. An $M(k \times k)$ PAM module can be programmed to behave in four different modes denoted as

$$M_{00}(k \times k); \quad M_{01}(k \times k); \quad M_{10}(k \times k) \text{ and } M_{11}(k \times k) \tag{6.32}$$

where the subscripts correspond to the values appeared on the U, V control terminals. With respect to these four operation modes, a PAM module $M(k \times k)$ can be used to evaluate the following four arithmetic expressions.

$M_{00}(k \times k)$:

$$\begin{aligned} \mathbf{P} &= \mathbf{A} \times \mathbf{B} + \mathbf{C} + \mathbf{D} \\ &= \left(\sum_{i=0}^{k-1} a_i \times 2^i\right) \times \left(\sum_{j=0}^{k-1} b_j \times 2^j\right) + \sum_{r=0}^{k-1} c_r \times 2^r + \sum_{r=0}^{k-1} d_r \times 2^r \\ &= \sum_{i=0}^{k-1} \sum_{j=0}^{k-1} a_i b_j \times 2^{i+j} + \sum_{r=0}^{k-1} (c_r + d_r) \times 2^r \end{aligned} \tag{6.33a}$$

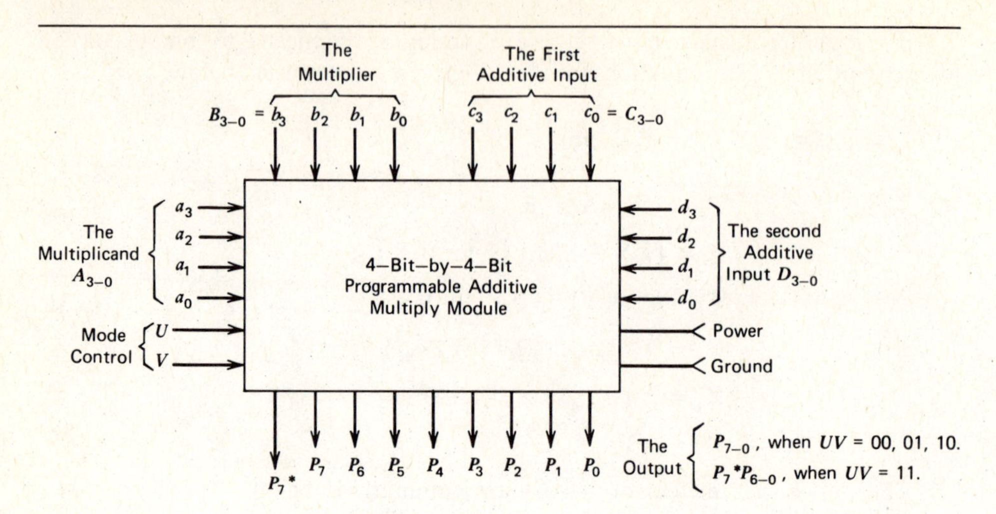

Mode Control U V	Operation Mode
0 0	M_{00}
0 1	M_{01}
1 0	M_{10}
1 1	M_{11}

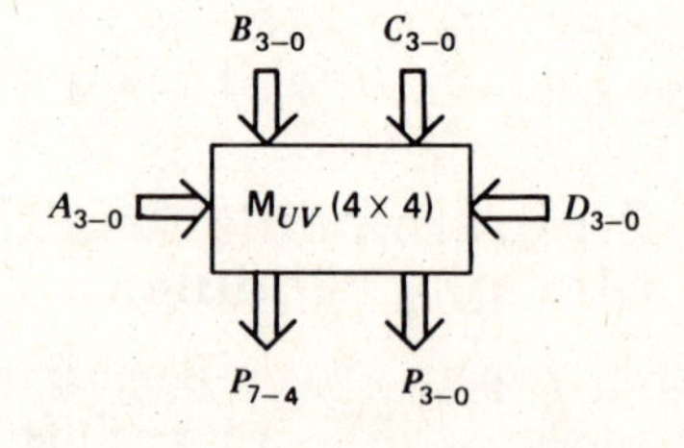

The simplified diagram of a PAM Module $M(4 \times 4)$ operated in Mode $\mathbf{M}_{UV}$

Figure 6.21 The input-output diagram of a 4-by-4 programmable additive multiply module and its simplified diagram $\mathbf{M}_{UV}(4 \times 4)$.

$M_{01}(k \times k)$:

$$\mathbf{P} = \left(\sum_{i=0}^{k-1} a_i \times 2^i\right) \times \left(\sum_{j=0}^{k-2} b_j \times 2^j\right) + \left(\sum_{i=0}^{k-1} \bar{a}_i \times 2^i\right) \times b_{k-1} \times 2^{k-1}$$

$$+ \sum_{r=0}^{k-1} c_r \times 2^r + \sum_{r=0}^{k-1} d_r \times 2^r$$

$$= \sum_{i=0}^{k-1} \sum_{j=0}^{k-2} a_i b_j \times 2^{i+j} + \sum_{i=0}^{k-1} \bar{a}_i b_{k-1} \times 2^{k+i-1}$$

$$+ \sum_{r=0}^{k-1} (c_r + d_r) \times 2^r \qquad \textbf{(6.33b)}$$

$M_{10}(k \times k)$:

$$\mathbf{P} = \left(\sum_{i=0}^{k-2} a_i \times 2^i\right) \times \left(\sum_{j=0}^{k-1} b_j \times 2^j\right) + a_{k-1} \times 2^{k-1} \times \left(\sum_{j=0}^{k-1} \bar{b}_j \times 2^j\right)$$

$$+ \sum_{r=0}^{k-1} c_r \times 2^r + \sum_{r=0}^{k-1} d_r \times 2^r$$

$$= \sum_{i=0}^{k-2} \sum_{j=0}^{k-1} a_i b_j \times 2^{i+j} + \sum_{j=0}^{k-1} a_{k-1} b_j \times 2^{k+j-1}$$

$$+ \sum_{r=0}^{k-1} (c_r + d_r) \times 2^r \tag{6.33c}$$

$M_{11}(k \times k)$:

$$\mathbf{P} = \sum_{i=0}^{k-2} \sum_{j=0}^{k-2} a_i b_j \times 2^{i+j} + a_{k-1} b_{k-1} \times 2^{2k-2}$$

$$+ \sum_{j=0}^{k-1} a_{k-1} \bar{b}_j \times 2^{k+j-1} + \sum_{i=0}^{k-1} a_i b_{k-1} \times 2^{k+i-1} + 1 \times 2^{2k-1}$$

$$+ (\bar{a}_{k-1} + \bar{b}_{k-1}) \times 2^{k-1} + \sum_{r=0}^{k-1} (c_r + d_r) \times 2^r \tag{6.33d}$$

where $\mathbf{A} = a_{k-1} \cdots a_1 a_0$ is the multiplicand input; $\mathbf{B} = b_{k-1} \cdots b_1 b_0$ is the multiplier input, and $\mathbf{C} = c_{k-1} \cdots c_1 c_0$ and $\mathbf{D} = d_{k-1} \cdots d_1 d_0$ are two additive inputs. The resulting output takes the following form

$$\mathbf{P} = \begin{cases} P_{2k-1} P_{2k-2} \cdots P_1 P_0, & \text{for } M_{00}(k \times k);\ M_{01}(k \times k) \text{ and } M_{10}(k \times k) \\ P^*_{2k-1} P_{2k-2} \cdots P_1 P_0, & \text{for } M_{11}(k \times k) \end{cases} \tag{6.34}$$

A matrix description of the four operating modes of a PAM module $M(k \times k)$ is given in Fig. 6.22 for the case of $k = 4$. Gated full-adder realization of this PAM module $M(4 \times 4)$ is shown in Fig. 6.23. The internal excitation logic is described by the following set of equations.

$$\begin{aligned} T_{i3} &= (v \oplus a_i) b_3 \quad \text{for } i = 0, 1, 2 \\ T_{3j} &= a_3 (b_j \oplus u) \quad \text{for } j = 0, 1, 2 \\ T_{33} &= \overline{(u \oplus v)} a_3 b_3 + u \bar{v} \bar{a}_3 b_3 + \bar{u} v a_3 \bar{b}_3 \\ R_a &= u v \bar{a}_3; \qquad R_b = u v \bar{b}_3 \\ S &= uv \end{aligned} \tag{6.35}$$

This PAM module $M(4 \times 4)$ requires an IC package of 29 external leads. The same development technique can be applied to produce standard PAM modules of different sizes, such as $M(2 \times 2)$, $M(4 \times 2)$, $M(8 \times 8)$, and so on. A summary of application parameters for PAM modules is given in Table 6.6.

					a_3	a_2	a_1	a_0
				×)	b_3	b_2	b_1	b_0
					a_3b_0	a_2b_0	a_1b_0	a_0b_0
				a_3b_1	a_2b_1	a_1b_1	a_0b_1	
			a_3b_2	a_2b_2	a_1b_2	a_0b_2		
		a_3b_3	a_2b_3	a_1b_3	a_0b_3			
		0			c_3	c_2	c_1	c_0
+)	0	0			d_3	d_2	d_1	d_0
	P_7	P_6	P_5	P_4	P_3	P_2	P_1	P_0

(a) Mode $\mathbf{M}_{00}(4 \times 4)$

					a_3b_0	a_2b_0	a_1b_0	a_0b_0
				a_3b_1	a_2b_1	a_1b_1	a_0b_1	
			a_3b_2	a_2b_2	a_1b_2	a_0b_2		
		$\bar{a}_3b_3$	$\bar{a}_2b_3$	$\bar{a}_1b_3$	$\bar{a}_0b_3$			
		0			c_3	c_2	c_1	c_0
+)	0	0			d_3	d_2	d_1	d_0
	P_7	P_6	P_5	P_4	P_3	P_2	P_1	P_0

(b) Mode $\mathbf{M}_{01}(4 \times 4)$

					$a_3\bar{b}_0$	a_2b_0	a_1b_0	a_0b_0
				$a_3\bar{b}_1$	a_2b_1	a_1b_1	a_0b_1	
			$a_3\bar{b}_2$	a_2b_2	a_1b_2	a_0b_2		
		$a_3\bar{b}_3$	a_2b_3	a_1b_3	a_0b_3			
		0			c_3	c_2	c_1	c_0
+)	0	0			d_3	d_2	d_1	d_0
	P_7	P_6	P_5	P_4	P_3	P_2	P_1	P_0

(*c*) Mode $\mathbf{M}_{10}(4 \times 4)$

					$a_3\bar{b}_0$	a_2b_0	a_1b_0	a_0b_0
				$a_3\bar{b}_1$	a_2b_1	a_1b_1	a_0b_1	
			$a_3\bar{b}_2$	a_2b_2	a_1b_2	a_0b_2		
		a_3b_3	$\bar{a}_2b_3$	$\bar{a}_1b_3$	$\bar{a}_0b_3$			
		$\bar{a}_3$			c_3	c_2	c_1	c_0
+)	1	$\bar{b}_3$			d_3	d_2	d_1	d_0
	P_7	P_6	P_5	P_4	P_3	P_2	P_1	P_0

(*d*) Mode $\mathbf{M}_{11}(4 \times 4)$

Figure 6.22 Summand matrix descriptions of the four operation modes of a Programmable Additive Multiply Module $\mathbf{M}(4 \times 4)$.

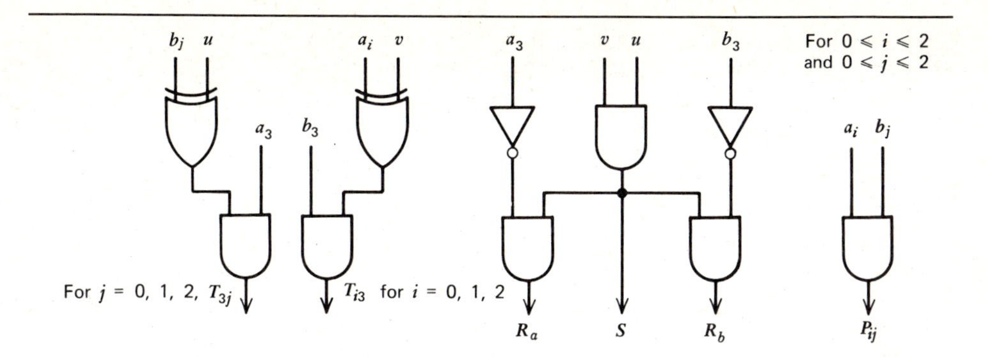

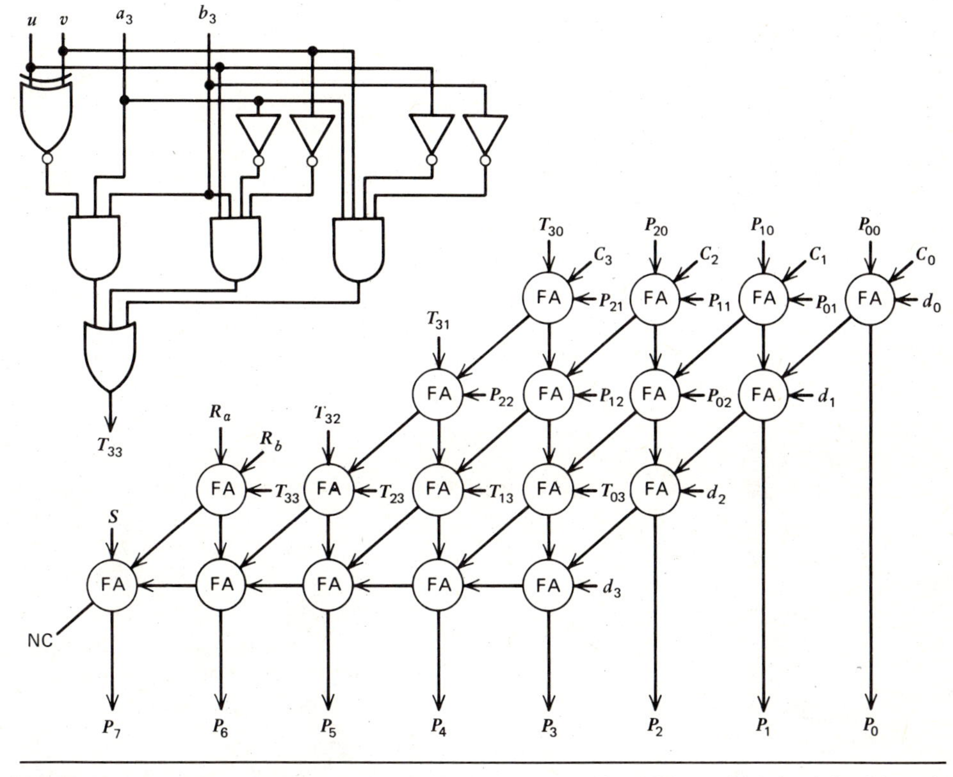

Figure 6.23 The schematic diagram of a **M**(4 × 4) programmable additive multiply module with mode control and excitation logic (Hwang [18]).

Table 6.6 Hardware Complexities and Application Parameters for Programmable Additive Multiply (PAM) Modules of Various Sizes

PAM Module $M(k \times k)$	Major Hardware Components: Number of Full Adders	Major Hardware Components: Number of Excitation Logic Gates	Number of External Leads Per Package	Carry Propagation Delay[a]: PAM Module M_{00}, M_{01}, M_{10}	Carry Propagation Delay[a]: PAM Module M_{11}
$M(2 \times 2)$	6	23	14 + 1	6Δ	8Δ
$M(4 \times 2)$	10	17	20 + 1	10Δ	12Δ
$M(4 \times 4)$	18	39	28 + 1	14Δ	16Δ
$M(8 \times 8)$	26	95	52 + 1	30Δ	32Δ
$M(k \times k)$	$k^2 + 2$	$k^2 + 2k + 15$	$6k + 5$	$(4k - 2)\Delta$	$4k\Delta$

[a] The overhead due to excitation logic is excluded in counting this delay, because the excitation logic is applied to all PAM modules simultaneously. The delay due to excitation logic will be counted only once for an entire network of PAM modules.

6.8 Universal Multiplication Networks

An iterative method is described to construct large *Universal Multiplication Networks* (UMN) using the small PAM modules developed in the preceding section. We shall denote an n-by-n UNM by UMN($n \times n$). Note that UMA($k \times k$) = UMN($k \times k$) when k is small like $k = 4$. Only $M(4 \times 4)$ PAM modules, programmed in four different modes $M_{00}(4 \times 4)$, $M_{01}(4 \times 4)$, $M_{10}(4 \times 4)$, or $M_{11}(4 \times 4)$, are used as the building blocks of UMN($4k \times 4k$). Similar techniques can be applied to designing with other sizes of standard PAM modules.

Two UMN(8×8) are constructed in Figs. 6.24 and 6.25. In each case, four PAM modules $M(4 \times 4)$ are used. The network has been programmed as an unsigned multiplication network using four $M_{00}(4 \times 4)$'s. We shall denote this unsigned UMN simply as $N_{US}(8 \times 8)$. The network, consisting of one each of the $M_{00}(4 \times 4)$, $M_{01}(4 \times 4)$, $M_{10}(4 \times 4)$, and $M_{11}(4 \times 4)$ PAM modules, has been programmed as a two's complement multiplication network, and denoted as $N_{TC}(8 \times 8)$.

This iterative scheme can be extended to designing any networks UMN($4k \times 4k$). In general, a UMN($4k \times 4k$) requires k^2 PAM modules $M(4 \times 4)$. For example, interconnecting 16 PAM modules $M(4 \times 4)$ yields a network $N_{TC}(16 \times 16)$ as shown in Fig. 6.26 in two's complement mode. Note that in these two network diagrams, the convention A_{3-0} for representing binary number $a_3 a_2 a_1 a_0$ is used. Among these 16 modules, we have programmed nine $M_{00}(4 \times 4)$, three $M_{01}(4 \times 4)$, three $M_{10}(4 \times 4)$, and one $M_{11}(4 \times 4)$ as illustrated. Of course, one can construct an unsigned network $N_{US}(16 \times 16)$ using 16 PAM modules of $M_{00}(4 \times 4)$.

The numbers of each mode of PAM modules $M(4 \times 4)$ required in a two's complement UMN $N_{TC}(4k \times 4k)$ are $(k - 1)^2$ $M_{00}(4 \times 4)$, $k - 1$ $M_{01}(4 \times 4)$, $k - 1$ $M_{10}(4 \times 4)$, and one $M_{11}(4 \times 4)$. The homogeneous additive structure, universal

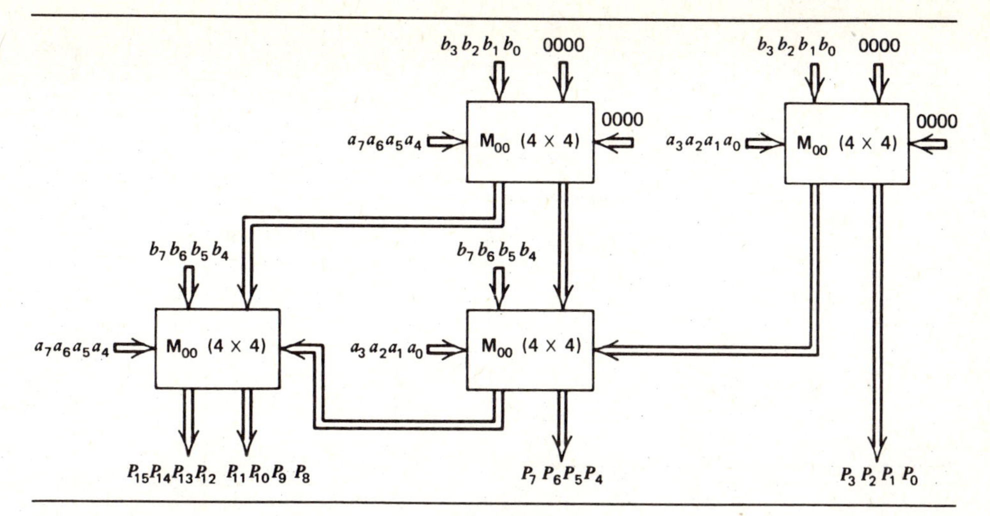

Figure 6.24 An $\mathbf{N}(8 \times 8)$ universal array multiplication network programmed as an unsigned array multiplier $\mathbf{N}_{US}(8 \times 8)$ consisting of four $\mathbf{M}_{00}(4 \times 4)$ **PAM** modules.

programmability, and limited number of required I/0 leads in PAM modules of small and moderate sizes make these programmable multiply modules attractive. Typical UMA($n \times n$) of sizes 16×16 or less can be made on a monolithic circuit with less than 64 pins. Standard PAM modules of size 4-by-4 or less can be mounted on single IC chips of less than 30 pins.

In what follows, we analyze the speed performance of the multiplication arrays and networks. The total multiply time required in a bipolar UMA($n \times n$) is obtained

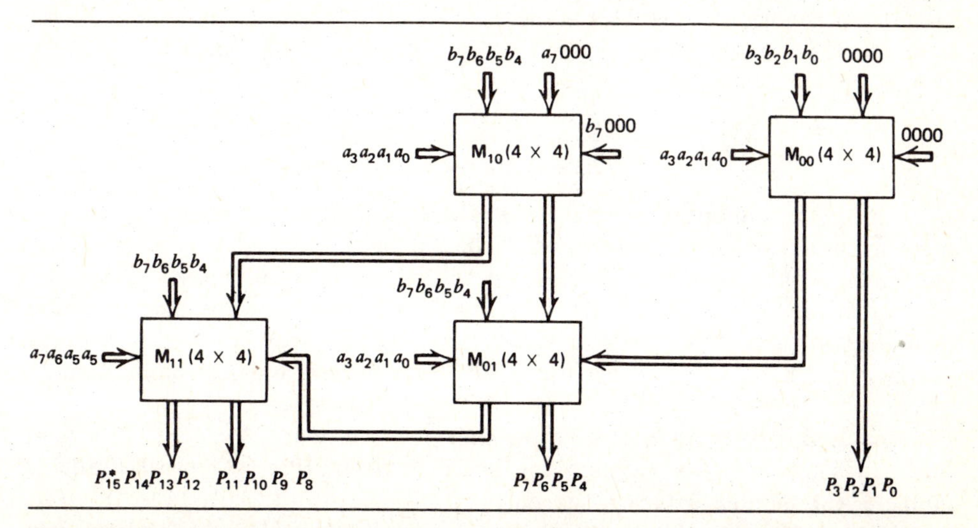

Figure 6.25 The same network $\mathbf{N}(8 \times 8)$ in Figure 6.10 reprogrammed as an $\mathbf{N}_{TC}(8 \times 8)$, a two's complement array multiplier.

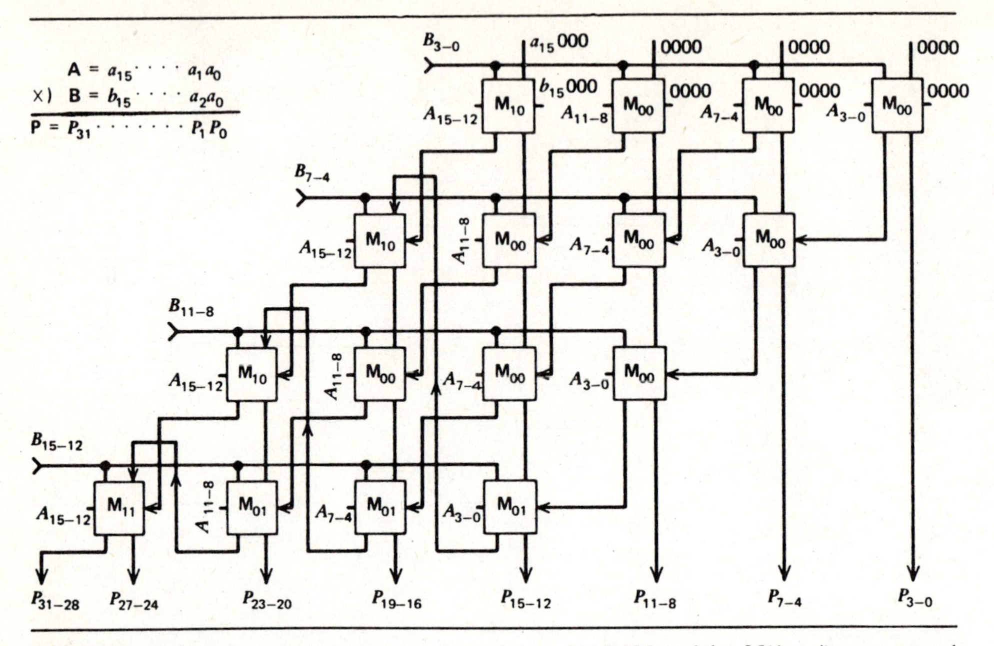

Figure 6.26 A 16-by-16 multiplication network consisting of 16 **PAM** modules, **M**(4 × 4), programmed in two's complement mode (Hwang [18]).

by examining the circuit shown in Fig. 6.20, 7Δ time delays for each of the two complement stages and 6Δ additional delays in the input excitation logic. Once entering the full-adder array, the longest carry propagation path (along the rightmost diagonal and the bottom row of full adders) give a delay of $[(n - 2) + (n + 1)]2\Delta = (4n - 2)\Delta$ units. Furthermore, if we use CLA in the last row, say 12Δ constant delay for a 2-level CLA adder, a reduced array delay of $(n - 2)2\Delta + 12\Delta = (2n + 8)\Delta$ units is yielded. To sum up, the total multiply time required in UMA$(n \times n)$ is

$$\Delta_{\mathrm{UMA}}(n \times n) = \begin{cases} (4n + 14)\Delta, & \text{without CLA in last row} \\ (2n + 22)\Delta, & \text{with CLA in last row} \end{cases} \tag{6.36}$$

Among the four operating modes of a UMA, the unsigned and the two's complement operations are more frequently used. The complementing logic can be completely eliminated if one's complement operation will not occur, resulting in a faster and cheaper circuit with only one mode control line to distinguish the unsigned operation from the two's complement operation. The total multiply time has been reduced by 12Δ less from those in Eq. 6.36 due to the removal of the pre- and postcomplementing circuits.

Let $\Delta_{rs}(4 \times 4)$ be the carry propagation delay in a PAM module $M(4 \times 4)$ operated in mode $M_{rs}(4 \times 4)$. Analyzing the circuit given in Fig. 6.23 reveals the delay in the PAM module $M(4 \times 4)$ as $\Delta_{00}(4 \times 4) = \Delta_{01}(4 \times 4) = \Delta_{10}(4 \times 4) = 14\Delta$ and $\Delta_{11}(4 \times 4) = 16\Delta$. Using these measures, we estimate below the total multiply time of any UMN$(4k \times 4k)$ made from the PAM modules $M(4 \times 4)$. Let $\mathbf{\Delta}_{\mathrm{TC}}(4k \times 4k)$

and $\Delta_{US}(4k \times 4k)$ be the delays in networks $N_{TC}(4k \times 4k)$ and $N_{US}(4k \times 4k)$, respectively. We have

$$\Delta_{US}(4k \times 4k) = (2k - 1) \times 14\Delta + 7\Delta = (28k - 7)\Delta \quad \textbf{(6.37)}$$

$$\Delta_{TC}(4k \times 4k) = \Delta_{US} + 2\Delta = (28k - 5)\Delta \quad \textbf{(6.38)}$$

where 7Δ accounts for the delay in the FA excitation logic and 2Δ due to the extra delay in the $M_{11}(4 \times 4)$ module. At present, one can apply the high-speed TTL or the current-mode ECL to construct the above array multipliers, PAM modules, or multiplication networks. With these two logic families, the unit delay Δ for one-level logic could be as low as a few subnanoseconds. For example, Pezaris' multiplier used an ECL circuit yielding a two-level delay of $2\Delta = 1.6$ nsec, and thus the unit delay $\Delta = 1.6$ nsec/2 = 0.8 nsec. With this unit delay, a universal array multiplier of size 16-by-16 with CLA in the last row can complete the multiplication in

$$\Delta_{UMA}(16 \times 16) = (2 \times 16 + 25)\Delta = 45.6 \text{ nsec.}$$

With little sacrifice in speed, we can use a universal multiplication network $N(16 \times 16)$ like the one shown in Fig. 6.26 to yield a 32-bit product in

$$\Delta_{TC}(16 \times 16) = (28 \times 4 - 5) \times 0.8 = 85.6 \text{ nsec}$$

or in $\Delta_{US}(16 \times 16) = (28 \times 4 - 7) \times 0.8 = 84$ nsec. As summarized in Table 6.7, the global array UMA approach offers faster multiply time than that of the modular networks. However, the modular approach is more appealing to practical implementation within the scope of current electronics and packaging technologies. Tradeoffs can alwys be made between the operating speed and the building-block module sizes. Optimal choice of the module size is affected by the logic family, by the fabrication method, and by the degree of universality and modularity desired in the system.

Table 6.7 Major Components and Operating Speed of UMA($n \times n$) and UMN($n \times n$) for Typical Array and Network Sizes

	UMA($n \times n$)		UMN($n \times n$)[b]	
Size $n \times n$	No. of Full Adders Required	Multiply Time[a] $\Delta_{UMA}(n \times n)$	No. of PAMs $M(4 \times 4)$, Required	Multiply Time $\Delta_{TC}(n \times n)$
8×8	53	47Δ	4	51Δ
16×16	237	79Δ	16	107Δ
32×32	989	143Δ	64	219Δ
$n \times n$	$n^2 - n - 3$	$(4n + 15)\Delta$	$\left(\frac{n}{4}\right)^2$	$(7n - 5)\Delta$

[a] Without CLA in the last row.
[b] $n = 4k$ for **UMN**($n \times n$).

6.9 Recoded Array Multiplication

A recoded two's complement array multiplier based on Booth's algorithm is presented in this section. Practical schemes for large array multiplications using ROM's, adders, and logarithmic tables are to be discussed in the next two sections. Majithia and Kita [19] have proposed an iterative logic array for directly multiplying two's complement numbers. The original presentation was based on fractional two's complement notation. With the introduction of a scaling factor, the array can be extended to process any two's complement numbers. We shall illustrate these array operations using the fractional notation.

Let $\mathbf{Y} = Y_0.Y_{-1}Y_{-2}\cdots Y_{(n-1)}$ be a fractional multiplier in two's complement notation. According to Eq. 6.14, $\mathbf{Y}$ has a value Y_v

$$Y_v = -Y_0 + \sum_{k=1}^{n-1} Y_{-k} \times 2^{-k} \tag{6.39}$$

and the resulting product $\mathbf{Z} = Z_0.Z_{-1}\cdots Z_{-2(n-1)}$ can be similarly evaluated. Note that the product $\mathbf{Z}$ has $2n - 1$ bits, if the two operands are n bits each including the sign.

The fractional multiplication algorithm is illustrated by the flow chart in Fig. 6.27. The successive bit scanning operations start with the most significant bit Y_0 down to the lower bits. The shifting of the current sum of partial products $S^{(k)}$ one place to the right is equivalent to shifting the multiplicand X one place to the right before the next ADD or SUBTRACT operation. The same end result is obtained by shifting the multiplicand, with the advantage of speeding up the multiplication process if the stored shifted technique is employed. It should be noted that when shifting a two's complement number to the right, an extended sign bit will be shifted into the vacated most significant position; that is, shifting in a "1", if the m.s.b. was a "1" and shifting in a "0" if otherwise.

The design of a *controlled add-subtract-shift* (CASS) cell with five inputs and five outputs is illustrated in Fig. 6.28. The logic equations characterizing this CASS cell are listed below

$$\begin{aligned} S &= A \oplus BP \oplus CP \\ T &= (A \oplus D)(B + C) + BC \\ U &= B;\ Q = P;\ R = D \end{aligned} \tag{6.40}$$

input lines A and B carry the two operand bits. P and D are the two mode control lines governing the same operation for all cells in the same row. When $P = 0$, the cell performs no arithmetic operation except to physically shift the extended multiplicand to the right one place by passing B to U. When $P = 1$ and $D = 0$, the cell acts as an adder. When $P = 1$ and $D = 1$, the cells act as a subtractor.

An iterative array for the direct multiplication of two 3-bit two's complement fractional numbers is shown in Fig. 6.29. In general, if both $\mathbf{X}$ and $\mathbf{Y}$ are n-bit numbers

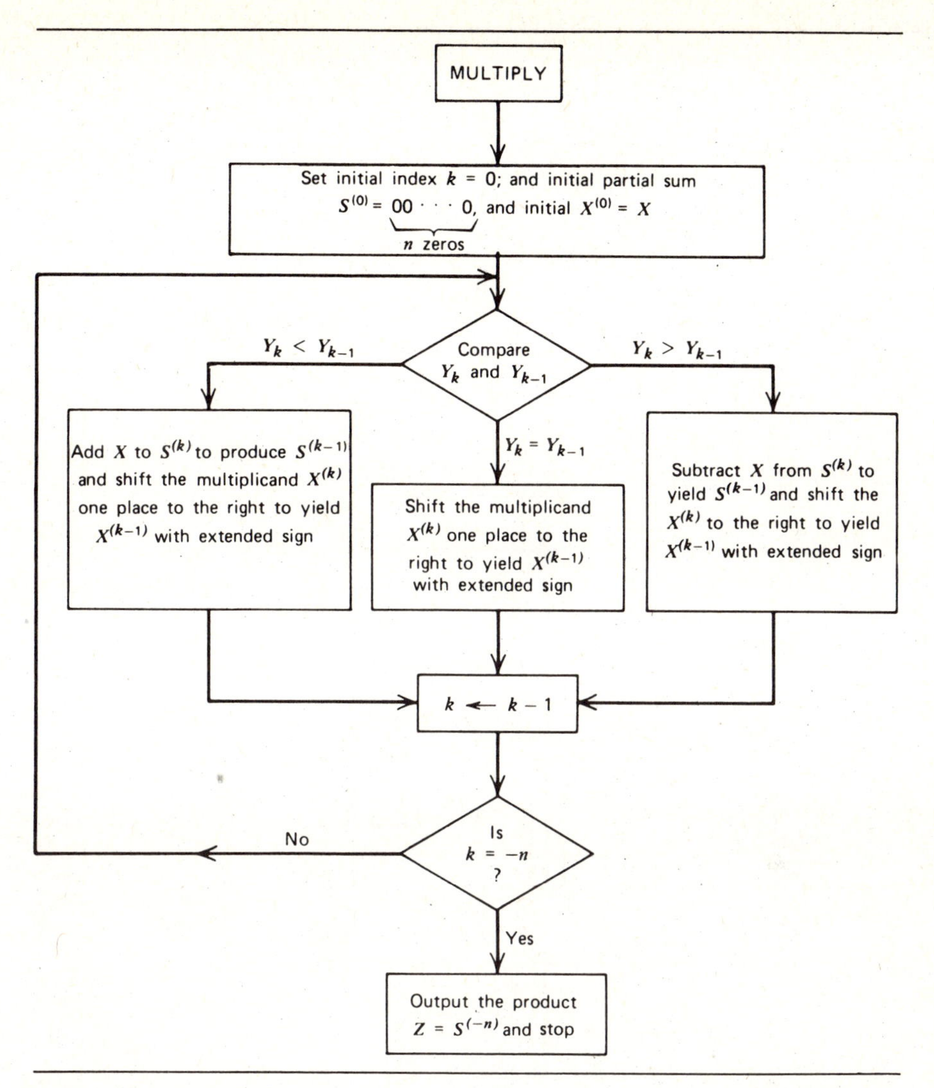

Figure 6.27 The Booth's recorded multiplication algorithm for two's complement array multiplication.

including the sign bit, the array requires $(3n^2 - 2)$ CASS cells and n 1-bit magnitude comparators. The design is verified below by multiplying $\mathbf{X} = 0.01$ by $\mathbf{Y} = -0.11$ in two's complement notation, $\mathbf{X} = X_0 X_{-1} X_{-2} = 001$ and $\mathbf{Y} = Y_0 Y_{-1} Y_2 = 101$ for the case of $n = 3$. The detailed intermediate steps are described below.

Step 1. Initialize $k = 0$, $S^{(0)} = 000$, and $Y^{(0)} = 001$.

Step 2. Compare $Y_0 = 1$, $Y_{-1} = 0$, subtract $\mathbf{X}^{(0)} = 001$ from $S^{(0)} = 000$ to give $S^{(-1)} = 1110$, and shift $X^{(0)}$ one place to give $X^{(-1)} = 0001$.

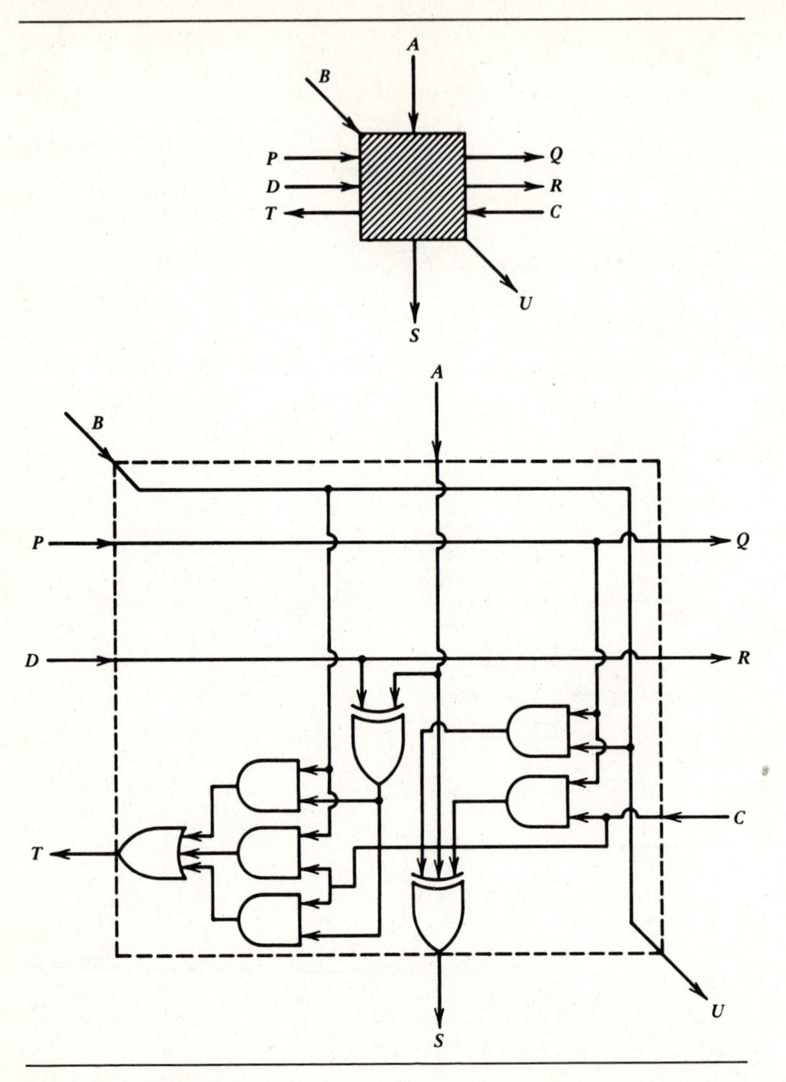

Figure 6.28 A schematic logic circuit diagram of the *Controlled Add-Subtract-Shift* (**CASS**) cell.

Step 3. Compare $Y_{-1} = 0$ and $Y_3 = 1$, add $X^{(-1)}$ to $S^{(-1)}$ to give $S^{(-2)} = 1111$, and shift $X^{(-1)}$ to give $X^{(-2)} = 00001$.

Step 4. Compare $Y_{-2} = 1$ and $Y_{-3} = 0$, subtract $X^{(-2)}$ from $S^{(-2)}$ to give $S^{(-3)} = 11101$, which is the desired product **P**, and then stop. The answer is **P** $= 1.1101$ which is equal to -0.0011 as expected.

The comparators have a delay of $\Delta_c = 3\Delta$ units each. This means the operation mode of each row is settled after Δ_c. The CASS cell has a worst-case delay of $\Delta_t = 8\Delta$

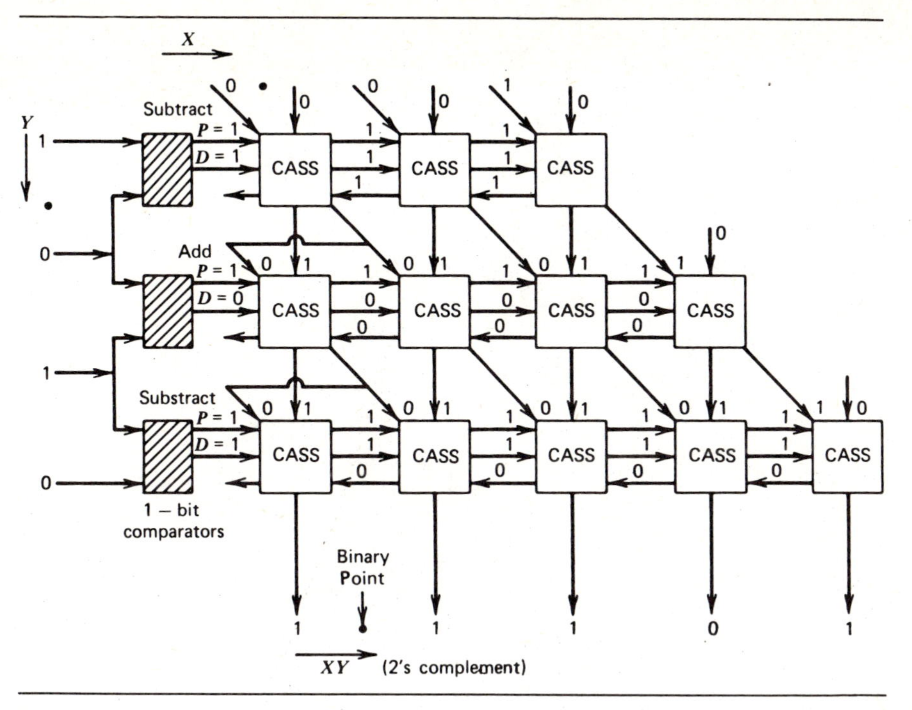

Figure 6.29 The interative logic array for recoded multiplication of two 3-bit two's complement fractions (Majithia and Kita [19]).

units to produce the output S. The first bit of the product appears after a time of $\Delta_c + \Delta_t$, the next bit after a delay of $\Delta_c + 2\Delta_t$, the third bit after a delay of $\Delta_c + 3\Delta_t$, and so on. Because there are $2n - 1$ diagonals of cells, the total delay in obtaining the complete product will be

$$\Delta_T = \Delta_c + (2n - 1)\Delta_t = 3\Delta + (2n - 1)8\Delta = (16n - 5)\Delta \qquad \textbf{(6.41)}$$

which is about four times slower than the speed of previous array multipliers, but still much faster than the conventional serial-parallel multipliers.

6.10 ROM-Adder Multiplication Networks

The availability of inexpensive and widely used ROM's has made the table lookup approach a viable means for producing the product of any two binary numbers. The multiplication table stored in a ROM contains results of all possible combinations of the input operands. The primary difficulty is the number of bits required in the ROM.

For direct multiplication, the combined m-bit multiplicand and n-bit multiplier define a unique address in the memory. The contents of the addressed word, out of the 2^{m+n} possible words, form the $(m + n)$-bit product, if no rounding is allowed.

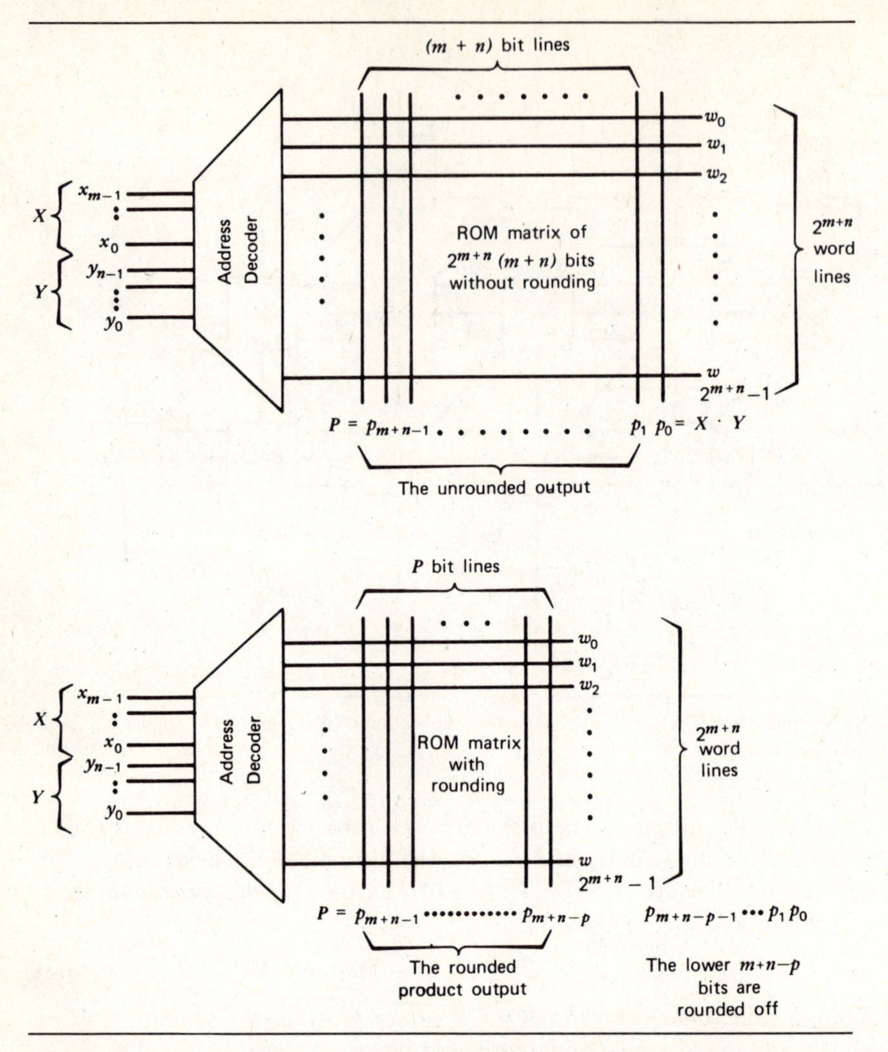

Figure 6.30 A ROM-implemented m-by-n multiplier with and without output rounding.

As illustrated in Fig. 6.30, the total bit capacity required in the ROM is

$$N = 2^{m+n} \times (m + n) \tag{6.42}$$

Even for moderate values of m and n, say $m = n = 8$, one may need a ROM with a capacity of $2^{8+8} \times 16 = 2^{20} = 1{,}048{,}576$ bits, exceeding one million bits to store the whole multiplication table. Therefore, in many practical applications, the product is rounded to p bits for some $p < m + n$ with the understanding that a scaling factor 2^{m+n+p} has to be imposed to the least significant bit of the rounded product. The bit capacity needed can thus be reduced to

$$N_r = 2^{m+n} \times p \tag{6.43}$$

This rounding may introduce an error ε which is bounded by

$$\frac{-q}{2} < \varepsilon < \frac{q}{2} \tag{6.44}$$

where q represents the value of the least significant bit of the rounded product. The bit capacities needed in a ROM for various values of q, p and $m = n = 4, 8, 10$ bits are enumerated in Table 6.8. In the case of $m = n = p$, the least significant n bits are

Table 6.8 Word Length Versus Bit-Capacity for Direct Multiplication Using **ROMs**

Input Range $m = n$	Word Length p	Value of the l.s.b. in the Rounded Product q	The Reduced Bit Capacity $N_r = 2^{2n} \times p$
4	8	0	2,048
4	4	2^4	1,024
8	16	0	1,048,576
8	14	2^2	917,504
8	12	2^4	786,432
8	8	2^8	524,288
10	20	0	20,971,520
10	16	2^4	16,777,216
10	10	2^{10}	10,485,759

rounded. The maximal error as a percentage of the range of the input quantities is given by

$$\varepsilon_{\max} = \frac{0.5}{2^n - 1} \times 100 \tag{6.45}$$

A plot of $\varepsilon_{\max}$ is given in Fig. 6.31 as a function of the input range n.

Presently, single-chip ROM modules up to 16K bits in capacity are available as standard off-the-shelf products from the electronic memory industry. For example, a standard ROM of capacity $256 \times 8 = 2048$ bits is manufactured by several companies, including American Microsystems, Electronic Arrays, Inc. and National Semiconductor. Such a 2048-bit ROM can be used to store the whole 4-by-4 multiplication table without rounding. Large tables can be built from a number of smaller memory modules. In order to use the smaller memories, one has to partition the given table into disjoint sections. Combining the three techniques of *table lookup*, ROMs, and *arithmetic function partition* produces a more practical way of implementing large multiplication networks with small ROMs and cheap adders. As indicated in the preceding section, an 8×8 multiplication table without rounding may require a ROM of over a million bits in capacity. Therefore, partitioning the whole table into

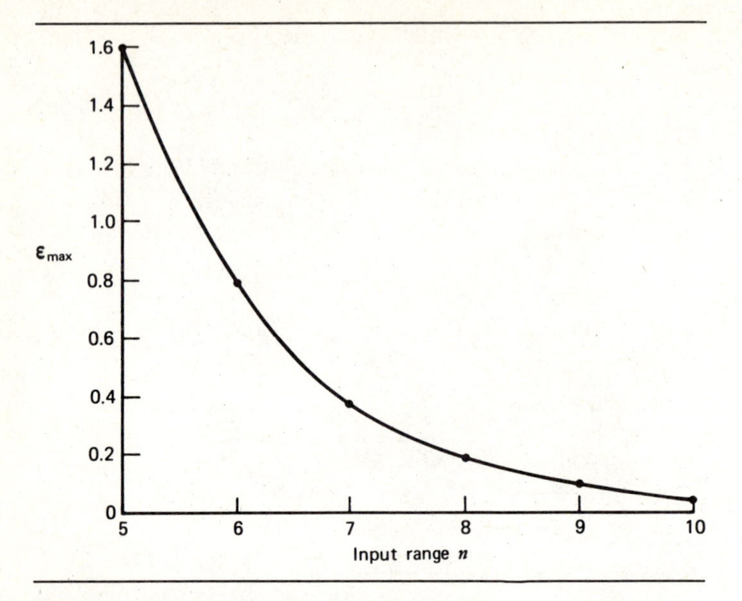

Figure 6.31 Maximal error curve for rounded products as a percentage of the input range *n*.

several smaller subtables is highly desirable from a practical implementation viewpoint. One can partition the given number $A_8 = a_7 a_6 a_5 a_4 a_3 a_2 a_1 a_0$ as the sum of two numbers: one number is simply the upper half of the number A_8 followed by four trailing zeros, and the other is formed by the lower half of the number A_8 with four leading zeros as shown below

$$A_8 = A_4^U + A_4^L \tag{6.46}$$

where $A_4^U = a_7 a_6 a_5 a_4 0000$ and $A_4^L = 0000 a_3 a_2 a_1 a_0$. $B_8 = B_4^U + B_4^L$ can be similarly defined. With these partitioned numbers, the original multiplication can be restructured as the sum of four 4-by-4 multiplications as follows:

$$A_8 \times B_8 = [A_4^U + A_4^L] \times [B_4^U + B_4^L] = A_4^U B_4^U + A_4^U B_4^L + A_4^L B_4^U + A_4^L B_4^L \tag{6.47}$$

These four products and their sum can be implemented with four 2048-bit **ROM**s and five 4-bit ripple-carry adders as sketched in Fig. 6.32. These **ROM**s and adders are available as standard off-the-shelf items such as the National Semiconductor MM523 256-by-8-bit **ROM** and the Texas Instruments SN 7483 4-bit full adders. The leading and the trailing zeros are not connected in the circuit. These zeros determine the columnwise relationships among all the sectioned product terms. External pull up resistors, not shown in the circuit, are required to make the output signals of the Metal-Oxide Semiconductor (MOS) memories compatible with input requirement of the TTL adders. Also, power supplies of ± 12 volts are required for the MOS ROMs, whereas only 5-volt power is required for the TTL devices. Even with all of these increased compatibility and power supply requirements, the ROM-adder implementation is still considered inexpensive and easy to construct.

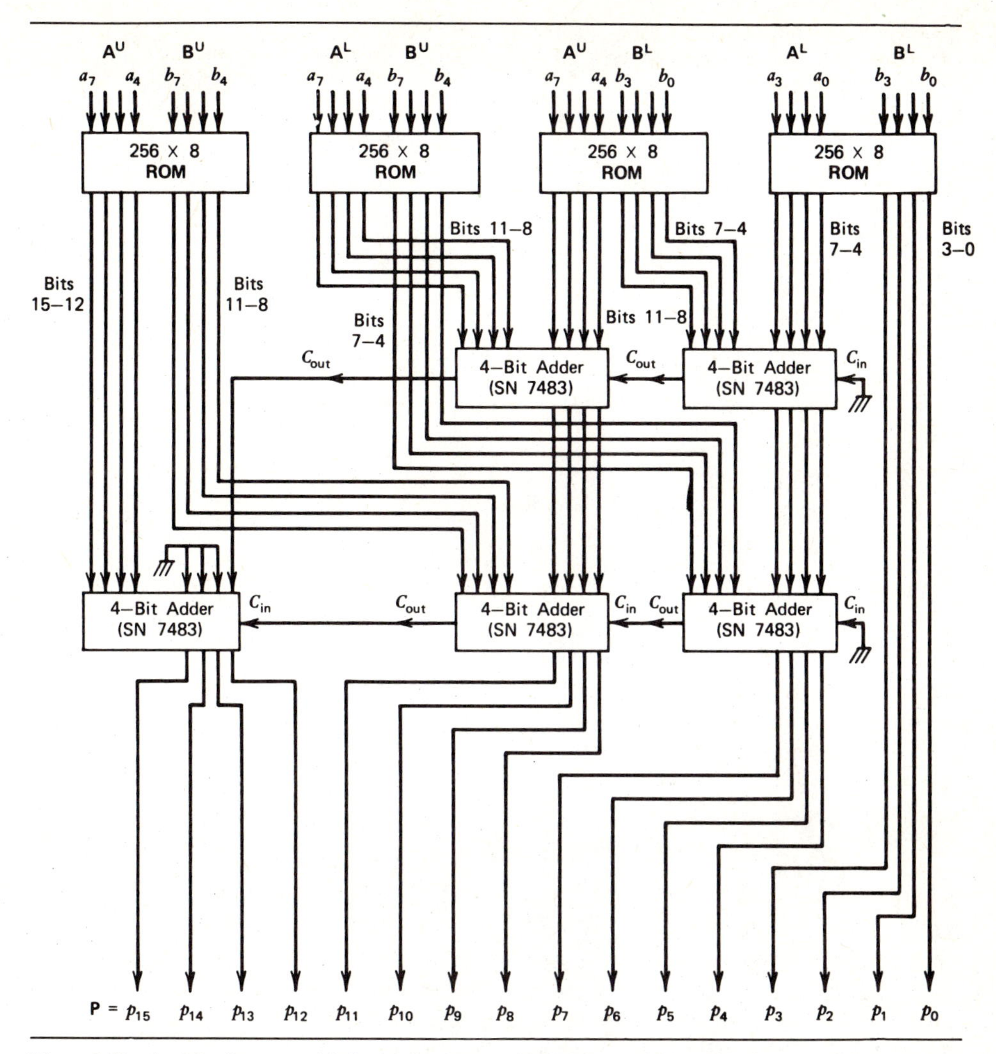

Figure 6.32 An 8-by-8 array multiplier implemented with 256-byte **ROM**s and 4-bit adders (Hemel [16]).

Table 6.9 provides a numerical account of the numbers of ROMs and adders required in multiplication networks with sizes equal to multiples of 4-bit slices. In general, a $4k$-by-$4k$ ROM-adder multiplication network, constructed similar to that given in Fig. 6.32, requires k^2 256-byte ROM modules and $k(3k - 1)/2$ 4-bit adders. The total delay in such a **ROM-adder** multiplication network will be

$$\Delta_{RA} = \Delta_r + (2k - 1)\Delta_a \tag{6.48}$$

where Δ_r is memory access time of the ROM and Δ_a is the total carry delay in the four-bit adder. According to present memory and adder technology, the speed of the 8-by-8 **ROM-adder** multiplication network appears less than 1 microsecond, a figure that is acceptable to many minicomputers and microprocessors.

Table 6.9 Hardware Requirements of Multiplication Networks Made from 4-bit Adders and 256-byte ROM's

Parameter k	Network Size $4k \times 4k$	Required No. of 256-Byte ROM's	Required No. of 4-Bit Adders
1	4×4	1	0
2	8×8	4	5
3	12×12	9	12
4	16×16	16	22
6	24×24	36	51
8	32×32	64	100

6.11 Logarithmic Multiplication/Division Schemes

Another method to reduce the required bit capacity of a **ROM** multiplication table is to convert the multiplication/division into addition/subtraction by means of logarithm and antilogarithm transformations. Instead of storing the whole table, one need store only the logarithm, antilogarithm, and addition tables, which usually require much smaller ROMs. Multiplication/division using logarithms is based on the following relationship

$$A \times B = \mathbf{antilog}(\mathbf{log}\ A + \mathbf{log}\ B) \qquad \textbf{(6.49)}$$

$$A/B = \mathbf{antilog}(\mathbf{log}\ A - \mathbf{log}\ B) \qquad \textbf{(6.50)}$$

where A and B are binary numbers and logarithms to the base 2 are used. Note that only the absolute magnitudes of A and B are considered and the correct sign of the resulting product is assumed to be generated externally.

The complete model for a logarithmic table lookup multiplier/divider, showing all the error sources, is depicted in Fig. 6.33. The subscripts within parentheses in expressions $X_{(n)}$, $Y_{(n)}$, $\mathbf{log}_{(m)}$, and $\mathbf{antilog}_{(n)}$ refer to the word lengths of the two input operands of the rounded logarithms and rounded antilogarithm, respectively. The errors, e_1, e_2, and e_3 come from rounded logarithms and rounded antilogarithms when implementing with ROMs. The addition/subtraction could be done by either a high-speed hardwired parallel adder or via table lookup. The maximal errors as a percentage of half range are plotted in Fig. 6.34 as a function of word length m of the rounded logarithm for a running parameter n, the number of product bits.

The entries in Table 6.10 indicate the number of bits required for various combinations of product and logarithm word lengths. The optimal size of ROMs can be determined by checking Fig. 6.34 and Table 6.10. For example, with an error of 0.5 percent, there are a variety of combinations of m and n that will work. The combination $m = n = 8$ bits gives a bit requirement of only 6144 bits, as compared with the 524,288 bits using an 8-by-8 direct multiplication table with half-range rounding, as

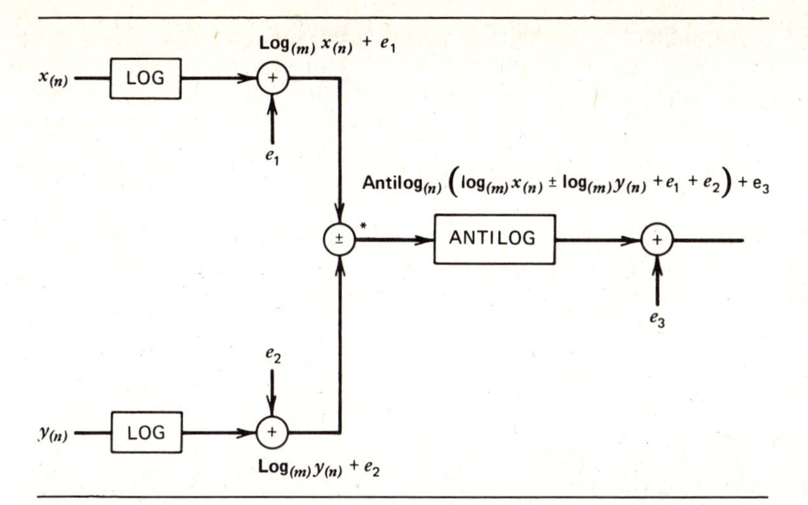

Figure 6.33 Model for logarithmic multiplication showing all possible error sources. (*Addition for multiply and subtraction for divide.)

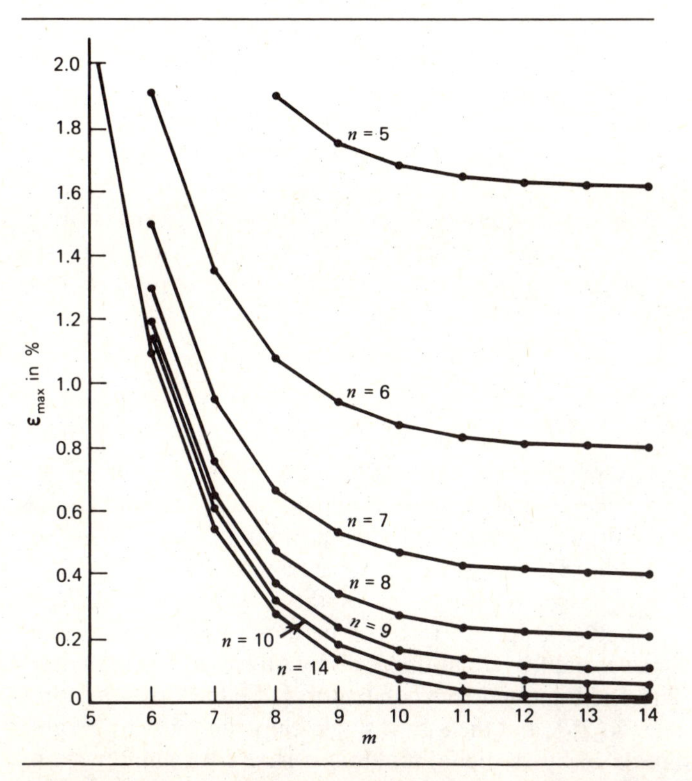

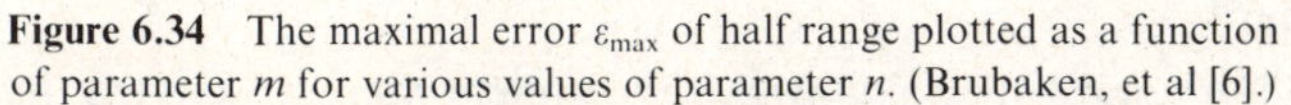

Figure 6.34 The maximal error ε_{max} of half range plotted as a function of parameter m for various values of parameter n. (Brubaken, et al [6].)

shown in Table 6.8, not to mention the over one million bits for direct multiplication without rounding. This significant memory reduction factor 524,288 ÷ 6144 = 85.3 is very attractive, especially for those applications in which minor errors can be tolerated.

Table 6.10 Total Number of ROM Bits Required for Logarithmic Multiplication as a Function of Rounded Logarithmic and Product Word Lengths

Number of Log Bits (m)	Number of Product Bits (n)								
	6	7	8	9	10	11	12	13	14
5	832	1,504							
6	1,152	1,984	3,584						
7	1,664	2,688	4,608	8,320					
8	2,560	3,840	6,144	10,496	18,944				
9	4,224	5,888	8,704	13,824	23,552	42,496			
10		9,728	13,312	19,456	30,720	52,224	94,208		
11			22,016	29,696	43,008	67,584	114,688	206,848	
12				49,152	65,536	94,208	147,456	249,856	450,560
13					108,544	143,360	204,800	319,488	540,672
14						237,568	211,296	442,368	688,128

A direct multiplication/division using ROM requires one memory access, whereas logarithmic multiplication requires two memory access times plus an addition time. Thus, the speed for logarithmic multiplication is about one-third the speed of direct multiplication via table lookup. In either case, the speed is governed by the memory technology. Current speed of logarithmic multiplication/division appears to be a few 100 nanoseconds.

6.12 Bibliographic Notes

The subject material of this chapter is presented as a broad-scope account of various iterative array and network schemes for high-speed parallel multiplication. Earlier treatments of integer array multiplication can be found in Braun [5] and in Chu [7]. Later, Wallace [28] and Dadda [8], among many other researchers [9, 14, 15, 22], extended the original Braun design. Descriptions of the existing basic multiply building modules and their network applications can be found in many electronic manufacturers' manuals, such as the 4-by-2 multiply cells by Advanced Micro Devices [1], Texas Instruments [26], and Fairchild Semiconductors [11]; the 4-by-4 multiply modules by Texas Instruments [26]; and the 8-by-8 single-chip multiplier by Hughes Aircraft [17]. LSI monolothic 16-by-16 digital multipliers have been announced by TRW [27] and McIver [20].

Indirect and direct sign-complemented array multipliers were discussed by

Deegan [10], Gibson and Gibbard [13], Mowle [23], and Toma [27]. A 17-by-17 array multiplier made with mixed types of full adders at the Lincoln Laboratory at MIT was reported by Pezaris [25] for direct multiplication of two's complement numbers. Baugh and Wooley [3] proposed a uniform direct two's complement multiplication algorithm. Possible modifications of Pezaris' original design, gated-full-adder implementation of the Baugh-Wooley algorithm, and the programmable universal parallel multiplication networks were introduced by Hwang [18]. The structure of a tenary array multiplier by Majithia and Kita [19] was based on Booth's algorithm [4]. The **ROM-adder** economical network implementation of parallel multipliers was selected from the work by Hemel [16]. Finally, logarithmic multiplication through the table-lookup method and discussions on implementing the scheme with ROMs are based on the papers by Mitchell [21] and by Brubaker and Becker [6]. More recent work on array multipliers can be found in Agrawal [2].

References

[1] Advanced Micro Dvices, "TTL/MSI AM2505 4-bit by 2-bit 2's Complement Multiplier," 901 Tompson Pace, Sunnyvale, CA.

[2] Agrawal, D. P., "Optimum Array-Like Structures for High-Speed Arithmetic," *Proc. of 3rd Symposium on Computer Arithmetic*, IEEE Computer Society, #75C1017-3C, Nov. 1975, pp. 208–219.

[3] Baugh, C. R. and Wooley, B. A., "A Two's Complement Parallel Array Multiplication Algorithm," *IEEE Trans. Computers*, Vol. C-22, No. 1–2, December 1973, pp. 1045–1047.

[4] Booth, A. D., "A Signed Binary Multiplication Technique," *Quart. J. Mech. Appl. Math.*, Vol. 4, Pt. 2, 1951, pp. 236–240.

[5] Braun, E. L., *Digital Computer Design*, Academic Press, New York, 1963.

[6] Brubaker, T. A. and Becker, J. C., "Multiplication Using Logarithms Implemented with Read-Only Memory," *IEEE Trans. Computers.*, Vol. C-24, 1975.

[7] Chu, Y., *Digital Computer Design Fundamentals*, McGraw-Hill, New York, 1962.

[8] Dadda, L., "Some Schemes for Parallel Multipliers," *Alta Frequenza*, Vol. 34, March 1965, pp. 349–356.

[9] Dean, K. J., "Design of a Full Multiplier," *Proc. IEEE*, Vol. 115, Nov. 1968, pp. 1592–1594.

[10] Deegan, I. D., "Cellular Multiplier for Signed Binary Numbers," *Electronic Letters*, Vol. 7, 1971, pp. 436–437.

[11] Fairchild Semiconductors, "TTL/MSI 9344 Binary (4-bit by 2-bit) Full Multiplier," 313 Fairchild Dr.: Mountain View, CA., 1971.

[12] Flores, I., *The Logic of Computer Arithmetic*, Prentice Hall, Englewood Cliffs, N.J., 1963.

[13] Gibson, J. A. and Gibbard, R. W., "Synthesis and Comparison of Two's Complement Parallel Multipliers," *IEEE Trans. Computers*, Vol. C-24, Oct. 1975, pp. 1020–1027.

[14] Guild, H. H., "Fully Iterative Fast Array for Binary Multiplication and Fast Addition," *Electronic Letters*, Vol. 5, May 1969, p. 263.

[15] Habibi, A. and Wintz, P. A., "Fast Multipliers," *IEEE Trans. Computers*, Vol. C-19, Feb. 1970, pp. 153–157.

[16] Hemel, A., "Making Small ROMs Do Math Quickly, Cheaply and Easily," *Electronic Computer Memory Technology*, W. B. Riley, (ed.), McGraw-Hill, New York, 1971, pp. 133–140.

[17] Hughes Aircraft Co., "Bipolar LSI 8-bit Multiplier H1002MC," 500 Superior Avenue, Newport Beach, CA., 1972.

[18] Hwang, K., "Global Versus Modular Two's Complement Array Multipliers," *IEEE Trans. Computers*, Vol. C-28, No. 4, April 1979, pp. 300–306.

[19] Majithia, J. C. and Kita, R., "An Iterative Array for Multiplication of Signed Binary Number," *IEEE Trans. Computers.*, Vol. C-20, Feb. 1971, pp. 214–216.

[20] McIver, G. W. et al., "A Monolithic 16 × 16 Digital Multiplier," *Dig. Tech. Paper Int. Solid State Circuits Conf.*, Feb. 1974, pp. 54–55.

[21] Mitchell, J. N.,"Computer Multiplication and Division Using Binary Logarithms," *IRE Trans. Elec. Computers*, Vol. EC-11, Aug. 1962, pp. 512–517.

[22] Mori, R. D., "Suggestion for an IC Fast Parallel Multiplier," *Electronic Letters*, Vol. 5, Feb. 1969, pp. 50–51.

[23] Mowle, F. J., *A Systematic Approach to Digital Logic Design*, Addison Wesley, Reading, Mass., 1976.

[24] National Semiconductor Corp. *Data Sheets* MM 521, MM 522, MM 523, and DM 8200.

[25] Pezaris, S. D., "A 40^{ns} 17-bit-by-17bit Array Multiplier," *IEEE Trans. Computers*, Vol. C-20, No. 4, April 1971, pp. 442–447.

[26] Texas Instruments, Inc., *TTL Databook and Supplement to TTL Databook*, T. I., Dallas, Texas 1974, pp. 496–498, S262–S270.

[27] TRW, "MPY-LSI Multipliers: AJ 8 × 8, 12 × 12 and 16 × 16," LSI Products, TRW, Redondo Beach, Calif., March 1977.

[28] Wallace, C. S., "A Suggestion for Fast Multipliers," *IEEE Trans. Electronic Computers,* Vol. EC-13, Feb. 1964, pp. 14–17.

Problems

Prob. 6.1 Show the schematic block diagram of an unsigned 16-by-16 Braun array multiplier, with two-level carry lookahead in the last row of the array. Determine the speed of the multiplier under the assumption that $\Delta = 5$ nsec.

Prob. 6.2 Build bit-slice Wallace trees of the following sizes using the carry-save full adders.

(a) A 9-bit-slice Wallace tree
(b) A 15-bit-slice Wallace tree

Prob. 6.3 Construct a 12-by-12 array multiplier with nine 4-by-4 Nonadditive Multiply Modules (NMMs), Wallace trees of up-to-5 slice inputs and two carry-propagate adders. Show the schematic block diagram of you design in a form similar to that in Fig. 6.6. Specify all the data-line widths as well as the inputs and outputs by their digit-position weights.

Prob. 6.4 Show the schematic design of a 32-by-32 multiplication network using sixteen 8-by-8 Additive Multiply Modules (AMMs). Estimate the total multiply time of your design, assuming that each AMM contributes 40 nsec delay.

Prob. 6.5 Prove that the negatively signed two's complement notation as defined in Eq. 6.14 is indeed a valid one such that $N_v + (-N_v) = 0$, where $-N_v$ is the value of $-\mathbf{N}$, the two's complement of a number $\mathbf{N}$.

Prob. 6.6 Construct an 8-by-8 Pezaris array multiplier with Type 0, Type 1, Type 3, and Type 4 full adders and verify your design by plugging in the following pair of two's complement numbers:

$$A = 10110111$$

$$B = 01110110$$

Show the input/output values of all cells in the array as well as the final product in two's complement form.

Prob. 6.7 Repeat Problem 6.6 for an 8-by-8 Tri-section array multiplier using only Type 0, Type 1, and Type 2 full adder cells.

Prob. 6.8 Repeat Problem 6.6 for an 8-by-8 bi-section array multiplier using only Type 0, Type 1, and Type 2 full adder cells.

Prob. 6.9 Repeat Problem 6.6 for an 8-by-8 Baugh-Wooley array multiplier using only Type 0 full adders. Show the schematic design including all excitation logic, and verify the design with the two given operands A and B for all intermediate stages in the array.

Prob. 6.10 Given 16-by-16 Programmable Additive Multiply (PAM), Modules $M(16 \times 16)$. Construct a 64-by-64 univerisal multiplication network with 16 such PAMs. Determine the multiply times $\Delta_{US}(64 \times 64)$ and $\Delta_{TC}(64 \times 64)$ of two operation modes of the resulting network.

Prob. 6.11 Construct a 12-by-8 array multiplier implemented with 256-byte ROMs and 4-bit full adders (SN 7483). Estimate the speed of your design under the assumption that $\Delta_{ROM} = 300$ nsec and $\Delta_{adder} = 20$ nsec. The schematic diagram of your **ROM-adder** network must be shown.

Chapter

7

Standard and High-Radix Dividers

7.1 Introduction

Although division is the inverse process of multiplication, it differs from multiplication in many aspects. First, division is a shift-and-subtract-divisor operation in contrast to multiplication, which is a shift-and-add-multiplicand operation. The results of one subtraction (comparison) determine the next operation in a division sequence. Therefore, division has an inherent serial dependency among the subsequent operation cycles. This problem does not occur in multiplication, because all summands are generated simultaneously. Second, division is not a deterministic process. Instead, it is a trial-and-error process. The successive quotient digits are selected from a digit set via a digit discrimination procedure.

Machine division process generally consists of three parts: *operand initialization*, *quotient generation*, and *remainder determination*. The initial step requires the standardization of the dividend and the divisor and the check of possible quotient overflow. The quotient digits are sequentially selected from the most significant digit to the least significant ones. The remainder is usually obtained automatically at the end of the quotient generation process. Efforts to improve division efficiency have focused on seeking fast and convenient ways to generate the quotient digits.

Division schemes can be categorized into four classes according to the permissible values of each quotient digit generated. In a *restoring division* with radix r, each quotient digit is selected from the conventional digit set

$$\{0, 1, 2, \ldots, r - 1\} \tag{7.1}$$

Nonrestoring division has quotient digits selected from the signed-digit set

$$\{-(r - 1), -(r - 2), \ldots, -1, +1, \ldots, r - 2, r - 1\} \tag{7.2}$$

with zero excluded. In the **SRT** *division* with radix 2, the signed digit set

$$\{-1, 0, 1\} \tag{7.3}$$

including zero is used. *Generalized SRT division* with radix r uses the quotient digit set

$$\{-m, \ldots, -1, 0, 1, \ldots, m\} \tag{7.4}$$

where

$$\frac{r-1}{2} \leq m \leq r - 1.$$

We shall study the principles and implementation of these four compare-shift division methods in this chapter. Unconventional approaches, using multiplicative division through convergence or reciprocation, and high-speed division using iterative cellular arrays will be studied in the next chapter.

7.2 Basic Subtract-Shift Division Properties

Division instructions are executed in most of today's digital computers via a recursive procedure. The time required for digital division is spent primarily in the repeated execution of this recursive procedure. Various division methods can be described by the following recursion formula.

$$R^{(j+1)} = r \times R^{(j)} - q_{j+1} \times D \tag{7.5}$$

where

$j = 0, 1, \ldots, n - 1$ is the *recursion index*,

D is the divisor

q_{j+1} is the $(j + 1)$th *quotient digit* to the right of the radix point

n is the word length of the quotient and q_0 is the sign

r is the radix

$r \cdot R^{(j)}$ is the *partial dividend* before the determination of the $(j + 1)$th quotient digit

$R^{(j+1)}$ is the *partial remainder* after the determination of the $(j + 1)$th quotient digit

$R^{(0)}$ is the *dividend* (initial partial remainder),

$R^{(n)}$ is the *final remainder*.

Without loss of generality, it will be assumed that both the dividend $R^{(0)}$ and the divisor D are fractions, and so is the generated quotient Q.

$$Q = q_0 . q_1 q_2 \cdots q_{n-1} q_n \tag{7.6}$$

where q_0 is the sign of the quotient determined by the following operation

$$q_0 = r_0^{(0)} \oplus d_0 \tag{7.7}$$

where $r_0^{(0)}$ and d_0 are the signs of the dividend and the divisor, respectively. The radix point is located between the sign q_0 and the most significant digit q_1. The final remainder could be either positive or negative depending on the method used. For the conventional restoring division, the sign of the remainder is identical with that of the dividend.

In order to simplify the discussion, both dividend and divisor are assumed positive fractions; that is, $r_0^{(0)} = d_0 = 0$ and so is the quotient $q_0 = 0$ according to Eq. 7.7.

When dividing a large dividend $R^{(0)}$ by a small divisor D, it is possible to generate a quotient Q, which exceeds the largest fractional value represented by an n-digit word. Such a *quotient overflow* is a result of the following condition

$$R^{(0)} \geq D \quad \textbf{(7.8)}$$

Overflow means that more than n digits are required to hold the quotient, which has a value no less than unity. The machine divider will issue an error signal when overflow does occur. Throughout the chapter, we assume $R^{(0)} < D$ and $D \neq 0$ implicitly. For the floating operations (to be described in Chapter 9), we shall further assume normalized division with a nonzero most significant digit $d_1 \neq 0$. The above assumptions can be always satisfied by appropriately shifting the dividend or the divisor.

The division process can be verified by applying the recursive Eq. 7.5 repeatedly. For $j = 0$

$$R^{(1)} = r \times R^{(0)} - q_1 \times D \quad \textbf{(7.9)}$$

For $j = 1$

$$R^{(2)} = r \times R^{(1)} - q_2 \times D = r^2 \times R^{(0)} - (r \times q_1 + q_2) \times D \quad \textbf{(7.10)}$$

For $j = n - 1$

$$R^{(n)} = r^n \times R^{(0)} - (r^{n-1} \times q_1 + r^{n-2} \times q_2 + \cdots + r \times q_{n-1} + q_n) \times D \quad \textbf{(7.11)}$$

The above iterative derivation shows that the division process consists of a sequence of additions, subtractions or shifts corresponding to the negative, positive or zero values of the successively generated quotients q_{j+1} for $j = 0, 1, \ldots, n - 1$. Eq. 7.11 can be rewritten as follows:

$$\frac{R^{(0)}}{D} = \sum_{j=1}^{n} r^{-j} \times q_j + \frac{r^{-n} \times R^{(n)}}{D} \quad \textbf{(7.12)}$$

where

$$Q = \sum_{r=1}^{n} r^{-j} \times q_j \quad \textbf{(7.13)}$$

represents the quotient, and the final remainder

$$R = r^{-n} \times R^{(n)} \quad \textbf{(7.14)}$$

is obtained by shifting the remainder $R^{(n)}$ n places to the right. The quotient selection procedure is based on one of the following arithmetic conditions. For restoring division, we use

$$0 \leq R^{(j+1)} < D \tag{7.15}$$

For nonrestoring division, we use

$$|R^{(j+1)}| \leq |D| \tag{7.16}$$

For generalized SRT division, we use

$$|R^{(j+1)}| \leq k \times |D| \tag{7.17}$$

where $\frac{1}{2} \leq k \leq 1$. Details of various quotient selection criteria will be described in subsequent sections.

7.3 Conventional Restoring Division

The conventional radix-r division uses the digit set $\{0, 1, 2, \ldots, r-1\}$. The value of each quotient digit q_{j+1} for $j = 0, 1, \ldots, n-1$ is selected to satisfy Eq. 7.15, provided overflow condition (Eq. 7.8) does not exist. This quotient-selection criterion can be implemented with repeated *subtraction* of the divisor D from the current partial dividend $r \times R^{(j)}$, until the difference becomes negative. The multiple $r \times R^{(j)}$ is obtained by simply shifting the current partial remainder $R^{(j)}$ one digital position to the left. The number of subtractions performed before the remainder turns negative determines the value of the quotient digit being selected. An additional step is required to restore the new partial remainder $R^{(j+1)}$. which must satisfy Eq. 7.15. The restoration is realized with one *addition* of the divisor to the negative difference obtained at the end of the subtracting process.

It takes $k+1$ subtractions and *one* addition to determine a quotient digit having a value $k \in \{0, 1, \ldots, r-1\}$. Therefore, in the worst case, r subtractions and *one* addition may be required to generate one quotient digit. The process is identical with the customary decimal division method using paper and pencil as illustrated in Fig 7.1.

The above restoring process can be greatly simplified for binary arithmetic, in which the radix $r = 2$. Binary restoring division requires at most one subtraction and one addition per each quotient digit. Equation 7.5 can be written for binary case as

$$R^{(j+1)} = 2R^{(j)} - q_{j+1} \times D \tag{7.18}$$

where each quotient digit $q_{j+1} \in \{0, 1\}$. The selection condition given in Eq. 7.15 implies that

$$q_{j+1} = \begin{cases} 0, & \text{if } 2R^{(j)} < D \\ 1, & \text{if } 2R^{(j)} \geq D \end{cases} \tag{7.19}$$

Given: Dividend $R^{(0)} = 0.1257$ and
Divisor $D = 0.39$ for $r = 10$
Find: Quotient $Q = 0.q_1q_2$ for $n = 2$,
Remainder $R^{(2)} = 0.r_1r_2 \times 10^{-2} = 0.00r_1r_2$

$R^{(0)} =$	0.1257		
$r \cdot R^{(0)} =$	1.257		
$-D =$	-0.39	1st Subtraction	$q_1 = 3$
	0.867	>0	
$-D =$	-0.39	2nd Subtraction	
	0.477	>0	
$-D =$	-0.39	3rd Subtraction	
	0.087	>0	
$-D =$	-0.39	4th Subtraction	
	-0.69	<0	
$+D =$	$+0.39$	Restoration Addition	
$R^{(1)} =$	0.087		
$r \cdot R^{(1)} =$	0.87		
$-D =$	-0.39	1st Subtraction	$q_2 = 2$
	0.48	>0	
$-D =$	-0.39	2nd Subtraction	
	0.09	>0	
$-D =$	-0.39	3rd Subtraction	
	-0.70	<0	
$+D =$	$+0.39$	Restoration Addition	

$R^{(2)} = 0.09 \times 10^{-2} = 0.0009 =$ The remainder

Quotient $Q = 0.q_1q_2 = 0.32$

Figure 7.1 A numerical example of the restoring division with the conventional decimal radix $r = 10$.

This discrimination procedure can be easily implemented with one subtraction to obtain the *tentative* partial remainder

$$\underline{R}^{(j+1)} = 2R^{(j)} - D \tag{7.20}$$

If the sign of $\underline{R}^{(j+1)}$ is positive, then $q_{j+1} = 1$ and $R^{(j+1)} = \underline{R}^{(j+1)}$. Otherwise, we obtain $q_{j+1} = 0$ and one addition is needed to restore the correct partial remainder

$$R^{(j+1)} = \underline{R}^{(j+1)} + D = 2R^{(j)} \tag{7.21}$$

A restoring binary division can be described by the Robertson diagram shown in Fig. 7.2. In this diagram, the horizontal axis shows the values of the current partial dividend $2R^{(j)}$. The vertical axis corresponds to the next partial remainder $R^{(j+1)}$. All values are measured in terms of multiples of the divisor D.

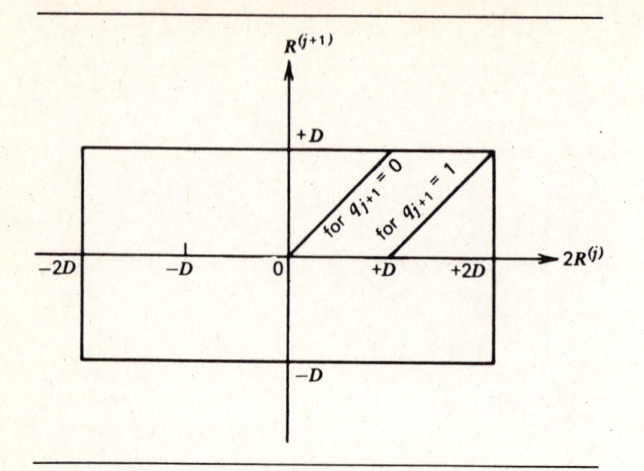

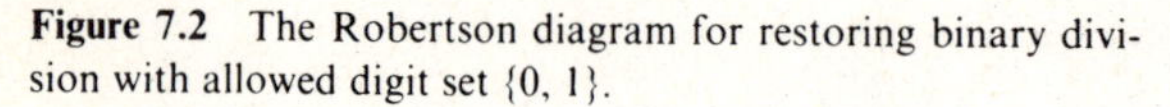
Figure 7.2 The Robertson diagram for restoring binary division with allowed digit set {0, 1}.

The straight line $2R^{(j)} - D = R^{(j+1)}$ on the right is designated to $q_{j+1} = 1$, because $D \leq 2R^{(j)} < 2D$ and the next partial remainder is $2R^{(j)} - D$. The left straight line $2R^{(j)} = R^{(j+1)}$ is designated to $q_{j+1} = 0$, because $0 \leq 2R^{(j)} < D$ and the remainder is $2R^{(j)}$. The above restoring binary division uses a brute-force choice of $q_{j+1} = 1$ at the beginning of each iteration. Penalty is paid in restoration steps, when the initial guess was not correct.

In summary, exactly n subtractions and, on the average, $n/2$ additions are required, if we assume equal probability of generating "0" or "1" quotient digits. Tc produce the successive partial dividend, $2R^{(j)}$ for $j = 0, 1, \ldots, n-1$, requires n one-digit left shifts. We shall show that with actual circuit implementation, the restoration addition can be bypassed and the old partial remainder can be saved by a stored shifting technique using shifters.

7.4 Binary Restoring Divider Design

The design of an arithmetic processor, which executes binary DIVIDE instructions using the restoring method just described is illustrated in this section. The unit shown in Fig. 7.3 consists of three n-bit working registers, namely, the *Accumulator* (**AC**), the *Auxiliary register* (**AX**), and the *Quotient-Multiplier* register (**QM**). Four flip-flops are used for storing Link (L), overflow (F), and Signs (S_a and S_x). A $2n$-bit dividend is initially stored in the cascaded register **AC** · **QM**. The n-bit divisor is stored in the **AX** register throughout the execution. The signs of the dividend and the divisor will be stored in S_a and S_x, respectively. The initial contents of L and F are zero.

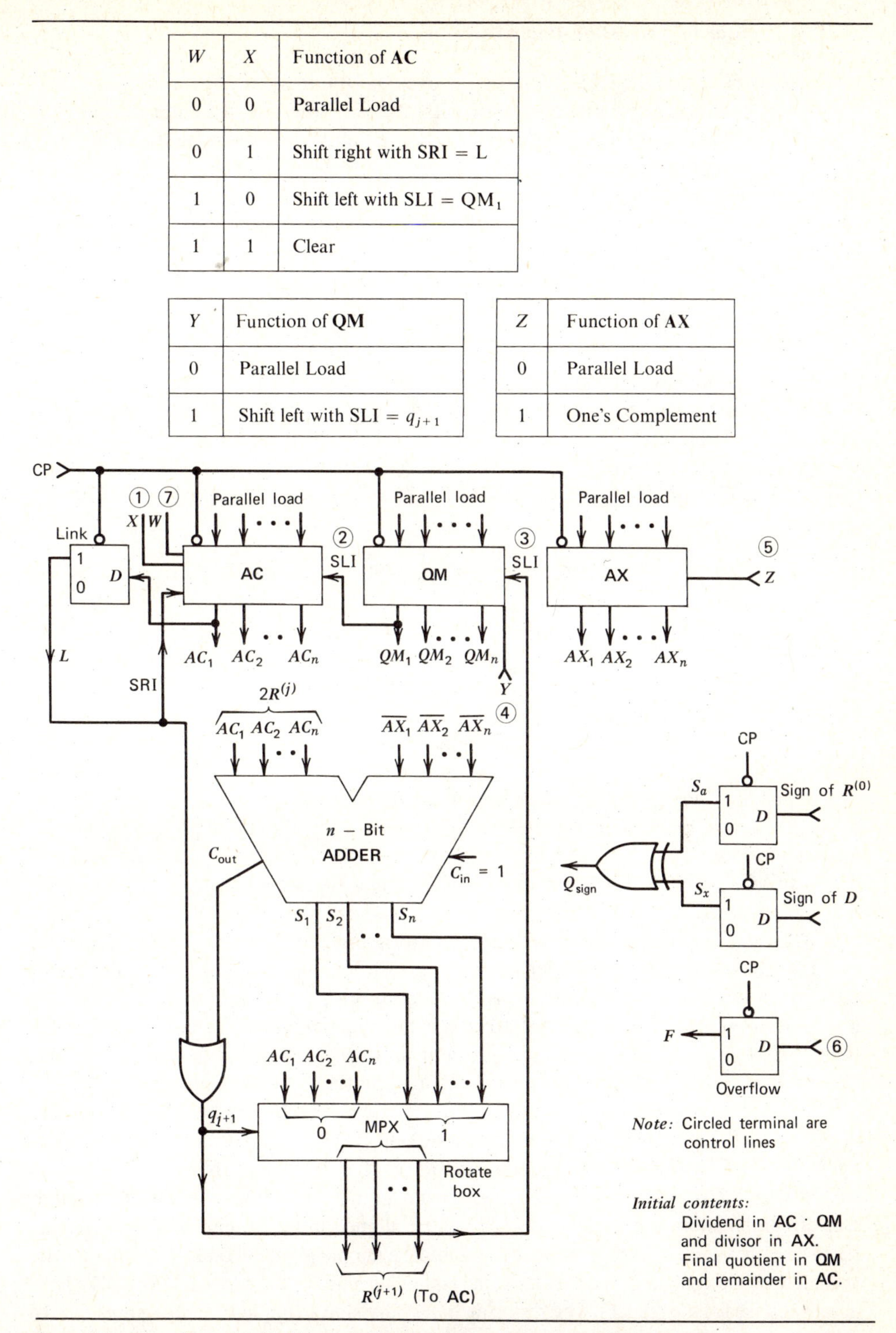

W	X	Function of **AC**
0	0	Parallel Load
0	1	Shift right with SRI = L
1	0	Shift left with SLI = QM_1
1	1	Clear

Y	Function of **QM**
0	Parallel Load
1	Shift left with SLI = q_{j+1}

Z	Function of **AX**
0	Parallel Load
1	One's Complement

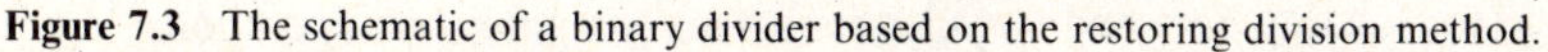

Figure 7.3 The schematic of a binary divider based on the restoring division method.

The partial dividend $2R^{(j)}$ for the jth cycle is obtained by shifting the cascaded register $L \cdot$ **AC** $\cdot$ **QM** one bit to the left with the new quotient bit entering the right end of the **QM** register. The bit pushed off the left end of **AC** is to be stored in the L flip-flop. The L flip-flop serves as a *buffer storage* to facilitate the following circular rotation operation

$$\begin{aligned} L &\leftarrow AC_1; \\ AC_1 AC_2 \cdots AC_n &\leftarrow AC_2 AC_3 \cdots AC_n QM_1; \\ QM_1 QM_2 \cdots QM_n &\leftarrow QM_2 QM_3 \cdots QM_n q_{j+1}. \end{aligned} \tag{7.23}$$

where q_{j+1} is the new quotient being generated.

The comparison operation specified in Eq. 7.19 is implemented with two's complement subtraction as shown in Fig. 7.3. The divisor in **AX** is subtracted from the shifted partial remainder in the extended $(n + 1)$-bit register $L \cdot$ **AC**. The input-output relationship of the n-bit adder is described by the following arithmetic equation

$$C_{\text{out}} \cdot S_{1-n} = AC_{1-n} - AX_{1-n} = AC_{1-n} + \overline{AX}_{1-n} + 1 \tag{7.24}$$

where $S_{1-n} = S_1 S_2 \cdots S_n$ is the sum outputs and C_{out} is the carry out of the n-bit adder. The adder and its extended link bit essentially implement Eq. 7.20. The $(j + 1)$st quotient digit is determined by the following logic equation:

$$q_{j+1} = L \vee C_{\text{out}} \tag{7.25}$$

This hardware circuit organization is verified below. When the most significant bit of $R^{(j)}$ was a "1", which after shift $2R^{(j)}$ is residing in the link L, then $(L \cdot \mathbf{AC}) > (\mathbf{AX})$ regardless of the subtraction result. On the other hand, if $L = 0$, then $C_{\text{out}} = 1$ implies $(\mathbf{AC}) \geq (\mathbf{AX})$. Both of these two cases lead to the selection $q_{j+1} = 1$. Thus Eq. 7.25 is verified.

The restoration addition can be avoided by using a multiplexer logic (rotate box) to retain the old partial remainder when $q_{j+1} = 0$. The hardware divide unit tends to retire the initial partial remainder from $2n$ to n bits and, at the same time, the quotient grows up to n bits. At the intermediate stage, the effective partial remainder occupies the leftmost $2n - j$ bits, whereas the quotient occupies the right j bits of the cascaded register **AC** $\cdot$ **QM** .The sum of the two word lengths is always $2n$ bits. The n-bit quotient enters the cascaded register from the right end bit by bit and pushes the upper half of the initial remainder (dividend) off the left end. Because of the initial cycle for overflow check, the final remainder $R^{(n)} = R \times 2^n$ will end up in $L \cdot AC_1 AC_2 \cdots AC_{n-1}$, where $R = R^{(n)} \times 2^{-n}$ is the desired remainder in $2n$ bits. With one right shift at the end of computation, the final remainder will be aligned in the accumulator. The above hardware operations are verified by the numerical example given in Table 7.1. In order to complete the design, all the control terminal equations must be determined. This will be left as an exercise for the readers. The restoring binary division so implemented requires $n + 1$ cycles, one cycle for loading data and n cycles for actual compare-shift operations.

Table 7.1 A Numerical Example of Binary Restoring Division Using the Hardware Unit Shown in Fig. 7.3

	Partial Remainder		Divisor		Adder Inputs		Adder Outputs			
	$R^{(j)}$		D	Link	$r \times R^{(j-1)}$	$\overline{D}$	Carry	Sum	Quotient	Remarks
Cycles j	AC	QM	AX	L	AC*	$\overline{\textbf{AX}}$	C_{out}	$S_1S_2S_3S_4$	q_j	$(r = 2, n = 4)$
0	0110	1101	1011	0	xxxx	xxxx	x	xxxx	sign q_0	Load data, check overflow. Determine sign by Eq. 7.7
1	0010	1011	1011	0	1101	0100	1	0010	1	$R^{(1)} = r \times R^{(0)} - D$ (Subtract-shift)
2	0101	0110	1011	0	0101	0100	0	1010	0	$R^{(2)} = r \times R^{(1)}$ (shift left)
3	1010	1100	1011	0	1010	0100	0	1111	0	$R^{(3)} = r \times R^{(2)}$ (shift left)
4	1010	1001	1011	1	0101	0100	0	1010	1	$R^{(4)} = r \times R^{(3)} - D$ (subtract-shift)
5	1010	.1001	1011	1	0101	xxxx	x	xxxx	x	Shift right L · **AC** to align remainder into **AC**
Results	Remainder	Quotient								$0.q_1q_2q_3q_4 = Q$ in **QM**

Note: $\frac{A}{B} = Q + \frac{R}{B}$; $\frac{0.01101101}{0.1011} = 0.1001 + \frac{0.00001010}{0.1011}$ for $n = 4$. **AC*** refers the left-shifted **AC**. "x" refers to don't-care condition.

7.5 Binary Nonrestoring Division

The restoring division is identical with the paper-pencil division method. It is simple to implement and easy to understand. The restoring addition steps (even using stored multiplexing as a bypass), however, may slow down the process especially when many zeros are generated in the quotient. In this section we shall study an improved method which will completely eliminate restoration steps without using the stored multiplexing logic.

Instead of a brutal choice of a "1" for the quotient digit at the beginning of each iteration, this nonrestoring method selects either $+1$ or -1 as the quotient digit as the situation dictates. Note that "0" is excluded as a legal choice. The quotient choice is made so that the error incurred in each selection is counterbalanced in later steps, and the remedy consumes no extra add, subtract, or shift delay time. The idea is based on the relaxation of the quotient selection criterion in Eq. 7.15 to the version specified in Eq. 7.16. We can rewrite Eq. 7.16 as follows under the assumption that the divisor $D > 0$.

$$|R^{(j+1)}| < D \tag{7.26}$$

The absolute value means that the successive partial remainders $R^{(j+1)}$ for $j = 0, 1, \ldots, n-1$, can be either *positive* or *negative*. There is no need to restore a negative remainder to positive, as long as its absolute magnitude is less than the divisor. At each iteration, the divisor is either added to or subtracted from the partial dividend. The specific operation performed in each step is determined by

$$R^{(j+1)} = \begin{cases} 2R^{(j)} - D, & \text{if } 2R^{(j)} > 0 \\ 2R^{(j)} + D, & \text{if } 2R^{(j)} < 0 \end{cases} \tag{7.27}$$

The corresponding quotient digit is generated accordingly

$$q_{j+1} = \begin{cases} 1, & \text{if } 0 < 2R^{(j)} < 2D, \\ -1, & \text{if } -2D < 2R^{(j)} < 0 \end{cases} \tag{7.28}$$

When $2R^{(j)} = 0$, the process can be terminated. The resulting quotient is represented by signed-digit code containing no zeros. Therefore, the quotient is neither canonical nor minimal. On the contrary, it is maximal in terms of number of nonzero digits. Conversion of the signed-digit quotient to conventional binary code may be necessary in order to match other operations in the arithmetic processor.

The Robertson diagram for binary nonrestoring division is given in Fig. 7.4. Note that the choice of the value $+1$ or -1 for the quotient digit is symmetrical with respect to the origin. As long as $D > R^{(0)}$ (no overflow), then $|R^{(j+1)}| < D$ and $|2R^{(j)}| < D$ and $q_{j+1} \in \{-1, 1\}$. The two 45°-slope straight lines correspond to the two equations given in Eq. 7.27. They match the choices of $q_{j+1} = 1$ and $q_{j+1} = -1$, respectively, as in Eq. 7.28. This nonrestoring method needs an accurate sign-determining mechanism which can detect signs of the successive partial remainders until the entire division process is completed.

The time required to compare two binary numbers is proportional to the word

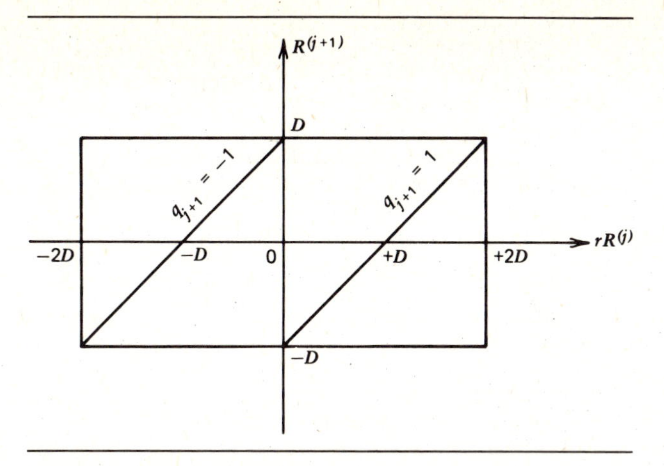

Figure 7.4 The Robertson diagram for nonrestoring binary division with permissible digit set $\{-1, 1\}$.

length. Absence of redundancy in the restoring division requires comparison of all the digits in the divisor against the partial remainder to determine the value of the quotient digit. The quotient generated in a nonrestoring division is certainly redundant. The imposed redundancy enables the selection of quotient digits on *estimates* of the values of the divisor and of the partial remainder. These estimates can be made cruder (shorter in length) when more redundancy is introduced. The required comparison time to select the quotient digit decreases as the estimate word length becomes shorter.

A negative quotient digit, say $q_j = -k$ for some positive integer k, implies that k additions have occurred. The same number of subtractions were performed for a positive quotient $q_j = k$. Because the adder is used in each step (addition or subtraction but no shifting alone), the average shift length in a nonrestoring division is *one* and provision for multiple shifts in nonrestoring division is ineffective. This is different from the case of restoring division, in which multiple shifts are allowed to bypass the adder when strings of zeros are to be generated in the quotient.

7.6 High-Radix Nonrestoring Division

Computer arithmetic is often facilitated by considering "groups" of bits rather than by each bit individually. Such grouping may be interpreted as use of digits of higher radix than two. For example, *pairs* of bits can be considered as radix-4 digits, and *trios* of bits as radix-8 digits, and so on. In general, a string of l bits is equivalent to m radix-r digits, where

$$m = \frac{l}{\lceil \log_2 r \rceil} \tag{7.29}$$

For practical cases, $r = 2^k$ for some integer $k > 0$ and $l = m \times k$.

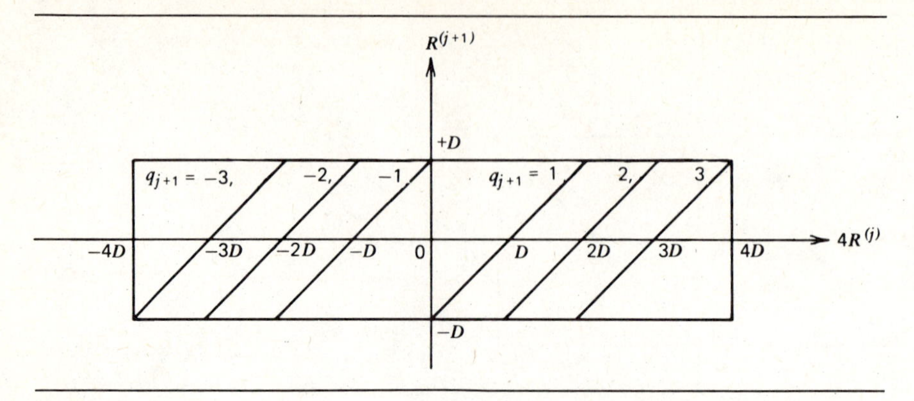

Figure 7.5 The Robertson diagram for a radix-4 nonrestoring division with permissible digit set $\{-3, -2, -1, +1, +2, +3\}$.

Both the binary restoring and nonrestoring methods can be generalized to designs with higher radix number systems. The number of iterations required to execute a DIVIDE instruction decreases rapidly as the radix increases. This means faster execution time may be expected when using higher radix division schemes. The tradeoffs lie in the increased hardware complexity associated with the generation of multiples of high-radix divisors, and the complicated procedures in selecting high-radix quotient digits.

The restoring division with arbitrary radix r has been described in section 7.3. In what follows, we explain the principles of radix-r nonrestoring division. The recursive process is again described in Eq. 7.5. Each quotient is selected from the digit set given in Eq. 7.2. For example, when radix $r = 4$, the quotient digit set $= \{-3, -2, -1, +1, +2, +3\}$. Overflow does not exist when $R^{(0)} < D$.

The $(j + 1)$st quotient digit q_{j+1} is chosen such that the absolute value of the partial remainder lies within the following range:

$$-D < R^{(j+1)} < D \tag{7.30}$$

The Robertson diagram in Fig. 7.5 depicts a nonrestoring division scheme with radix $r = 4$. The straight lines correspond to the six possible values of each quotient digit. It should be noted that multiple choices of the value of quotient q_{j+1} appear in some regions. This can be seen by the overlapped projection of the lines on the $r \times R^{(j)}$ axis. These multiple choices are listed below:

$$q_{j+1} = \begin{cases} -3, & \text{if } -4D < 4R^{(j)} < -3D, \\ -3 \text{ or } -2, & \text{if } -3D < 4R^{(j)} < -2D, \\ -2 \text{ or } -1, & \text{if } -2D < 4R^{(j)} < -D, \\ -1, & \text{if } -D < 4R^{(j)} < 0, \\ +1, & \text{if } 0 < 4R^{(j)} < D, \\ +1 \text{ or } +2, & \text{if } D < 4R^{(j)} < 2D, \\ +2 \text{ or } +3, & \text{if } 2D < 4R^{(j)} < 3D, \\ +3, & \text{if } 3D < 4R^{(j)} < 4D \end{cases} \tag{7.31}$$

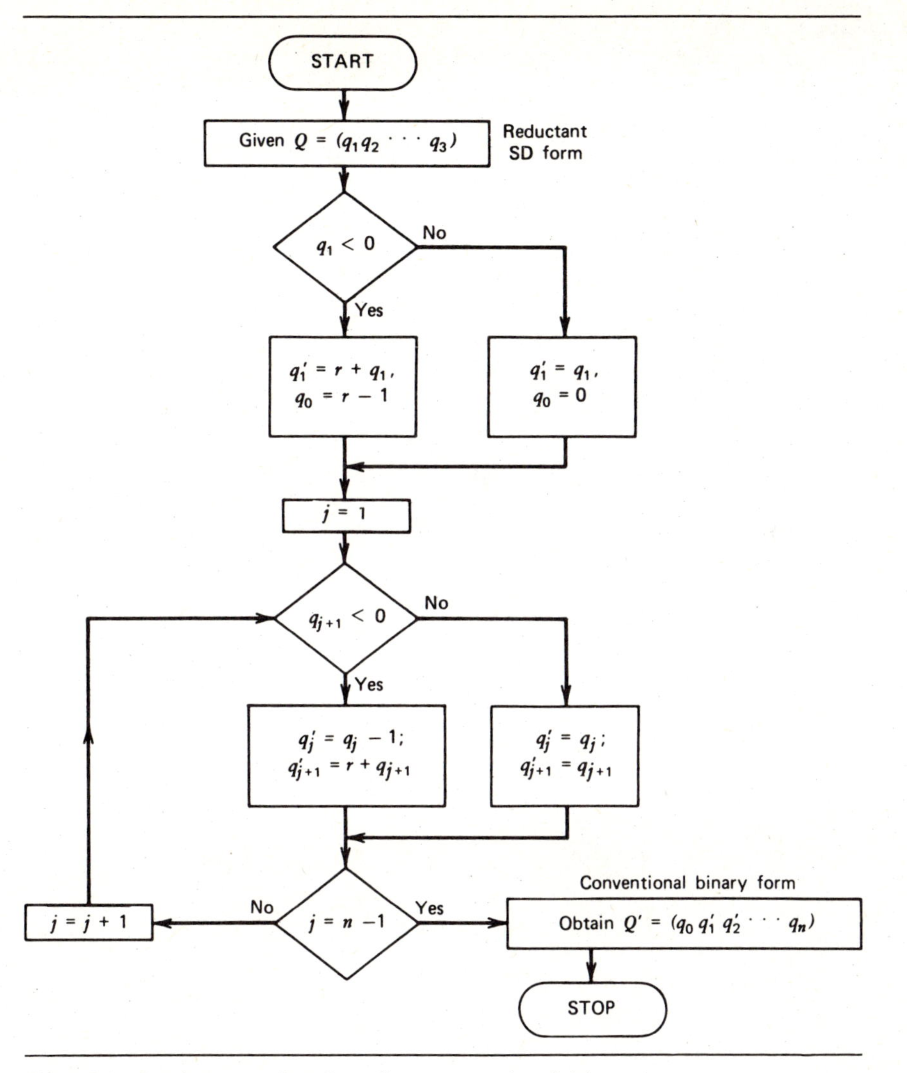

Figure 7.6 Quotient converison for radix-r nonrestoring division.

Described in the flow chart of Fig. 7.6 are the general rules for converting a radix-r redundant signed-digit quotient $(q_1 q_2 \cdots q_n)$, with each

$$q_i \in \{-(r-1), \ldots, -1, +1, \ldots, +(r-1)\}$$

for all i, into the conventional number in radix-complement (r's complement) notation $q_0.q_1'q_2' \cdots q_n'$ with each $q_i' \in \{0, 1, \ldots, r-1\}$ for all i and the sign bit $q_0 = 0$ for positive number and $q_0 = r - 1$ for negative number.

This conversion procedure requires a serial inspection of the quotient digits, the most significant digit first. The digit q_{j+1} is unchanged when it is positive. When a negative digit is encountered, it is added to the radix, and a unit is borrowed from the adjacent higher order digit. Note that because zero is not a permissible digit, there is no need for borrow propagation. A numerical example is given below for the case of $r = 4$ and $n = 3$.

Example. Given $Q = (q_1 q_2 q_3) = (\bar{2}\ 1\ \bar{3})$. The natural value of Q is

$$Q = \left(-2 \times \frac{1}{4}\right) + \left(1 \times \frac{1}{4^2}\right) + \left(-3 \times \frac{1}{4^3}\right)$$

$$= \frac{-2}{4} + \frac{1}{16} + \frac{-3}{64} = \frac{-32 + 4 - 3}{64} = \frac{-31}{64}$$

This redundant quotient Q can be converted to a 4's complement number $Q' = (q_0 . q'_1 q'_2 q'_3)$. The detailed steps are shown below:

Initial step. The fact that $q_1 = \bar{2} < 0$ leads to a negative sign

$$q_0 = r - 1 = 4 - 1 = 3$$
$$q'_1 = r + q_1 = 4 + \bar{2} = 2 \qquad \textbf{(7.32)}$$

For $j = 1$. $q'_1 = 2$ and $q_2 = 1$ implies that

$$q'_2 = q_2 = 1 \qquad \textbf{(7.33a)}$$

For $j = 2$. $q'_2 = 1$ and $q_3 = \bar{3}$ implies that q'_2 should be modified to be

$$q'_2 = q'_2 - 1 = 1 - 1 = 0 \qquad \textbf{(7.33b)}$$

and

$$q'_3 = r + q_3 = 4 + \bar{3} = 1 \qquad \textbf{(7.34)}$$

Equations 7.32 through 7.34 give the resulting binary code in radix-complement form

$$Q' = (q_0 . q'_1 q'_2 q'_3)$$
$$= (3.201)_4$$

Q' has a natural value of $-31/64$, which can be easily verified by recomplementing Q' back to its positive version

$$\bar{Q} = (0 \cdot 133)_4 = 1 \times \tfrac{1}{4} + 3 \times \tfrac{1}{16} + 3 \times \tfrac{1}{64}$$
$$= \tfrac{31}{64}$$

7.7 Principle of SRT Division

The **SRT** method of binary division was discovered independently at about the same time by Sweeney, Robertson, and Tocher [11, 14, 15]. It was proposed to upgrade the binary floating-point arithmetic. The method involves a normalized divisor, and the successive partial dividend is also normalized within the following ranges:

$$\tfrac{1}{2} < |D| < 1 \tag{7.35}$$

$$\tfrac{1}{2} < |2R^{(j)}| < 1 \tag{7.36}$$

In nonrestoring division, "0" is a prohibited quotient digit. This means shifting-alone operations are not utilized. This is not a desirable omission because shifting usually requires much less time than addition/subtraction. SRT division improves the nonrestoring division by allowing the quotient digit set $\{-1, 0, 1\}$, with zero included as a legitimate choice.

The divisor is either *added* to or *shifted*, or *subtracted* from the partial dividend or neither, depending on the outcome of comparing the partial dividend against the divisor. The specific operation performed at each cycle is characterized by

$$R^{(j+1)} = \begin{cases} 2R^{(j)} + D, & \text{if } 2R^{(j)} < -D, \\ 2R^{(j)}, & \text{if } -D \leq 2R^{(j)} \leq D, \\ 2R^{(j)} - D, & \text{if } D < 2R^{(j)} \end{cases} \tag{7.37}$$

The rules for selecting the quotient digit q_{j+1} are

$$q_{j+1} = \begin{cases} -1, & \text{if } 2R^{(j)} < -D, \\ 0, & \text{if } -D \leq 2R^{(j)} \leq D, \\ 1, & \text{if } D < 2R^{(j)} \end{cases} \tag{7.38}$$

The above rules are illustrated graphically in Fig. 7.7. Because both the divisor and the partial dividend are normalized fractions as specified in Eqs. 7.35 and 7.36, we can eliminate the full comparison of $2R^{(j)}$ with D or $-D$ as specified in Eqs. 7.37 and

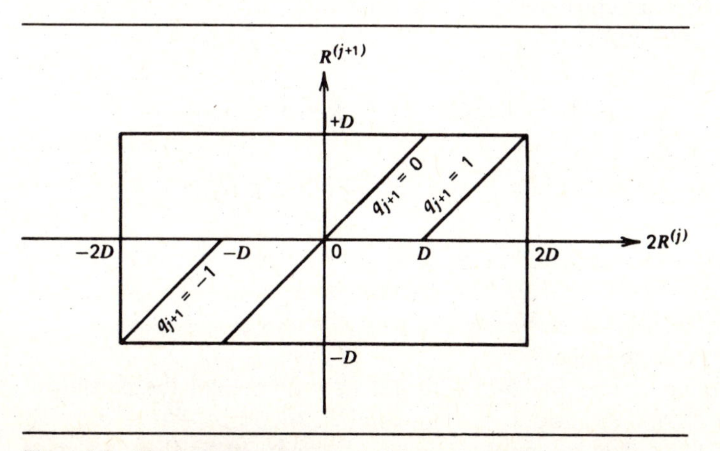

Figure 7.7 The Robertson diagram for binary nonrestoring division with allowed digit set $\{-1, 0, 1\}$.

7.38. This is done by using the minimum value of the divisor as the threshold. The comparison operation can therefore be reduced to the following simplified form

$$R^{(j+1)} = \begin{cases} 2R^{(j)} + D, & \text{if } 2R^{(j)} \leq \dfrac{-1}{2} \\ 2R^{(j)}, & \text{if } \dfrac{-1}{2} < 2R^{(j)} < \frac{1}{2} \\ 2R^{(j)} - D, & \text{if } \frac{1}{2} \leq 2R^{(j)} \end{cases} \tag{7.39}$$

and the quotient selection rule can be accordingly simplified as

$$q_{j+1} = \begin{cases} -1, & \text{if } 2R^{(j)} < \dfrac{-1}{2}, \\ 0, & \text{if } \dfrac{-1}{2} \leq 2R^{(j)} \leq \frac{1}{2}, \\ 1, & \text{if } \frac{1}{2} < 2R^{(j)} \end{cases} \tag{7.40}$$

The advantage of using the set of rules in Eq. 7.40 lies in the fact that only comparison of $2R^{(j)}$ against the constant $\frac{1}{2}$ or $-1/2$ is required. This can be easily implemented in a binary system. The procedure just described has the effect of normalizing the partial remainder by shifting over leading *zeros* if the partial remainder is positive, and by shifting over leading *ones* if the partial remainder is negative.

The quotient digits so generated will be in redundant signed-digit form. The process can be further simplified if the divisor is known or has been converted to positive at the preliminary standardization operation. Under such circumstances, the following set of rules should be used:

$$q_{j+1} = \begin{cases} -1, & \text{if } |2R^{(j)}| > \frac{1}{2}, \text{ and Sign } (2R^{(j)}) = \text{negative}, \\ 0, & \text{if } |2R^{(j)}| \leq \frac{1}{2}, \\ +1. & \text{if } |2R^{(j)}| > \frac{1}{2}, \text{ and Sign } (2R^{(j)}) = \text{positive} \end{cases} \tag{7.41}$$

The above choices are summarized in the Robertson diagram shown in Fig. 7.8. The range of $2R^{(j)}$ is restricted to $-1 < 2R^{(j)} < 1$ and that of $R^{(j+1)}$ is $-\frac{1}{2} < R^{(j+1)} < \frac{1}{2}$. The q-lines are shown in three cases corresponding the values $D = \frac{1}{2}, \frac{3}{4}$, and 1, two extremes plus an intermediate value.

For example, if the dividend $R^{(0)} > 0$ with a divisor $D = \frac{1}{2}$, all the partial remainders will be positive and each quotient digit can be either 0 or $+1$ as dictated by the line $R^{(j+1)} = 2R^{(j)} - \frac{1}{2}$ in the first quadrant. The quotient, in this particular case, is in conventional binary form with an average shift of 2 by Table 5.7. On the

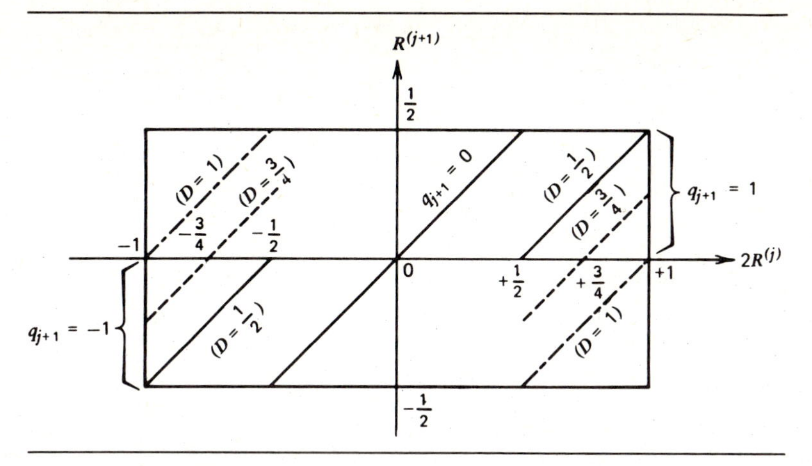

Figure 7.8 The Robertson diagram describing binary **SRT** division for three possible values of the divisor $\frac{1}{2}$, $\frac{3}{4}$, and 1.

other hand, for $D = \frac{3}{4}$, the quotient will be in canonical minimal signed-digit form with an average shift length of 3. The quotient representation for the case $D = 1$ has the feature that successive nonzero digits are of opposite signs. This can be seen by the two q-lines residing in the second and the fourth quadrants of Fig. 7.8. A redundant code with average shift length of 2 will occur for $D = 1$.

The above analyses lead to the following conclusion: The efficiency of an **SRT** division system at one time depends on the magnitude of the normalized divisor. The average shift length over all possible values of the normalized divisor is obtained as 2.6 by Freiman [5]. This implies that further improvement can be made to upgrade the performance of **SRT** division by controlling its divisor ranges. Implementation of **SRT** division will be demonstrated in later sections.

7.8 Modified Binary SRT Divisions

Since the discovery of **SRT** division in late 1950s, many researchers have investigated improved methods to optimize the original **SRT** division as described in the preceding section. The analysis of the efficiency of **SRT** division proved to be rather complicated. Freiman [5] has shown that the **SRT** division yields a quotient in minimal, but not necessarily in canonical, form with an average shift length of 3 for divisors in the range

$$\tfrac{3}{5} \leq |D| \leq \tfrac{3}{4} \tag{7.42}$$

In fact, Robertson [11] has shown that for a divisor with precise value

$$|D| = \tfrac{3}{4} \tag{7.43}$$

the quotient will be in canonical minimal form. An elegant and complete proof of this range was given by Shively [13].

The first variant of the **SRT** division to possess a uniform shift length of 3 for $\frac{1}{2} < |D| < 1$ was given by Wilson and Ledley [18]. Their method proposes a simplified approach to dividing positive and normalized fractions. The approach is based on an inverse process of the rapid multiplication, in which the bit patterns are decomposed into strings of units and zeros with isolated zeros and units, as illustrated by the example in Fig. 7.9.

The Wilson-Ledley division method is described by the flow chart in Fig. 7.10, under the assumption that the divisor D is a positive, normalized fraction, and the dividend N is either normalized or with (at most) a single zero after the binary point. The process starts with forming the first partial remainder

$$N^{(s)} = N^{(s-1)} - D \tag{7.44}$$

Because $N^{(s)}$ is negative under the above assumption, the negative loop of Fig. 7.10 is entered. If $N^{(s)}$ has α zero to the right of binary point, then the quotient Q has at least $\alpha - 1$ units to the right of the binary point. (In this initial step, $q_i = 0$ for $i = 0$ means merely that $Q < 1$). Now $N^{(s)}$ is normalized and the second partial remainder is formed as

$$N^{(s+1)} = N^{(s)} + D \tag{7.45}$$

If $N^{(s+1)}$ is negative, then $q_{0+\alpha} = 0$ and the following bits are units. If $N^{(s+1)}$ is positive, then $q_{0+\alpha} = 1$ and the following bits are zeros. The procedure continues in this fashion, differencing and normalizing each time, and determining q_i and q_{i+1} through $q_{i+\alpha-1}$ at each step. Note that, at each step, a multiple number of quotient digits may be determined. The process ends when the recursive index reaches the number of bits desired in the quotient.

A numerical example is given in Fig. 7.11 for the division of

$$N = 0.1001111100001100$$

by $D = 0.1101$. This example verifies the above procedure nicely. This method saves, on the average, two-thirds the number of addition or subtractions required in a normal binary division. It also yields a quotient in minimal but not necessarily in canonical form for all normalized divisors. The method demonstrates a design in which full-length comparison may not be required at each step.

Another class of modified **SRT** division methods was proposed by Metze [10] based on Freiman's statistical analysis of **SRT** algorithms. The key idea behind Metze's approach lies in the fact that the number of quotient digits generated per each iteration is increased by normalizing the partial remainder. Partial remainders need not be examined with high precision, yet the quotient is represented in minimal but not necessarily canonical form.

The **SRT** division specified in Eq. 7.39 involves the comparison of the shifted partial remainder $2R^{(j)}$ with the constant $K = \frac{1}{2}$. A precision of 2^{-1} obviously suffices for this comparison. As long as the divisor lies within the range of Eq. 7.42,

0	1 2 3 4	5 6 7 8	9 10 11 12 13	14 15 16 17 18	Bit positions
0.	1 1 1 1	0 0 0 0	1 1 0 1 1	0 0 1 0 0	Binary number
	string of units	string of zeros	isolated zero in string of units	isolated unit in string of zeros	

Figure 7.9 An illustration of the decomposition of a binary number into four types of substrings.

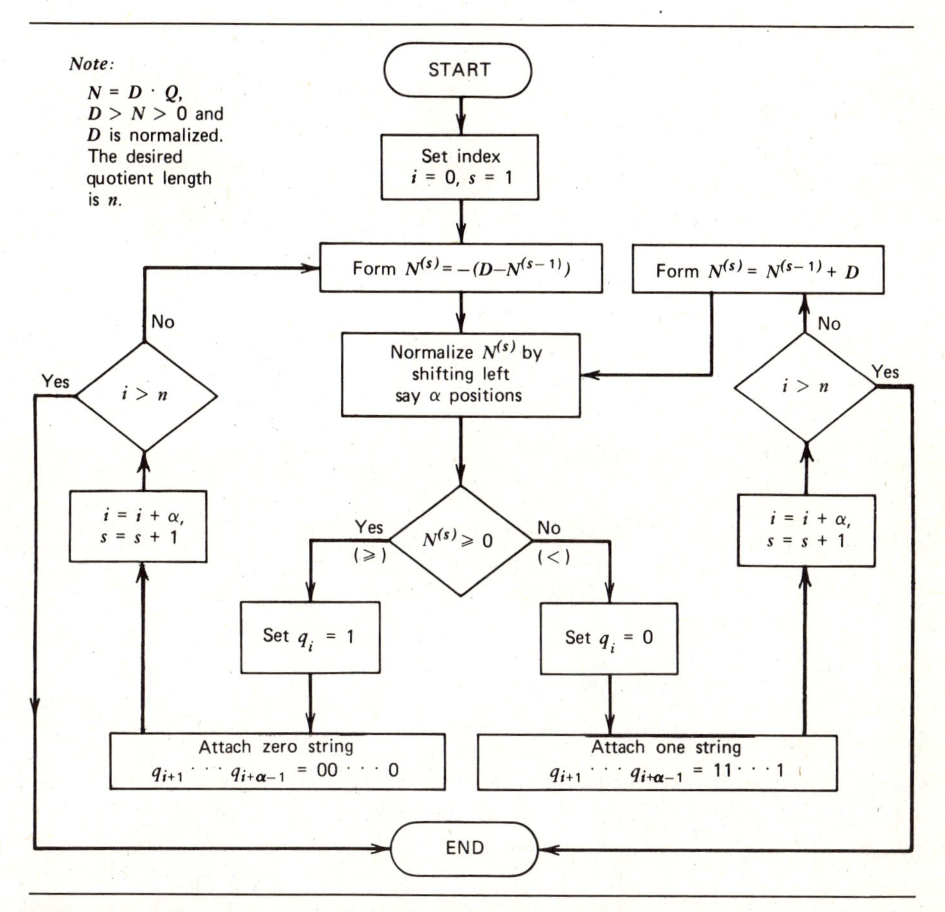

Figure 7.10 Wilson and Ledley's algorithm for rapid binary division.

Given $N = 0.1001111100001100$ and $D = 0.1101$
Find $Q = N/D = q_0.q_1 \cdots q_n$ for $n = 10$.

			Strings of quotient bits generated
$+D$		0.1101	
$N^{(0)} = N$:	$-)$	0.1001 1111 0000 1100	$q_0.q_1q_2q_3q_4q_5q_6q_7q_8q_9q_{10}$
$N^{(1)} = -(D - N^{(0)})$:		−0.0011 0000 1111 0100	$(\alpha = 2)$
Normalized:		−0.11 0000 1111 0100	0.1?
$+D$:	$+)$	0.11 01	
$N^{(2)} = N^{(1)} + D$:		+0.00 0011 0000 1100	$(\alpha = 4)$
Normalized:		+0.11 0000 1100	0.11000?
$-D$:	$-)$	0.11 01	
$N^{(3)} = -(D - N^{(2)})$:		−0.00 0011 0100	$(\alpha = 4)$
Normalized:		−0.11 0100	0.110000111?
$+D$:	$+)$	0.11 01	
$N^{(4)} = N^{(3)} + D$:		0.00 0000	$(\alpha > 6)$
Normalized:		$(\alpha = 6)$	0.1100001111
			$q_0.q_1q_2q_3q_4q_5q_6q_7q_8q_9q_{10}$

Figure 7.11 A numerical example illustrating the Wilson–Ledley rapid division method described in Fig. 7.10.

the comparison constant $K = \frac{1}{2}$ permits a *local* quotient range defined by finding the extremes of K/D, that is

$$\tfrac{2}{3} \le |Q| \le \tfrac{5}{6} \tag{7.46}$$

where

$$\left(\frac{K}{D}\right)_{\min} = \frac{\frac{1}{2}}{D_{\max}} = \frac{\frac{1}{2}}{\frac{3}{4}} = \tfrac{2}{3} \tag{7.47}$$

and

$$\left(\frac{K}{D}\right)_{\max} = \frac{\frac{1}{2}}{D_{\min}} = \frac{\frac{1}{2}}{\frac{3}{5}} = \tfrac{5}{6} \tag{7.48}$$

This means that Metze's method is identical to the **SRT** division whenever the divisor lies within the range of Eq. 7.42. For divisors outside the range, Metze's method supersedes the **SRT** division by choosing a different comparison constant, which improves the quotient selection process. The comparison constant for different divisor region should be chosen so as to maintain the local quotient range as specified in Eq. 7.46.

To achieve this dynamic choice of comparison constants, one simply chooses the constant K_i for the ith divisor region so as to satisfy the following bounds

$$\tfrac{6}{5}K_i \le |D_i| \le \tfrac{3}{2}K_i \tag{7.49}$$

where D_i is the divisor in the ith region and the following limits are used

$$D_{\min} = \frac{K_i}{Q_{\max}} = \frac{K_i}{\frac{5}{6}} = \frac{6}{5}K_i \tag{7.50}$$

$$D_{\max} = \frac{K_i}{Q_{\min}} = \frac{K_i}{\frac{2}{3}} = \frac{3}{2}K_i \tag{7.51}$$

In general, a large number of regions permits low precision recursive comparisons and vice versa. This modified **SRT** division method therefore consists of two phases. Initially, the appropriate comparison constant K_i is determined by applying Eq. 7.49 through Eq. 7.51. At each recursive step, the rules of **SRT** division are obeyed, except making $K = K_i$.

Table 7.2 A Set of Comparison Constants and Corresponding Divisor Regions for Metze's Modified **SRT** Division

Region	Comparison Constants	Theoretical Divisor Regions	Practical Divisor Regions Without Overlapping
1	$K_1 = \frac{3}{8}$	$\frac{9}{20} \leq \lvert D_1 \rvert \leq \frac{9}{16}$	$\lvert D_1 \rvert < \frac{9}{16}$
2	$K_2 = \frac{7}{16}$	$\frac{21}{40} \leq \lvert D_2 \rvert \leq \frac{21}{32}$	$\frac{9}{16} \leq \lvert D_2 \rvert < \frac{5}{8}$
3	$K_3 = \frac{1}{2}$	$\frac{3}{5} \leq \lvert D_3 \rvert \leq \frac{3}{4}$	$\frac{5}{8} \leq \lvert D_3 \rvert < \frac{3}{4}$
4	$K_4 = \frac{5}{8}$	$\frac{3}{4} \leq \lvert D_4 \rvert \leq \frac{15}{16}$	$\frac{3}{4} \leq \lvert D_4 \rvert < \frac{15}{16}$
5	$K_5 = \frac{3}{4}$	$\frac{9}{10} \leq \lvert D_5 \rvert \leq \frac{8}{9}$	$\frac{15}{16} \leq \lvert D_5 \rvert$

Note: The maximum precision of initial determination of divisor region and of recursive comparisons equals 2^{-4}.

Table 7.2 lists five divisor regions and corresponding comparison constants chosen to satisfy Eq. 7.46 and Eq. 7.49 simultaneously. Recursive comparisons with the constant K_i require a precision of at most 2^{-4}. The last column in Table 7.2 shows the practical divisor regions without overlapping. The regions are selected so as to ensure a discrimination precision of up to 2^{-4}. Three numerical examples are shown in Fig. 7.12 and Fig. 7.13 which compare the original **SRT** division and Metze's **SRT** division methods applied to the same set of numerical data. Example **A** deals with a divisor in the range $\frac{3}{5} \leq |D| \leq \frac{3}{4}$, for which the **SRT** division yields minimally represented quotients. Example **B** shows an **SRT** division by a divisor outside the above range. The quotients so generated do not yield a minimal representation. Example **C** in Fig. 7.13 shows the modified **SRT** division applied to the same data as in Example **B**. A minimally represented quotient is obtained using Metze's approach.

Example A. Divisor in region $\frac{3}{5} \leq |D| \leq \frac{3}{4}$
Given $N = \frac{77}{256}$ and $D = \frac{11}{16}$ and $K = \frac{1}{2}$

$$
\begin{array}{llll}
R^{(0)} = N = & 0.01001101 & \text{Quotients} & \\
2R^{(0)} = & 0.1001101 & 2R^{(0)} > \frac{1}{2} \rightarrow q_1 = 1 & \\
-D = & -)\ 0.1011 & & Q = 0.q_1q_2q_3q_4 \\
R^{(1)} = 2R^{(0)} - D = & -0.0001001 & & = 0.100\bar{1} \\
R^{(2)} = 2R^{(1)} = & -0.001011 & |2R^{(1)}| < \frac{1}{2} \rightarrow q_2 = 0 & = \frac{7}{16} \\
R^{(3)} = 2R^{(2)} = & -0.01011 & |2R^{(2)}| < \frac{1}{2} \rightarrow q_3 = 0 & \text{(Minimal form)} \\
R^{(4)} = 2R^{(3)} = & -0.1011 & |2R^{(3)}| < \frac{-1}{2} \rightarrow q_r = \bar{1} & \\
+D = & +0.1011 & & \\
 & 0.0000 & &
\end{array}
$$

Example B. Divisor outside region $\frac{3}{5} \leq |D| \leq \frac{3}{4}$
Given: $N = \frac{63}{256}$ and $D = \frac{9}{16}$

$$
\begin{array}{llll}
R^{(0)} = N = & 0.00111111 & \text{Quotients} & \\
R^{(1)} = 2R^{(0)} = & 0.0111111 & |2R^{(0)}| \leq \frac{1}{2} \rightarrow q_1 = 0 & \\
2R^{(1)} = & 0.111111 & 2R^{(1)} > \frac{1}{2} \rightarrow q_2 = 1 & \\
-D = & -)\ 0.1001 & & Q = 0.q_1q_2q_3q_4 = 0.0111 = \frac{7}{16} \\
R^{(2)} = 2R^{(1)} - D = & 0.011011 & & \text{(Nonminimal)} \\
2R^{(2)} = & 0.11011 & 2R^{(2)} > \frac{1}{2} \rightarrow q_3 = 1 & \\
-D = & -)\ 0.1001 & & \\
R^{(3)} = 2R^{(2)} - D = & 0.01001 & & \\
2R^{(3)} = & 0.1001 & 2R^{(3)} > \frac{1}{2} \rightarrow q_4 = 1 & \\
-D = & -)\ 0.1001 & & \\
 & 0.0000 & &
\end{array}
$$

Figure 7.12 Numerical examples describing the **SRT** division applied to two cases of different divisor regions.

Example C. Modified (Metze) **SRT** division applied to the same operands as in Example B.

Given: $N = \frac{63}{256}$ and $D = \frac{9}{16}$ with comparison constant
$K = \frac{7}{16}$ (from Table 7.2)

			Quotients
$R^{(0)} = N =$		0.00111111	
$2R^{(0)}$	=	0.0111111	$2R^{(0)} < \frac{7}{16} \to q_1 = 1$
$-D$	= −)	0.1001	
$R^{(1)} = 2R^{(0)} - D =$		−0.0001001	
$R^{(2)} = 2R^{(1)}$	=	−0.001001	$\|2R^{(1)}\| < \frac{7}{16} \to q_2 = 0$
$R^{(3)} = 2R^{(2)}$	=	−0.01001	$\|2R^{(2)}\| < \frac{7}{16} \to q_3 = 0$
$R^{(4)} = 2R$	=	−0.1001	$2R^{(3)} < \frac{-7}{16} \to q_4 = \bar{1}$
$+D$	= +)	0.1001	
		0.0000	

Answer: $Q = 0.q_1q_2q_3q_4 = 0.100\bar{1}$ minimal form yielded.

Figure 7.13 Metz's modified **SRT** division method yielding minimal **SD** quotient.

7.9 Robertson's High-Radix Division

Robertson originally proposed the **SRT** division in a generalized form [11] for arbitrary radix r with a quotient digit set as shown in Eq. 7.4. The same recursive formula (Eq. 7.5) for nonrestoring division can be used to describe Robertson's generalized method. However, successive quotient digits are selected such that Eq. 7.17 is satisfied. Several discrete values of the constant k used in $|R^{(j+1)}| \leq k \times |D|$ for known division methods exist in the range

$$\tfrac{1}{2} \leq k \leq 1 \tag{7.52}$$

In particular, $k = 1$ corresponds to nonrestoring division (Eq. 7.16).

The mechanization of Robertson's division method requires three distinct steps:

Step 1. The partial remainder $R^{(j)}$ is left shifted *one* digital position (in radix r, this means $[\log_2 r]$ binary-digit shifts) to obtain $r \times R^{(j)}$, the same as that described in section 7.6.

Step 2. One of several permissible arithmetic procedures is selected, such that the maximum absolute value of $r|R^{(j)}|$, namely, $rk|D|$ as shown in Eq. 7.17, is reduced by the amount of $1/r$, so that the next partial remainder $R^{(j+1)}$ satisfies Eq. 7.17.

Step 3. A quotient digit is generated corresponding to the arithmetic procedure selected.

The key to the success of this high-radix division lies in the analysis of arithmetic procedures, which enable Step 2 as stated above. Attention is focused on arithmetic procedures, which transform $r \cdot R^{(j)}$ into $R^{(j+1)}$ according to the following set of equations:

$$R^{(j+1)} = r \times R^{(j)} - i \times |D| \tag{7.53}$$

for $i = -m, \ldots, -2, -1, 0, 1, 2, \ldots, m$ and $(r - 1)/2 \leq m \leq r - 1$.

We have called each equation in Eq. 7.53 a ***q*-line**. The set of ***q*-lines** can be interpreted as a plot of $R^{(j+1)}$ versus $r \times R^{(j)}$ with running parameter q_{j+1}. We have called this plot a Robertson diagram. The $2m + 1$ ***q*-lines** are confined within the following rectangle centered at the origin with vertices

$$(\pm rk \times |D|, \pm k \times |D|) \tag{7.54}$$

as shown in Fig. 7.14. The projections on the $r \times R^{(j)}$ axis of the ***q*-lines** within the rectangle cover that portion of the $r \times R^{(j)}$ axis within the rectangle. The ***q*-lines** have slope r, whereas the vertices (Eq. 7.54) lie on the lines through the origin of slope $\pm(1/r)$. The rectangle must be sufficiently large to ensure that $k \geq \frac{1}{2}$.

Two considerations governing the choice of the values of k and m are given below:

1. The number of ***q*-lines** should be minimized, because the number of multiples of the divisor which must be formed during division is proportional to the number of ***q*-lines**.
2. The overlap of projections of the ***q*-line** on the $r \times R^{(j)}$ axis should be maximized. The precision necessary in the quotient-selection process decreases as the overlap increases. Remember that the less quotient precision required, the faster the selection will be.

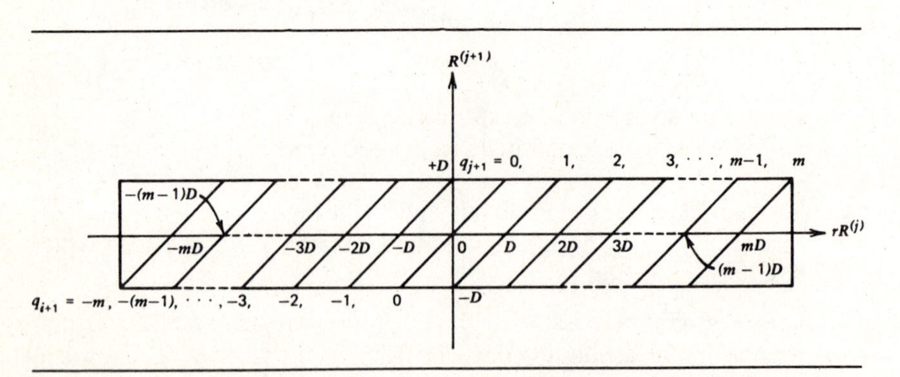

Figure 7.14 The Robertson diagram for the Robertson's high-radix **SRT** division with digit set $\{-m, \ldots, -2, -1, 0, 1, 2, \ldots, m\}$.

These two conditions are actually contradictory to each other. Therefore, the choice of the constant k, confined in the range of Eq. 7.52, should be taken from a discrete set of values such that the upper right vertex

$$(r \times k \times |D|, k \times |D|) \tag{7.55}$$

of the rectangle lies on the rightmost ***q*-line**

$$R^{(j+1)} = r \times R^{(j)} - m \times |D| \tag{7.56}$$

The values of k as a function of r and m can be found by solving for the point of intersection of line $R^{(j+1)} = r \times R^{(j)}$ and the ***q*-line** specified in Eq. 7.56. The value of $R^{(j+1)}$ at the intersection is k and equals $m/(r-1)$. Therefore, we obtain

$$k = \frac{m}{r-1} \tag{7.57}$$

With the condition $k \geq \frac{1}{2}$, we obtain

$$m \geq \frac{r-1}{2} \tag{7.58}$$

With the obvious upper bound of $r - 1$, we have

$$\frac{r-1}{2} \leq m \leq r-1 \tag{7.59}$$

Given a radix r, the optimal choices of m and thus k depend heavily on the detailed design tradeoffs, such as time and equipment cost.

In the binary **SRT** division with $r = 2$, we have used $m = r - 1 = 2 - 1 = 1$ and the upper limit $k = m/(r-1) = 1/(2-1) = 1$. In the following example case of radix $r = 4$ with digit set $\{-2, -1, 0, 1, 2\}$, we choose $m = 2$. All the necessary divisor multiples $\pm 2D$, and $\pm D$ can be formed by shifting and complementation, which are much simpler compared with the generation of multiples D, $2D$, and $3D$ in a conventional radix-4 restoring division with digit set $\{0, 1, 2, 3\}$.

A radix-4 Robertson's division scheme is illustrated in Fig. 7.15. The unit consists of a simple binary adder, a conditional doubling and complement circuit, and a selection logic. The selection circuit compares $r \times R^{(j)} = 4R^{(j)}$ with $0.5D$ and $1.5D$ to a precision of 7 bits, provided

$$\tfrac{1}{4} \leq |D| < 1 \tag{7.60}$$

Table 7.3 provides the correspondence of the selected quotient digit values and the operations performed in the three selection circuits in Fig. 7.15.

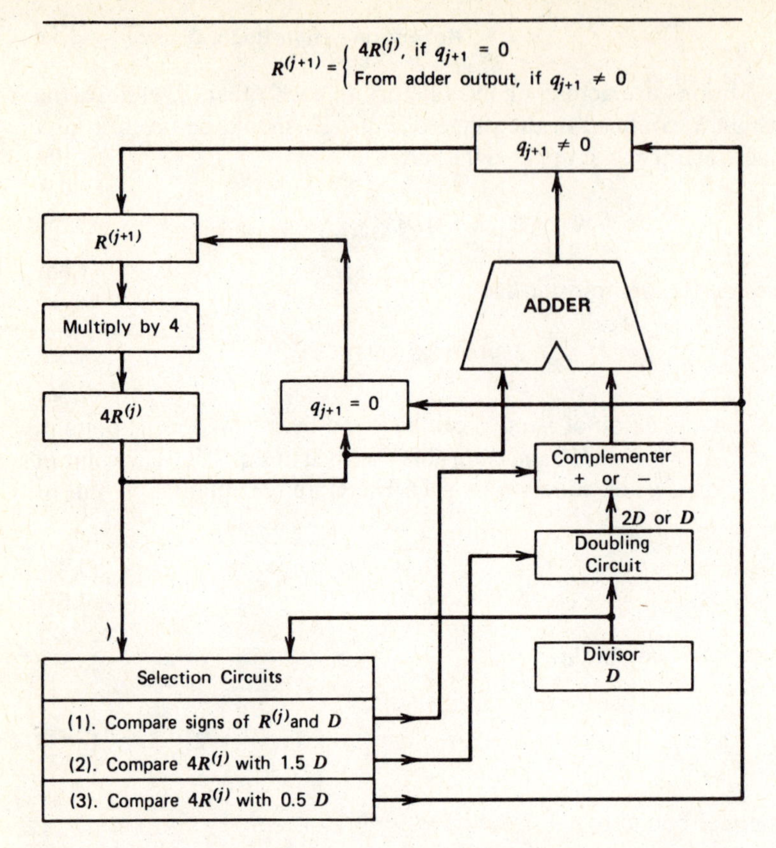

Note: ① When signs of $R^{(j)}$ and D agree, the complementer is set to subtract, otherwise add.

② The doubling circuit outputs $2D$ if $4\,|R^{(j)}| \geq \frac{5}{3}\,|D|$;
D if $4\,|R^{(j)}| \leq \frac{4}{3}\,|D|$, and either way if $\frac{4}{3}\,|D| < 4\,|R^{(j)}| < \frac{5}{3}\,|D|$.

Figure 7.15 The radix-4 **SRT** division scheme proposed by Robertson [11].

Table 7.3 The Quotient-Selection and Selection-Logic Correspondence in Fig. 7.15

Selection Circuit (1)	Selection Circuit (2)	Selection Circuit (3)	Value of q_{j+1}
Subtract	2D	$q_{j+1} \neq 0$	+2
Subtract	1D	$q_{j+1} \neq 0$	0
Subtract	1D	$q_{j+1} \neq 0$	+1
Add	2D	$q_{i+1} = 0$	−2
Add	1D	$q_{j+1} = 0$	0
Add	1D	$q_{j+1} \neq 0$	−1

7.10 Case Study II: The ILLIAC-III Arithmetic Processor

An arithmetic processor, which adopts the signed-digit (**SD**) redundancy and high-radix methods, including the multiplier recoding and **SRT** division, is illustrated in this section. The design of such a radix-256 arithmetic processor has been developed in the ILLIAC III Computer at the University of Illinois. This case study elaborates the hardware demands of functional blocks required. This presents an actual implementation of those theoretical models described in previous and present chapters for high-speed multiplication and division.

It should be noted that the design is not technology dependent. The main effort focuses on the organization of the functional subblocks into an efficient arithmetic processor. A high-speed arithmetic unit usually includes a substantial investment in hardware to accelerate the execution of MULTIPLY; we hope much of this investment may be also used to speed up the execution of DIVIDE instruction.

The strategy adopted in this arithmetic unit is to design a high-speed multiplication scheme, and to embed the generalized high-radix **SRT** division within it. In order to emphasize the design of major recursive steps, the preliminary and terminal steps such as overflow detection and floating-point provisions are not included in the illustrations. Seven byte (56 bits) fractional data are assumed. With one additional byte for the exponent, the unit can be expanded to handle 60-bit (8 bytes) floating-point operations.

Figure 7.16 shows the functional block diagram of the ILLIAC III arithmetic processor. The conventions used in the figure are described below:

1. Working registers are denoted by 7-byte rectangles with the subdivisions indicating bytes. In other words, registers are each 56 bits. Each 56-digit **SD** number is stored in two 56-bit registers (two bits per each signed digit).
2. Combinational logic modules are shown in circles or rectangles with rounded corners. Logic gating is represented in terms of AND($\wedge$), OR($\vee$), and XOR($\oplus$).
3. Thick bus lines represent data in **SD** form. Thin bus lines correspond to conventional data with 1 bit per digit. The pairing of registers for holding redundant **SD** numbers is indicated by abbreviating the pair, such as **US** and **UM**, by **USM**.
4. Register-to-register transfer gating signals are indicated as follows

$$\begin{aligned} &\mathbf{S}\ D\ \mathbf{T}, \\ &\mathbf{S}\ Rn\ \mathbf{T}, \qquad (7.61) \\ &\mathbf{S}\ Ln\ \mathbf{T}, \end{aligned}$$

where **S** and **T** are the *source* and *destination* registers, respectively. The center-positioned "*D*" means "*D*irect transfer" of the contents of **S** to **T**, and "*Rn*" and "*Ln*" mean that the contents of the source register is shifted *n* places to the "*R*ight" and "*L*eft," respectively, during the transfer to the destination register.

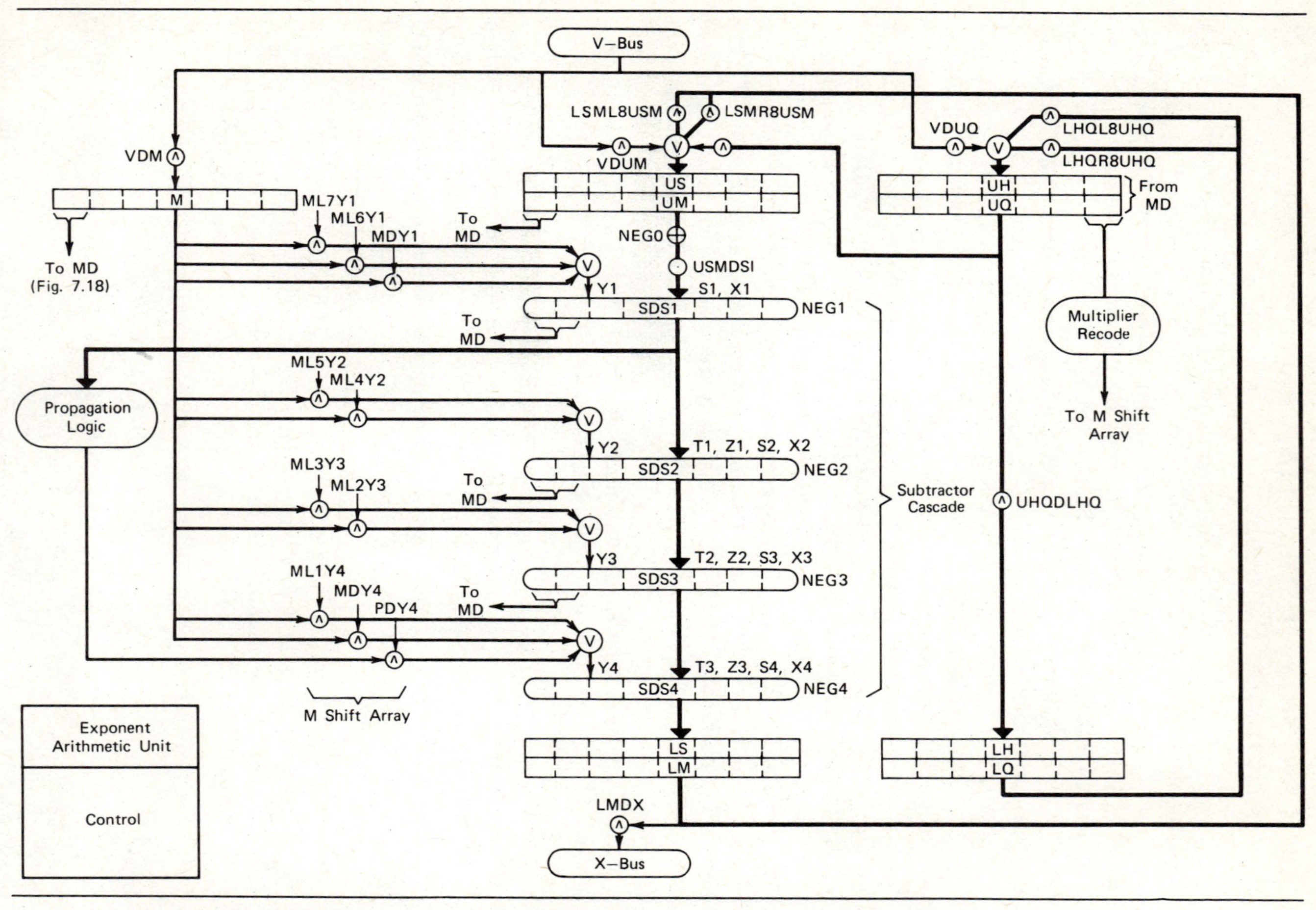

Figure 7.16 The functional block diagram of the ILLIAC-III arithmetic processor. (Atkins [3, 4]).

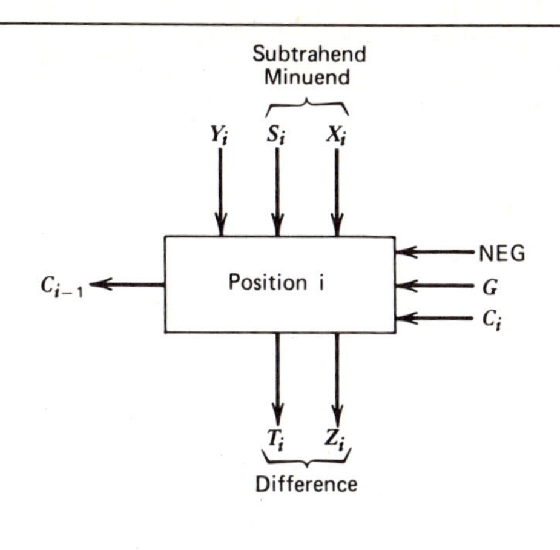

S_i = sign of minuend digit
X_i = magnitude of minuend digit
Y_i = subtrahend in conventional binary form
T_i = sign of difference digit
Z_i = magnitude of difference digit
NEG = control to complement T_i
If NEG = 1, then T_i is complemented, or else not
G = gate on interpositional connections
C_i = interpositional connection

$$T_i = C_i \oplus \text{NEG}$$
$$Z_i = C_i \oplus X_i \oplus Y_i$$
$$C_{i-1} = (S_i X_i \vee X_i Y_i)G$$
$$C_i = (S_{i+1} X_{i+1} \vee X_{i+1} Y_{i+1})G$$

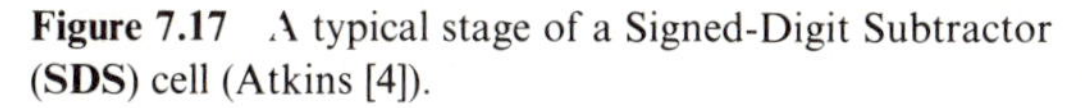

Figure 7.17 A typical stage of a Signed-Digit Subtractor (**SDS**) cell (Atkins [4]).

The add/subtract logic is formed with four *signed-digit subtractors* (**SDS**) in cascade, as denoted by SDSi for $i = 1, 2, 3, 4$ in the block diagram. The basic **SDS** cells are described in Fig. 7.17. The **SDS** cascade will perform addition or subtraction on the data stored in register **M** in conventional form and on the data stored in **USM** registers in **SD** form. The cascade incorporates the facilities to postpone the carry/borrow propagation, until the terminal step. The control signals NEG$_i$ for $i = 1, 2, 3, 4$ govern the required "add" or "subtract" operations.

Each line Y_i for $i = 1, 2, 3, 4$ is a bit of the multiplicand or of the divisor from the **M** register. Note that the following operations are to be performed by the **SDS** cascade.

$$\text{(old partial product)} \quad + \quad \text{(multiplicand)} \\ = \quad \text{(new partial product);} \qquad \textbf{(7.62)}$$

$$\text{(old partial remainder)} \quad - \quad \text{(divisor)} \\ = \quad \text{(new partial remainder)} \qquad \textbf{(7.63)}$$

Each digit of the old partial product or remainder consists of two bits, $S_i M_i$, stored in the pair of registers **USM**, where M_i is the magnitude bit and S_i is the associated sign bit. The outputs of SDS*i* serve as inputs to SDS($i + 1$) for $i = 1, 2, 3$. The final outputs of SDS4 go to the pair register **LSM**.

This **SDS** cascade provides multiplication in radix 256; that is, 8 bits of the multiplier are scanned simultaneously. The multiplicand is loaded from the V-bus to the **M** register, the multiplier into **UQ**. The low-order byte of **UQ** drives recoding logic, which couples to the control lines in the shift array. The recoding requires plus and minus multiples of 1, 2, 4, 8, 16, 32, 64, and 128 times the multiplicand. The multiples are formed by shift array as shown by the AND logic controls. During the execution of a MULTIPLY, the contents of **LSM** and **LHQ** (partial product and multiplier) are shifted right 8 bits back into the **USM** and **UQ** registers. This continues 8 cycles, and the 9th cycle converts the product in **SD** form through the conversion logic. The outputs of the conversion logic are routed to SDS4, and consequently converted to conventional representation.

For DIVIDE execution, the divisor is placed in register **M**, and the dividend and subsequent partial remainders in **SD** form are in **USM** registers. The quotient digits in redundant **SD** form are stored in **UHQ** registers, the sign bits in **UH**, and magnitude bits in **UQ**. The same conversion logic, as used in the execution of MULTIPLY, will be used for DIVIDE. The contents of **UHQ** are gated to **USM** by the control signal **UHQ*D*USM** and then converted into the conventional binary form at the final cycle. The hardware will require an 8-bit left shift **LSM** to **USM** under the control signal **LSM*L*8USM**, and from **LHQ** to **UHQ** under the control signal **LHQ*R*8UHQ**, and so on.

A radix-4 *model division* (**MD**) is sketched in Fig. 7.18, in which quotient digits are selected based on inspection of only the 6 high-order bits of the current divisor and partial remainders. The model division determines which multiples of the divisor are to be subtracted from the partial remainder through the **SDS** cascade. The next partial remainder is stored in **LSM** and then shifted left 8 bits into **USM**. These cycles continue until the full precision quotient is generated. The quotient is then gated directly from **UHQ** into **USM**, converted, and gated into **LM**, where it is available to the control processing unit. The radix-256 division is obtained by means of four successive applications of radix-4 model division.

Table 7.4 provides a summary of the functions of all subblocks of the arithmetic processor shown in Fig. 7.16 and of the **MD** unit shown in Fig. 7.18. Note that a radix-4 multiplication takes place at each SDS stage. The values of the multiples selected per each SDS stage are as follows:

$$\begin{array}{ll} \text{SDS1} & 0, \pm 128A, \pm 64A \\ \text{SDS2} & 0, \pm 32A, +16A \\ \text{SDS3} & 0, \pm 8A, +4A \\ \text{SDS4} & 0, \pm 2A, \pm A \end{array} \tag{7.64}$$

where A is the multiplicand. The recoding is performed in 3-bit overlapping groups according to the specification in Table 7.5. The model division can be viewed as quotient recoder, which determines which multiples of the divisor are to be subtracted.

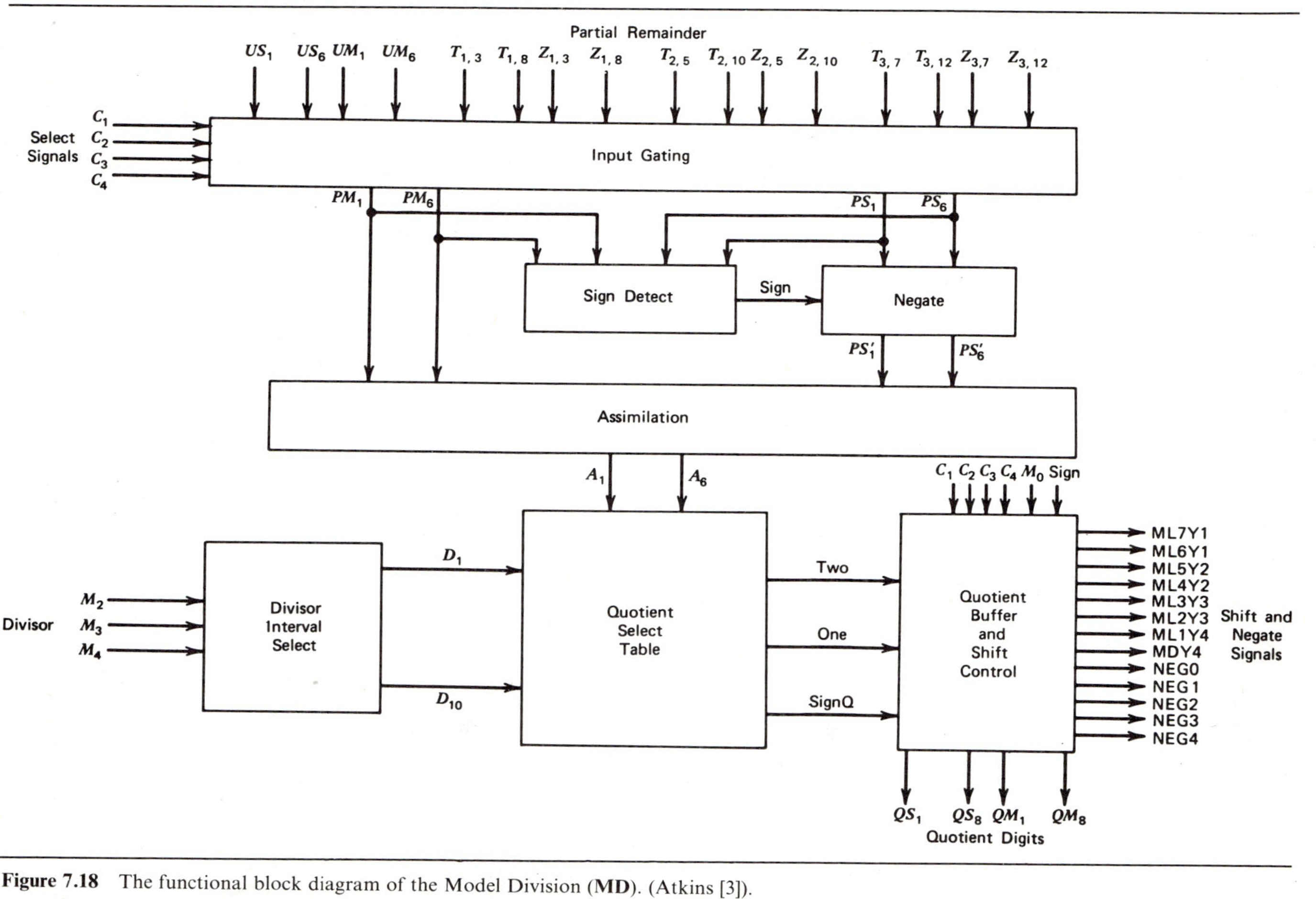

Figure 7.18 The functional block diagram of the Model Division (**MD**). (Atkins [3]).

Table 7.4 A Summary of the Functions of all Subblocks in Fig. 7.16 and in Fig. 7.17

Abbreviated Name	Name	Description[a]
M	Multiplicand register	Holds a number in conventional binary form to be added or subtracted from the redundantly represented number in **US-UM**.
US	Upper sign register	Part of the primary rank of the double rank registers associated with the **SDS** array. Holds the sign bits of a binary number in **SD** format to be used as input to SDS1.
UM	Upper magnitude register	Part of the primary rank of the double rank register associated with the **SDS** array. Contains the magnitude of binary numbers in **SD** format to be used as input to SDS1.
LS	Lower sign register	Secondary rank for the **US** register.
LM	Lower magnitude register	Secondary rank for the **UM** register.
UH	Upper **H** register	The primary rank of a double rank register used to hold the sign bits of a quotient and in aligning and normalizing operands.
LH	Lower **H** register	The secondary rank for the **UH** register.
UQ	Upper **Q** register	The primary rank of double rank register used to hold the magnitude bits of a quotient and in aligning and normalizing operands. Holds the multiplier during multiplication.
LQ	Lower **Q** register	The secondary rank for **UQ**.
V-BUS	In bus (the "**V**" denotes an arrowhead point into the system)	Provides input (to **AU**) interface between exchange net and **AU**. Includes cable terminators and parity checking.
X-BUS	Exit bus	Provides output (from **AU**) interface between **AU** and exchange net. Includes cable drivers and parity generation.
MR	Multiplier recode	Recodes the lower order 9 bits of **UQ** register into signals which operate the gates of the **M** shift array during multiplication.
SDS1–SDS4	Signed-digit subtracter 1 through 4	Signed-digit subtracters.
PL	Propagation logic	Produces the P_i bits from the T_i and Z_i outputs of S1. These P bits are then combined with Z_i bits in S4 to produce the assimilated result. This unit is similar to borrow lookahead logic.
MSA	**M** shift array	Selects multiples of the contents of the **M** register to be added in SDS1, SDS2, SDS3, SDS4 of the subtracter cascade.
MD	Model division	A radix-4 table-lookup division which generates quotient digits to be stored in **UH-UQ**, and to control the **M** shift array in forming full precision partial remainders.

(**Table 7.4**, continued)

EAU	Exponent arithmetic unit	Performs arithmetic on the 7-bit exponents of floating-point operands.
IG	Input gating	AND-OR gating configuration to gate the partial remainder inputs selected by the control signal C_i to subsequent stages of the model.
SDT	Sign detect	To determine the sign of the selected inputs, i.e. the sign of the leading nonzero digit. Used to control NEGATE and in forming the sign of the quotient digits.
NEG	Negate	To negate the selected inputs by complementing all of the **PS** bits. With this feature the quotient select table need only implement the first quadrant of the Robertson diagram.
ASM	Assimilation	Converts partial remainder in **SD** format into a conventional binary number. Uses borrow look-ahead technique to accelerate this step.
DISL	Divisor interval select logic	DECODES divisor, i.e., M_1 to M_4. Because $M_1 = 1$, it may be eliminated.
QST	Quotient select table	To logical implementation of the Robertson diagram. It may be constructed with diode matrix logic.
QBSC	Quotient buffer and shift control	Stores the quotient digits unit all 8 are formed and gated to the lower order byte of **UH-UQ**. Produces the **M**-shift **ARRAY** gate signals and the NEGI signals which control whether the **SDS** adds or subtracts the selected multiple of the divisor.

[a] **AU** = Arithmetic Unit

Table 7.5 The Multiplier Recoding Scheme Used in ILLIAC III (Wallace)

Triplet of Multiplier Bits			
X_{i-1}	X_i	X_{i+1} [b]	**Recoded Digit/ Multiple Selected** [a]
0	1	1	$+2A$
0	1	0	$+A$
0	0	1	–
0	0	0	0
1	1	1	–
1	1	0	$-A$
1	0	1	–
1	0	0	$-2A$

[a] A is the multiplicand.
[b] $X_{i+1} = 0$ initially.

Redundancy has been introduced into the representation of the quotient similar to that in multiplier recoding. The *Divisor Interval Selection Logic* (**DISL**) is specified in Table 7.6. The range of the quotient for the division of two nonzero fractions f_1 and f_2 is given by $\frac{1}{4} < f_1/f_2 < 16$. Division by zero is detected during the preliminary steps of the operation.

Table 7.6 The Specification of the *Divisor Interval Selection Logic* (**DISL**) of the Model Division

Interval Name	Logic Equations	Range of Divisor d Represented
D_1	$\overline{M}_{10}\overline{M}_{11}\overline{M}_{12}$	$\frac{1}{2} \leq d < \frac{9}{16}$
D_2	$\overline{M}_{10}\overline{M}_{11}M_{12}$	$\frac{9}{16} \leq d < \frac{5}{8}$
D_3	$\overline{M}_{10}M_{11}\overline{M}_{12}$	$\frac{5}{8} \leq d < \frac{11}{16}$
D_4	$\overline{M}_{10}M_{11}M_{12}$	$\frac{11}{16} \leq d < \frac{3}{4}$
D_5	$M_{10}\overline{M}_{11}$	$\frac{3}{4} \leq d < \frac{7}{8}$
D_6	$M_{10}M_{11}$	$\frac{7}{8} \leq d < 1$
D_7	$D_1 \vee D_2 \vee D_3$	$\frac{1}{2} \leq d < \frac{11}{16}$
D_8	$D_4 \vee D_5 \vee D_6$	$\frac{11}{16} \leq d < 1$
D_9	$D_4 \vee D_5$	$\frac{11}{16} \leq d < \frac{7}{8}$
D_{10}	$(D_5 \vee D_6) = M_{11}$	$\frac{3}{4} \leq d < 1$
D_{11}	$(D_1 \vee D_2 \vee D_3 \vee D_4) = \overline{M}_{11}$	$\frac{1}{2} \leq d < \frac{3}{4}$

7.11 Bibliographic Notes

The material included in this chapter follows the ideas of Robertson [11] and his followers. Excellent surveys on the subjects have been contributed by MacSorley [14], Garner [6], and Tung [17]. The invention of **SRT** division is attributed to Sweeney [14], Robertson [11], and Tocher [15]. The improvements of **SRT** division were made by Metze [10] and Wilson and Ledley [18]. High-radix division methods were originally due to Robertson [11] and later by Atkins [2, 3, 4] and Kalaycioglu [8]. Freiman [5] and Shively [13] have evaluated the performance of **SRT** division via statistical analysis. The case study of **ILLIAC III** arithmetic design is selected from the work of Atkins, especially from his M.S. thesis [3] and reference [4]. The Robertson diagram was named after the pioneering work by Robertson. Interested readers are urged to study the more recent work of Robertson [12], which establishes the correspondence between methods of digital division and multiplier recoding parameters. Such correspondence has been used in the **ILLIAC-III** arithmetic architecture as described in Section 7.10.

References

[1] Avizienis, A., "The Recursive Division Algorithm," *Class Notes*, Engr. 225A, UCLA, Los Angeles, California, 1971.

[2] Atkins, D. E., "Higher-Radix Division Using Estimates of the Divisor and Partial Remainders," *IEEE Trans. Comput.*, Vol. C-17, No. 10, October 1968, pp. 925–934.

[3] Atkins, D. E., "The Theory and Implementation of SRT Division," *Tech. Report No. 230*, Dept. of Computer Science, University of Illinois, Urbana, Illinois, 1967.

[4] Atkins, D. E., "Design of Arithmetic Units of ILLIAC III: Use of Redundancy and Higher Radix Methods," *IEEE Trans. Computer*, August 1970, pp. 720–733.

[5] Freiman, C. V., "Statistical Analysis of Certain Binary Division Algorithms," *Proc. of IRE*, Vol. 49, No. 1, January 1961, pp. 91–103.

[6] Garner, H. L., "Number Systems and Arithmetic," in *Advances in Computers*, Vol. 6, 1965, pp. 168–177.

[7] Gilman, R. E., "A Mathematical Procedure for Machine Division," *Comm. of Ass. for Comp. Mach.*, Vol. C, No. 4, April 1959, pp. 10–12.

[8] Kalaycioglu, V. U., "Analysis and Synthesis of Generalized-Radix Additive Normalization Division Techniques," *Tech. Report*, Dept. of Elec. & Comp. Engr., University of Michigan, Ann Arbor, Michigan, May 1975.

[9] Krishnamurthy, E. V., "On Range-Transformation Techniques for Division," *IEEE Trans. Comput.*, Vol. C-19, No. 3, March 1970, pp. 227–231.

[10] Metze, G., "A Class of Binary Divisions Yielding Minimally Represented Quotients," *IRE Trans.*, Vol. EC-11, No. 6, December 1962, pp. 761–764.

[11] Robertson, J. E., "A New Class of Digital Division Methods," *IEEE Trans. Comput.*, Vol. C-7, September 1958, pp. 218–222.

[12] Robertson, J. E., "The Correspondence Between Methods of Digital Division and Multiplier Recoding Procedures," *IEEE Trans. Comput.*, Vol. C-19, No. 8, August 1970, pp. 692–701.

[13] Shively, R. R., "Stationary Distribution of Partial Remainders in SRT Digital Division," *PH.D. Thesis*, University of Illinois, Urbana, Illinois, 1963.

[14] MacSorley, O. L., "High-Speed Arithmetic in Binary Computers," *Proc. of IRE*, Vol. 49, January 1961, pp. 67–91.

[15] Tocher, K. D., "Techniques of Multiplication and Division for Automatic Binary Computers," *Quart. J. Mech. Appl. Math.*, Vol. XI, Pt. 3, 1958, pp. 364–384.

[16] Tung, C., "A Division Algorithm for Signed-Digit Arithmetic," *IEEE Trans. Comput.*, Vol. 17, 1968, pp. 887–889.

[17] Tung, C., "Arithmetic," Chap. 3 in *Computer Science*, A. F. Cardenas, et al. (eds.), Wiley-Interscience, N.Y., 1972.

[18] Wilson, J. B. et al., "An Algorithm for Rapid Binary Division," *IRE Trans.*, Vol. EC-10, No. 4, December 1961, pp. 662–670.

Problems

Prob. 7.1 Apply the following data to a 16-bit restoring divider (Fig. 7.3) and list the successive contents of the partial remainder, of the divisor, of the link, of the adder inputs and outputs, and the quotient bits being generated. Arrange the listing in a tabular form similar to Table 7.1.

$$\text{Dividend } A = 0.0110110111100111$$
$$\text{Divisor } B = 0.10100111$$

Prob. 7.2 Convert the following radix-r redundant signed-digit numbers into the conventional binary form using the algorithm given in Fig. 7.6.

(a) $Q = (7\ \underline{6}\ \underline{9}\ \underline{2})_{10}$ for $r = 10$,
(b) $Q = (\underline{7}\ 6\ 5\ \underline{4})_8$ for $r = 8$,
(c) $Q = (\underline{1}\ 1\ \underline{1}\ \underline{1}\ 1\ 1)_2$ for $r = 2$

Verify your answers by checking the values represented.

Prob. 7.3 Explain why the efficiency of a binary **SRT** divider is sensitive to the range of the divisors being used. Prove that when $D = \frac{3}{4}$, the quotient will be in canonical minimal signed-digit form.

Prob. 7.4 Apply the Wilson-Ledley algorithm to work out the quotient $Q = N/D$ with given $N = 0.0101100010110111$ and $D = 0.1001$ for $n = 10$.

Prob. 7.5 Prove that Metze's modified **SRT** method is identical to the original **SRT** division, whenever the divisor lies within the range $\frac{3}{5} \leq |D| \leq \frac{3}{4}$.

Prob. 7.6 Prove that Metze's modified **SRT** division always yields a minimal signed-digit quotient, even when the divisor is outside the range specified in Prob. 7.5.

Prob. 7.7 Prove that the parameter m used in Eq. 7.55 must be chosen within the range $(r - 1)/2 \leq m \leq r - 1$ in order to satisfy the requirements for radix-r **SRT** division.

Prob. 7.8 Propose one possible practical realization of the Signed-Digit Subtractor (**SDS**) cell defined in Fig. 7.17. Show the complete schematic of the 7-byte **SDS** subtractor design with 56 such **SDS** cells. Then illustrate with numerical example how to subtract a conventional binary number from a binary coded signed-digit number, using this unit.

Prob. 7.9 Compare the binary restoring division and the binary nonrestoring division methods with respect to their processing speeds and hardware requirements. In the case of nonrestoring division, the hardware needed to convert the signed-digit quotient to the conventional binary form should be included in the comparison.

Prob. 7.10 Explain the merits of binary SRT division as compared with other fast standard division methods, from both designer's and user's viewpoints. Explore the relative increase in hardware complexity of higher-radix SRT division over the binary SRT division.

Chapter

8

Convergence Division and Cellular Array Dividers

8.1 Introduction

Most conventional computers use the trial subtraction division schemes described in the preceding chapter. Several unconventional division methods which differ from the trial subtraction division are described in this chapter.

Division by repeated multiplications shall be described. There are two types of multiplicative division methods. The first method treats the dividend and divisor as numerator and denominator of a fraction. It accomplishes the division by multiplying both the numerator and denominator with the same sequence of convergence factors until the denominator approaches unity. The resulting numerator then becomes the desired quotient. The second method achieves division by finding the reciprocal of the divisor, and then multiplying the dividend with this reciprocal to obtain the quotient. Both of these two methods converge to the quotient quadratically. The convergence division has been used in some large computers such as the CDC **6600** and **7600**, and the IBM System/**360** Model **91**. The main advantage of this type of division schemes is in the sharing of hardware required to multiply. A fast hardware multiplier is crucial to the success of multiplicative division via quadratic convergence.

The second half of this chapter is devoted to the design and construction of various high-speed iterative cellular arrays for parallel or semiparallel binary divisions. Large combinational logic arrays are used to realize the comparisons required in division. Shifting is realized by physical wiring in the array. This cellular array approach is especially attractive to LSI fabrication. The cellular array dividers need less control circuitry and still offer satisfactory speed as compared with the standard dividers. Three classes of division arrays are illustrated: the restoring, nonrestoring, and augmented division arrays. The Cappa-Hamacher [3] array divider, using carry save and lookahead techniques to speed up carry propagation in the array, will be described in detail. A cost-effectiveness analysis of all existing cellular array dividers is provided. In order to be consistent with the previous divider illustrations, all the numbers to be processed are assumed to be positive fractions.

8.2 Convergence Division Methods

A class of division methods entirely different from those described in the last chapter is presented below. These methods employ the multiplication hardware to carry out the division through a convergence approach. Iterative multiplications are performed to generate the desired quotient.

Let us consider the division operation

$$\frac{N}{D} = Q \tag{8.1}$$

where the quotient can be treated as a fraction formed by a numerator N and a denominator D. On each iteration, a constant factor R_k is chosen to multiply both the numerator and the denominator without changing the natural value of the ratio. The sequence of multiplying factors, R_k for $k = 0, 1, \ldots, n$, are chosen such that the resulting denominator converges quadratically toward unity, and the resulting numerator then converges quadratically toward the desired ratio or quotient Q.

This concept can be stated mathematically as

$$\frac{N}{D} = \frac{N}{D} \times \frac{R_0}{R_0} \times \frac{R_1}{R_1} \times \cdots \times \frac{R_n}{R_n} \tag{8.2}$$

and for sufficient large n

$$D \times R_0 \times R_1 \times \cdots \times R_n \to 1$$

then we obtain

$$N \times R_0 \times R_1 \times \cdots \times R_n \to Q \tag{8.3}$$

To match the design of floating-point arithmetic, we assume both N and D be positive and normalized fractions within the range

$$\frac{1}{r} \leq N, \quad D < 1 \tag{8.4}$$

The general rule of thumb is to choose the successive multipliers R_i for $i = 0, 1, \ldots, n$ such that

$$D_{i-1} < D_i \tag{8.5}$$

for all $i = 1, 2, \ldots, n$ where

$$D_0 = D \times R_0 \tag{8.6}$$

and

$$D_i = D \times R_0 \times R_1 \times \cdots \times R_i = D_{i-1} \times R_i \tag{8.7}$$

The process is continued with the multiplication of D_i by a new factor R_{i+1} until for some integer $i = n$ the denominator

$$D_n \to 1.0 \tag{8.8}$$

within the maximum machine precision. The normalized range in Eq. 8.4 implies that the divisor can be expressed as

$$D = 1 - \delta \tag{8.9}$$

for some δ in the range

$$0 < \delta \leq \frac{r-1}{r} \tag{8.10}$$

We can choose the first multiplying factor

$$R_0 = 1 + \delta, \tag{8.11}$$

then

$$\begin{aligned} D_0 &= D \times R_0 \\ &= (1 - \delta) \times (1 + \delta) \\ &= 1 - \delta^2 \end{aligned} \tag{8.12}$$

Clearly, $1 - \delta^2$ is larger than $1 - \delta$ and closer to unity. For the first iteration, select the multiplying factor

$$R_1 = 1 + \delta^2 \tag{8.13}$$

then

$$\begin{aligned} D_1 &= D_0 \times R_1 \\ &= (1 - \delta^2) \times (1 + \delta^2) \\ &= 1 - \delta^4 \end{aligned} \tag{8.14}$$

In general, at the ith iteration,

$$R_i = 1 + \delta^{2^i} \tag{8.15}$$

and

$$\begin{aligned} D_i &\doteq D_{i-1} \times R_i \\ &= (1 - \delta^{2^i}) \times (1 + \delta^{2^i}) \\ &= 1 - \delta^{2^{i+1}} \end{aligned} \tag{8.16}$$

Equation 8.5 is satisfied, because δ is a proper fraction and

$$\delta^{2^i} > \delta^{2^{i+1}} \tag{8.17}$$

thus

$$D_{i-1} = 1 - \delta^{2^i} < 1 - \delta^{2^{i+1}} = D_i \tag{8.18}$$

In the binary case, $\delta < \frac{1}{2}$, therefore,

$$\delta^{2^i} < 2^{-2^i} \tag{8.19}$$

and thus

$$1 - D_i = \delta^{2^{i+1}} < 2^{-2^{i+1}} \tag{8.20}$$

For sufficiently large i, say $i = n$, the error shown in Eq. 8.20 can be made very small and negligible.

It is important to note that the factor R_i specified in Eq. 8.15 can be directly obtained by taking the radix complement (2's complement in binary case) of the D_{i-1} as follows:

$$\begin{aligned} R_i &= 2 - D_{i-1} \\ &= 2 - (1 - \delta^{2^i}) \\ &= 1 + \delta^{2^i} \end{aligned} \tag{8.21}$$

The simplicity of obtaining R_i from D_{i-1} is the key merit of this method. After n iterations, the denominator

$$D_n = 1 - \delta^{2^n} \tag{8.22}$$

falls within the machine precision chosen to represent an one, say the all-one vector. Because the error, $\delta^{2n} < 2^{-2^{n+1}}$, in Eq. (8.22) becomes less than the least significant bit of the full word length, the numerator then becomes the desired quotient with an explicit value

$$\begin{aligned} Q = \frac{N}{D} &= N \times R_0 \times R_1 \times \cdots \times R_n \\ &= N \times (1 + \delta) \times (1 + \delta^2) \times \cdots \times (1 + \delta^{2^n}) \end{aligned} \tag{8.23}$$

The smaller the value of δ in the input divisor D, the faster the convergence rate will be. This quadratic convergence division method appeals especially to binary number system due to the ease in obtaining two's complement of binary numbers. It can be also applied to high-radix system, if the divisor is allowed to be expressed in bit-normalized form. Cubic and high-order convergence division methods will be left as exercises for the readers.

8.3 Multiplicative Divider Design

The convergence division scheme described in the preceding section has actually been implemented in several existing computers. The IBM System/**360** model **91** arithmetic execution unit is a good example. We shall show how the multiplicative convergence method is applied to design efficient dividers in the model **91** computer.

The IBM machine division deals with fractions of 56 bits long. The divisor D is a bit-normalized, floating-point fraction of the following form

$$D = 0.\underbrace{1\text{xxx}\cdots\text{x}}_{56\text{ bits}} \tag{8.24}$$

The convergence constant δ is therefore upper bounded by 0.5

$$\delta = 1 - D \leq 0.5 \tag{8.25}$$

With a 56-bit precision, the internal representation of unit (1) is normally approached from below, that is

$$1 = 0.\underbrace{111 \cdots 1}_{56 \text{ bits}} \tag{8.26}$$

Whenever the convergence factor δ^{2^k} becomes small enough such that

$$\delta^{2^k} < 2^{-56} \tag{8.27}$$

where 2^{-56} corresponds to the least significant bit of the divisor, the resulting denominator will approach 1 as indicated in Eq. 8.26.

The multiplying factor R_i for the ith iteration is obtained by taking the two's complement of the denominator D_{i-1} obtained at the $(i - 1)$st iteration as specified in Eq. 8.21. For a 56-bit fraction, five multiplying factors, R_0, R_1, R_2, R_3, R_4, are sufficient with the two's complement operations inserted in between, and the quotient

$$Q = N \times R_0 \times R_1 \times R_2 \times R_3 \times R_4 \tag{8.28}$$

where $R_4 = 2 - D_3$ and $D_3 = D \times R_0 \times R_1 \times R_2 \times R_3$.

The time for each multiplication can be reduced if the number of bits in the multiplier is reduced. The initial multiplier R_0 is truncated to t bits as $1 + \delta_t$, where $\delta_t - \delta < 2^{-t}$. It can be shown that the resulting denominator is equivalent to

$$\begin{aligned} D_1 &= (1 - \delta) \times (1 + \delta_t) \\ &= 1 - \delta^2 + \Delta_t \end{aligned} \tag{8.29}$$

where the factor Δ_t for $0 < \Delta_t < 2^{-t}$ is due to truncation. Because the additional Δ_t is always positive, the denominator can converge toward unity from above or below in two forms

$$D_i = \begin{cases} 0.11111 \cdots \text{xxxx} \cdots \\ 1.0000 \quad \cdots \text{xxxx} \cdots \end{cases} \tag{8.30}$$

The multiplier can be reformed as substrings of 0's or 1's according to the rate of convergence desired. The string bits, because they are all zero or all one, can be skipped in the multiplication process. Thus the multiply time can be improved considerably as can the divide time in IBM **360/91**.

In order to improve the initial minimum string length, and thus reduce the number of iterations, the first multiplier $R_0 = 1 + \delta$ is generated by table-lookup, addressed by the first seven bits of the divisor. The first multiply guarantees a result which has seven similar bits to the right of the binary point; that is, $1 \pm \delta$ has the form $\bar{a}.aaaaaaa \cdots$. The above operations associated with the execution of a DIVIDE are summarized by the flow chart in Fig. 8.1.

The design is intended to implement the operations specified in the flow chart

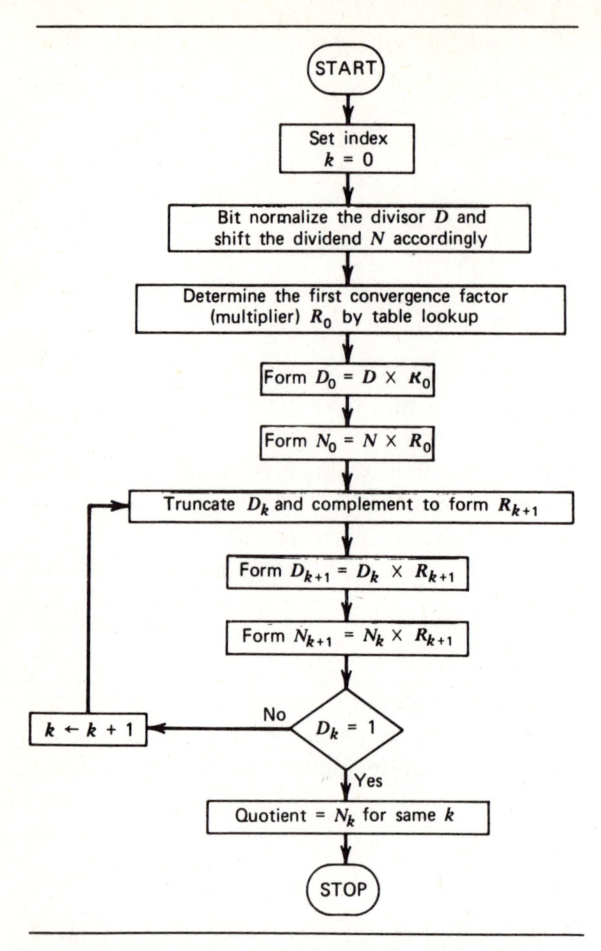

Figure 8.1 The sequence of operations in the execution of a DIVIDE using the multiplicative convergence method.

utilizing the MULTIPLY hardware sketched in Fig. 8.2. The carry-save adder loop is used to perform the six-operand addition required in each multiply step. Three major problems should be considered towards a complete design.

1. The multiplier R_i has a variable-length on each iteration. The first multiplier, determined by table-lookup, is 10 bits long and yields a minimum 0 or 1 string length of seven; the second multiplier is 14 bits; the third multiplier is 28 bits, and so on. The minimum 0 or 1 string length is doubled on each iteration.

2. The result of one iteration is the multiplicand for the next iteration. A carry propagate adder must be used to merge the sum and carry outputs of the carry-save adders in the divide loop.

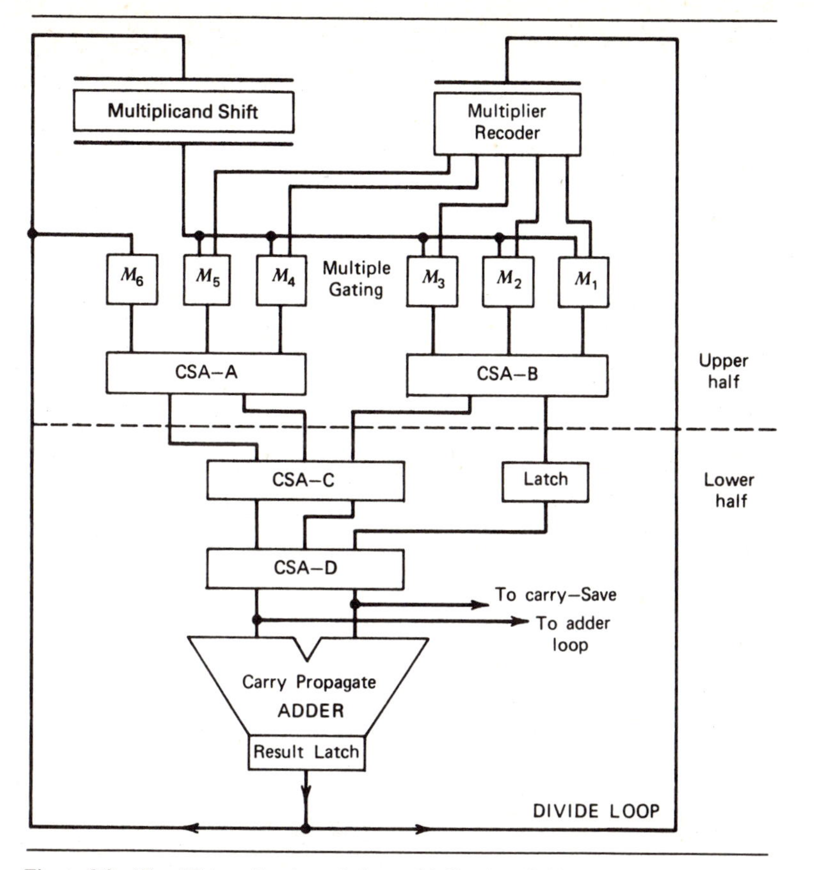

Figure 8.2 The **CSA** realization of the multiplicative division scheme (Anderson, et al. [1]).

3. It seems that two multipliers are required in each iteration—one for the numerator multiply and one for the denominator multiply. However, efforts are made to overlap the two multiplies and execute them by one multiply circuit. This will save both hardware and time.

The entries in Table 8.1 show the formats of the successive denominators and their multiplier for five iterations. The entries in Table 5.4 show that the leading 1's or 0's in a multiplier can be skipped, because they result in a zero multiple out of the multiplier recoder. The sign of the output changes without affecting its magnitude, if the input of the multiplier recoder is complemented. This property will be used to produce $\pm\delta^2$ at the output of the recoder.

To begin the execution of a DIVIDE the divisor D is multiplied by the first multiplier R_0 and the first denominator D_0 is generated at the outputs of the CSA tree. These two outputs are combined by the carry propagate adder and looped back to

Table 8.1 The Formats of the Denominators and Their Multipliers in the Multiplicative Convergence Division Process. (Reprint with permission, Anderson, et al. [1])

Digit	0	1	2	3	4	5	6	7	8	9	10	11	12	13	14	15	16
D	0	1xxx	xxxx	xxxx	xxxx	xxxx	xxxx	xxxx	xxxx	xxxx	xxxx	xxxx	xxxx	xxxx	xxxx	0000	0000
R_0	[01	xxxx	xxxx	xx0]	Determined by table lookup of denominator												
$D \times R_0 = D_0$	{0	1111	111x	xxxx	xxxx	xxxx	xxxx	xxxx	xxxx	xxxx	xxxx	xxxx	xxxx	xxxx	xxxx	xxxx	xxxx
	{1	0000	000x	xxxx	xxxx	xxxx	xxxx	xxxx	xxxx	xxxx	xxxx	xxxx	xxx	xxxxx	xxxx	xxxx	xxxx
R_1	{1	0000	[000x	xxxx	xx11	1]	Determined by complementing denominator										
	{0	1111	[111x	xxxx	xx11	1]											
$D_0 \times R_1 = D_1$	{0	1111	1111	1111	11xx	xxxx	xxxx	xxxx	xxxx	xxxx	xxxx	xxxx	xxxx	xxxx	xxxx	xxxx	xxxx
	{1	0000	0000	0000	00xx	xxxx	xxxx	xxxx	xxxx	xxxx	xxxx	xxxx	xxxx	xxxx	xxxx	xxxx	xxxx
R_2	{1	0000	0000	000[0	00xx	xxxx	xxx1]										
	{0	1111	1111	111[1	11xx	xxxx	xxx1]										
$D_1 \times R_2 = D_2$	{0	1111	1111	1111	1111	1111	111x	xxxx	xxxx	xxxx	xxxx	xxxx	xxxx	xxxx	xxxx	xxxx	xxxx
	{1	0000	0000	0000	0000	0000	000x	xxxx	xxxx	xxxx	xxxx	xxxx	xxxx	xxxx	xxxx	xxxx	xxxx
R_3	{1	0000	0000	0000	0000	0000	[000x	xxxx	xxxx	1]							
	{0	1111	1111	1111	1111	1111	[111x	xxxx	xxxx	1]							
$D_2 \times R_3 = D_3$	{0	1111	1111	1111	1111	1111	1111	1111	1111	xxxx	xxxx	xxxx	xxxx	xxxx	xxxx	xxxx	xxxx
	{1	0000	0000	0000	0000	0000	0000	0000	0000	xxxx	xxxx	xxxx	xxxx	xxxx	xxxx	xxxx	xxxx
R_4	{1	0000	0000	0000	0000	0000	0000	0000	0[000	xxxx	xxxx	x|x|xx	xxxx	xxxx	x|x|xx	xxxx	xxxx 11]
	{0	1111	1111	1111	1111	1111	1111	1111	1[111	xxxx	xxxx	x|x|xx	xxxx	xxxx	x|x|xx	xxxx	xxxx 11]
D_4	{0	1111	1111	1111	1111	1111	1111	1111	1111	1111	1111	1111	1111	1111	1111	1111	1111
(not formed)	{1	0000	0000	0000	0000	0000	0000	0000	0000	0000	0000	0000	0000	0000	0000	0000	0000

Note: Short precision divide result is $N_4 = N \times R_0 \times R_1 \times R_2 \times R_3$.
Long precision divide result is $N_5 = N_4 \times R_4$.

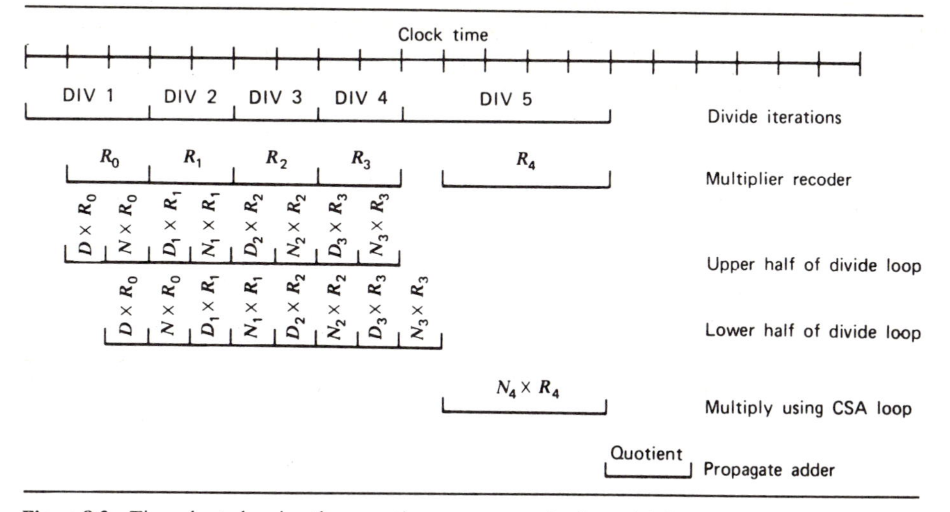

Figure 8.3 Time chart showing the execution concurrency in the multiplicative divide loop of Figure 8.2 (Anderson, et al. [1]).

the input as the new multiplicand. The truncated and complemented output forms the new multiplier. Two temporary storage latches are used at the outputs of the two halves of the divide loop. Thus, as soon as $D \times R_0$ is gated into CSA-C, the next multiply $N \times R_0$ can be started. Now $D_0 = D \times R_0$ advances to the result latch and loops back to start the next multiply $D_0 \times R_1$. At this time $N \times R_0$, which is latched in CSA-C, advances through the adder tree to the result latch. So the two multiplications follow each other around the divide loop. The first determines how the second should be multiplied by converging eventually to the quotient.

The chain operation continues until the last multiplier R_4 has been used. Because the denominator D_4 is equivalent to *one*, the multiplication is completed. The multiplier is gated 12 bits at a time as was shown in Table 8.1. The result of the final multiply $(N \times R_0 \times R_1 \times R_2 \times R_3) \times R_4$, is the final quotient Q. Figure 8.3 shows the timing chart that reveals the over-lapped concurrency in the divide loop. The multiplier recoder latch is changed each time a denominator multiply is completed. There are always two multiplications in execution,one in the upper half of the divider and one in the lower half of the divider as separated by the dash line in Fig. 8.2.

8.4 Division Through Divisor Reciprocation

Another multiplicative convergence method for binary division is given in this section. The method generates the *reciprocal* of the divisor by an iterative process, and then obtains the quotient by multiplying the dividend with the divisor reciprocal. A convergence algorithm is presented below to perform the required reciprocation. The method proves especially attractive to binary reciprocation, because of its easy implementation with combinational logic circuits. An efficient hardware multiplier is also needed in this division scheme.

Let Q as defined in Eq. 8.1 be the quotient to be evaluated. The divisor D is assumed to be a positive and normalized fraction within the range specified by Eq. 8.4. The reciprocal of D, denoted $1/D$, is therefore bounded in the range

$$1 < \frac{1}{D} \leq 2 \tag{8.31}$$

An initial approximation p_0 to $1/D$ can be found by either a combinational logic unit or by an ROM lookup table. Such an initial approximation is depicted by the block diagram in Fig. 8.4. To reduce the hardware complexity, only the first k digits to the right of digit $d_1 = 1, d_2 d_3 \cdots d_{k+1}$, of the divisor D are needed as the inputs to the approximator box, and the outputs of the box form the binary approximation p_0 in the form

$$p_0 = 1.s_1 s_2 \cdots s_t \tag{8.32}$$

where the values of k and t are determined by hardware and precision demanded. A partial table specifying the conversion logic of the approximator for the case of $k = 2$ and $t = 4$ is also given in Fig. 8.4.

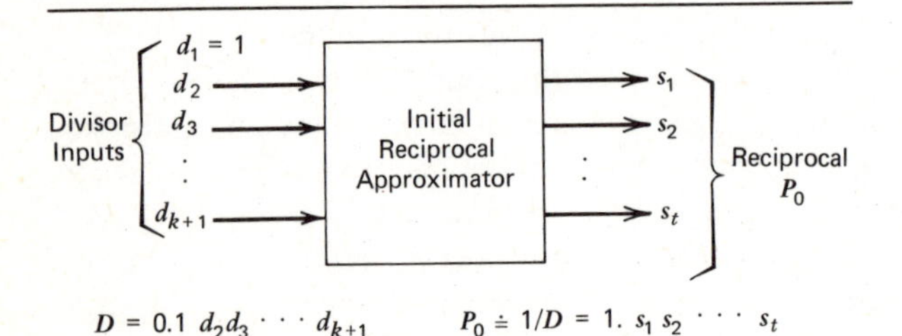

Inputs		Outputs			
d_2	d_3	S_1	S_2	S_3	S_4
0	0	1	1	1	1
0	1	1	0	0	1
1	0	0	1	0	1
1	1	0	0	1	0

Figure 8.4 Block diagram and partial table description of the initial reciprocal approximator.

Evidently, p_0 is a stepwise approximation to $1/D$, being constant in each of 2^j equal intervals between $\frac{1}{2}$ and 1. It can easily be shown that the optimal value of p_0 for the hth interval ($h = 1, 2, \ldots, 2^j$) is the reciprocal of the number corresponding to its midpoint, that is,

$$p_0(h) = \frac{2^{j+1}}{2^j + h - \frac{1}{2}} \tag{8.33}$$

A plot of the initial approximation p_0 versus the values of the divisor D is given in Fig. 8.5 for the case of $j = 2$.

The final reciprocal is obtained by the following recursive process with initial conditions p_0 and $a_0 = p_0 \times D$

$$p_i = p_{i-1} \times (2 - a_{i-1}) \tag{8.34}$$

where

$$a_i = a_{i-1} \times (2 - a_{i-1}) \tag{8.35}$$

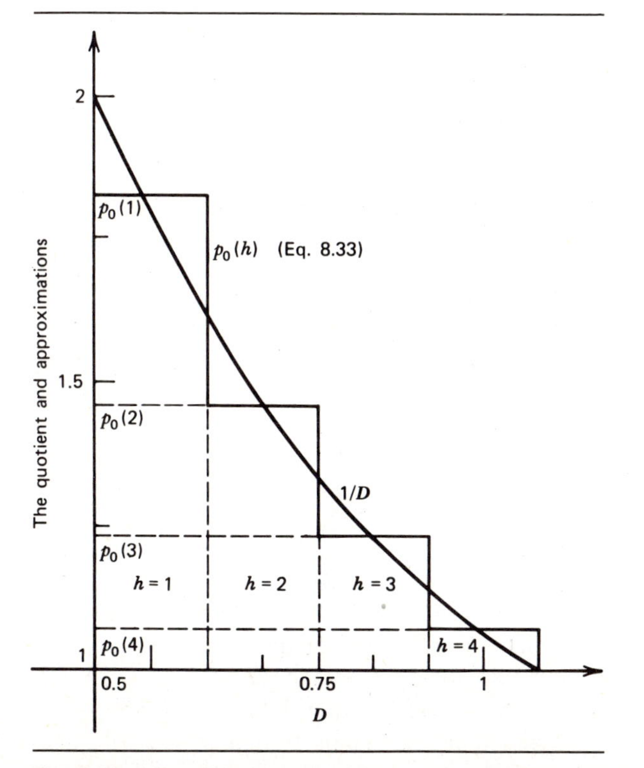

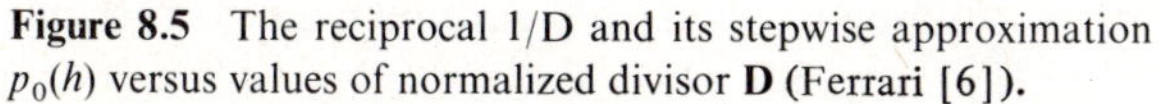
Figure 8.5 The reciprocal 1/D and its stepwise approximation $p_0(h)$ versus values of normalized divisor **D** (Ferrari [6]).

and $i = 1, 2, \ldots$ is the recursion index. Theoretically, the successive approximations p_i for all i approach the final reciprocal quadratically as the number of iteration increases, that is,

$$\lim_{i \to \infty} p_i = \frac{1}{D} \tag{8.36}$$

and

$$\lim_{i \to \infty} a_i = 1 \tag{8.37}$$

The actual number of iterations is determined by the machine precision. Let ε be a very small number reflecting the order of magnitude of the least significant digit of the machine word. The required n iterations satisfy

$$|1 - p_n \times D| < \varepsilon \tag{8.38}$$

where p_n is the final approximation of the calculated reciprocal. Circuit implementation of this reciprocal algorithm will be given in the next section.

8.5 Binary Reciprocator Using CSA Tree

Several arithmetic circuits have been proposed to implement the above reciprocal algorithm. Wallace [18] has suggested the use of a multilevel carry-save adder (**CSA**) tree for carrying out the iterative dual multiplications required in the reciprocal-converging process. Stefanelli [16] has proposed a different iterative array to generate the divisor reciprocal in redundant SD code, which may require a code converter to transform back into the conventional binary code. Only the Wallace tree approach is studied in this section.

The two multiplications specified in Eqs. 8.34 and 8.35 are to be executed by the **CSA** tree shown in Fig. 8.6. The **CSA** tree allows the overlapped execution of these two multiplications at each iteration of the convergence process. It is interesting to note that both operations share the same multiplier $(2 - a_{i-1})$ at the ith iteration. The multiplicands p_{i-1} and a_{i-1} of the two multiply operations are fed into the CSA tree at different levels; this makes the overlapping possible.

Multiplier recoding using redundant **SD** code was suggested by Wallace to reduce the total number of summands (multiples of the multiplicands) required to be added by the tree. The addition of these summands is completed in a time which is proportional to the logarithm of the number of summands. The recoding scheme requires only multiples of the multiplicand that are obtainable by shifting and complementing. It is also a *local* recoding, in which each recoded digit depends only on a small group of original multiplier bits. For example, base-4 recoded multiplier digits belong to $\{-2, -1, 0, +1, +2\}$ and each is determined entirely by three adjacent original bits. This will reduce the number of summands by half as described in Chapter 5. Negative multiples can be obtained with two's complement correction circuits.

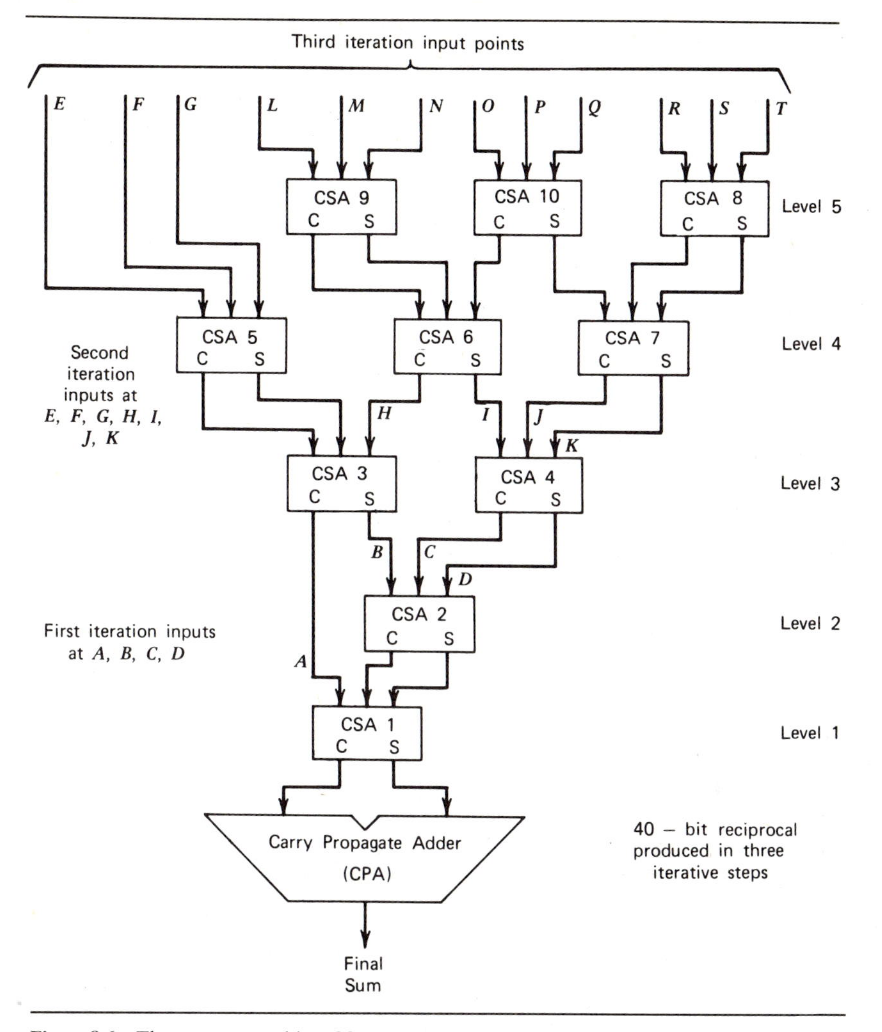

Figure 8.6 The carry-save adder (**CSA**) tree suggested by Wallace [18] for division using iterative binary reciprocation.

The initial approximation p_0 can be recoded to give only three summands, based on the choice of $k = 6$ inputs and $t = 4$ outputs for the approximator box shown in Fig. 8.4. In Wallace's original design,

$$|1 - a_0| \leq 2^{-5} \tag{8.39}$$

and a_0 has the form

$$a_0 = p_0 \times D = \bar{x}.xxxxxefghjk. \tag{8.40}$$

The first iteration step should result in 10 similar digits immediately after the binary point. It can be shown that the next $a_1 = a_0 \times (2 - a_0)$ of the form in Eq. 8.40

is obtainable by using an approximated multiplier, which is not exactly $2 - a_0$, but rather

$$2 - a_0 \simeq 1 - 2^{-5} \times (\bar{x}.efghj) \qquad \textbf{(8.41)}$$

where the number in parentheses is regarded as a signed two's complement fractional number. This approximated multiplier can be obtained directly from a_0 as shown in Eq. 8.40. Only four summands are required after recoding this multiplier. Similarly, an approximated multiplier can be used in the next iteration step requiring seven summands, and one in the third step requiring 12 summands. Only three iteration steps are needed to produce a 40-bit reciprocal.

Two distinct advantages of using the approximated multipliers are described below:

1. The small number of recoded multiplier digits allows the multiplication to be done by only part of the CSA tree, and the multiply time, especially in earlier steps, is reduced. For example, the four summand for $p_0 \cdot D$ can be introduced at points A, B, C, and D in the first iteration step. Those seven for the second step at points E, F, G, H, I, J, K.

2. The number of digits in p_i is small in the early steps. This, plus the fact that some of the leading digits of the a's need not be formed, implies that, for a 40-bit word length, both multiplications in each of the first and the second iterative steps can be performed simultaneously by splitting the CSA tree into two shorter-word length sections, similar to that in the divide loop of IBM **360/91** Execution unit.

The number of the levels on the **CSA** tree is determined by the word length; longer word length requires more levels. The last step "a" multiplication, that is, $a_n = a_{n-1} \times (2 - a_{n-1})$ if n is the last step, need not be done, because, $Q = p_n$ does not depend on it. Using the Wallace tree (Fig. 8.6), a 40-bit reciprocal can be obtained with only four passes, and at least the first three are faster than a full multiplication due to the overlapped use of the recoded approximated multipliers.

8.6 Restoring Cellular Array Divider

Restoring division requires two operations per each cycle. The first operation is always a trial subtraction of the divisor from the dividend starting from the most significant quotient digit. If the subtraction is unsuccessful by producing a negative difference, a quotient bit "0" is generated, and the second operation requires that the divisor be added back to restore the old partial remainder, because there has been an overdraw. The effects of the trial subtraction and restoration addition cancel each other, resulting in a correct partial remainder. The present cycle is completed by simply shifting the remainder one bit left, ready for the next trial cycle. On the other hand, if the current trial subtraction is successful, a quotient bit of "1" is selected and

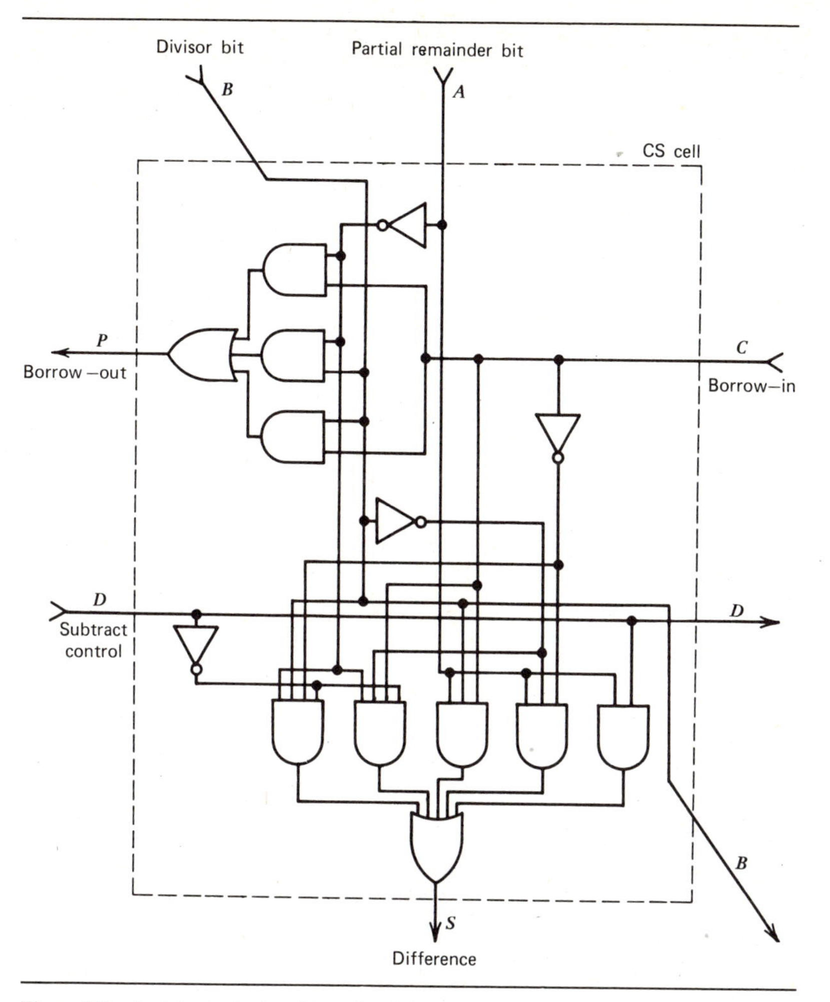

Figure 8.7 An interior logic schematic of the Controlled Subtractor (**CS**) cell.

the difference is equal to the new partial remainder. In this case, no restoring addition is needed.

These conditional restoring operations can be implemented with cellular iterative arrays. The basic cell used in the array construction is a *Controlled Subtractor* (**CS**) with four inputs and four outputs as shown in Fig. 8.7. The internal logic of the **CS** cell is specified by the following equations

$$P = \bar{A}B + \bar{A}C + BC \tag{8.42}$$

$$S = AD + A\bar{B}\bar{C} + ABC + \bar{A}B\bar{C}\bar{D} + \bar{A}\bar{B}C\bar{D} \tag{8.43}$$

where A, B, C are the remainder, divisor, and borrow-in input bits, respectively; P is the borrow-out signal and D is the operation control signal for all the cells in the same row. The next partial remainder digit S can be logically described by

$$S = \begin{cases} A \oplus B \oplus C, & \text{if } D = 0 \\ A, & \text{if } D = 1 \end{cases} \tag{8.44}$$

In other words, the **CS** cell behaves like a subtractor when $D = 0$ and maintains the old remainder as the next remainder, if $D = 1$. The subtraction is performed by subtracting the vector B (the divisor) from the vector A (the old partial remainder). Figure 8.8 shows the layout of the complete restoring division array, where the dividend being a 6-bit fraction, $N = .n_1 n_2 \cdots n_6$, is fed from the vertical input lines on the top row and on the rightmost diagonal, and the divisor, being a 3-bit fraction $D = .d_1 d_2 d_3$, enters the array along the diagonal lines. The quotient, being a 3-bit fraction $Q = .q_1 q_2 q_3$ if $D > N$, is generated to the left of the array. Each row determines one quotient bit, which is equal to the complement of the borrow-out from the leftmost **CS** cell in that row. The initial borrow-in to all rows is set as "0" for true magnitude subtraction. The left shifting of the partial remainder required in division is now

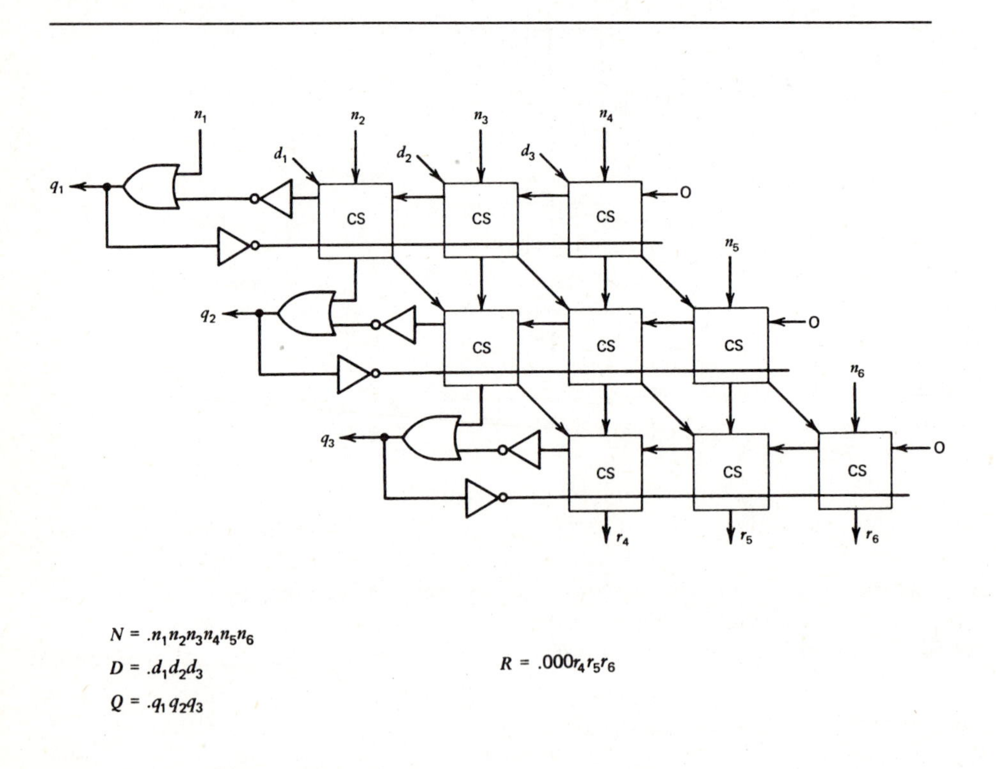

Figure 8.8 The schematic logic diagram of a 3-by-3 restoring array divider (Dean's version with redundant cells removed).

replaced by the equivalent operation of maintaining the remainder fixed and shifting right the divisor along the diagonal lines.

A value "1" is generated for the i-th quotient digit, q_i, if either the leading digit of the last remainder is an "1" or the carry-out line of the i-th row is a zero indicating that the result of the trial subtraction is positive. In this case ($q_i = 1$), the difference of the trial subtraction is passed on to the next row as the new partial remainder. On the other hand, whenever $q_i = 0$ (if none of the above two conditions is true), the trial subtraction is bypassed by passing the old partial remainder onto the next row. The final remainder, $R = 0.000r_4r_5r_6$, appears at the outputs of the bottom row with three leading zeros corresponding to the length of the quotient length ($n = 3$ shown in Fig. 8.8).

Equations 8.42 and 8.43 show that each **CS** cell can be implemented with a 3-level logic with 3Δ gate delays. In general, a binary cellular array divider which receives a $2n$-bit dividend and a n-bit divisor will produce an n-bit quotient and a $2n$-bit remainder with n leading zeros. We shall call such an array an n-by-n divider. The n-by-n restoring array divider ($n = 3$ in Fig. 8.8) requires n^2 **CS** cells and $2n$ inverters. Because of the sequential nature of the successive row operations, an n-by-n restoring divider should have a DIVIDE execution time on the order of $\mathbf{O}(n^2)$. A detailed study of speed performance of such array dividers will be given later when they are compared with other types of division arrays.

One important observation is mentioned at this point. The old partial remainder in a restoring division is retained to the next row, when the quotient bit generated in a row has a value "0". This is made possible by the multiplexer logic embedded within each **CS** cell. Therefore, no actual restoration addition is performed on any row of the restoring array. On the contrary, only when the old partial remainder is ruined in a trial subtraction, do the restoring steps become necessary. Such a stored shifting technique using a multiplexer has been used in the design of standard restoring binary dividers, as discussed in section 7.4. This feature makes the restoring division competent with respect to its nonrestoring counterparts. Without the stored multiplexing logic, the restoring division is usually 33 percent slower than nonrestoring division.

8.7 Nonrestoring Cellular Array Divider

The restoration addition is not required in a nonrestoring division. The operation to be performed per each row of a nonrestoring array is either *addition* or *subtraction*, depending on whether the sign of the previous row outputs agrees with the dividend sign. When an unsuccessful subtraction occurs, the partial remainder changes sign with respect to the dividend. A "0" quotient should be generated and the divisor is first shifted right along the diagonals, and then added to the partial remainder in the next row. When the partial remainder does not change its sign, "1" quotient digit is generated and the operation at the next row should be a subtraction.

Before we show the complete array structure let us check a numerical example given in Fig. 8.9. It is instructive to note that the right shift (divide-by-2) and add operation in case of unsuccessful trial subtraction results in a "net" operation of

Example:

Dividend $N = (0.101001)_2 = (41/64)_{10}$
Divisor $D = (0.111)_2 = (7/8)_{10}$
TC: Two's Complement. PR: Partial Remainder

	Sign Bit ↓		Quotient Bits
Dividend N	0.101001		
Subtract D in TC	1.001		
Negative PR	1.110001	<0	$q_0 = 0$
Shift PR	1.10001		
Add D	0.111		
Positive *PR*	0.01101	>0	$q_1 = 1$
Shift *PR*	0.1101		
Subtract D	1.001		
Negative *PR*	1.1111	<0	$q_2 = 0$
Shift *PR*	1.111		
Add D	0.111		
	0.110	>0	$q_3 = 1$

Quotient $Q = q_0 . q_1 q_2 q_3 = (0.101)_2 = 5/8)_{10}$
Remainder $R = (0.00r_3r_4r_5r_6)_2 = (0.000110)_2$
$= (6/64)_{10}$

Figure 8.9 A numeric example of the conventional nonrestoring array division using the cellular array divider given in Fig. 8.10.

Table 8.2 Correspondence Between Restoring and Nonrestoring Array Divisions

	Restoring Array Divisoin	**Nonrestoring Array Division**
First cycle	$-D$ $+D$ (restoration)	$-D$
Second cycle	$-D/2$	$+D/2$
Net operation	$-D/2$	$-D/2$

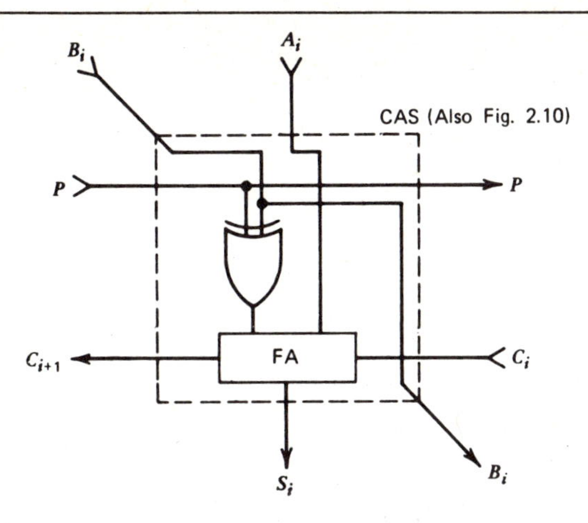

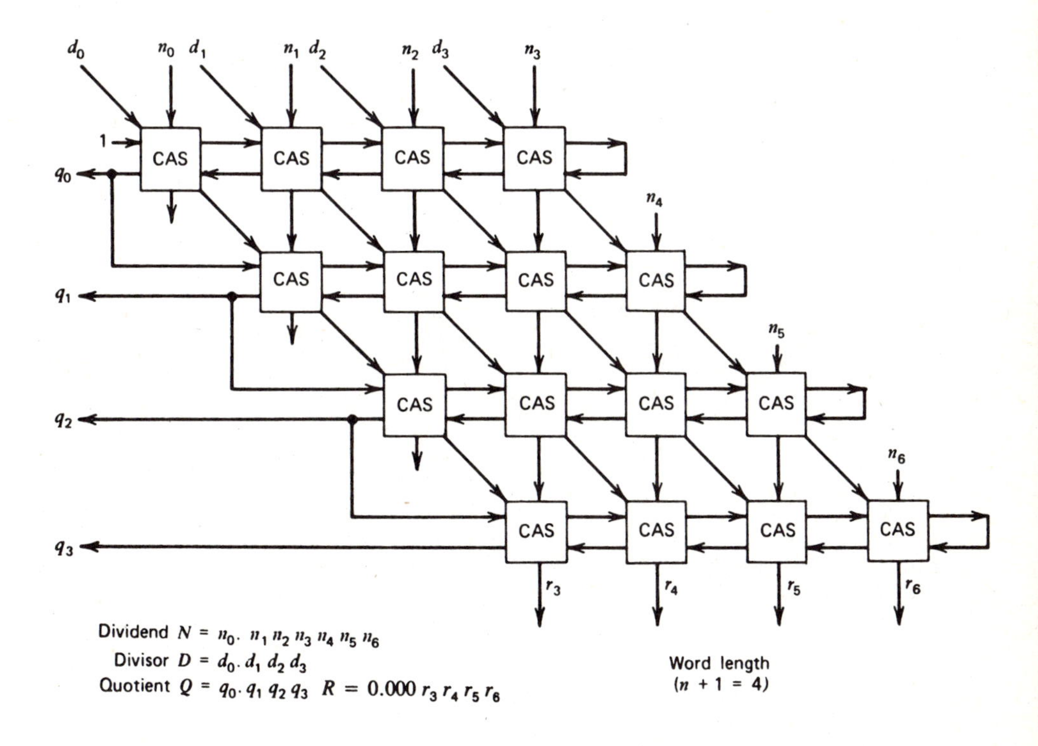

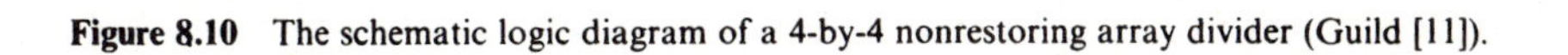

Figure 8.10 The schematic logic diagram of a 4-by-4 nonrestoring array divider (Guild [11]).

adding one-half of the divisor. This is the exact result by adding back the full divisor, shifting right the divisor, and then subtracting the half divisor in restoring binary division. The correspondence is demonstrated in Table 8.2.

A nonrestoring binary divider can be implemented with an iterative array composed of *Controlled Add/Subtract* (**CAS**) cells, which were described in section 2.3. An $(n + 1)$-by-$(n + 1)$ nonrestoring division array consists of $(n + 1)^2$ **CAS** cells. The cell interconnection is illustrated by the array in Fig. 8.10 for the case of $n = 3$. The terminal assignments for the input operands are the same as before. The initial operation performed on the top row is always a subtraction. Therefore, the control line P is permanently set to be "1" on the top row. The subtraction is performed in two's complement arithmetic, in which the feedback lines on the right-end cells serve as the initial carry-in's. Note that this initial carry-in is zero when addition is performed in the row ($P = 0$). The carry-out of the leftmost cell in each row determines the value of the quotient. By feeding back the current quotient to the next row, we can decide the operation for the next row. This carry-out signal indicates the sign of the current partial remainder, which in turn, decides the next row operation as described earlier.

Using the above nonrestoring division array will also require an execution time of order $\mathbf{O}(n^2)$, because of the carry (or borrow) propagation along each row, and because all the rows are serially connected on their carry lines. In order to estimate the DIVIDE execution time by this nonrestoring array, we have to specify the actual internal circuit realization of the **CAS** cells. The sum and carry equations of a **CAS** cell can be modified in the following form from Eq. 2.3.

$$
\begin{aligned}
S_i &= A_i \oplus (B_i \oplus P) \oplus C_i \\
&= A_i B_i \bar{C}_i P + A_i \bar{B}_i \bar{C}_i \bar{P} + \bar{A}_i B_i C_i P \\
&\quad + A_i \bar{B}_i C_i P + A_i B_i C_i \bar{P} + \bar{A}_i \bar{B}_i \bar{C}_i P
\end{aligned} \qquad \mathbf{(8.45)}
$$

$$
\begin{aligned}
C_{i+1} &= (A_i + C_i)(B_i \oplus P) + A_i C_i \\
&= A_i B_i \bar{P} + A_i \bar{P}_i P + B_i C_i \bar{P} + \bar{B}_i C_i P \\
&\quad + A_i C_i
\end{aligned} \qquad \mathbf{(8.46)}
$$

These two equations can each be realized with a 3-level combinational logic circuit (including the inverters). Therefore, the delay per each basic **CAS** cell is 3Δ units. We shall use this cell delay in determining the exact divide time in section 8.10.

8.8 Carry-Save Cellular Array Division

All the division arrays we have learned so far require the carry (borrow) to be fully propagated in each row of the array. This renders the total division time to be *quadratically* proportional to the length of the quotient. A viable approach, which eliminates this ripple carry propagation, is described in this and the next sections using the carry-save and lookahead techniques. The improved scheme is implementable with an augmented cellular array, which is purely combinational allowing for MSI/LSI implementation with minimum requirement of external timing and control logic.

Each of the successive partial remainders in a nonrestoring division can be decomposed into a combination of two vectors: the *sum vector* and the *carry vector*. This decomposition makes the carry anticipation possible by the conventional look-ahead circuit. Therefore, the sign digit of each partial remainder at the left end of each row can be determined in minimum carry delay. Apparently, carry generate and propagate functions should be embedded within each adding cell in the array to make the carry lookahead possible. This method results in a total division time, which increases almost *linearly* with the quotient length. The improvement in speed is obtained at only a moderate increase in hardware.

There are three types of logic cells used in this array construction. These cells are described by the schematic logic circuits in Fig. 8.11. The actual implementation of these cells may use the minimum level circuits. These logic diagrams do not mean to be minimum in circuit complexity. They simply demonstrate the logic of those cells. The **A** cells in the array are essentially *controlled 3-to-2 carry-save full adder/subtractors*. The superscripts $i = 0, 1, \ldots, n$ indicate the row designations from top to bottom, and the subscripts $j = 0, 1, 2, \ldots, n$ designate the digital positions of each row. The left-most bit is always the sign.

The sum digit output S_j^i and the left-shifted carry out C_{j-1}^i of the ith row are logically expressed as

$$S_j^i = S_j^{i-1} \oplus C_j^{i-1} \oplus (D_j \oplus K^i) \quad \textbf{(8.47)}$$

$$C_{j-1}^i = (D_j \oplus K^i)(S_j^{i-1} + C_j^{i-1}) + S_j^{i-1}C_j^{i-1} \quad \textbf{(8.48)}$$

where S_j^{i-1} and C_j^{i-1} are the jth sum and carry output digits of the $(i-1)$st row, D_j is the jth divisor bit, and K^i is the operation control signal of the ith row. The operation performed on the ith row is determined by

$$K^i = \begin{cases} 0 & \text{for addition} \\ 1 & \text{for subtraction} \end{cases} \quad \textbf{(8.49)}$$

Each **A** cell also contains and AND and an OR gate, which generates the carry generate and carry propagate functions required for row carry lookahead

$$G_j^i = C_j^i S_j^i \quad \textbf{(8.50)}$$

$$P_j^i = C_j^i + S_j^i \quad \textbf{(8.51)}$$

where C_j^i comes from the carry-out of the immediate right hand A cell on the same row.

The **S** cells are located at the sign position of each row. It generates a *pseudo sign* S_0^i expressed by

$$S_0^i = (S_0^{i-1} \oplus C_0^{i-1} \oplus \overline{K^i}) \oplus C_0^i \quad \textbf{(8.52)}$$

The complemented quotient digit $\bar{Q}_i$ at the ith row is generated by the XOR gate

$$\bar{Q}_i = S_0^i \oplus C_i^* \quad \textbf{(8.53)}$$

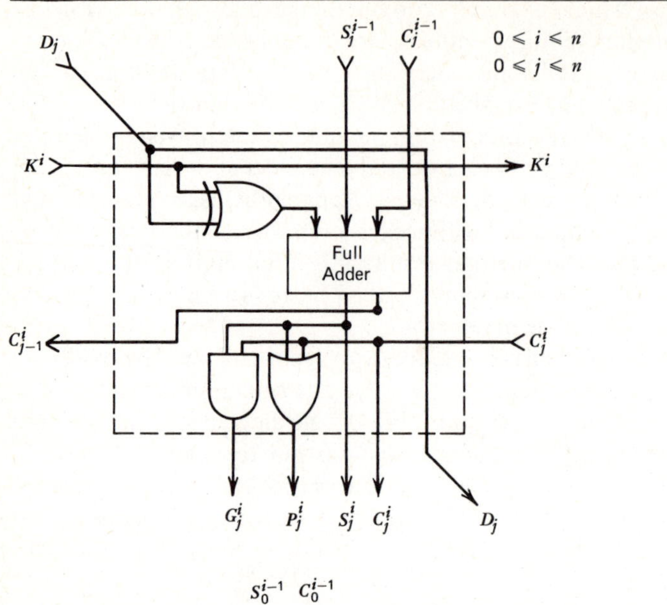

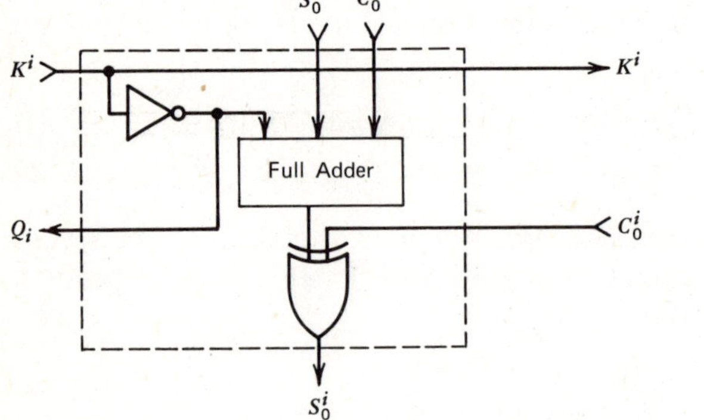

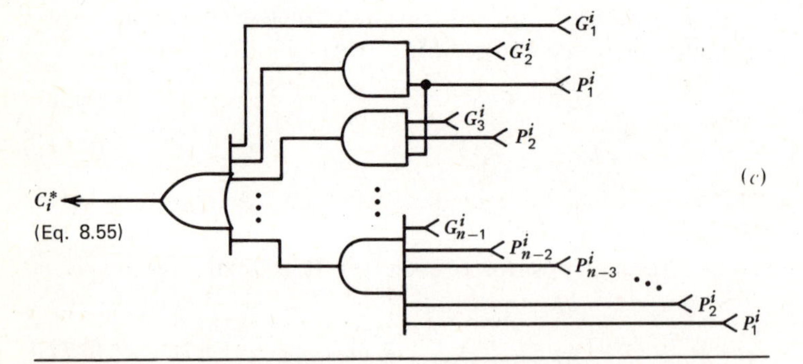

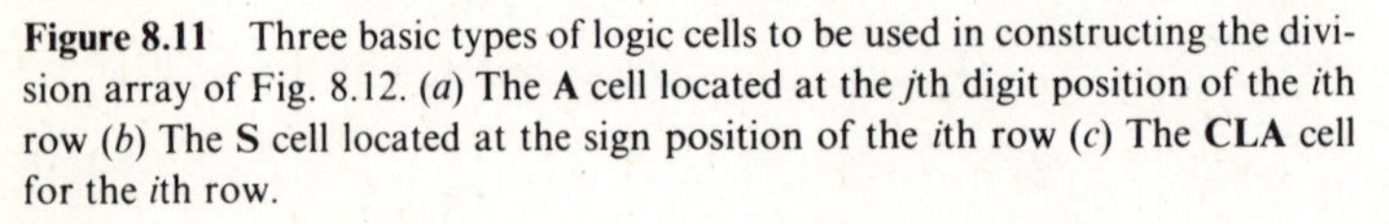

Figure 8.11 Three basic types of logic cells to be used in constructing the division array of Fig. 8.12. (*a*) The **A** cell located at the *j*th digit position of the *i*th row (*b*) The **S** cell located at the sign position of the *i*th row (*c*) The **CLA** cell for the *i*th row.

where C_i^* is the lookahead carry generated at the ith row. The signal $\bar{Q}_i$ will be used to control the add-subtract operation on the next row. In notation, we have for $i = 0, 1, \ldots, n-1$

$$K_{i+1} = \bar{Q}_i \tag{8.54}$$

and $K_0 = 0$ at the top row.

The output of the *Carry Lookahead* (**CLA**) cell, C_i^*, can be written as

$$\begin{aligned} C_i^* = G_1^i + P_1^i G_2^i + P_1^i P_2^i G_3^i + \cdots \\ + P_1^i P_2^i \cdots P_{n-2}^i G_{n-1}^i \end{aligned} \tag{8.55}$$

Present technology allows this lookahead carry to be generated by a circuit with 2Δ delay, when n is less than 10.

8.9 Carry-Lookahead Cellular Array Divider

With the building components specified, we are now ready to describe the structure of the cellular array which performs high-speed division. The circuit is sketched in Fig. 8.12. It uses some **A** cells, **S** cells, and **CLA** cells, in addition to a number of XOR gates and inverters. The first row always performs a subtraction by enforcing a "1" along the control line K^0. The initial carry-in at the right end of each row is connected to the quotient digit Q_{i-1}, generated at the row above it. This means that subtraction is accomplished in the form of two's complement addition.

A $2n$-bit dividend $N = N_0 \cdot N_1 \cdots N_n$ is fed from the top input lines, S_j^{-1}, of those cells on the top row and on the rightmost diagonal. The initial carry vector C_j^{-1} for $j = 0, 1, \ldots, n$ should be zero. The complemented divisor is fed through the diagonal input lines. The array division operations are described by the flow chart in Fig. 8.13. The nonrestoring operation in each step can be illustrated by the following aligned formulation:

$$\begin{array}{rcll}
S^{i-1} &=& S_0 \,.\, S_1 S_2 \cdots S_{n-1} \quad N_{n+i} & = \text{Old sum vector} \\
C^{i-1} &=& C_0 \,.\, C_1 C_2 \cdots 0 \qquad Q_{i-1} & = \text{Old carry vector} \\
\pm D &=& D_0 \,.\, D_1 D_2 \cdots D_{n-1} D_n & = \text{Add/subtract divisor} \\
\hline
S^i &=& S_0' \,.\, S_1' S_2' \cdots S_{n-1}' \quad S_n' & = \text{New sum vector} \\
C^i &=& C_0' \,.\, \underbrace{C_1' C_2' \cdots C_{n-1}' 0}_{\substack{\text{To determine the} \\ \text{CLA bit } C_i^*}} & = \text{New carry vector}
\end{array}$$

A numerical example is given in Fig. 8.14 to illustrate the operating procedures. Readers are encouraged to verify the numeric entries against the array circuit in Fig. 8.12. The carry speedup logic shown in Fig. 8.12 has a single-level lookahead configuration. This is feasible at present only for word length $n < 10$. For longer word lengths, two or more levels of carry lookahead may be required per each row.

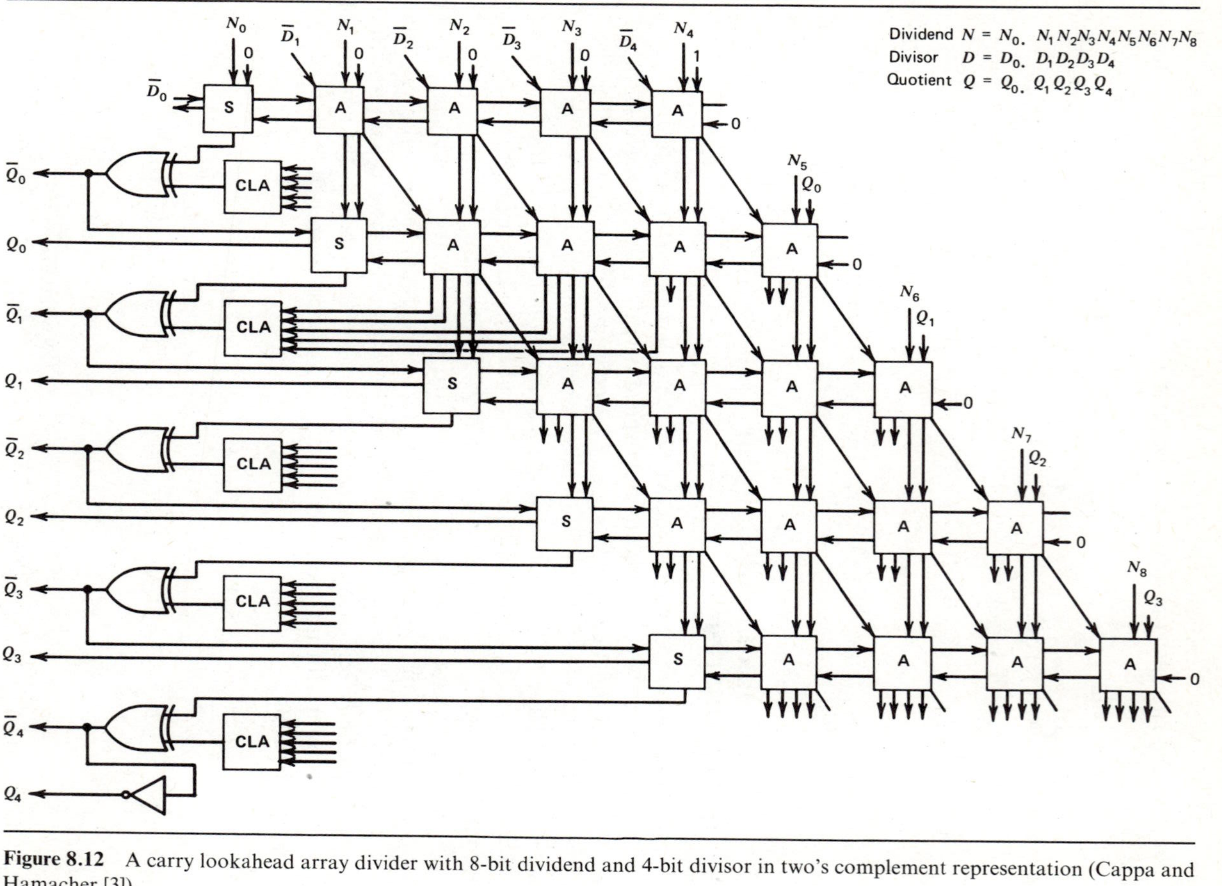

Figure 8.12 A carry lookahead array divider with 8-bit dividend and 4-bit divisor in two's complement representation (Cappa and Hamacher [3]).

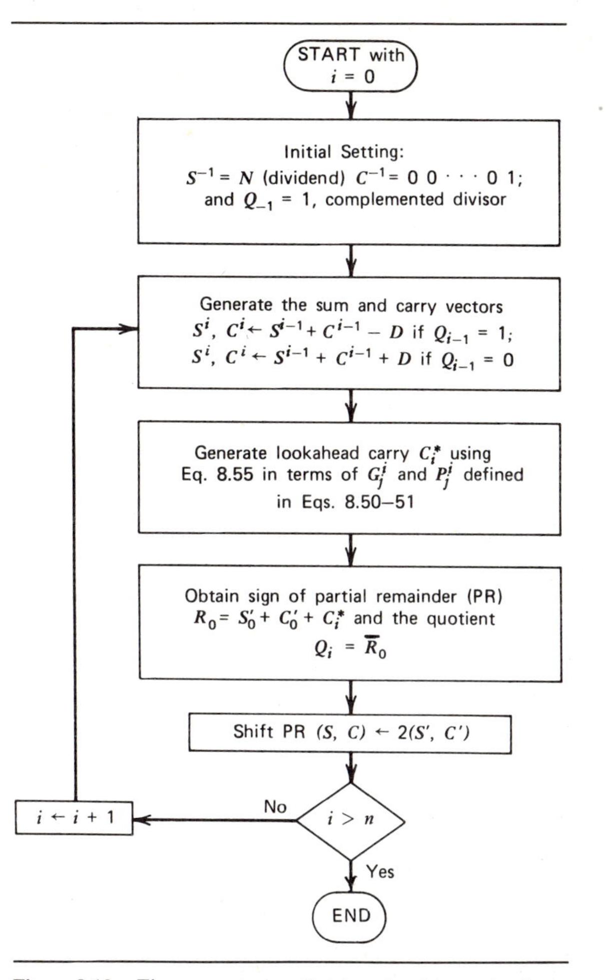

Figure 8.13 The nonrestoring division algorithm using lookahead and carry-save techniques for describing the Cappa–Hamacher array divider shown in Fig. 8.12.

In summary, the above division array implemented with carry-save cells and augmented with carry lookahead logic per each row may require $n(n + 1)$ **A** cells, $n + 1$ **S** cells, $n + 1$ **CLA** circuits, and $n + 1$ XOR gates plus one inverter, where n is the quotient length (excluding the sign). The total division time depends on the number of levels of carry lookahead used. It is quite clear, however, that the time delay per each row is a constant, independent of the word length n. Using this array, the total divide time should be $(n + 1)$ times the delay per each row.

$$\Delta_{CLA}(\text{array divider}) = (n + 1) \times \Delta_{\text{row}} + \Delta \tag{8.56}$$

where Δ_{row} is the row delay to be analyzed in the next section and the Δ is attributed to the inverter delay at the bottom row for producing the last quotient digit.

Example: Dividend = $(0.10100100)_2 = (\frac{164}{256})_2$
Divisor = $(0.1110) = (\frac{14}{16})_{10}$

Dividend	0.10100100	
Forced carry	0.0001	
Divisor (1's complement)	1.0001	
Sum vector	1.10100100	Quotient Bit
Carry vector shifted	0.001	↓
Lookahead carry bit	0←	
True sign	1	$Q_0 = 0$
Shift sum vector	1.0100100	
Shift carry vector	0.0100	
True divisor	0.1110	
Sum vector	1.1110100	
Carry vector	0.100	
Lookahead	1←	
True sign	0	$Q_1 = 1$
Shift sum	1.110100	
Shift carry plus 1	1.001	
Complemented divisor	1.0001	
	1.110100	
	0.001	
Lookahead	0←	
True sign	1	$Q_2 = 0$
Shift sum	1.10100	
Shift carry	0.0010	
True divisor	0.1110	
	1.01100	
	1.010	
Lookahead	0←	
True sign	0	$Q_3 = 1$
Shift sum	1.0100	
Shift carry plus 1	0.0101	
Complemented divisor	1.0001	
	0.0000	
	0.101	
Lookahead	0←	
True sign	0	$Q_4 = 1$

Quotient = $Q_0 \cdot Q_1 Q_2 Q_3 Q_4 = (0.1011)_2 = (\frac{11}{16})_{10}$

Figure 8.14 Example of the Cappa–Hamacher nonrestoring array division with carry lookahead and carry-save.

Equation 8.56 implies that the total divide time is a linear function $\mathbf{O}(n)$, of the word length, which is a big improvement over the $\mathbf{O}(n^2)$ delay for the noncarry-lookahead division arrays presented in previous sections. Strictly speaking, the row delay Δ_{row} should rise logarithmically as $\mathbf{O}(\log_m n)$, where n is the word length and m is the maximum fan-in of the gate function used. Present TTL logic has $m = 10$ typically. This number, $\log_m n$, reflects the number of lookahead levels. Therefore, the total divide time of the carry-lookahead array divider should have an order of magnitude $\mathbf{O}(n \log_m n)$.

For practical applications, the word length n is usually small, say $n \leq 100$. Therefore, the second factor $\log_m n$ is one or two for $m = 10$, which can be treated as constant. The delay shown in Eq. 8.56 is pretty accurate for practical designs. At least, the delay can be considered as piecewise, linearly increasing with a break point at $n = 10$ bits, then it increases linearly with a higher slope beyond the break point. More details of hardware and speed analyses of array division schemes are to be discussed in the next section.

8.10 An Evaluation of Various Cellular Array Dividers

The three classes of division arrays described in the previous sections are compared in this section in terms of operating speed and hardware requirements. The speed will be measured in terms of the number of basic gate delays, and the hardware demands will be evaluated in terms of the numbers of logic gates required in the array construction.

The **CS** cells used for constructing restoring division arrays require 10 gates to implement, each with a 3Δ delay per cell. The gate count includes only the basic gate types (AND, OR, NOT) needed. By the same token, the **CAS** cells used for constructing nonrestoring arrays, as specified in Eq. 8.45 and 8.46, require 12 gates per cell, also with a 3Δ delay per cell.

The worst-case signal delay in an n-by-n restoring array is the time required to generate the least significant quotient digit q_n at the bottom row of the array. In a restoring array, the sum outputs (new partial remainder) of each row are not considered stabilized until a stable signal appears on the leftmost borrow-out line. This means that each row itself must take $n \times 3\Delta$ time delay to generate a correct partial remainder to be used by the next row. Because the successive partial remainders must be generated row by row in a serial manner, the total time required to generate the last quotient digit at bottom row of Fig. 8.8 is

$$\Delta_{\text{restoring}} = (3n^2 + 1)\Delta \tag{8.57}$$

where the 1Δ is attributed to the inverter on the last row.

For a $2n$-by-n nonrestoring array divider, we obtain similar result if the top row required to determine the quotient sign is considered as an extra requirement

$$\Delta_{\text{nonrestoring}} = 3(n + 1)^2\Delta \tag{8.58}$$

The above analysis shows that restoring and nonrestoring division arrays offer about the same speed for large n. For short word length, the restoring array is faster.

To evaluate the hardware demands of a $2n \times n$ carry-lookahead array divider, we have to determine the gate counts in the basic cells. Instead of following the logic diagrams given in Fig. 8.11, we use the minimum-delay 3-level circuits to estimate the gate count and the delays for the **A** cells and **S** cells used in an augmented array divider. These circuits are obtained by fully expanding the terminal equations for the **A** cell and **S** cell. For clarity, all the subscripts and superscripts are removed. Instead, outputs of the cells are "primed," whereas the inputs are not. After expansion, Eq. 8.47 becomes

$$\begin{aligned} S' = {} & \bar{S}\bar{C}\bar{D}\bar{K} + \bar{S}\bar{C}D\bar{K} + \bar{S}C\bar{D}\bar{K} \\ & + \bar{S}CDK + S\bar{C}\bar{D}\bar{K} + S\bar{C}DK \\ & + SCD\bar{K} + SC\bar{D}K \end{aligned} \tag{8.59}$$

Equation 8.48 becomes

$$C' = \bar{D}KS + \bar{D}KC + D\bar{K}S + D\bar{K}C + SC \tag{8.60}$$

Equations 8.50 and 8.51 become

$$G = C \wedge (S' \text{ in Eq. 8.59}) \tag{8.61}$$

$$P = C \vee (S' \text{ in Eq. 8.59}) \tag{8.62}$$

Note that the C in Eqs. 8.61 and 8.62 comes from the carry-out of the right-hand A cell on the same row, whereas the C in Eqs. 8.59 and 8.60 comes from that of the immediate above row. Equation 8.52 becomes

$$\begin{aligned} S_0^i = {} & \bar{S}C\bar{K}\bar{C}^i + \bar{S}CKC^i + S\bar{C}\bar{K}\bar{C}^i \\ & + S\bar{C}KC^i + \bar{S}\bar{C}\bar{K}C^i + \bar{S}\bar{C}K\bar{C}^i \\ & + SC\bar{K}C^i + SCK\bar{C}^i \end{aligned} \tag{8.63}$$

As a result of these minimal-level circuits, each **A** cell contains 17 gates. The delays for $S'(S_j^i$ in Eq. (8.47) and $C'(C_{j-1}^i$ in Eq. 8.48) are each 3Δ. The delays for the G and P functions are each $3\Delta + \Delta = 4\Delta$. The **S** cell requires 9 gates and its time delay is 3Δ. Because it takes 3Δ delay to generate C^i, the carry-in signal enters the **S** cell from the carry-out of the leftmost **A** cell on the same row. Therefore, it takes totally $3\Delta + 3\Delta = 6\Delta$ to generate the pseudo sign S_0^i of each row, after the input signals arrive from the outputs of the row immediately above.

The gate cost of the **CLA** cell varies with the word length. For $n < 10$, one-level **CLA** requires at most $n - 1$ gates. For $10 \leq n \leq 64$, two-level **CLA** require approximately $n + \sqrt{n}$ gates. Summing up these gate counts, we obtain the gate count for a $2n$-by-n division array with two-level **CLA**.

$$\begin{aligned} \text{Gate Count} &= 17n(n + 1) + (n + 1) \times (n + \sqrt{n} + 9 + 1) + 1 \\ &= (n + 1)(18n + \sqrt{n} + 10) + 1 \end{aligned} \tag{8.64}$$

The time delay per each row Δ_{row} as was mentioned in Eq. 8.56 is now determined as follows: Δ_{row} is determined by the time to generate each quotient digit $\bar{Q}_i$ which equals the sum of several circuit delays, 3Δ from the output of XOR gate plus 6Δ from either the pseudo-sign generation logic (one A cell plus one S cell) or the carry lookahead logic, which for single-level lookahead equals $4\Delta + 2\Delta = 6\Delta$ also. Therefore, we have

$$\Delta_{row}(\text{single-level CLA}) = 3\Delta + 6\Delta = 9\Delta \quad \textbf{(8.65)}$$

For two-level carry lookahead, the second delay will be $6\Delta + 2\Delta = 8\Delta$. Therefore,

$$\Delta_{row}(\text{two-level CLA}) = 3\Delta + 8\Delta = 11\Delta \quad \textbf{(8.66)}$$

Substituting the row delays obtained above into Eq. 8.56, we obtain the following total divide time of the augmented **CLA** array divider

$$\Delta_{CLA}(\text{array divider}) = \begin{cases} (9n + 10)\Delta \text{ for single-level CLA;} \\ (11n + 12)\Delta \text{ for two-level CLA} \end{cases} \quad \textbf{(8.67)}$$

Equation 8.67 is sufficient for use in practical design problems. In a strict sense, for a division array with $k = \lceil \log_m n \rceil$ levels of lookahead logic, we obtain the following formula for estimating the divide time.

$$\begin{aligned} \Delta_{\substack{k\text{-level}\\ \text{CLA}}}(\text{Array Divider}) &= (n + 1) \times [9\Delta + (k - 1) \times 2\Delta] + \Delta \\ &= [2(n + 1) \times k + 7n + 8]\Delta \end{aligned} \quad \textbf{(8.68)}$$

where m is the maximum gate fan-in allowed in the circuit.

Table 8.3 summarizes the total gate counts and divide times associated with each of the three classes of iterative array dividers. Graphical plots of the speeds and gate costs are given in Fig. 8.15. From these plots, we conclude with the following remarks: The carry lookahead array dividers cost, on the average, 50 percent more than the nonrestoring division arrays and 70 percent more than the restoring division arrays.

Table 8.3 Gate Counts and Time Delay of Three Types of Array Dividers in Terms of Quotient Word Length n and Unit Gate Delay Δ

Type of Array Divider	Gate Count[a]	Divide Time[a]
Restoring (Dean)	$14n^2$	$(3n^2 - 1)\Delta$
Nonrestoring (Guild)	$17(n + 1)^2$	$3(n + 1)^2\Delta$
Nonrestoring with 2-level CLA and carry-save (Cappa-Hamacher)	$(n + 1) \times (18n + \sqrt{n} + 10) + 1$	$(11n + 12)\Delta$
Nonrestoring with 1-level CLA and carry-save (Cappa-Hamacher)	$18n^2 + 28n + 11$	$(9n + 10)\Delta$

[a] The word length n does not include the extra requirement to cover the sign bit. This implies that $n + 1$ is the actual word length for nonrestoring arrays.

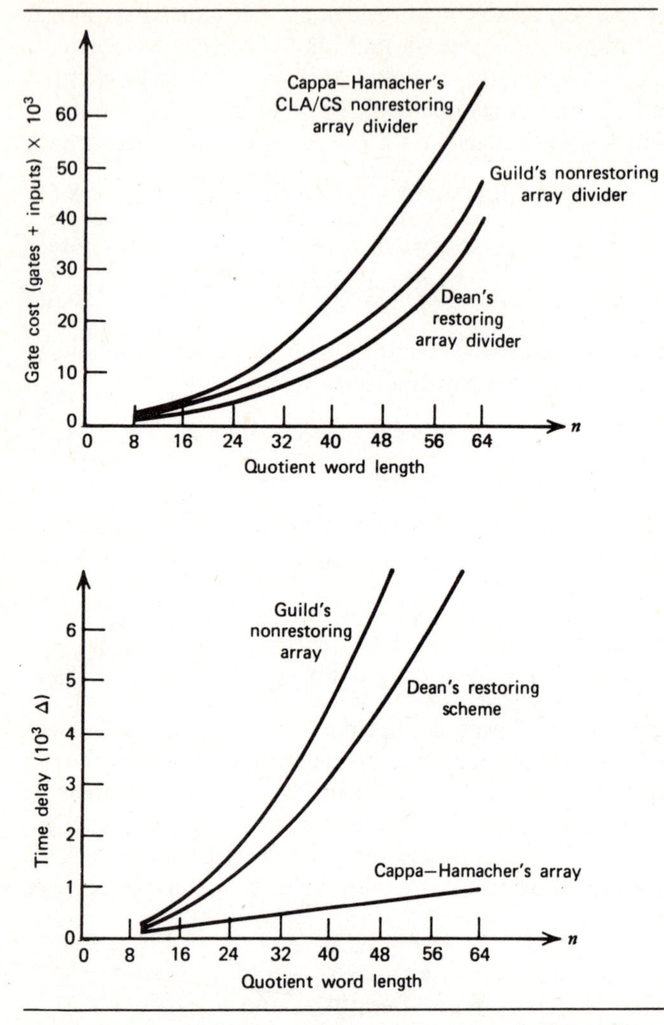

Figure 8.15 Cost and dealy versus quotient word length of three classes of iterative array dividers (Cappa [2]).

On the average the **CLA** array dividers are 5 times faster than the two array divisions, without using carry lookahead logic. The speed curves corresponding to restoring and nonrestoring arrays almost overlap each other. Among the three classes, the restoring array divider uses the least amount of hardware equipment.

8.11 Bibliographic Notes

The multiplicative convergence division was first proposed by Goldschmidt [10]. The scheme has been adopted in the IBM **360/91** floating-point execution unit design [1]. Reciprocal division was first described by Burks, et al in [19]. The theory of reciprocal convergence is presented following the lines of Wallace [18] and Ferrari [6]. Stefanelli [16] has proposed a reciprocal division scheme using iterative array with signed-digit code. Socemeantu [15] also studied this new approach. Iterative cellular array division has been proposed by many authors. The restoring array divider was proposed by Dean [4]. Nonrestoring division arrays were proposed by Majithia [13] and Guild [11]. Comparison of restoring and nonrestoring cellular array dividers was studied by Gardiner, et al [8]. The carry-save division and carry lookahead array divider was proposed by Cappa and Hamacher [3]. Cappa's thesis [2] presented a comprehensive treatment of high-speed array arithmetic using the carry-lookahead and carry-save approaches. Iterative arrays for fast multiply and divide were also proposed in recent works of Deverell [5], Gex [9], and Hamacher, et al [12]. A potential advantageous approach is to realize the signed-digit arithmetic with large-scale cellular array logic.

References

[1] Anderson, S. F. et al., "The IBM System/360 Model 91: Floating-Point Execution Unit," *IBM Journal*, January 1967, pp. 34–53.

[2] Cappa, M., "Cellular Iterative Arrays for Multiplication and Division," *M.S. Thesis*, Dept. of Elec. Eng., Univ. of Toronto, Canada, October 1971.

[3] Cappa, M., and Hamacher, V. C., "An Augmented Iterative Array for High-Speed Binary Division," *IEEE Trans. on Computers*, Vol. C-22, February 1973, pp. 172–175.

[4] Dean, K. J., "Binary Division Using a Data Dependent Iterative Arrays," *Electronics Letters*, Vol. 4, July 1968, pp. 283–284.

[5] Deverell, J., "The Design of Cellular Arrays for Arithmetic," *The Radio and Electronic Engineer*, Vol. 44, No. 1, January 1974, pp. 21–26.

[6] Ferrari, D., "A Division Method Using a Parallel Multiplier," *IEEE Trans. Comput.*, Vol. EC-16, April 1967, pp. 224–226.

[7] Flynn, M. J., "On Division by Functional Iteration," *IEEE Trans. Comput.*, Vol. C-19, August 1970, pp. 702–706.

[8] Gardiner, A. B. and Hont, J., "Comparison of Restoring and Nonrestoring Cellular Array Dividers," *Electronics Letters*, Vol. 7, April 1971, pp. 172–173.

[9] Gex, A., "Multiplier-Divider Cellular Array," *Elec. Letters*, Vol. 7, July 1971, pp. 442–444.

[10] Goldschmidt, R. Z., "Applications of Division by Convergence," *M.S. Thesis*, M.I.T. Cambridge, Mass., June 1964.

[11] Guild, H. H., "Some Cellular Logic Arrays for Nonrestoring Binary Division," *The Radio and Elec. Engr.*, Vol. 39, June 1970, pp. 345–348.

[12] Hamacher, V. C. and Gavilan, J., "High-Speed Multiplier/Divider Iterative Arrays," *Proc. of 1973 Sagamore Computer Conf. on Parallel Processing*, 1973, pp. 91–100.

[13] Majithia, J. C., "Nonrestoring Binary Division Using a Cellular Array," *Electronics Letters*, Vol. 6, May 1970, pp. 303–304.

[14] Robertson, J. E., "Theory of Computer Arithmetic Employed in the Design of New Computer at the University of Illinois," *Tech. Rept.*, No. 319, Dept. of Computer Science, University of Illinois, Urbana, 1960.

[15] Socemeantu, A., "Cellular Logic Array for Redundant Binary Divisions," *Proc. IEE* (London), Vol. 119, No. 10, October 1972, pp. 1452–1456.

[16] Stefanelli, R., "A Suggestion for High-Speed Parallel Binary Divider," *IEEE Trans. Comp.*, Vol. C-21, January 1972, pp. 42–55.

[17] Svoboda, A., "An Algorithm for Division," *Inf. Proc. Machines*, No. 9, Prague, Czechoslovakia, 1963, pp. 25–34.

[18] Wallace, C. S., "A Suggestion for a Fast Multiplier," *IEEE Trans. Comp.*, Vol. EC-13, February 1964, pp. 14–17.

[19] Wilkes, M. V., Wheeler, D. J., and Gill, S., *Preparation of Programs for an Electronic Digital Computer*, Addison-Wesley, Reading, Mass., 1951.

Problems

Prob. 8.1 The multiplicative division algorithm described in section 8.2 converges to the desired quotient quadratically using the set of convergence factors

$$\{\delta^{2^i} \mid i = 0, 1, 2, \ldots, n\},$$

where $\delta = 1 - D$ and D is the given divisor.

Extend the quadratic convergence division to a *cubic convergence division* scheme using the successive convergence factors $\{\delta^{3^i} \mid i = 0, 1, \ldots, k\}$, where k is the number of iterations required. Prove that $k < n$ and show the improvement of convergence rate as a function of p, where p is the fraction length of the operands.

Prob. 8.2 Explain and specify the necessary arithmetic operations required to generate the successive multipliers for the cubic convergence division method developed in *Prob.* 8.1. Note that there are three arithmetic operations associated with the generation of each multiplier, which means that longer cycle time per iteration may be required. Comment on the tradeoffs between convergence rate (reciprocal of the number of iterations) and the iteration cycle time in both quadratic and cubic convergence division systems.

Prob. 8.3 Prove that the optimal choice of the initial approximation, $p_0(h)$ for the hth interval, using the Ferrari's stepwise approach (Fig. 8.5), is the reciprocal of the number corresponding to its midpoint, that is Eq. 8.33.

Prob. 8.4 Extend the levels of the Wallace tree shown in Fig. 8.6 so that it can be used to produce a 72-bit reciprocal with few iterations. Determine also the minimum number of iterations required to generate the 72-bit reciprocal.

Prob. 8.5 Construct a 16-by-16 restoring array divider using the Controlled Subtractor (CS) cells for the division of a 31-bit unsigned fraction by a 16-bit unsigned fraction. Test your design with the dividend

$$N = 0.0000111001000110111111110101101010$$

and divisor

$$D = 0.1101010110101111$$

Estimate the total division time using this array in terms of number of unit gate delays.

Prob. 8.6 Construct a 16-by-16 nonrestoring array divider using the Controlled Add/Subtract (CAS) cells for the division of a 32-bit two's complement number by a 16-bit two's complement number. Test your design with the dividend

$$N = 0.0110111101111010100110110011110$$

and divisor $= 0.101100100111011$ in two's complement form. Estimate the total division time in terms of unit gate delays.

Prob. 8.7 Construct a 16-by-16 carry lookahead array divider with 32-bit dividend and 16-bit divisor in two's complement arithmetic. Test your design with the same pair of two's complement operands, N and D, given in *Prob.* 8.6. Estimate the total division time of your design, assuming that two-level carry lookahead is to be used per each row of the array construction.

Prob. 8.8 Compare three types of 32-by-32 array dividers, similar to those constructed in the above problems, with respect to the following two aspects:

(a) Total divide times, assuming that $\Delta = 4$ nsec.
(b) Total gate counts.

Evaluate these array dividers by using the reciprocal of the product of the divide time and the gate count to determine their cost-effectiveness. Which of these three 32-by-32 array designs is most cost effective in your evaluation?

Chapter

Normalized Floating-Point Arithmetic Processors

9.1 Rationales of Floating-Point Arithmetic

Most of the early computers used fixed-point arithmetic, which essentially handles "small" integers in an exact form for business or commercial applications. For scientific computing, we must round the numbers constantly in order to reduce the number of digits to a manageable amount. Fixed-point arithmetic presents some difficulties in handling scientific and engineering computations. The problems are caused primarily by the *range*, *precision* and *significance* of digital numbers represented in a machine.

The most commonly used range of fixed-point numbers is the unity interval from -1 to $+1$. When the range of numbers becomes very large or very small during a computation, it will be cumbersome for the programmer to keep track of the radix points of all the intermediate numbers. The out-of-range numbers are usually handled by means of *scaling* with software, firmware, or hardware means. Obviously, numbers used in scientific computing do not fall naturally into the unity interval. In most cases, the given numbers must be scaled up or down to fit the unity interval properly, and at the end of the computation the results will be transformed back to the user's domain. Without these transformations, the fixed-point hardware may produce meaningless results.

The scaling problem includes the selection of appropriate *scaling factors*, and for some complicated cases, the setup of appropriate *scaling loops*, which modify the scaling factors in the loop as circumstances dictate. Any n-digit, base-r, fixed-point number having an absolute value less than r^k bears a maximum error of r^{k-n}. The constant r^k is a common scaling factor shared by the entire set of numbers. In particular $k = 0$ and $k = n$ correspond to FXP fractions and FXP integers, respectively.

The natural precision of an n-digit fixed-point fraction number is rather restricted with a maximum error of r^{-n}. In order to increase the precision, multiple-precision fixed-point arithmetic was suggested. But to use more than one word for each fixed-point number automatically implies more programming overhead and high waste on data and instruction storage spaces. Nevertheless, multiple-precision fixed-point arithmetic may provide exact results that do not require much error analysis.

For complicated computational problems, these scaling and precision extension procedures involve extensive mathematical analyses and side computations for keeping track of the scaling factors or for controlling variable word lengths. Usually, the common scaling factor of an entire set of numbers is the maximum scaling factor used. The introduction of scaling may create a *significance loss* problem. For example, the actual difference between a common scaling factor, r^p and the order of magnitude r^t of a number for some $t < p$, may create $p - t$ leading zeros in the fixed-point number, leaving only a maximum of $n - p + t$ instead of n significant digits presented. In a chain of calculations, successive losses of significance precision may result in a singularity situation, which the fixed-point hardware cannot handle except through programmer intervention.

Floating-point arithmetic was proposed in early 1940s to overcome the above difficulties associated with fixed-point arithmetic. Even though it may double or triple the hardware demand with more complicated rounding schemes, floating-point arithmetic has become universally accepted for high-speed scientific computations. There are two types of floating-point arithmetic: *unnormalized* and *normalized.* A normalized FLP processor operates only with normalized floating-point numbers and enforces postnormalization steps on all intermediate and final results. Unnormalized FLP arithmetic refers to FLP operations that do not necessarily operate on normalized operands. The normalized operations have the advantages of procedural convenience, unique FLP number representation, and resulting in maximum significance in the mantissa at each stage of calculation. Floating-point arithmetic is the outgrowth of previous efforts to develop automatic scaling procedures to overcome the limited range and rigid precision problems associated with fixed-point arithmetic. Most modern general-purpose digital computers are equiped with both *fixed-point* and *floating-point* arithmetic processors.

9.2 Base Choices and FLP Singularities

Most computers were designed with binary floating-point arithmetic, such as the Digital Equipment PDP-11 family, until the advent of the IBM System/360, which chooses base 16 for floating-point arithmetic. Other machines such as ILLIAC II used base 4, and the Burroughs machines used base 8. ILLIAC III used base 256 for multiplication and division but numbers were stored in base 16. Sweeney [14] has shown that as the base increases substantially less alignment and normalization shifting are required. Binary and octal based floating-point machines were used for scientific computation with few complaints about their numerical properties. The hexadecimal machines, however, have drawn many unfavorable comments from users who have applied them to scientific problems. This stimulated research on the numeric properties of floating-point number systems as a function of the base value. These studies have dealt with the analysis of representational error, round off simulation, the determination of rigorous bounds for the relative error in the evaluation of mathematical functions, and the corresponding machine design considerations.

The arguments and discussions of the relative merits of binary, octal, hexadecimal, and other floating-point systems have centered primarily upon numerical characteristics such as the *exponent range*, *density of numbers*, and the *maximum relative error* of machine representation. Three important parameters (r, p, q) are used to characterize floating-point representations, where r is the base (radix), p is the mantissa length, and $q+1$ is the exponent length. Note that $p+q+2=n$ equals the word length including the sign bit. We shall wait until Chapter 10 to analyze FLP arithmetic errors against base choices.

Before we define normal FLP arithmetic operations on legitimate floating-point numbers, let us consider first the abnormal situations. Singular results may be generated by floating-point hardware; even when legitimate operations are performed on legitimate floating-point data. We use the symbols $+\infty$ and $-\infty$, to refer to the positive and negative *quasi-infinites* corresponding to FLP numbers whose exponent exceeds the largest allowable positive value. Two *infinitesimals*, $+\varepsilon$ and $-\varepsilon$, are similarly defined as FLP numbers whose exponent exceeds the largest negative value. The subscripts "u" and "n" are used to distinguish the unnormalized infinitesimals $\pm\varepsilon_u$, from the normalized infinitesimals $\pm\varepsilon_n$.

Let p be the number of digits in the mantissa field and $q+1$ be that of the exponent of a floating-point number. These infinities and infinitesimals satisfy the following properties for a FLP machine with base $r = 2^k$, where $k = \log_2 r$ is the number of bits required to represent each base-r digit. Note that the smallest nonzero binary fraction 2^{-p} equals the smallest nonzero base-r fraction $r^{-p/k}$.

$$\begin{aligned} +\infty &> (1 - 2^{-p}) \times r^{2^q - 1} \\ -\infty &< -(1 - 2^{-p}) \times r^{2^q - 1} \end{aligned} \tag{9.1}$$

$$\begin{aligned} 0 < +\varepsilon_u &< 2^{-p} \times r^{-(2^q - 1)} \\ -2^{-p} \times r^{-(2^q - 1)} &< -\varepsilon_u < 0 \end{aligned} \tag{9.2}$$

$$\begin{aligned} 0 < +\varepsilon_n &< r^{-1} \times r^{-(2^q - 1)} \\ -r^{-1} \times r^{-(2^q - 1)} &< -\varepsilon_n < 0 \end{aligned} \tag{9.3}$$

The above ranges are illustrated in Fig. 9.1 for both normalized and unnormalized FLP arithmetic systems.

Now we are ready to define the singular results that may occur during the execution of floating-point instructions in a digital computer.

Exponent overflow This refers to the condition that the exponent of the resulting number exceeds the upper limit or the lower limit as shown in Eq. 9.1. In other words, both $+\infty$ and $-\infty$ are considered overflow conditions.

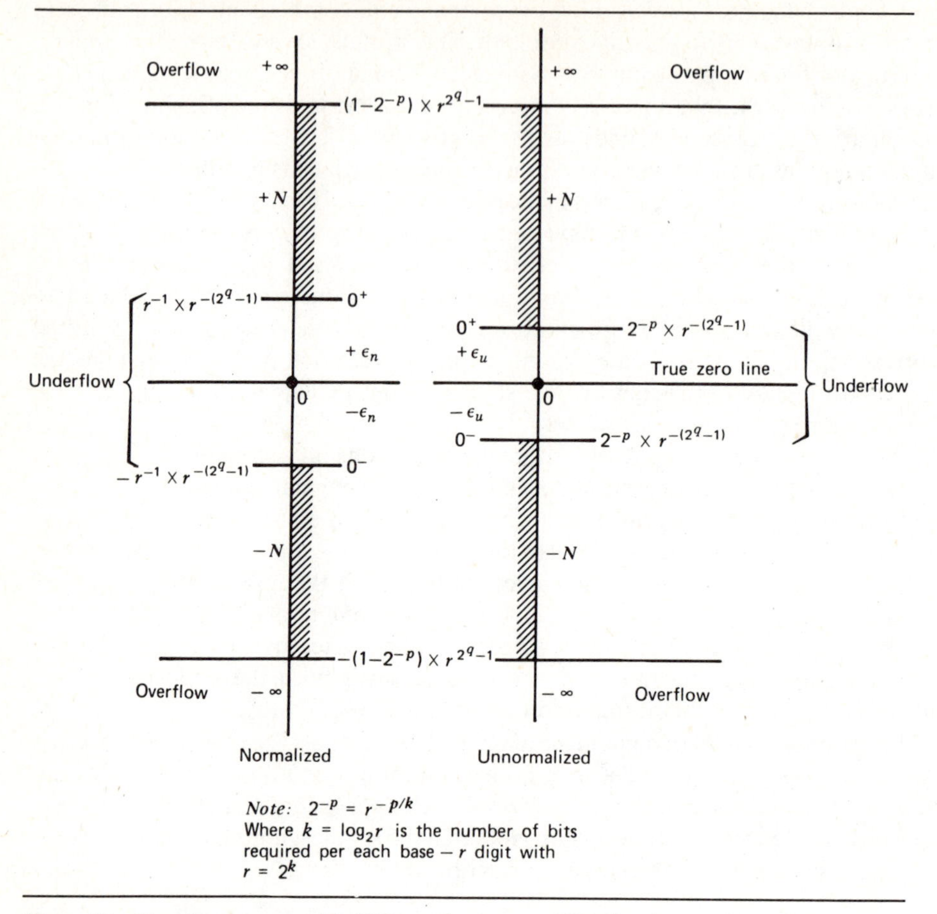

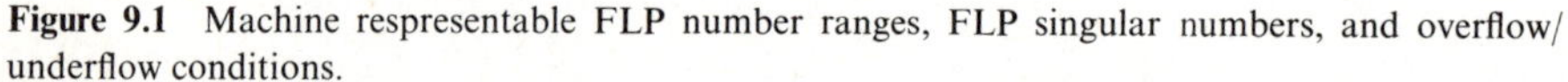
Figure 9.1 Machine respresentable FLP number ranges, FLP singular numbers, and overflow/underflow conditions.

Exponent underflow This refers to the condition that the exponent of the resulting number exceeds the minimal allowable value and falls into the following intervals $(-\varepsilon, 0_-)$ and $(0_+, +\varepsilon)$ as shown in Eqs. 9.2 and 9.3. It should be noted that the true zero value at the origin of the real axis is an exception. In other words, zero does not fall into the infinitesimal intervals $(-\varepsilon, 0_-)$ and $(0_+, +\varepsilon)$. Whenever an exponent overflow or underflow occurs, a corresponding indicator signal will be issued by the floating-point hardware. Sometimes, overflow/underflow signals can be bypassed without interrupting the computation; at other times, programmer intervention may be required to preserve meaningful computation.

Unlike fixed-point arithmetic, the zero in a floating-point notation may have more than one meaning. A much wider meaning of zero is introduced below.

Order-of-magnitude zero Any floating-point number having a zero mantissa, that is, $(m, e) = (0, e)$, is called an *Order-of-Magnitude Zero* (OMZ), where the exponent e can assume any legitimate value representable in the exponent field. Such an OMZ can be the result of the subtraction of a number by itself.

$$(m, e) - (m, e) = (0, e) \tag{9.4}$$

The values represented by $(0, e)$ imply a wide range of indeterminate numbers that satisfy the following inequality

$$-r^{e-p} < (0, e) < r^{e-p} \tag{9.5}$$

where p is the number of mantissa digits and r is the implied radix.

In particular, an OMZ $(0, e)$ with $e = 0$ denoted as $(0, 0)$ is called a *True Zero*. Recall that a true zero is not a normalized number, nor is it a member of the infinitesimal set.

Legitimate floating-point operands (in-range numbers) are indicated in Fig. 9.1 as $+N$ or $-N$. It can be seen that

$$\begin{aligned} +\varepsilon < +N < +\infty \\ -\infty < -N < -\varepsilon \end{aligned} \tag{9.6}$$

The true zero corresponds to the dividing line between positive and negative numbers. OMZ's are not representable on the real axis. Table 9.1 lists all the possible

Table 9.1 FLP Arithmetic Operations Involving Infinities, Infinitesimals, OMZs, and Normal FLP Numbers

Addition + and Subtraction −	**Remarks**
$\infty \pm N = \infty$; $N \pm \varepsilon = N$ $N - \infty = -\infty$; $\varepsilon - N = -N$ $\infty + \infty = \infty$; $\infty \pm \varepsilon = \infty$ $\varepsilon \pm \varepsilon = \varepsilon$; $\varepsilon - \infty = -\infty$ $\infty - \infty = \infty$; $\varepsilon - \varepsilon = \varepsilon$ $N \pm z = N$; $z - N = -N$ $z_1 \pm z_2 = z_3$;	N: Normal FLP Number ∞: Infinity ε: Infinitesimal $z_i = (0, e_i)$: An OMZ
Multiplication *	
$\infty * N = \infty$; $N * \varepsilon = \varepsilon$, $\infty * \infty = \infty$ $\varepsilon * \varepsilon = \varepsilon$; $\infty * \varepsilon = \infty$ (or $\infty * \varepsilon = \varepsilon$) $z_1 * z_2 = z_3$; $N * z_1 = z_2$,	Most machines choose $\infty * \varepsilon = \infty$ or $\varepsilon/\varepsilon = \infty$ instead of $\infty * \varepsilon = \varepsilon$ or $\varepsilon/\varepsilon = \varepsilon$ to emphasize the alarming cases of overflow. N/z or $N/0$ will set the Zero Divisor flag
Division /	
$\infty/N = \infty$; $\varepsilon/N = \varepsilon$; $N/\infty = \varepsilon$; $N/\varepsilon = \infty$; $\infty/\varepsilon = \infty$; $\varepsilon/\infty = \varepsilon$ $\infty/\infty = \infty$; $\varepsilon/\varepsilon = \infty$ (or $\varepsilon/\varepsilon = \varepsilon$) $z_1/N = z_2$; $N/z = N/0$ ↑ ↑ (both suppressed)	

operations defined over a legitimate operand $\pm N$, and an infinity $\pm\infty$, or an infinitesimal $\pm\varepsilon$. Operations involving two singular numbers or indeterminate OMZ's are also specified in this table. Division by an OMZ is suppressed and a *zero divisor indicator* will be turned on.

Those operations defined in Table 9.1 are presented only as a reference choice. In fact, the system designer can define his or her own set of indeterminate operations to suit the special application requirements. The general rule is to handle the indeterminate singular numbers in such a manner that FLP computation can be meaningfully continued without external interruption, and hopefully, the final results will not be seriously contaminated and traceable error analysis can be made.

9.3 Normalized FLP Arithmetic Operations

The four standard arithmetic operations—addition, subtraction, multiplication, and division—can be performed by floating-point hardware with wider operating range and better precision control. We shall characterize these operations over normalized floating-point numbers, $x_1 = (m_1, e_1)$ and $x_2 = (m_2, e_2)$, where $x = m \times r^e$ and r is the implied radix. The mantissa m is a signed fraction with p significant digits (excluding the sign) lying within the following normalized range

$$\frac{1}{r} \le |m| \le 1 - r^{-p} < 1 \quad \textbf{(9.7)}$$

and the exponent e is a signed integer with q significant digits (excluding the sign) such that

$$0 \le |e| \le r^q - 1 \quad \textbf{(9.8)}$$

The exponent is a variable which determines the real position of the radix point. FLP *addition/subtraction*, $x_1 \pm x_2$, is formally defined as follows:

$$(m_1, e_1) \pm (m_2, e_2) = \begin{cases} ((m_1 \pm m_2 \times r^{-(e_1 - e_2)}), e_1), & \text{if } e_1 > e_2 \\ ((m_1 \times r^{-(e_2 - e_1)} \pm m_2), e_2), & \text{if } e_1 \le e_2 \end{cases} \quad \textbf{(9.9)}$$

The above equation shows that radix point of the two numbers x_1 and x_2 must be aligned before meaningful addition/subtraction can be performed. This is done by comparing the relative magnitude of the two exponents and shifting the mantissa with a smaller exponent $|e_1 - e_2|$ places to the right. The operation is also reflected in Eq. 9.9 by multiplying the mantissa with a shifting factor $r^{-|e_1 - e_2|}$. The addition/subtraction of the mantissas then proceeds with the larger exponent serving as the resulting exponent. Note that the absolute value of the resulting mantissa, say $|m|$, is always bounded in the following range.

$$0 \le |m| < 2 \quad \textbf{(9.10)}$$

This compare-shift-add process is essentially sequential in nature, so floating-point add/subtract requires longer execution time than its fixed-point counterpart. Even after alignment, there may still exist two possible complications. The first complication corresponds to the situation that $1 \leq |m| < 2$, when adding two numbers of the same sign or when subtracting two oppositely signed numbers. In these cases, the absolute value $|m|$ of the sum or of the difference exceeds the unity. This is sometimes termed the problem of *mantissa overflow*. We shall not consider it a real overflow. However, the problem can be easily solved by shifting the overflown mantissa one bit to the right and simultaneously incrementing the exponent by one

$$(m_1, e_1) \pm (m_2, e_2) = \begin{cases} (r^{-1} \times (m_1 \pm m_2 \times r^{-(e_1 - e_2)}), e_1 + 1), & \text{if } e_1 > e_2 \\ ((r^{-1} \times (m_1 \times r^{-(e_2 - e_1)} \pm m_2), e_2 + 1), & \text{if } e_1 \leq e_2 \end{cases} \qquad \textbf{(9.11)}$$

The second complication comes when the resulting mantissa equals zero as described in Eq. 9.4. An OMZ has been created in this case. The final step in floating-point add/subtract is to normalize the resulting mantissa, if it has leading zeros. When OMZ has occurred, however, the postnormalization will not be possible. A special signal must be generated to indicate the existence of OMZ to the user.

The *FLP multiplication* $x_1 \times x_2$, and *FLP division* x_1/x_2 are defined by the following two arithmetic equations.

$$(m_1, e_1) \times (m_2, e_2) = (m_1 \times m_2, e_1 + e_2) \qquad \textbf{(9.12)}$$

$$(m_1, e_1)/(m_2, e_2) = (m_1/m_2, e_1 - e_2) \qquad \textbf{(9.13)}$$

The mantissa multiply/divide and the corresponding exponent add/subtract can be executed simultaneously. Therefore, for the same fractional length, these floating-point multiply/divide take essentially the same amount of execution time as corresponding fixed-point operations. The time saved by not executing fixed-point scaling instructions is offset by the time required to initialize the operands and to postnormalize the resulting product or quotient.

Several observations should be mentioned. In general, if Eqs. 9.12 and 9.13 are used, the value of the resulting mantissa falls within the following intervals.

$$\frac{1}{r^2} \leq |m_1 \times m_2| < 1; \qquad \textbf{(9.14)}$$

$$\frac{1}{r} < |m_1/m_2| < r \qquad \textbf{(9.15)}$$

provided $m_1 \neq 0 \neq m_2$.

When $1/r \leq |m_1 \times m_2| < 1$, no correction is needed, and the resulting product is already in normalized form. However, when

$$\frac{1}{r^2} \leq |m_1 \times m_2| < \frac{1}{r} \qquad \textbf{(9.16)}$$

the resulting product is not normalized. One-digit left shift is sufficient to normalize it. In this case, Eq. 9.12 should be replaced by the following equation:

$$(m_1, e_1) \times (m_2, e_2) = (r \times m_1 \times m_2, e_1 + e_2 - 1) \qquad \textbf{(9.17)}$$

No postnormalization is needed for floating-point division, because the resulting quotient m_1/m_2 is always normalized for $m_1 < m_2$ (Eq. 9.15). However, when $m_1 \geq m_2 \neq 0$, *division overflow* (or *quotient overflow*) occurs with

$$1 \leq |m_1/m_2| < r \tag{9.18}$$

One-digit right shift is sufficient to suppress the quotient overflow. Eq. 9.13 should now be replaced by Eq. 9.19 if quotient overflows

$$(m_1, e_1)/(m_2, e_2) = (m_1 \times r^{-1}/m_2, e_1 - e_2 + 1) \tag{9.19}$$

Floating-point arithmetic rules for unnormalized numbers will be discussed in the next chapter. In subsequent sections we shall detail the above procedures with systematic flow-charts and present the associated floating-point arithmetic hardware processors. A case study of an existing floating-point machine will be given in Section 9.8.

9.4 Basic FLP Arithmetic Hardware

The hardware configuration of a basic floating-point arithmetic unit is described in this section. The unit, which consists of a number of registers, two adders, control and timing logic, will be used as the hardware host for illustrating the execution of standard FLP operations, ADD, SUBTRACT, MULTIPLY, and DIVIDE in subsequent sections. We assume a floating-point data format having a 32-bit word length as shown in Fig. 9.2. Radix $r = 2$ is assumed with an exponent bias constant $b = 128$. The 23-bit mantissa plus the sign bit is a normalized fraction in sign-magnitude form with an implied binary point immediately to the right of the sign bit M_s. The sign convention is as usual

$$M_s = \begin{cases} 0 & \text{for positive number} \\ 1 & \text{for negative number} \end{cases} \tag{9.20}$$

The 23-bit magnitude (fraction) $\mathbf{M} = M_{1-23}$ of the mantissa lies within the range

$$0.5 \leq \mathbf{M} < 1 \tag{9.21}$$

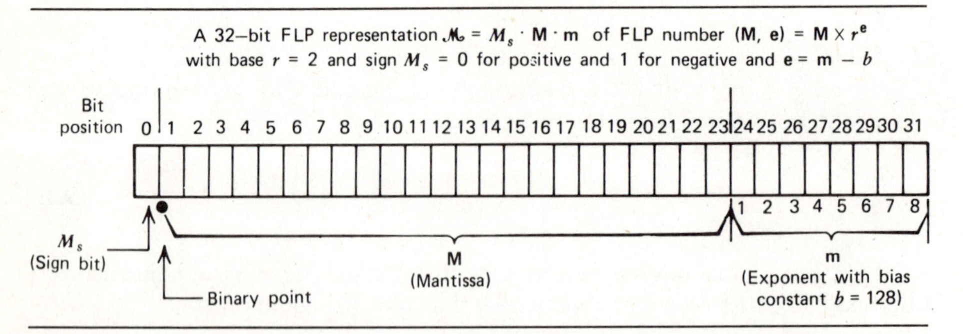

Figure 9.2 A 32-bit FLP data format to be used in illustrating normalized FLP arithmetic.

This implies that the most significant bit $M_1 = 1$. The exponent is *biased*, denoted by lowercase $\mathbf{m} = \mathbf{e} + b = m_{1-8}$. In other words, $\mathbf{m}$ is an 8-bit unsigned integer within the decimal range

$$0 \leq \mathbf{m} \leq 255_{10} \tag{9.22}$$

where the *true exponent* $\mathbf{e}$ can be retrieved from the biased value by

$$\mathbf{e} = \mathbf{m} - b = \mathbf{m} - 128 \tag{9.23}$$

Note that $\mathbf{e}$ is a signed integer in two's complement form within the range $-128 \leq \mathbf{e} \leq 127$. The true value $\mathbf{e}$ is used by the programmers, whereas the biased value $\mathbf{m}$ is used by the designers. The exponent adder processes only the biased exponents.

From now on, we shall reserve the scripted letters, $\mathcal{M} = M_s \cdot \mathbf{M} \cdot \mathbf{m}$, to represent a 32-bit floating-point data word fetched from the main memory, where M_s is the sign, the uppercase $\mathbf{M} = M_{1-23}$ is the normalized mantissa, and the lowercase $\mathbf{m} = m_{1-8}$ is the biased exponent.

Figure 9.3 shows the hardware organization of a 32-bit floating-point arithmetic processor. There are six registers in this processor denoted as **A**, **a**, **B**, **b**, **Q**, **q**. Registers with uppercase labels **A**, **B**, **Q** are each 23 bits long for storing the mantissas. Lowercase labeled registers **a**, **b**, **q** are each 8 bits long for storing the exponents. Subscripts are used to designate the bit position of these registers. For example, $\mathbf{A} = A_{1-23}$ and A_1 refers to the most significant bit of register **A**, and $\mathbf{a} = a_{1-8}$, and so on. Three flip-flops, A_s, B_s, Q_s, are used to store the signs of the numbers residing in the appropriate registers. Three cascaded registers of length 32 bits can be formed as follows:

$$\begin{aligned} \mathcal{A} &= A_s \cdot \mathbf{A} \cdot \mathbf{a}, \\ \mathcal{B} &= B_s \cdot \mathbf{B} \cdot \mathbf{b}, \\ \mathcal{Q} &= Q_s \cdot \mathbf{Q} \cdot \mathbf{q} \end{aligned} \tag{9.26}$$

Cascaded registers, $\mathcal{A}$, $\mathcal{B}$, $\mathcal{Q}$, are called the *Accumulator*, *Auxiliary Register*, and *Quotient-Multiplier* register, respectively. When a floating-point number (data) is moved from a memory cell $\mathcal{M}$ to a register $\mathcal{B}$, we simply write $\mathcal{B} \leftarrow \mathcal{M}$, which implies actually three separate register transfers of data as shown below.

$$\begin{aligned} B_s &\leftarrow M_s; \\ \mathbf{B} &\leftarrow \mathbf{M}; \\ \mathbf{b} &\leftarrow \mathbf{m}. \end{aligned} \tag{9.25}$$

In addition, there are two special flip-flops, denoted as C and c, for storing the carry out of the most significant bit of the *mantissa adder* and that out of the *exponent adder*, respectively. The carry bits being generated during the addition/subtraction of two mantissa or of two exponents can be used to discriminate the relative magnitude of two numbers, or to determine the overflow or underflow conditions. Details of how to use these carry conditions will appear in later sections.

Apparently, two parallel binary adders are required, one for processing the mantissas and the other for the exponents. The *Mantissa Adder* (**MA**) is 24 bits long and the *Exponent Adder* (**EA**) is 9 bits long. Let $Z = Z_0 \cdot Z_{1-23}$ and $z = z_0 \cdot z_{1-8}$

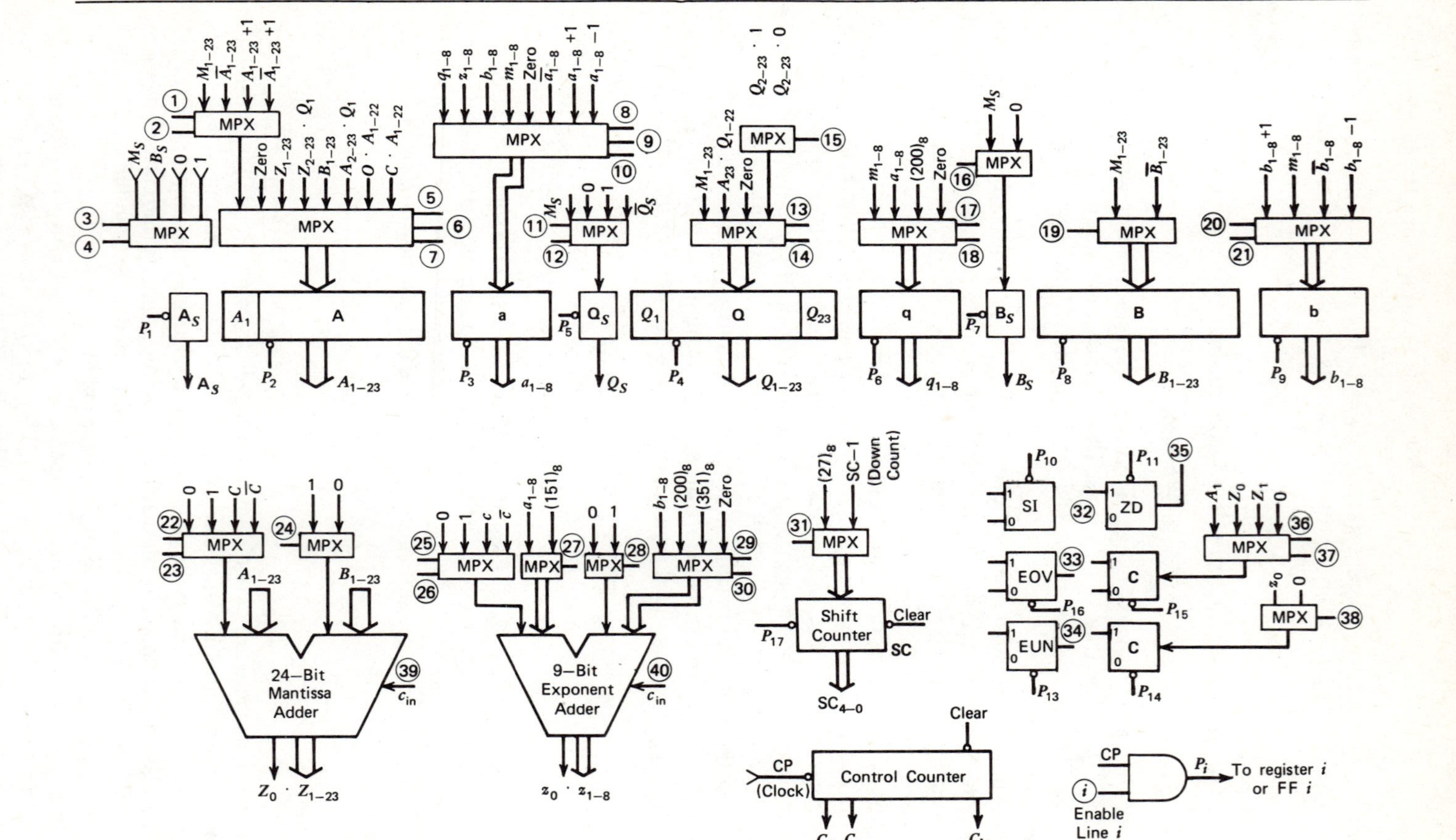

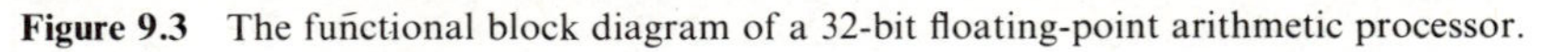
Figure 9.3 The functional block diagram of a 32-bit floating-point arithmetic processor.

be the output labels of the **MA** and **EA**, respectively. The input and output relationships of these two adders are described by the following two equations, where the operators "+" and "·" refer to arithmetic addition and string concatenation, respectively.

$$Z_0 \cdot Z_{1-23} = A_0 \cdot A_{1-23} + 0 \cdot B_{1-23} + C_{\text{in}} \tag{9.26}$$

$$z_0 \cdot z_{1-8} = a_0 \cdot a_{1-8} + 0 \cdot b_{1-8} + c_{\text{in}} \tag{9.27}$$

where A_{1-23}, a_{1-8}, B_{1-23}, and b_{1-8} stand for the contents of registers **A**, **a**, **B** and **b**, respectively. The leading bit A_0 (or a_0) can assume one of the four values 0, 1, C, and $\bar{C}$ (or 0, 1, c, and $\bar{c}$) through a 4-input multiplexer. The last terms C_{in} and c_{in} stand for the carry-in to the two adders **MA** and **EA**, respectively.

Multiplexers are used to select the appropriate data inputs to each register and to the adders at each processor cycle. Shifting is implemented with physical wiring embedded in the multiplexer logic. Several flip-flops are used to indicate the singularity conditions, such as *Exponent OVerflow* (EOV), *Exponent UNderflow* (EUN), *Zero Divisor* (ZD), *Order-of-Magnitude Zero* (OMZ), and a *Sign Indicator* (SI) for the result, and so on.

A mode control decoder, which is not shown in Fig. 9.3, maybe used to select among the four standard floating-point arithmetic operations. A *Control Counter* (CC) with decoded counter states is used to regulate the processor operations coupled with the master clock. Enable lines are used to control the state changes in all registers and flip-flops. In addition, a *Shift Counter* (**SC**) is used to record the number of shifts during multiplication, division, and normalization procedures. All the control lines are labeled by circled numbers. The detailed hardware operations associated with each arithmetic function are described separately in subsequent sections.

9.5 Normalized FLP Addition/Subtraction

Before a floating-point add/subtract instruction is executed, the two operands, the augend/minuend and the addend/subtrahend in normalized form, should be loaded into the accumulator $\mathscr{A}$ and the auxiliary register $\mathscr{B}$, respectively. The resulting sum/difference will be normalized and loaded back into the accumulator $\mathscr{A}$. Conceptually, these operations can be written as

$$\mathscr{A} \leftarrow \mathscr{A} \pm \mathscr{B} \tag{9.28}$$

There are four major steps which must be executed in order to complete the addition/subtraction of two floating-point numbers.

1. Check for zero operands.
2. Align the mantissas by equalizing their exponents.
3. Add/subtract the mantissas.
4. Normalize the resulting sum/difference.

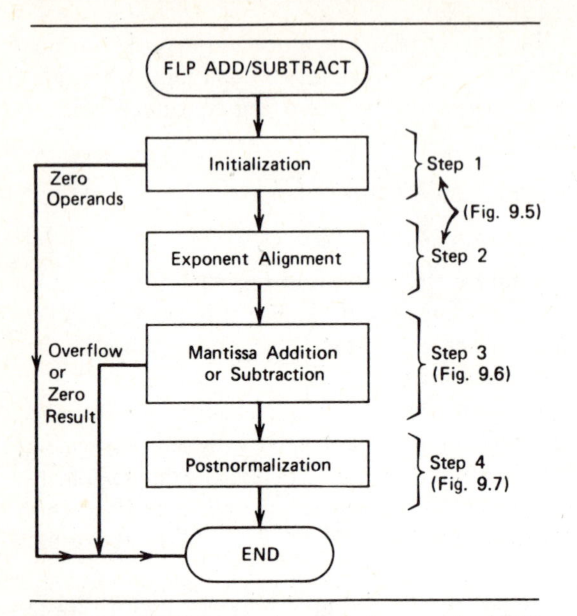

Figure 9.4 The four execution steps of a normalized *FLP Addition* or *FLP Subtraction* instruction.

These four steps are depicted in Fig. 9.4. Detailed hardware operations within each step are illustrated by separate flow charts. The first two steps of zero-checking and exponent equalization are described in Fig. 9.5. When the second operand (addend/subtrahend in register $\mathscr{B}$) is zero, no execution is needed. When the first operand (augend/minuend in $\mathscr{A}$) is zero, the result is equal to the addend for FLP addition and to the negative version of the subtrahend for FLP subtraction. The second step is entered, if and only if both operands are nonzero.

The exponent comparison as specified in Eqs. 9.9 and 9.10 are realized with two's complement subtraction. Register cascading (concatenation) is indicated by the "dot" operator between them. The arrow "$\leftarrow$" has been used as a replacement operator. Operators "$+$" and "$-$" used in the flow chart refer to arithmetic addition and subtraction, respectively, unless otherwise noted. The overbar means bitwise complement. The shift counter starts with a value of 27_8, which is 23_{10}. When the two exponents **a** and **b** are not equal, the mantissa of the smaller one will be shifted right sufficient number of times, so that the smaller exponent can be incremented to match the magnitude of the larger exponent.

Then we proceed to the third step of mantissa addition/subtraction. There are cases that the smaller exponent is too small to match with the larger exponent, in which the mantissa alignment sequence must exit after a maximum of 23 shifts. The operand with the smaller exponent now becomes an OMZ with a zero mantissa. The addition of such an OMZ to a legitimate operand is done in step 3 with the assumption that the two operands now have equal exponents but the OMZ has a

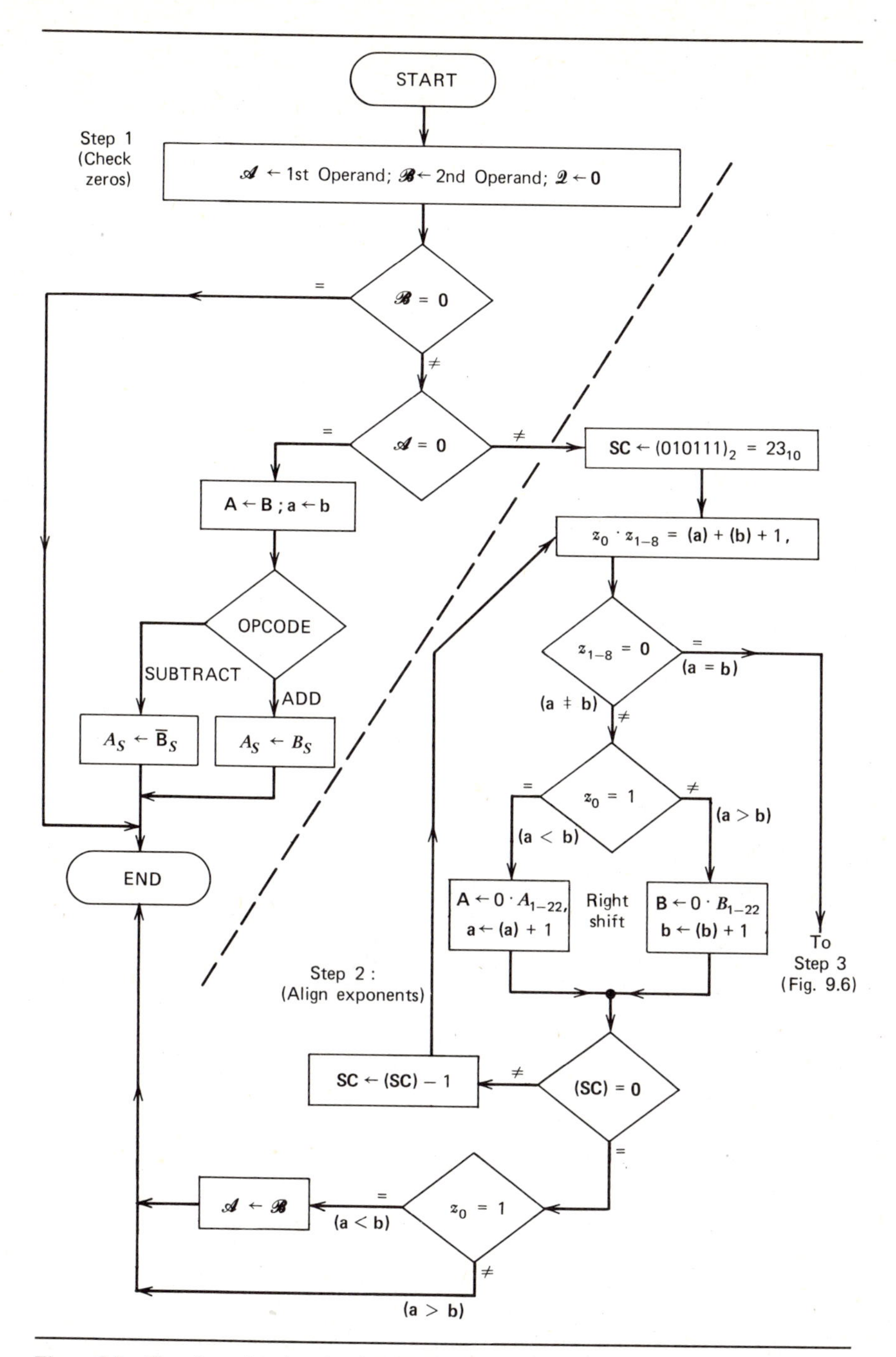

Figure 9.5 Flowchart showing the detailed microoperations performed during the first two execution steps of a FLP Add/Subtract instruction.

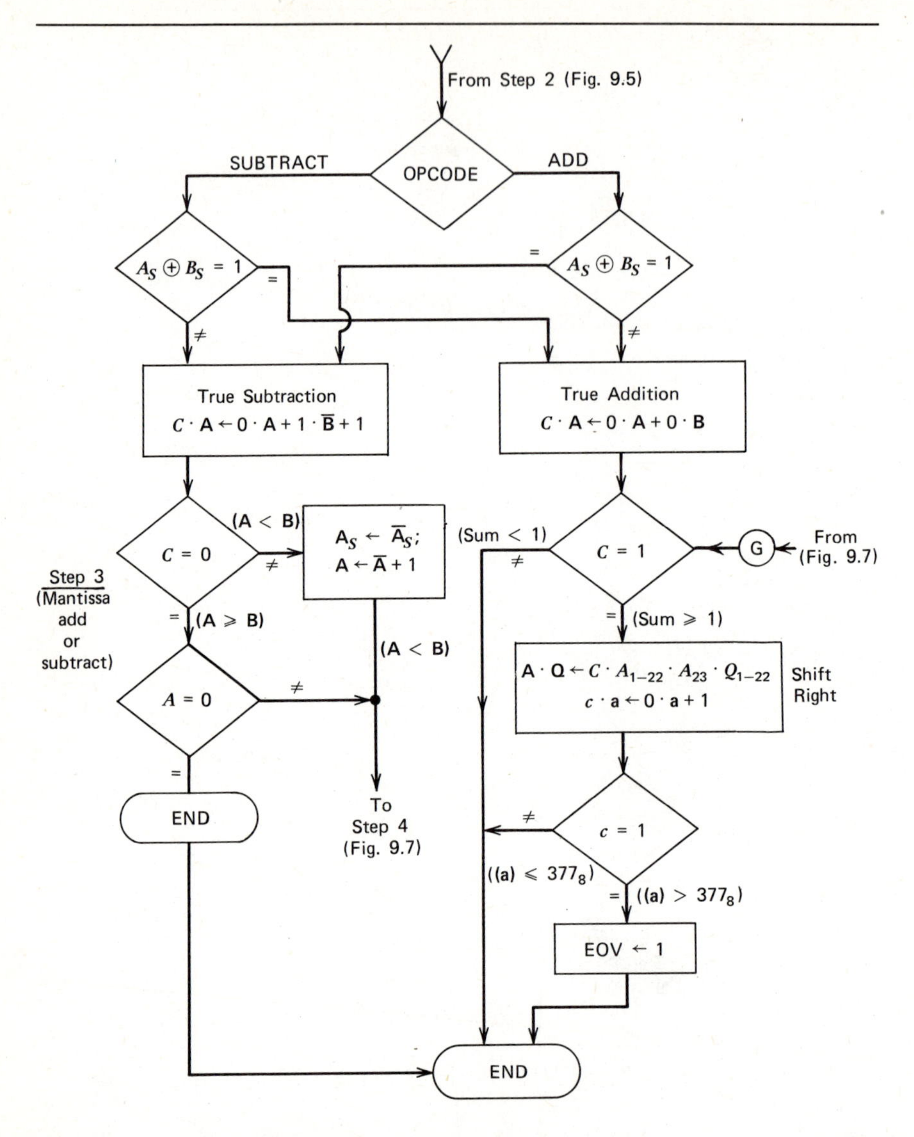

Figure 9.6 Flow chart illustrating the mantissa add/subtract sequence during the execution step 3 of a FLP Add or FLP Subtract instruction.

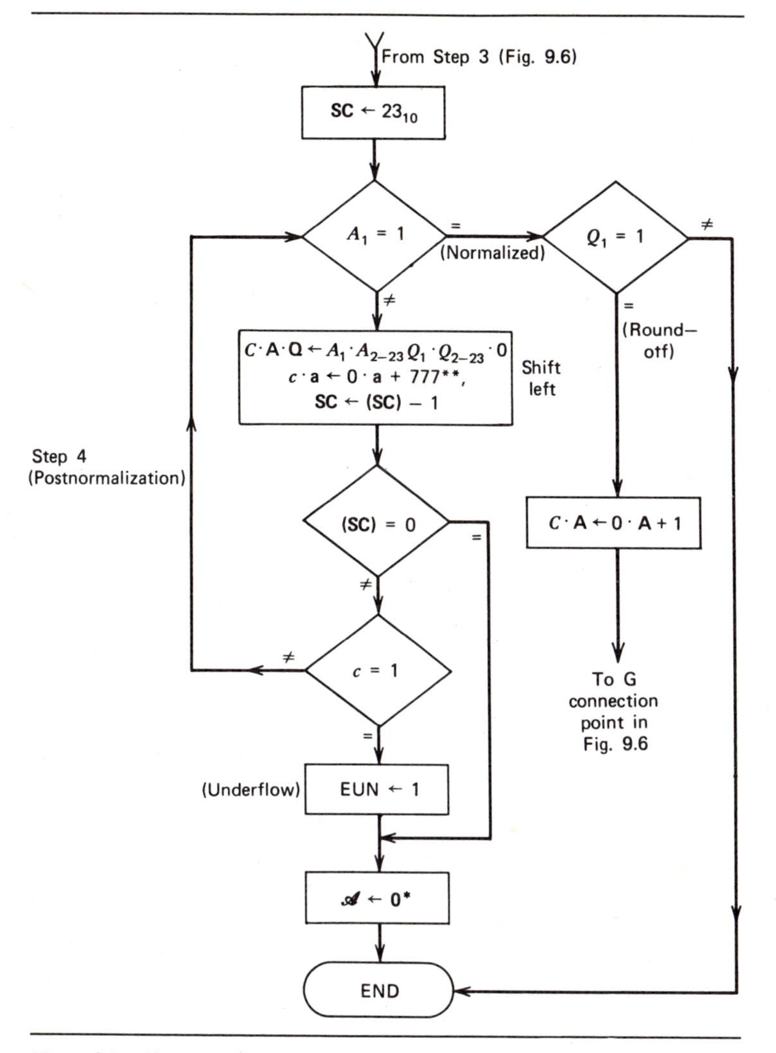

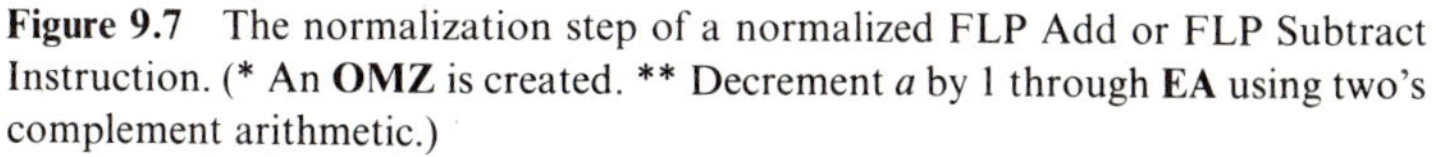
Figure 9.7 The normalization step of a normalized FLP Add or FLP Subtract Instruction. (* An **OMZ** is created. ** Decrement a by 1 through **EA** using two's complement arithmetic.)

zero mantissa, which still produces a legitimate result equal to either the positive or the negative version of the larger number involved.

Figure 9.6 illustrates the mantissa addition or subtraction operations. Again, two's complement arithmetic is assumed in both adders. The right half of the flow chart in Fig. 9.6 describes a *true addition* corresponding to one of the following four cases.

$$(+A) + (+B); (+A) - (-B);$$
$$(-A) + (-B); (-A) - (+B). \quad \textbf{(9.29)}$$

When the sum of a true addition exceeds unity, the resulting mantissa is shifted one bit to the right with a "one" (the carry) entering the most significant bit and, at the same time, the exponent is incremented by one according to Eq. 9.11. Note that the least significant bit A_{23} of register **A** has been shifted into the most significant position Q_1 of register **Q**. It is possible that such an exponent increment (even by one) may cause an exponent overflow, that is, (**a**) > 377_8. The flip-flop EOV should have been set when overflow occurred and there is no need to enter the normalization sequence after exponent overflows.

The left half of the flow chart in Fig. 9.6 corresponds to a *true subtraction* for one of the following four cases:

$$\begin{aligned} &(+A) + (-B); (+A) - (+B); \\ &(-A) + (+B); (-A) - (-B). \end{aligned} \tag{9.30}$$

The true subtraction is performed through two's complement addition as shown. The resulting difference may be either positive, negative, or zero. When it is negative ($A < B$), the result in two's complement form should be recomplemented back to its sign-magnitude form as shown. Again, the operation is ended when the difference is zero. The normalization step will be entered only when a nonzero difference results.

The above rulings on true addition and on true subtraction indicate the fact that a carry from the most significant bit position does not necessarily indicate an overflow as it does in fixed-point operation. Instead, the exponent is incremented by one, which means the binary point has been moved one digit to the left so that the resulting mantissa can be held properly. This is actually done by shifting the mantissa one bit right as specified in Eq. 9.11. The least significant bit being lost may be rounded. Overflow occurs only when the exponent exceeds the limit during true addition.

The normalization operations following the mantissa addition/subtraction are described in Fig. 9.7. The cascaded register $C \cdot \mathbf{A} \cdot \mathbf{Q}$ will be shifted left one bit at a time, until the leading bit A_1 in register **A** becomes a "1". Then a "roundoff" 1 is added to the mantissa, if $Q_1 = 1$. After a maximum of 23 left shifts, if the result is still unnormalized, the final result should be considered either an OMZ or a zero. During this left shifting operation, it is possible to create an exponent underflow. When underflow occurs, the final result can either be assigned a zero or treated as an OMZ. Only zero value was assigned in the flow chart description under such circumstances.

9.6 Normalized FLP Multiplication

As specified in Eq. 9.12, floating-point multiplication is accomplished by multiplying the mantissas of the two operands and adding their corresponding exponents. Exponent overflow or underflow may occur when true addition is performed on two exponents of the same sign. There are four major operations associated with floating-point multiplication. These operations can be executed in three sequential stages as illustrated in Fig. 9.8. The initial stage checks for zero operands and sets the product sign. The steps of mantissa multiplication and exponent addition can be executed

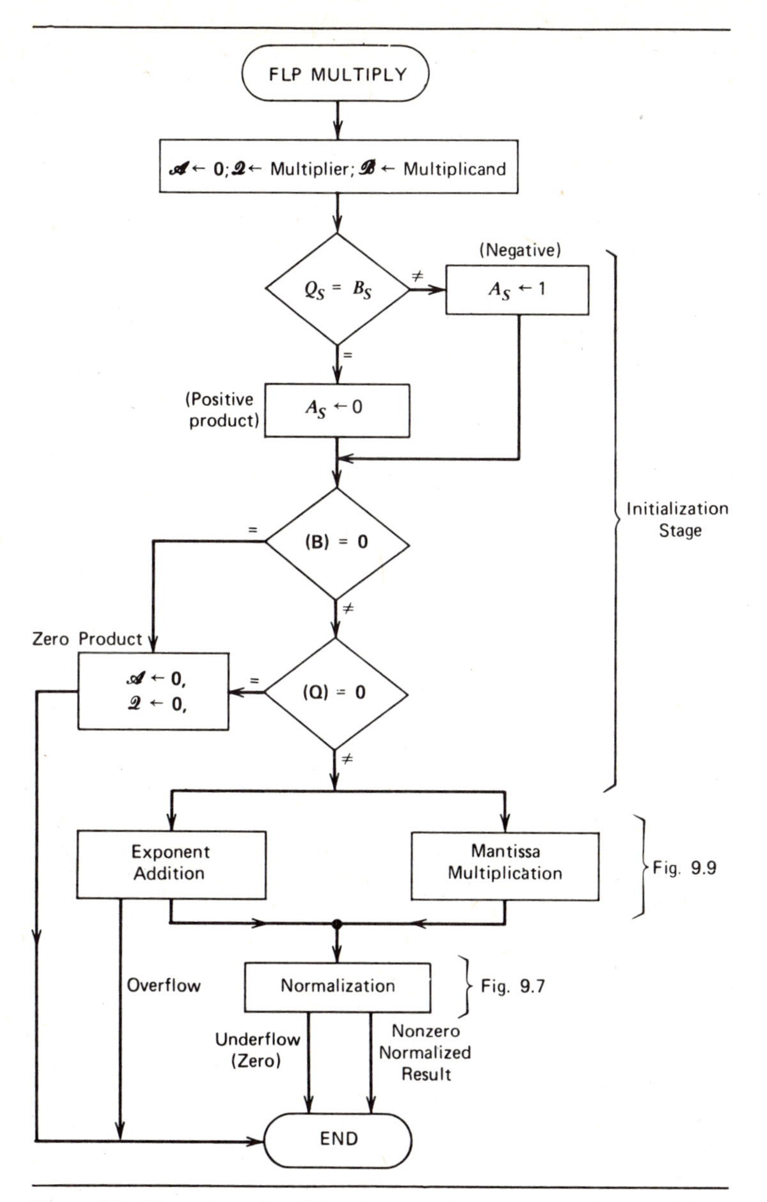

Figure 9.8 Flow chart describing the execution sequence of a normalized FLP Multiply instruction.

simultaneously. However, these two parallel steps must be properly synchronized before the normalization step is initiated. The final step of normalization is required only when the condition in Eq. 9.16 is produced.

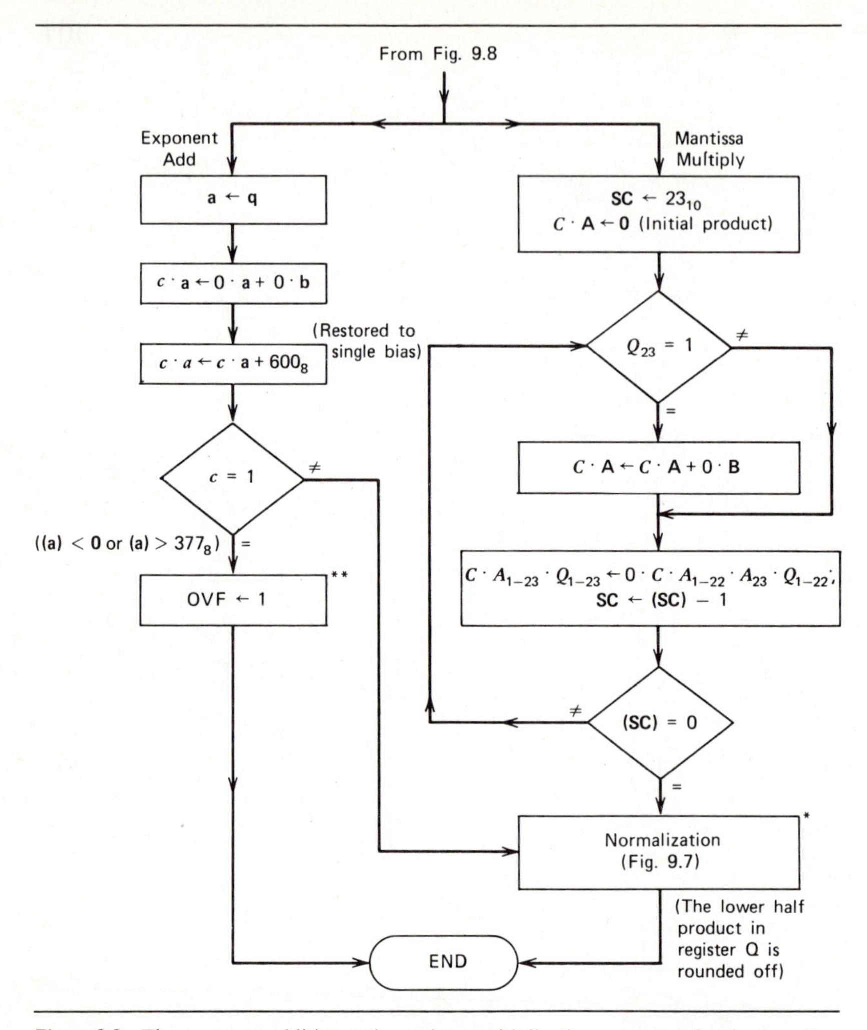

Figure 9.9 The exponent-addition and mantissa-multiplication sequences for the execution of a FLP Multiply instruction. (*At most one left shift is required to normalize the product. ** OVF is a flag signaling either EOV or EUN.)

Initially, the multiplicand is loaded into the $\mathscr{B}$ register, the multiplier in $\mathscr{Q}$ register, and zero in the accumulator $\mathscr{A}$. At the end of execution, the normalized product resides in the accumulator. The detailed micro-operations in the initialization stage are shown in Fig. 9.8. The product is positive when the two operands have same sign and negative otherwise. When either of the operands is zero after checking the corresponding mantissa subregisters **B** and **Q**, the resulting product should be set to zero. We then proceed to clear accumulator $\mathscr{A}$ and terminate the execution. Should neither operand contain a zero, we continue on to the intermediate step.

The flow chart in Fig. 9.9 illustrates the micro-operations associated with mantissa multiply and exponent add. These two operations can be executed in parallel. In some inexpensive machines with only one adder available for arithmetic operations, they may be executed in serial, one after the other even though they are mutually independent. The exponent addition follows a loop-free sequence as described at the left of the flow chart. The sequence starts with loading the multiplier exponent from **q** register into the subregister **a**. The exponent adder produces the sum of the two biased exponents; therefore, the resulting sum is doubly biased. The next operation required is to subtract one copy of the bias constant $b = 128_8 = 200_8$ from the doubly biased sum of exponents. The subtraction of 200_8 is actually implemented with the addition of the two's complement of 200_8, which is 600_8. When the carry out of the **EA** is a "1", an exponent overflow or underflow has occurred for (**a**) $> 377_8$ or (**a**) $<$ 0, respectively. Otherwise, the sequence proceeds to the normalization step.

The mantissa multiply can be implemented by any of the fixed-point multiply schemes described in Chapters 5 and 6. With the basic floating-point hardware host machine declared in section 9.4, we can carry out the FLP multiply with a simple add-shift scheme as illustrated on the right of the flow chart. The multiply sequence starts with loading the shift counter with $27_8 = 23_{10}$ and setting a zero as the initial partial product in mantissa register $C \cdot$ **A**. Then we check the value of the least significant bit of the multiplier mantissa Q_{23}. If it is a one, addition of the multiplicand mantissa **B** to the partial product is performed through the **MA**. Otherwise, no addition is required. Then the contents of the cascaded register C · A · Q is shifted one bit right and the shift counter is decremented by one. The process is completed when the SC becomes zero. Otherwise, inspection of the next high-order multiplier bit is conducted at the next iteration. This process continues until all multiplier bits are exhausted. As stated in Eq. 9.14, at most one leading zero may exist in the final product, in which case only one left shift is necessary to normalize it. In fact, the above operation results in a double-length product in the cascaded register **A** · **Q**. The sequence shows that the lower half of the resulting product residing in **Q** can be rounded off by advancing a "1" to the least significant bit A_{23} of **A** register, when $Q_1 = 1$. The operations in the normalization step implement the mantissa shifting and exponent decrement as was specified in Eq. 9.17 and illustrated in Fig. 9.7.

9.7 Normalized FLP Division

The way a floating-point DIVIDE instruction is executed is analogous to that of floating-point MULTIPLY, except the mantissa multiplication is replaced by mantissa division and the exponent addition by exponent subtraction. Exponent overflow or underflow may occur when true addition is performed on the two exponents of *opposite* signs. The scheme must avoid the situation of having a divisor which is smaller than the dividend mantissa, including the special case of a zero divisor. With this constraint, the postnormalization is unnecessary in floating-point division as long as prenormalization was conducted to avoid quotient overflow.

There are four major operations associated with floating-point division: *Initialization*, *Mantissa Alignment*, *Exponent Subtraction*, and *Mantissa Division*.

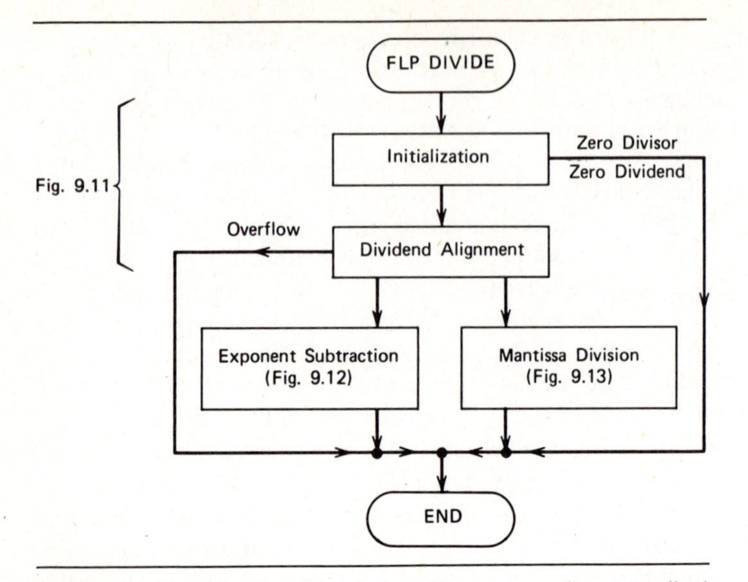

Figure 9.10 Flow chart describing the execution sequences for a normalized FLP Divide instruction.

Again, the exponent and mantissa operations can be executed in parallel, resulting in a process consisting of only three sequential stages as illustrated in Fig. 9.10. Appropriate synchronization between these two parallel processes must be established. The initial and final register assignments before and after the execution of the FLP Divide are depicted below:

Initial Contents	**Register Assignment**			**Final Contents**
Dividend	A_s	**A**	**a**	Remainder
Divisor	B_s	**B**	**b**	Divisor
Zero	Q_s	**Q**	**q**	Quotient

The initialization and mantissa-alignment sequences are demonstrated in Fig. 9.11. It is assumed that both the dividend and divisor are given in normalized form. The quotient sign Q_s is positive when the dividend and divisor have identical signs, and negative otherwise. When the divisor is zero, a quotient overflow is resulted and the ZD flag is set. When the divisor is nonzero and the dividend is zero, the quotient is set to be zero and so is the remainder. When both operands are nonzero, we proceed to align the mantissas.

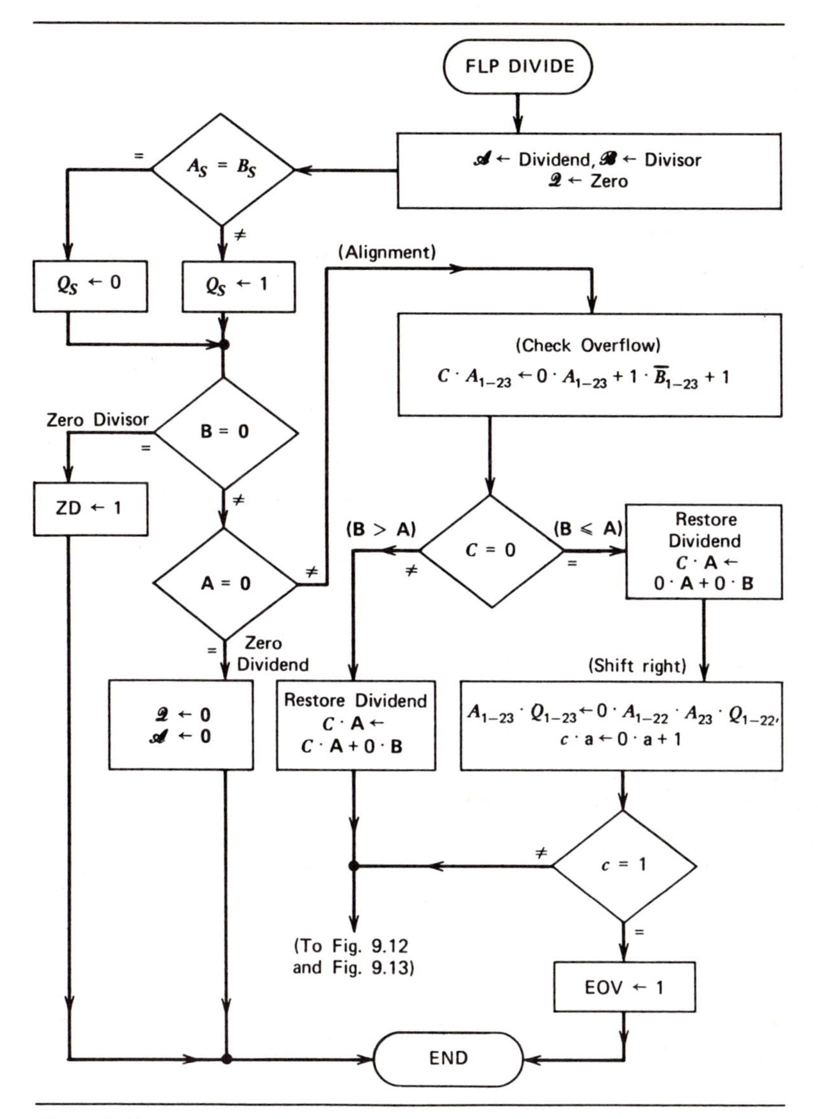

Figure 9.11 The initialization and dividend alignment sequences in FLP division.

Mantissa alignment is essentially an overflow prevention operation, ensuring that the dividend mantissa is smaller than the divisor mantissa. Otherwise, meaningful mantissa division cannot proceed. The alignment always results in a normalized quotient, as therefore, no postnormalization is required. The relative magnitude of the two mantissas is revealed by a comparison operation. Appropriate actions follow the comparison, depending on the outcome of the comparison. The magnitude comparison can be quickly executed by a combinational magnitude comparator described in Chapter 2. With the available hardware components in Fig. 9.3 we can accomplish the comparison with the addition of the two's complement of the divisor mantissa in register **B** to the dividend mantissa in register **A**. Note that two's complement addition is formed by bitwise complementing B and feeding a "1" into the carry-in terminal of the **MA** and then adding them to A.

When carry C from the above comparative subtraction in **MA** is a *zero*, the dividend (in **A** register) is greater than or equal to the divisor (in **B** register), denoted simply as ($\mathbf{A} \geq \mathbf{B}$). Therefore, we must add the divisor back to restore the original dividend, that is, $\mathbf{A} \leftarrow \mathbf{A} - \mathbf{B} + \mathbf{B}$, and then shift the cascaded register $\mathbf{A} \cdot \mathbf{Q}$ right one position with a zero entering from the left end. This will make the dividend smaller than the divisor. Simultaneously, the dividend exponent is incremented by "1" as required in Eq. 9.17. It is possible to have an exponent overflow associated with this incrementing operation. The leading exponent carry c can be used to detect the situation. On the other hand, C being *one* after the magnitude comparison reveals the opposite case ($\mathbf{A} < \mathbf{B}$), which will not cause overflow. Still, one addition is required to restore the original dividend, but no shifting is required. This completes the alignment sequence.

Both the exponent subtraction and mantissa division can be initiated after the dividend mantissa is properly aligned. The exponent subtraction operation is illustrated in Fig. 9.12. For inexpensive designs, these two operations can be executed sequentially by sharing the same hardware facility. The exponent subtraction is realization by two's complement addition through the **EA**.

The exponent carry c can be used to distinguish the case ($\mathbf{a} \geq \mathbf{b}$) (for $c = 0$) from the case ($\mathbf{a} < \mathbf{b}$) (for $c = 1$). In both cases, we have to restore the bias constant 200_8, being lost during the exponent subtraction. After restoring the bias constant back to the exponent register **a**, we can simply check the leading bit a_1 in register **a** to reveal the existence of any singularity conditions; EOV = 1 for exponent overflow and EUN = 1 for exponent underflow. If there is no singular condition, then we move the exponent difference from register **a** to register **q**, where the final quotient exponent should be stored.

The remainder exponent should be set $23_{10} = 27_8$ less than the dividend exponent. This is done by subtracting 27_8 from the contents of register $c \cdot \mathbf{a}$ through the **EA**. This operation is equivalent to adding 751_8 to the contents of register $c \cdot \mathbf{a}$. Note that, when $\mathbf{a} < \mathbf{b}$, the exponent difference $\mathbf{a} - \mathbf{b}$ is negative. By subtracting 23_{10} from this negative difference may result in an exponent underflow for the remainder. In this case, we simply assume a zero value (the most negative exponent in biased form) to the remainder exponent, which should be residing in register **a** at the end of computation.

The sequence for mantissa division is described in Fig. 9.13. The **SC** should be initially set to a value $23_{10} = 27_8$ corresponding to the mantissa length. The stored-shift restoring division similar to that shown in section 7.4 is implemented. The cascaded register $C \cdot \mathbf{A} \cdot \mathbf{Q}$ has the hardware shifting capability. In other words, the separate shifter (rotate box) used in Fig. 7.3 is not required with the embedded stored shifting feature in the register. The $C \cdot \mathbf{A} \cdot \mathbf{Q}$ register is similar to the $L \cdot \mathbf{AC} \cdot \mathbf{QM}$ register used in section 7.4. Again, the comparison operation is implemented with two's complement arithmetic. The 24-bit **MA** handles the required subtraction. The

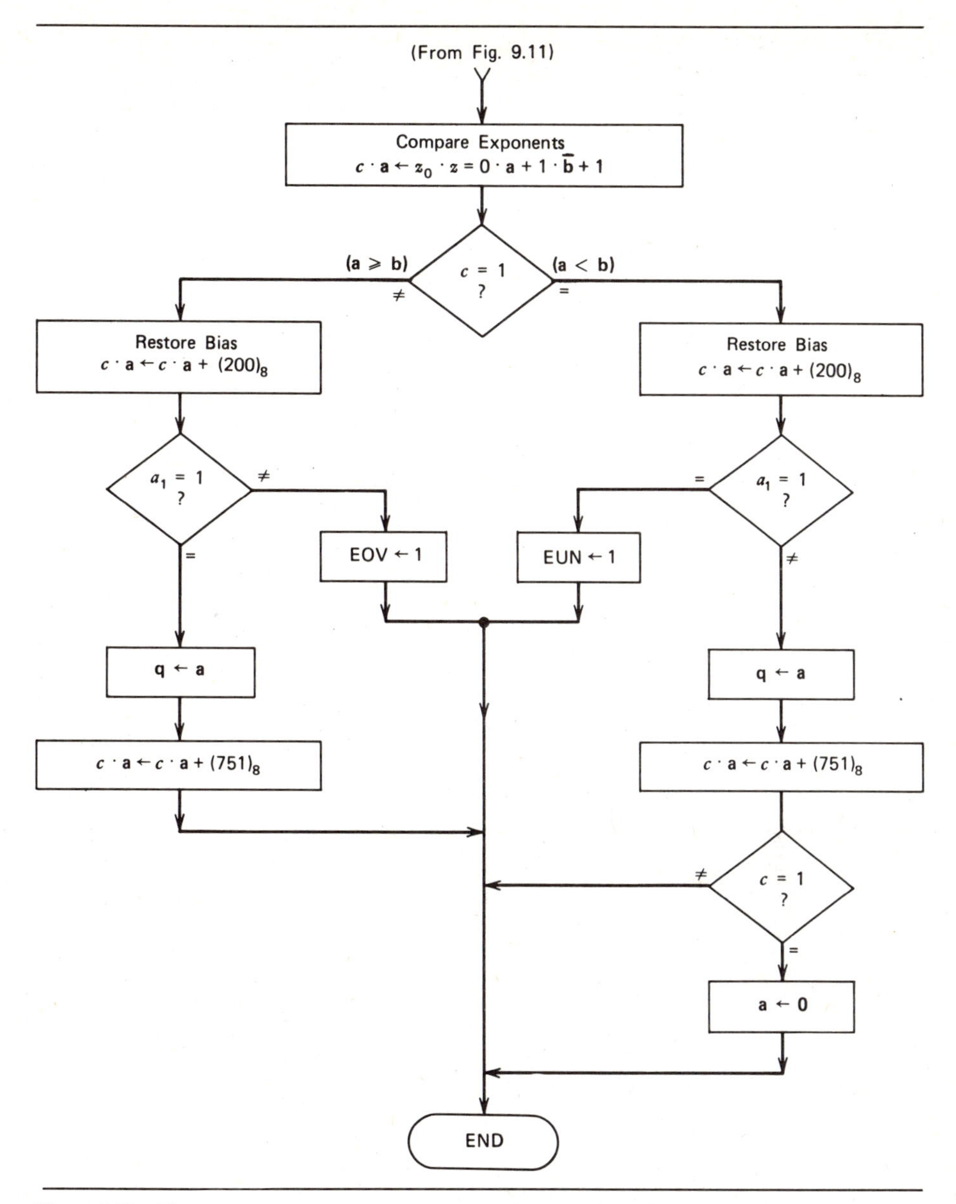

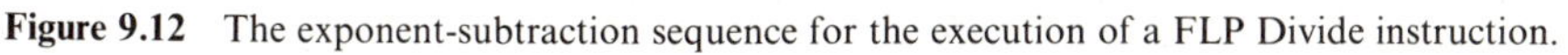

Figure 9.12 The exponent-subtraction sequence for the execution of a FLP Divide instruction.

operations specified in Eq. 7.24 are replaced by

$$Z_0 \cdot Z_{1-23} = C \cdot A_{1-23} + 1 \cdot \overline{B}_{1-23} + 1 \quad \textbf{(9.32)}$$

The quotient being generated is shifted in from the right end of the cascaded register $C \cdot \mathbf{A} \cdot \mathbf{Q}$. The **SC** decrements by one after each iteration until the value "0" is reached.

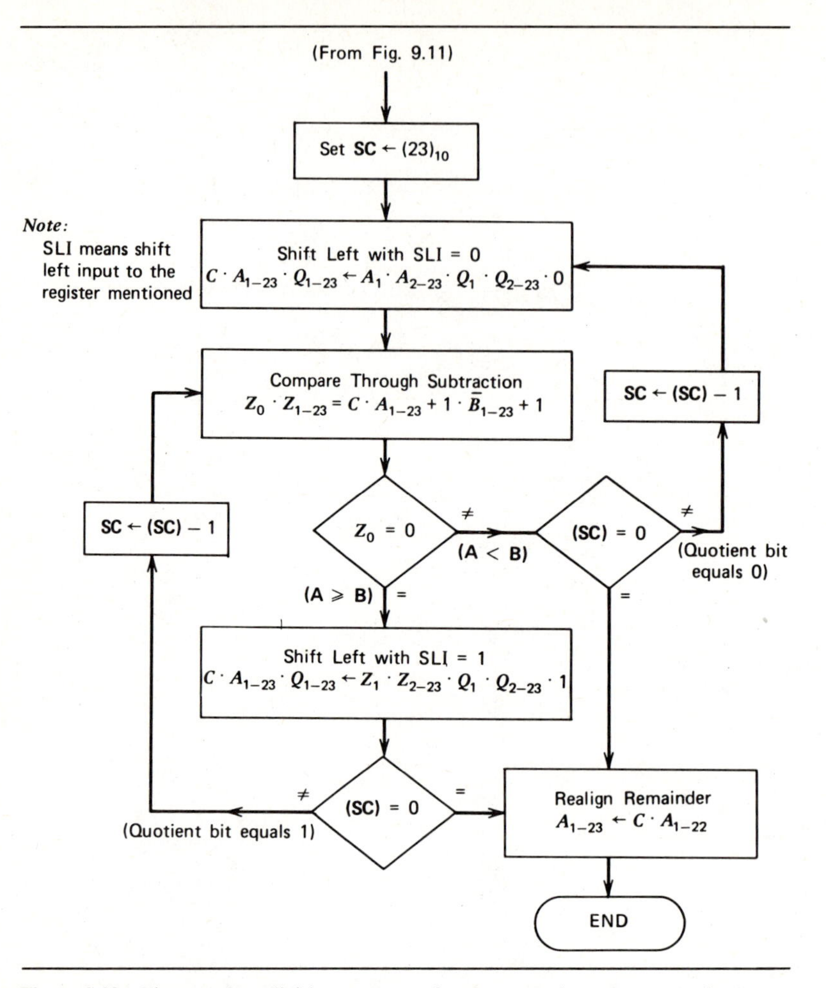

Figure 9.13 The mantissa division sequence for the execution of a normalized FLP Divide instruction.

A final right shift is necessary to realign the remainder mantissa into register **A**. This completes all FLP division sequences.

9.8 Case Study III: The IBM System/360 Model 91 FLP Arithmetic Processor

In this section, a practical FLP processor is shown. The IBM System/360 Model 91 computer has an FLP arithmetic processor (Fig. 9.14) which employs separate instruction-oriented algorithms for the ADD function and MULTIPLY/DIVIDE functions. Linked together by an FLP instruction unit, the multiple FLP execution units provide instruction execution at the burst rate of one instruction per cycle.

Figure 9.14 The main frame of the IBM System/**360** Model **91** Computer. (Courtesy of IBM Corporation, 1977.)

We present below the FLP processor configuration, FLP data format, FLP instruction instruction definitions, FLP ADD Unit, and FLP Multiply/Divide Unit contained in the Model **91** FLP processor. Provisions of adding temporary storage platforms in the system to enable pipelined arithmetic operations will be elaborated in Chapter 11.

A prime concern in designing the *FLP Execution Units* (**FLEU**) in Model **91** is to develop an overall organization that will match the performance of the *FLP Instruction Unit* (**FLIU**) in Model **91**. The execution time of FLP instructions is usually long compared with the issuing rate of these instructions by the **FLIU**. The most obvious approach is to apply a faster technology with special design techniques to reduce the FLP execution time per FLP instruction. However, even with state-of-the-art algorithms, no existing **FLEU** can match the one- or two-cycle performance of the **FLIU**. Another approach to closing the speed gap is to provide execution concurrency among instructions; this obviously requires two or more **FLEU**s. The general organization of an FLP arithmetic processor capable of concurrent execution is shown in Fig. 9.15. One **FLIU** and two **FLEUs** are shown in the figure. The primary function of the **FLIU** is to sequence the operands from storage to the proper **FLEU**. It must buffer the instructions and assign each instruction to a nonbusy **FLEU**. Because the execution time is not the same for all FLP instructions, there is a possibility for out-of-sequence execution. Dependence

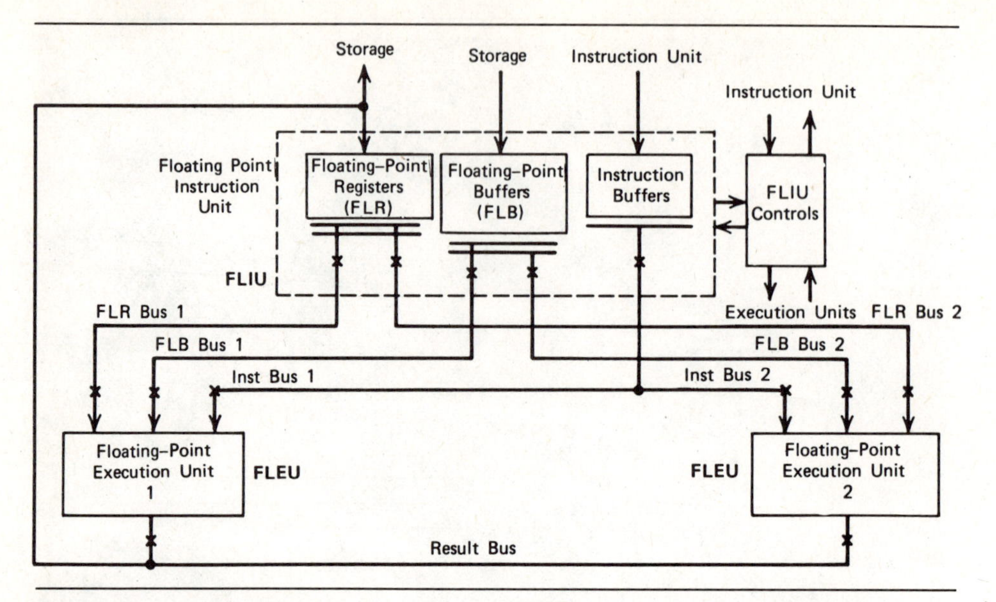

Figure 9.15 The organization of a FLP arithmetic processor capable of concurrent execution of FLP instructions (Anderson, et al. [2]).

Table 9.2 FLP Instructions of the IBM System/360 Model 91, Executed by Various Arithmetic Units of the System

FLP Instruction	Arithmetic Unit	Machine Cycles
Load (S/L)	FLIU	1
Load and Test (S/L)	FLIU	1
Store (S/L)	FLIU	1
Load Complement (S/L)	ADD Unit	2
Load Positive (S/L)	ADD Unit	2
Load Negative (S/L)	ADD Unit	2
Add Normalized (S/L)	ADD Unit	2
Add Unnormalized (S/L)	ADD Unit	2
Subtract Normalized (S/L)	ADD Unit	2
Subtract Unnormalized (S/L)	ADD Unit	2
Compare (S/L)	ADD Unit	2
Halve (S/L)	ADD Unit	2
Multiply	M/D Unit	6
Divide	M/D Unit	18

Abbreviations: S = short data format; L = Long data format; FLIU = Floating-Point Instruction Unit; M/D = Multiply/Divide.

among instructions also must be controlled. If the (**N** + 1)st instruction is dependent on the result of the **N**th instruction, instruction **N** + 1 must not be allowed to start until instruction **N** is completed. Buffering and sequence control of all instructions, storage operands, and FLP accumulators are the responsibilities of the **FLEU**s. Each **FLEU** is capable of executing all FLP instructions.

The Model **91** computer separates its FLP instruction set into two subsets: the set of ADD-type instructions and the set of MULTIPLT/DIVIDE instructions. Table 9.2 lists these instructions and identifies the unit in which each instruction is executed. With this separation, an **FLEU** called *FLP* **ADD Unit**, which executed all the add-type instructions (add or subtract) in two cycles, and an **FLEU** called the *FLP Multiply/Divide Unit* (**M/D** Unit), which executed the FLP MULTIPLY in six cycles and the FLP DIVIDE in 18 cycles, were designed in the Model **91** machine. The system organization of the IBM **360/91** FLP arithmetic processor is shown in Fig. 9.16. The FLP **ADD** Unit has three reservation stations and is treated as three separate **ADD** Units $\mathbf{A}_1$, $\mathbf{A}_2$, and $\mathbf{A}_3$. The FLP **M/D** Unit has two reservation stations—**M/D1** and **M/D2**. A detailed design of these functional units will be given in the next section following a brief sketch of the System/**360** architecture and FLP instructions.

FLP data occupy a fixed-length format in Model **91**, which may be with either a full-word short format or with a double-word long format:

SHORT FLP DATA FORMAT

Sign	Exponent	Mantissa
0	1 2 ··· 7	8 9 ··············· 31

LONG FLP DATA FORMAT

Sign	Exponent	Mantissa
0	1 2 ··· 7	8 9 ··· 63

The leftmost position is the sign bit: "0" for positive and "1" for negative. The subsequent seven positions are occupied by the *characteristic* (biased exponent). The *fraction* (mantissa) consists of 6 hexadecimal true magnitude digits for the short format or 14 for the long format. The radix-point is assumed to be immediately before bit position 8. Base (radix) 16 is assumed for the exponent with a bias constant of 64. Therefore, the range of the exponent is from −64 to +63 corresponding to the biased binary values from 0 to 127. Normalized SUBTRACT, MULTIPLY, and DIVIDE are performed preserving the maximum precision. Unnormalized ADD or SUBTRACT are also implemented in the system as shown in Table 9.2. Either a normalized or unnormalized sum or difference can be obtained by issuing the proper instruction.

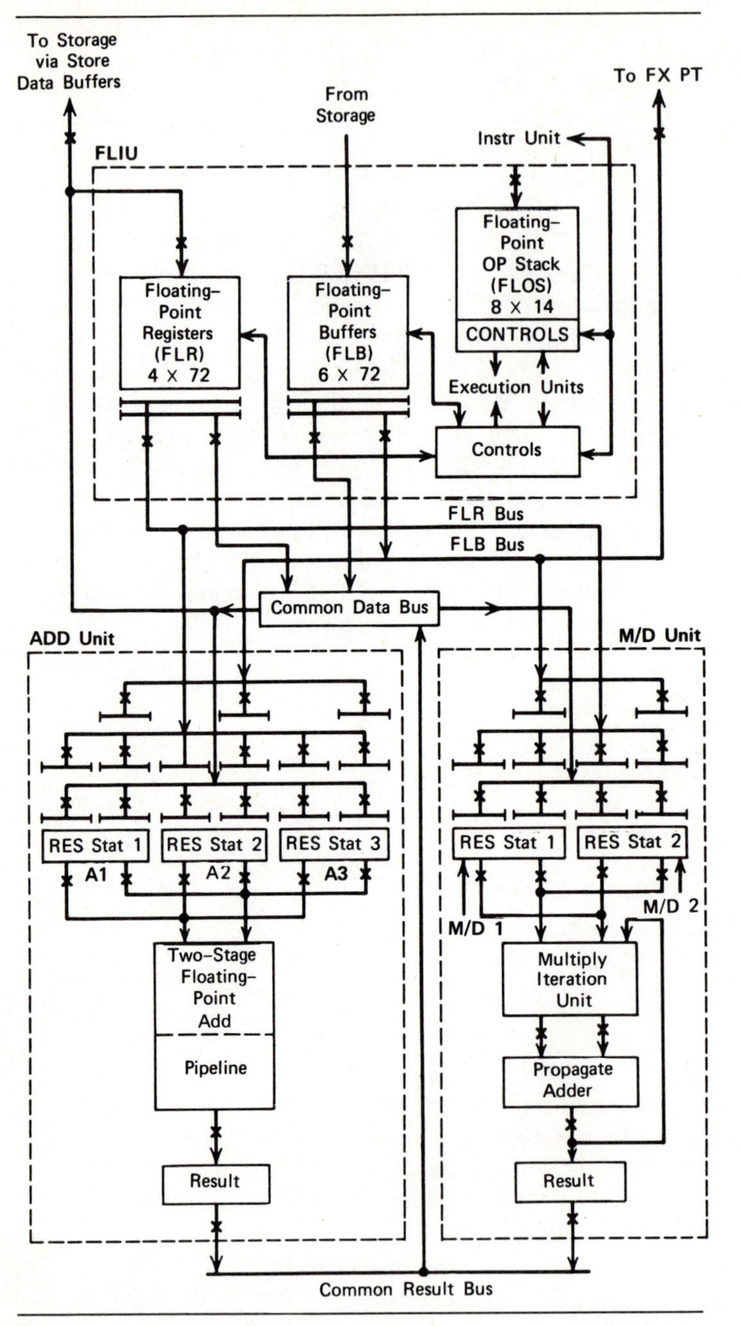

Figure 9.16 The schematic block diagram of the IBM System/360 Model 91 FLP arithmetic processor with three functional units (Anderson, et al. [2]).

The Floating-Point ADD Unit in IBM 360/91

The challenge in the design of the FLP **ADD** Unit in Model **91** was to minimize the number of logical levels in the longest delay path. The FLP add/subtract operations in Model **91** are much more involved compared with those described in section 9.5. The **ADD** Unit algorithm is separated into three parts: (1) *characteristic comparison and preshifting* (**CCP**), (2) *fraction addition*, and (3) *postnormalization*. The data flow in the FLP **ADD** Unit is illustrated in Fig. 9.17. The unit performs the following sequence of operations:

(a) Compare the two characteristics and establish their difference.

(b) Decode the difference into shift amount and shift right the fraction with the smaller characteristic and equalize it with the larger characteristic.

(c) Pass the second fraction through the *True-Complement* (T/C) logic so that subtraction can be performed by two's complement addition.

(d) Add the two fractions in a parallel adder.

(e) Provide the true sum and complemented sum depending on the high-order carry.

(f) Normalize (shift left) the resultant fraction, if required by the normalized instruction.

(g) Reduce the characteristic by the amount of left shift necessary to normalize the resultant fraction.

(h) Store the resultant operand in the proper accumulator.

Operations (a), (b), and (c) are performed by the first section of **CCP** in Fig. 9.17. The second section of the fraction adder merges operations (d) and (e). The third section of postnormalization conducts the last three operations (f), (g), and (h). A *Characteristic Difference Adder* (**CDA**) is used in the **CCP** section. If $C_A \geq C_B$, there is a carry out of the high-order position of the **CDA** and the carry is used to gate the fraction B to the preshifter. If $C_B > C_A$, there is no carry out of the **CDA**, and fraction A is gated to the preshifter. The preshifter is a parallel digit-shifter which shifts right each of the 14 hexadecimal digits any amount from zero to 15.

The *Fraction Adder* is a two-level **CLA** adder 56 bits long and divided into 14 groups. The *Zero-Digit Checker* (**ZDC**) in the third section is simply a large decoder which detects the number of leading zero digits and provides the shift amount to the postshifter. The implementation of the *Digit Postshifter* is the same as the *Digit Preshifter* except that the postshift is a left-shift. The *Characteristic Update Adder* (**CUA**) is executed in parallel with the fraction shift. The **ZDC** provides the **CUA** with the two's complement of the amount by which the resulting characteristic must be reduced. A total of *three* high-speed parallel adders and *two* shifters were used in this FLP **ADD** Unit. The large increase in hardware over the basic design in section 9.4 enables high-speed execution of the double-word FLP addition in two processor cycles, almost matching the instruction-issuing rate of the central processor.

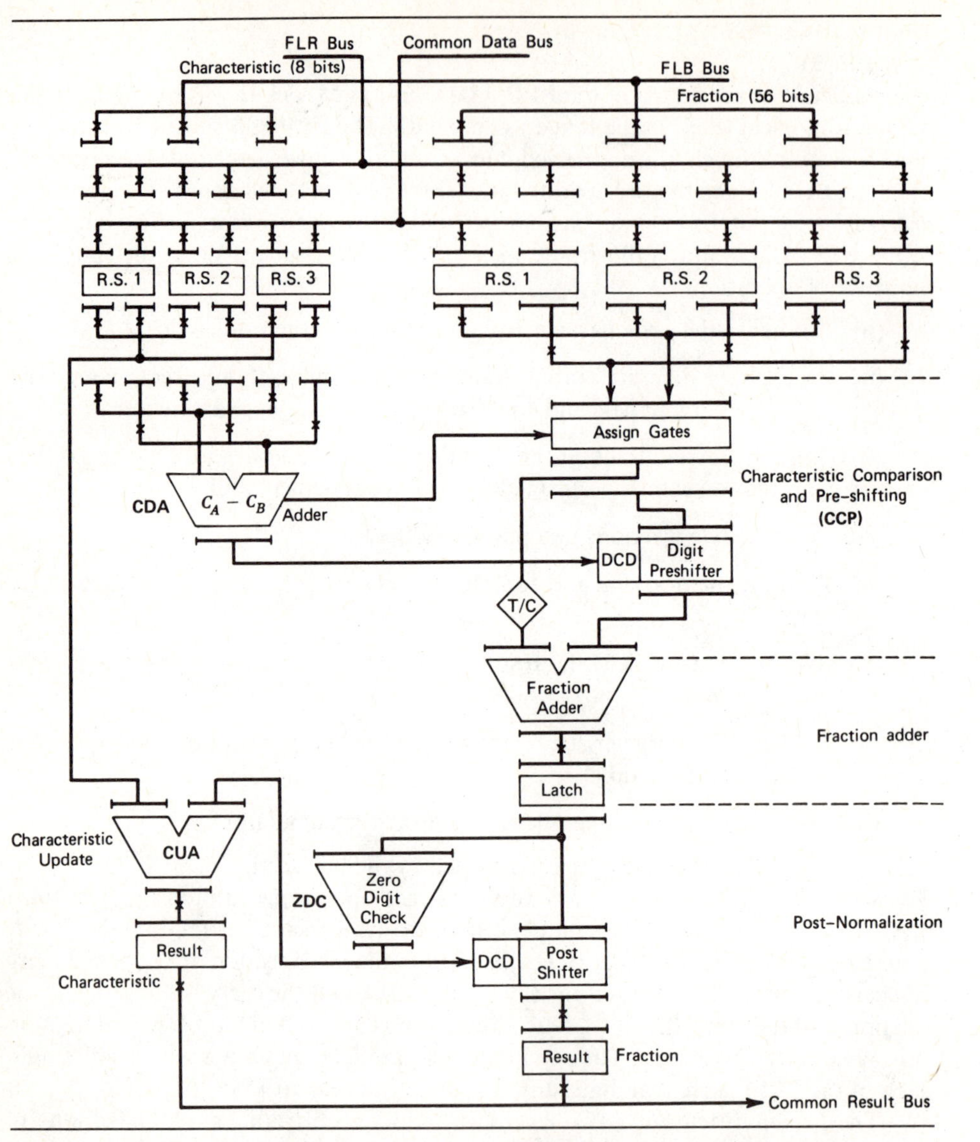

Figure 9.17 The schematic block diagram of the FLP Add Unit in IBM System/360 Model 91 computer (Anderson, et al. [2]).

The Floating-Point Multiply/Divide Unit in IBM 360/91

The algorithms developed for FLP **Multiply** and FLP **Divide** in Model **91** share essentially the same hardware. The multiply algorithm based on the overlapped 3-bit scanning has already been described in section 5.5 and implemented in section 5.6. The scanning pattern for the 56-bit multiplier (fraction portion only) was illustrated in Fig. 5.7. The five overlapped iterations (with 12 bits per iteration) and the generation of six multiples of the multiplicand per each iteration were illustrated in Fig. 5.8.

The multiplier recoding rules were specified in Table 5.4, generating one multiple after each 3-bit window is scanned. Two carry-save adders trees were suggested in Fig. 5.9 for the addition of the multiplicand multiples. The necessary multiply hardware based on this iterative algorithm was shown in Fig. 5.10.

On the other hand, the divide algorithm based on multiplicative quadratic convergence has been described in section 8.2 and implemented in section 8.3. Figure 8.1 illustrated the exact sequence of FLP operations in Model **91**. The hardware divide loop employing **CSAs** and **CPA** was described in Fig. 8.2. The formats of the denominators and their multipliers in the convergence process were specified in Table 8.1. The timing chart that showed concurrency in the divide loop was illustrated in Fig. 8.3. We shall now show how the above unit designs are combined into a unified system which handles the FLP multiply or FLP divide instructions of the Model **91** Computer.

The complete schematic diagram of the FLP **Multiply/Divide** Unit is shown in Fig. 9.18. The diagram illustrates the data flow for the execution of either a normalized MULTIPLY or a normalized DIVIDE. The data flow can be separated into two parts, the iterative hardware and the peripheral hardware (that hardware which is peripheral to the iterative hardware). The iterative hardware has already been described in section 5.6 and in section 8.3. The peripheral hardware includes the input reservation stations, the prenormalizer, the postnormalizer, the carry-propagate adder, the result register, and the characteristic arithmetic. The two reservation stations appear to the **FLIU** for assignment purpose as two distinct Multiply Units. If the input operands are unnormalized, they must be gated to the prenormalizer, normalized, and then returned to the originating reservation station. The amount of left shifting necessary to normalize an operand is gated to the characteristic arithmetic logic, where the characteristic is updated for this shift.

The required characteristic addition is overlapped with the execution of the fraction multiply. The postshift can never be more than one digit because the inputs are normalized. Because the product is accumulated as two operands, the output of the iterative hardware is gated to a **CPA** to form the final product. Detection of the need for postnormalization is done in parallel with the **CPA** and the result is gated to the common data bus, either shifted left one digit or unshifted. The iterative operations of the quadratic convergence division were described in detail in section 8.3. The necessary characteristic subtraction in FLP divide is carried out in parallel with the fraction division. No postnormalization is required for division as described in section 9.7.

Concurrency is the key to high performance in Model **91** FLP arithmetic. The processor organization we have studied exhibits several levels of concurrency as summarized below:

(a) Concurrent execution among instruction classes.

(b) Concurrent execution among instructions in the same class (**ADD** Unit).

(c) Concurrent execution within an instruction (multiply iterative hardware and divide loop).

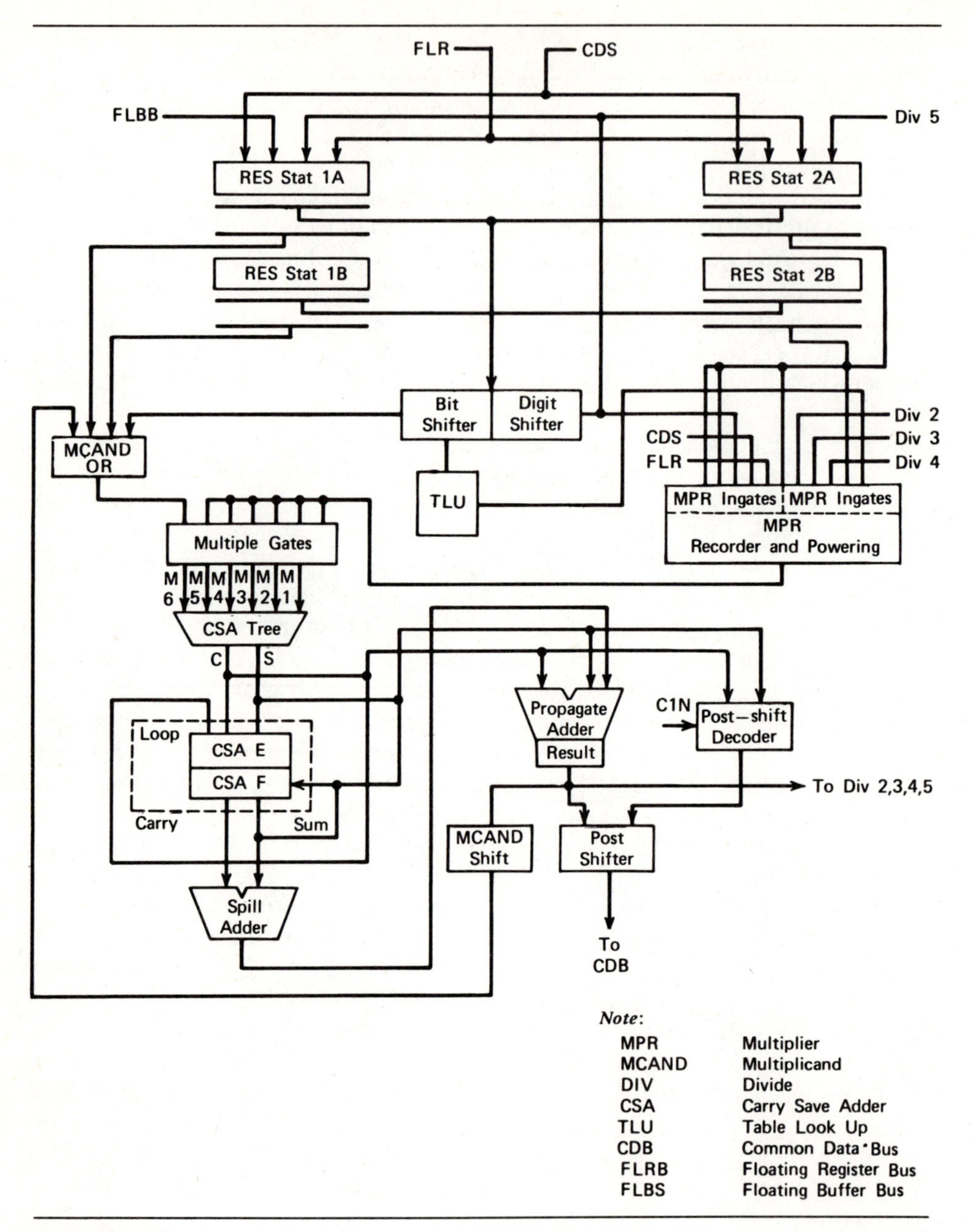

Figure 9.18 The schematic block diagram of the FLP Multiply/Divide Unit in the IBM System/360 Model 91 computer (Anderson, et al. [2]).

The common data busing and register tagging schemes in Model **91** permit the simultaneous execution of independent instructions while preserving the essential precedences inherent in the instruction stream. The hardware looks ahead about eight instructions and automatically optimizes the program execution on a local basis. The application of these techniques is not limited to FLP arithmetic or System/**360** architecture. It may be used in almost any computer that has multiple execution units and multiple accumulators.

9.9 Bibliographic Notes

This chapter presents the basic issues associated with FLP arithmetic design. In addition to the rationals given in section 9.1, readers may find additional justifications from Knuth [10] and Sterbenz [15]. Rigorious mathematical treatment of FLP arithmetic can be found in Knuth. Sterbenz's book describes the IBM 360 Systems in detail including comprehensive application requirements for users. The handling of FLP singularities, in particular in the Project STRETCH (IBM 7030 System), has been reported in Campbell [3]. He gave an excellent sketch of the FLP features of the 7030. Readers who wish to review more case studies of FLP processors should enjoy reading the book edited by Buckholz [3]. The FLP operations in the IBM System/**360** are described in [1, 8]. The normalized FLP instructions specified in section 9.3 are presented without concern for the rounding or truncation errors. Chu [4] and Mano [12] have given a brief account of normalized FLP arithmetic processors. Chu uses the Computer Design Language (CDL) to illustrate the FLP hardware, which can be easily simulated by a general-purpose computer. Such CDL simulators are now available for various types of machines. Base choices of FLP arithmetic were studied by Kahan [9], Matula [11], and Richman [13]. Numerical characteristics of FLP arithmetic were studied in Cody [5], Gary and Harrison [7], and Sweeney [14]. More advanced topics of FLP operations will be presented in the next chapter.

The case study of IBM System/**360** Model **91** FLP Instruction Unit, FLP **ADD** Unit, and the FLP **Multiply/Divide** Unit is based on the report of Anderson et al. [2]. Flynn and Low [6] have also presented an interesting report of the same system. The article by Tomasulo [16] should be of interest to those who wish to explore the maximum concurrency in systems with multiple arithmetic units.

References

[1] Amdahl, G. M., "The Structure of System/360, Part III, Processing Unit Considerations," *IBM System Journal*, Vol. 3, No. 1964, pp. 144–164.

[2] Anderson, S. F. et al., "The IBM System/360 Model 91: Floating-Point Execution Unit," *IBM Journal*, January 1967, pp. 34–53.

[3] Campbell, S. G., "Floating-Point Operation," *Planning a Computer System*, W. Buchholz (ed.), McGraw-Hill, New York, 1962, Chapter 8.

[4] Chu, Y., *Computer Organization and Microprogramming*, Prentice-Hall, Englewood Cliffs, N.J., 1972, Chapter 5.

[5] Cody, W. J., Jr., "Static and Dynamic Numerical Characteristics of Floating-Point Arithmetic," *IEEE Trans. Computers*, Vol. C-28, June 1973, pp. 596–601.

[6] Flynn, M. J. and Low, P. R., "The IBM System/360 Model 91: Some Remarks on System Development," *IBM Journal*, January 1967, pp. 2–7.

[7] Gary, H. L. and Harrison, C. Jr., "Normalized Floating-Point Arithmetic with an Index of Significance," *Proc. of the Eastern Joint Computer Conference*, 1959, pp. 244–248.

[8] IBM Staff, "Floating-Point Arithmetic," in *IBM System/360 Principles of Operation*, IBM System Ref. Lib. From A22-6821-7, September 1968, pp. 41–50.3.

[9] Kahan, W., "What Is the Best Base for Floating-Point Arithmetic, Is Binary Best?," *Lecture Notes*, Dept. of Computer Sci., University of California, Berkeley, California, 1970.

[10] Knuth, D. E., *The Art of Computer Programming: Seminumerical Algorithms*, Vol. 2, Addison-Wesley, Reading, Mass., 1969, Chapter 4.

[11] Matula, D. W., "A Formalization of Floating-Point Numeric Base Conversion," *IEEE Trans. Comput.*, Vol. C-19, August 1970, pp. 681–692.

[12] Mano, M. M., *Computer System Architecture*, Prentice-Hall, Englewood Cliffs, New Jersey, 1976, Chapter 10.

[13] Richman, P. L., "Floating-Point Number Representations: Base Choice Versus Exponent Range," *Tech. Rep. No. CS-64*, Dept. of Computer Sci., Stanford University, Stanford, California, 1967.

[14] Sweeney, D. W., "An Analysis of Floating-Point Addition," *IBM System Journal*, Vol. 4, No. 1, 1965, pp. 31–42.

[15] Sterbenz, P. H., *Floating-Point Computation*, Prentice-Hall, Englewood Cliffs, New Jersey, 1974.

[16] Tomasulo, R. M., "An Efficient Algorithm for Exploiting Multiple Arithmetic Units," *IBM Journal*, January 1967, pp. 25–33.

Problems

Prob. 9.1 Compare fixed-point arithmetic with floating-point arithmetic and describe their advantages and disadvantages with respect to number range, precision, programming, and design considerations.

Prob. 9.2 Consider a 48-bit FLP processor with 37-bit sign-magnitude mantissa, 11-bit exponent in biased two's complement form with base $r = 16$. Indicate the ranges of legitimate FLP numbers representable by this machine for unnormalized and normalized operations, respectively. What is the numerical gap between adjacent FLP numbers and explain the difference between normalized infinitesimals $\pm\varepsilon_n$ and the unnormalized infinitesimals $\pm\varepsilon_u$ of this machine. Justify the singular FLP operations specified in Table 9.1.

Prob. 9.3 Consider the 32-bit FLP arithmetic processor specified in Fig. 9.3 with 40 control terminals and 16 enable lines for registers or flip-flops. List logic and arithmetic conditions for which the control lines 5, 15, and 39 will be set to logic "1" by examining the flow charts given from Fig. 9.4 to Fig. 9.13.

Prob. 9.4 What are "true additions" and "true subtractions"? Verify the exponent overflow condition associated with true addition in Fig. 9.6, and the exponent underflow condition with postnormalization in Fig. 9.7. Explain why the normalized division described in section 9.7 requires no postnormalization.

Prob. 9.5 During normalized FLP add/subtract, the exponent alignment sequence illustrated in Fig. 9.5 shows that when the difference of two exponents exceed the maximum mantissa length p, the mantissa with the smaller exponent has been shifted entirely out of the register (leaving a zero mantissa or an OMZ). What is the maximum error that may occur if one uses only the larger number to determine the result under the above circumstances? Can you suggest a method to minimize this error?

Prob. 9.6 Describe separately how two's complement subtraction and one's complement subtraction can be used to compare the relative magnitude of two sign-magnitude numbers. Explain also how exponent overflow (EOV) or exponent underflow (EUN) can be detected using these two sign-complement arithmetic systems. From designers' viewpoint, which of these two systems requires less hardware investment and why?

Prob. 9.7 Show step-by-step numerical computations of the following FLP numbers using the basic 32-bit FLP arithmetic hardware described in this chapter. List the successive contents of the working registers, the inputs and outputs of the two adders, and some indicator values.

(a) $(+0.75 \times 2^{-5}) + (-0.625 \times 2^{+15})$

(b) $(+0.5625 \times 2^{+251}) - (-0.75 \times 2^{247})$

(c) $(+0.75 \times 2^{-5}) \times (-0.625 \times 2^{+15})$

(d) $(+0.5625 \times 2^{+251}) \times (+0.75 \times 2^{+18})$

(e) $(+0.75 \times 2^{-5}) \div (-0.625 \times 2^{+15})$

(f) $(+0.5625 \times 2^{-251}) \div (+0.75 \times 2^{+18})$

Prob. 9.8 Suppose that the following data format is suggested to extend the 32-bit machine to double-precision FLP computations. Explore the necessary hardware components (in block diagrams similar to Fig. 9.3) required to perform the double-precision FLP arithmetic. Note that each operand is now 64 bits long and occupies two of 32-bit memory words

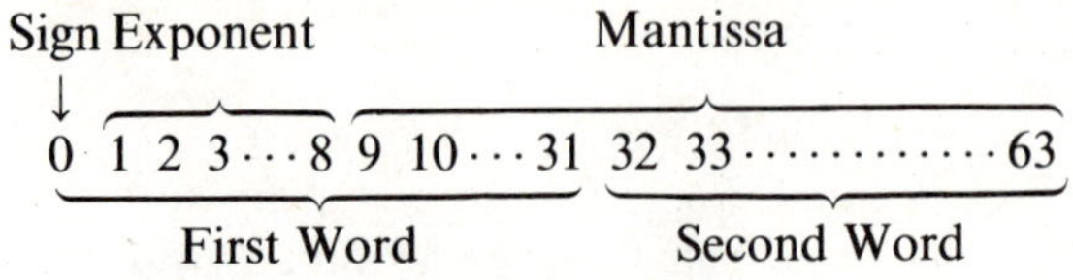

$$\begin{array}{c} \text{Sign} \quad \text{Exponent} \qquad\qquad\qquad \text{Mantissa} \\ \downarrow \\ 0\ \overbrace{1\ 2\ 3\cdots 8}\ \overbrace{9\ 10\cdots 31\ \ 32\ 33\cdots\cdots\cdots 63} \\ \underbrace{0\ 1\ 2\ 3\cdots 8\ 9\ 10\cdots 31}_{\text{First Word}}\ \underbrace{32\ 33\cdots\cdots\cdots 63}_{\text{Second Word}} \end{array}$$

Prob. 9.9 Consider the 32-bit FLP arithmetic processor described in the text with biased exponent and normalized mantissa. Let $\mathcal{M} = (M, m)$ and $\mathcal{N} = (N, n)$ be two nonzero, normalized FLP numbers with the format specified in Fig. 9.2. Describe the conditions which cause each of the following singular results:

(a) Overflow after $\mathcal{M} + \mathcal{N}$

(b) Overflow after $\mathcal{M} \times \mathcal{N}$

(c) Overflow after $\mathcal{M} \div \mathcal{N}$

(d) Underflow after $\mathcal{M} + \mathcal{N}$

(e) Underflow after $\mathcal{M} \times \mathcal{N}$

(f) Underflow after $\mathcal{M} \div \mathcal{N}$

Chapter

10

Advanced Topics on Floating-Point Arithmetic

10.1 Introduction

Only normalized FLP arithmetic operations and the associated processor designs were studied in Chapter 9. In this chapter, we explore several advanced topics in relation to efficient FLP arithmetic design. We start with the specification of unnormalized FLP arithmetic operations. Because postnormalization is not required in unnormalized arithmetic, there are some advantages and some disadvantages as well. We shall compare the two FLP arithmetic systems, *normalized* versus *unnormalized*, and indicate their design differences.

Errors incurred in FLP computations result primarily from normalization, truncation, or roundoff operations. The basic rounding schemes such as chopping and nearest-neighborhood rounding are explained first. A more rigorous treatment of axiomatic rounding theory is presented in section 10.4 in terms of algebraic mappings from the set of real numbers to the set of machine representable numbers. Five different rounding schemes are modeled with these mappings and are compared from designer's viewpoint. A new rounding implementation scheme is presented in section 10.7 using ROMs or PLAs. This ROM-rounding is then compared with several adder-based rounding schemes. Statistical analysis of the precision of FLP machines is presented. Arithmetic error bounds are derived in the light of optimal base choice for efficient FLP system design and applications. The results can be used by both the designers and the users in search of better rounding schemes and better error analysis methods. A theoretical study of various categories of FLP instruction algorithms is given in section 10.8. The comparative study is based on mathematical mapping defined for each category of FLP instructions. These mappings are used to partition the data space of operand pairs into disjoint classes, which convey valuable information for predicting the effects of normalization, truncation, and rounding in FLP operations.

We then describe the special requirements for multiple-precision FLP arithmetic, which provides better machine accuracy at the expense of more storage space and more sophisticated program control. In particular, double-precision addition, subtraction, multiplication, and division algorithms are shown for a computer system with 48-bit word length and 96-bit double words. Readers should have a thorough understanding of the normalized FLP operations before studying the material presented in this chapter, which includes formal proofs, mathematical descriptions, and advanced algorithms for FLP operations. The studies provide readers with tools necessary for developing efficient FLP arithmetic systems. Bibliography indicates directions to further research work that should be done.

10.2 Unnormalized Floating-Point Arithmetic

Floating-point arithmetic operations described in Chapter 9 dealt with normalized operands. In this section, we present the advantages and disadvantages of using unnormalized FLP arithmetic operations. A good example of unnormalized FLP arithmetic is the Control Data CDC **6600** Computer, which has all its instructions performing unnormalized arithmetic. We specify below the four basic arithmetic operations on unnormalized data and indicate their design differences from the normalized design.

A base-r unnormalized FLP number (f, e) has a mantissa f within the unity range

$$|f| < 1 \tag{10.1}$$

This unnormalized mantissa f may have value less than r^{-1}. This means that an unnormalized FLP number may have many internal representations. Each legitimate FLP representation (f, e) corresponds to a mantissa shifted in different number of positions and an exponent appropriately scaled up or down. Let us assume the convention that the mantissa f is a signed fraction with p true magnitude digits to the right of the radix point, and the exponent e is a sign-magnitude integer with $q + 1$ digits, including the sign and satisfying Eq. 9.8. Although sign-magnitude form is assumed for both fields, all the development could well be carried out in terms of complemented representation. Of course, one can always normalize an unnormalized operands before it is processed by the normalized floating-point hardware. This normalization operation may slow down the processing rate and require more hardware. The computer system CDC **6600** allows arithmetic operations to be directly performed on unnormalized numbers, and transform them into normalized form, if desired, using a separate NORMALIZE instruction. IBM System **360** has both normalized and unnormalized arithmetic hardware which executes either type of FLP instructions.

Arithmetic rules associated with processing two unnormalized numbers (f_1, e_1) and (f_2, e_2) are defined below with the assumption of base $r = 2$. FLP addition and FLP subtraction can be comprehended as one type of operations due to the fact that the mantissas may be either positive or negative. Without loss of generality (addition is commutative), we assume that the exponents $e_1 \geq e_2$. The following FLP addition rule has been proposed for unnormalized arithmetic

$$(f_1, e_1) + (f_2, e_2) = \begin{cases} (f_1 + f_2 \times 2^{e_2 - e_1}, e_1), & \text{if } S < 1 \\ ((f_1 + f_2 \times 2^{e_2 - e_1}) \times 2^{-1}, e_1 + 1), & \text{if } S \geq 1 \end{cases} \quad \textbf{(10.2)}$$

where $S = |f_1 + f_2 \times 2^{e_2 - e_1}|$. $S \geq 1$ means the resulting sum has a mantissa exceeding the fraction limit. Therefore, one right shift is needed. This definition is similar to normalized FLP addition except the final normalization step is omitted. The operands are not required to be normalized, overflow may be expected to occur less frequently in unnormalized arithmetic.

Let z be defined as the number of leading zeros in the mantissa field of an unnormalized FLP number (f, e). Note that $f = 0$, when $z = p$, where p is the mantissa length. Let (m, e') be obtained from (f, e) as follows:

$$\begin{cases} m = f \times 2^z \\ e' = e - z \end{cases} \quad \textbf{(10.3)}$$

In other words, (m, e') is the equivalent normalized FLP number. We can write the above identities simply in the form

$$(m, e') = (f \times 2^z, e - z) \quad \textbf{(10.4)}$$

FLP multiplication and FLP division with operands (f_1, e_1) and (f_2, e_2) are defined in terms of a variable λ, which is a function of m_1 and m_2. Suppose $|f_1| \leq |f_2|$, which ensures that $z_1 \geq z_2$; the unnormalized multiplication is defined by

$$(f_1, e_1) \times (f_2, e_2) = (m_1 \times m_2 \times 2^{-z_1 + \lambda}; e_1 + e_2 - z_2 - \lambda) \quad \textbf{(10.5)}$$

If one takes $\lambda = 0$, this operation may be accomplished by first normalizing the operand with the larger mantissa (absolute value) and then multiplying the mantissas and adding the two exponents. We can rewrite Eq. 10.5 as follows under the assumption $\lambda = 0$.

$$(f_1, e_1) \times (f_2, e_2) = (f_1 \times m_2, e_1 + e_2 - z_2) \quad \textbf{(10.6)}$$

The product mantissa $f_1 \times m_2$ has approximately the same magnitude as f_1 for $\frac{1}{2} \leq |m_2| < 1$. Furthermore, $\frac{1}{4} \leq |m_1 \times m_2| < 1$ for nonzero operands, clearly in all cases either $z = z_1$ or $z_1 + 1$ for $\lambda = 0$, where z is the number of leading zeros in the resulting product mantissa. If one chooses $\lambda = 1$, the product mantissa will be $2f_1 \cdot m_2$ and in all cases either $z = z_1 - 1$ or z_1. In general, we have the following values assigned to the parameter z after FLP multiplication with $\lambda = 0$ or 1.

$$z = z_1 \quad \text{or} \quad z_1 \pm 1 \quad \textbf{(10.7)}$$

Similarly, we define the unnormalized FLP division as

$$(f_1, e_1)/(f_2, e_2) = \begin{cases} \left(\dfrac{m_1}{m_2} \times 2^{-z_1 - \lambda}, e_1 - e_2 + z_2 + \lambda\right), & \text{if } |f_1| \leq |f_2|; \\ \left(\dfrac{m_1}{m_2} \times 2^{-z_2 - \lambda}, e_1 - e_2 - z_1 + 2z_2 + \lambda\right), & \text{if } |f_1| > |f_2|. \end{cases}$$

(10.8)

In particular, when a zero divisor is encountered with $f_2 = 0$ and $m_2 = 0$ for $z_2 = p$, then the quotient becomes

$$(0, e_1 - e_2 - z_1 + 2p + \lambda) \tag{10.9}$$

The quotient obtained in Eq. 10.8 has a mantissa with z leading zeros, where z is given below corresponding to the choice of $\lambda = 0$ or 1.

$$z = \max(z_1, z_2) \quad \text{or} \quad \max(z_1, z_2) \pm 1 \tag{10.10}$$

The above operations show that the number of significant digits in a p-digit product or p-digit quotient is approximately the same as that in the operand with fewer such digits. For this reason, unnormalized arithmetic is sometimes called *significance FLP arithmetic*. The number $p - z$ reflects the number of significant digits in a resultant FLP number.

Comparing unnormalized FLP arithmetic with normalized FLP arithmetic, we observe that the latter may be less subject to roundoff error buildup because rounding always occurs at the full p digits from the leading nonzero bit. The MANIAC III computer, built at the University of Chicago, has arithmetic units performing all three different types of arithmetics: fixed-point arithmetic, and normalized and unnormalized floating-point arithmetic.

The main advantages of unnormalized arithmetic are the gradual underflow by rescaling and the use of significance arithmetic. The gradual underflow allows an underflow result be replaced by an unnormalized number. Significance arithmetic, although producing unnormalized results, eliminates the pre- and postnormalization steps. Analysis of the tradeoffs between the two types floating-point arithmetic systems involve many factors such as degree of precision, error analysis, arithmetic hardware design and mechanization, application demand, programming ease, and so on. What follows may provide some answers to these issues.

10.3 Chopping and Rounding Operations

In many cases, the results of floating-point operations may exceed p digits, where p is the maximum mantissa length. For example, $p + 1$ digits may result from adding two p-digit normalized mantissas when the resulting sum exceeds value 1. Another example indicates that a $2p$-digit product may be expected from multiplying two p-digit mantissas.

The way we handled the case of mantissa sum overflow in the first example was to normalize the resulting sum by shifting it right one position and pushing the old least significant bit off the right end regardless of its value. Such an operation is called *chopping*. In a *chopped arithmetic*, the resulting mantissa is first normalized and then its low-order digits are discarded and its high-order p digits are retained unchanged. Some authors call this *truncation*. We shall treat positive and negative numbers equally in case of chopping or truncation as stated above.

If x is a real fraction (positive) in radix-r true magnitude form, we shall use $(x)_p$ to denote the quantity of x *chopped* to p radix-r digits.

$$
\begin{aligned}
x &= 0.130581 \\
(x)_4 &= 0.1305 \\
(x)_2 &= 0.13
\end{aligned}
\tag{10.11}
$$

We shall define machine arithmetic operations $\oplus$, $\ominus$, $\otimes$, $\oslash$ with chopping as follows by first performing the corresponding real arithmetic operations $+$, $-$, $\times$, $/$ and then chopping the result to p digits

$$
\begin{aligned}
a \oplus b &= (a + b)_p, \\
a \ominus b &= (a - b)_p, \\
a \otimes b &= (a \times b)_p, \\
a \oslash b &= (a / b)_p.
\end{aligned}
\tag{10.12}
$$

For example, if one applies the multiplication rule in Eq. 9.42, we obtain only a p-digit product chopped from the $2p$-digit real product, if the resulting product was already in normalized form. If the resulting product contains a leading zero as warranted by Eq. 9.16, one left shift is necessary to normalize it; and then drop the $p - 1$ lower bits.

Let us recall the way the mantissa multiply was implemented in section 9.6, the lower p bits of the product in **Q** register are dropped, but a "1" is advanced into the upper p bits of the product in **A** register whenever the leading bit Q_1 in register **Q** has a "1". (Otherwise, "0" is advanced.) This shows one way of rounding off the insignificant digits. Rounding in this way means that the result is rounded to the closest p-digit number. This rounding method approaches the natural value of the result from either *above* or *below*. When the low-order portion being rounded off has a value, which is greater than or equal to one-half of its maximum value

$$\varepsilon = r^{-p}/2 \tag{10.13}$$

the advanced "1" will make the machine representation approach the natural value from above; otherwise from below.

The threshold for rounding can be easily implemented by checking only the most significant bit of the portion being rounded off. If the digit has a value in

$$\left\{\frac{r}{2}, \frac{r}{2} + 1, \ldots, r - 1\right\},$$

where $r = 2k$ was assumed, the rounding is completed by advancing a "1" into the high-order portion. If the digit value is among $\{0, 1, \ldots, (r/2) - 1\}$, the rounding is equivalent to chopping. Therefore, chopping can be considered as a special case of rounding, in which the real value is always approached from below. Other rounding schemes such as always forcing a "1" into the high-order partition regardless of the value of the rounded portion is one that always approaches the natural value from above.

We shall use $(x)_p^*$ to represent a real number x which is *rounded* to retain p radix-r digits. For the same numeric example in Eq. 10.11, we have

$$\begin{aligned} x &= 0.130581 \\ (x)_4^* &= 0.1306 \\ (x)_2^* &= 0.13 \end{aligned} \tag{10.14}$$

Note that $(x)_4^* \neq (x)_4$ but $(x)_2 = (x)_2^*$ in this example. This means different results may be expected from applying different rounding schemes. The threshold used in $(x)_4^*$ is equal to $\varepsilon = 10^{-4}/2 = 0.000050$. Because $0.000081 > 0.000050$, an "1" has been rounded into the upper 4 decimal digits.

Similarly, we can define rounded machine arithmetic as the natural operations followed by rounding the result to p significant digits.

$$\begin{aligned} a \oplus b &= (a + b)_p^*, \\ a \ominus b &= (a - b)_p^*, \\ a \otimes b &= (a \times b)_p^*, \\ a \oslash b &= (a / b)_p^* \end{aligned} \tag{10.15}$$

Intuitively, one can see that the chopped arithmetic is simpler than rounded arithmetic both in terms of design and programming efforts. Some computer systems use a floating-point system which is close to the above rounded arithmetic. These computers implement rounded arithmetic using a *guard-digit register*, which is w digits long. This guard-digit register holds the low-order digits to the right of the first p significant digits. For example, IBM System/**360** uses one guard digit ($w = 1$), and IBM **7090** has 27 guard digits. These guard registers function similarly to the **Q** register used in MULTIPLY execution in section 9.6. These guard registers can be used to implement error-control schemes. Some other machines are more flexible, such as CDC **6600**, which uses a subset of operation codes executing chopped arithmetic, whereas another subset performs a version of rounded arithmetic. This gives the programmers more flexibility to control arithmetic errors at the expense of more hardware facilities.

10.4 Axiomatic Rounding Theory

The rounding techniques described in the preceding section considered only the absolute values of mantissas. In an enlarged sense, a good rounding scheme should distinguish positive numbers from negative numbers. For this reason, more rigorous algebraic descriptions are needed for modeling various machine-implementable rounding schemes. It is hoped that these formal approaches will summarize the existing rounding methods and stimulate more versatile schemes for future machine arithmetic design.

Let **R** be the real number system and let **M** be the set of machine representable numbers. Clearly, we have $\mathbf{M} \subseteq \mathbf{R}$. A *rounding* $\boldsymbol{\rho}$ is a mapping

$$\boldsymbol{\rho}: \mathbf{R} \rightarrow \mathbf{M} \tag{10.16}$$

defined for all $a, b, \in \mathbf{R}$ such that

$$\rho(a) \leq \rho(b) \quad \text{whenever } a \leq b \tag{10.17}$$

A rounding is called *optimal* if for all $a \in \mathbf{M}$

$$\rho(a) = a \tag{10.18}$$

In practice, this must be true for any reasonable machine representation of numbers. Optimal rounding implies that if $a \in \mathbf{R}$ and m_1, m_2 are two consecutive members of $\mathbf{M}$ with $m_1 < a < m_2$, then either $\rho(a) = m_1$ or $\rho(a) = m_2$, It is desired to have hardware-implemented roundings be always optimal.

A rounding is *downward-directed* or *upward-directed* if, for all $a \in \mathbf{R}$, we have, respectively,

$$\rho(a) \leq a \tag{10.19}$$

or

$$\rho(a) \geq a \tag{10.20}$$

A rounding is *symmetric* if for all $a \in \mathbf{R}$,

$$-\rho(-a) = \rho(a) \tag{10.21}$$

The *optimal downward-directed* rounding $\nabla : \mathbf{R} \to \mathbf{M}$ is defined by

$$\nabla(a) = \text{Max}\{m \in \mathbf{M} \mid m \leq a\} \tag{10.22}$$

and *optimal upward-directed* rounding $\Delta : \mathbf{R} \to \mathbf{M}$ is defined by

$$\Delta(a) = \text{Min}\{m \in \mathbf{M} \mid m \geq a\} \tag{10.23}$$

The above definitions imply that Δ is a method that rounds away from zero when the number is positive, and toward zero when it is negative. On the contrary, ∇ is one which does exactly the opposite. These two roundings, Δ and ∇, have been used in the implementation of interval arithmetic.

There are three symmetric roundings that are of interest to us: *Truncation* **T**, *Augmentation* **A**, and *Proximity* **P**. These roundings are formally defined below for all $a \in \mathbf{R}$ in terms of the above mappings Δ and ∇.

$$\mathbf{T}(a) = \begin{cases} \nabla(a), & \text{if } a \geq 0, \\ \Delta(a), & \text{if } a < 0 \end{cases} \tag{10.24}$$

$$\mathbf{A}(a) = \begin{cases} \Delta(a), & \text{if } a \geq 0, \\ \nabla(a), & \text{if } a < 0 \end{cases} \tag{10.25}$$

$$\mathbf{P}(a) = \begin{cases} \nabla(a), & \text{if } \nabla(a) \leq a < \dfrac{\nabla(a) + \Delta(a)}{2}; \\ \Delta(a), & \text{if } \dfrac{\nabla(a) + \Delta(a)}{2} \leq a < \Delta(a) \end{cases} \tag{10.26}$$

The truncation rounding **T** always rounds towards zero. The augmentation rounding **A** always rounds away from zero. The proximity rounding **P** always selects the closest machine number, and in the case of a tie from above and under, selects the one whose magnitude is larger. The proximity rounding **P** is most frequently used because it provides maximum accuracy. Existing machines implement either **T** or some approximated version of **P**. A common approximation for **P** is a rounding before the postnormalization shift.

A summary of the above five rounding modes **Δ**, **∇**, **T**, **A**, and **P** is given in Table 10.1. It is clear that the **P** rounding offers the most dynamic choices at the

Table 10.1 Five Rounding Schemes and their Implications for FLP Arithmetic with p Fraction Digits (Yohe [28])

Rounding Mode	Operation Symbol	Numerical Implications	
		Positive Number	Negative Number
Optimal Upward-directed	**Δ**	Round away from zero	Round toward zero
Optimal Downward-directed	**∇**	Round toward zero	Round away from zero
Truncation	**T**	Round toward zero	Round toward zero
Augmentation	**A**	Round away from zero	Round away from zero
Proximity	**P**	$\pi = \|\text{result}\|$ and $\theta = (p + 1)$st fraction digit.	
$\pi \geq \min$ and $\theta < r/2$		Round toward zero	
$\pi \geq \min$ and $\theta \geq r/2$		Round away from zero	
$\pi < \frac{1}{2}\min$		Round toward zero	
$\frac{1}{2}\min \leq \pi < \min$		Round away from zero	
exponent overflow		Round away from zero	

expense of increased hardware demand. The chopped and rounded schemes described in section 10.3 are similar to the **T** rounding and **P** rounding, respectively. The **P** mode is, in general, superior to the **T** mode, but the rates of error growth for both cases are essentially the same after a sequence of computations. Thus rounding tends to delay error growth as predicted.

Yohe [28] suggested that every computer should be capable of executing the five roundings modes as listed in Table 10.1 under program control. The designer of a floating-point arithmetic system must make decisions that affect both computational speed and accuracy. Recently, Garner [9] concluded that better accuracy is obtained with small base values and with sophisticated round-off algorithms, whereas computational speed is increased with large base values and crude round-off procedures such as truncation. A good rounding scheme may greatly enhance the precision performance of a floating-point processor, but it may also slow down the overall system speed.

10.5 Error Analysis of FLP Arithmetic

Error analysis of floating-point operations has proved to be rather complicated. Without a good understanding of the FLP error characteristics, one cannot achieve a design with optimal performance in terms of precision and speed. In this section, we describe several error analysis models proposed in recent years. In particular, we choose three specific FLP representations $(r, q, p) = (2, 9, 22), (4, 8, 23)$, and $(16, 7, 24)$ to present the representational error properties. These parameters are chosen to give essentially the same range of number representations for a 32-bit word as is found in most single-precision, FLP arithmetic on current machines. The comparative studies on these three floating-point systems can easily be modified to analyze other systems as well.

Representational error is a consequence of the fact that the set $\mathbf{M}$ of machine representable numbers is only a subset of the set $\mathbf{R}$ of real numbers. Each machine representation $m \in \mathbf{M}$ of a number can be viewed as an equivalence class of real numbers

$$m = \{x | x \in \mathbf{R} \quad \text{and} \quad x \equiv m\} \tag{10.27}$$

where $x \equiv m$ denotes the equivalence relation up to machine precision. The *representational error* is made in representing an element x of the equivalence class by m. For a (r, q, p) FLP system, the value of the gap between *adjacent* normalized floating-point numbers is equal to

$$2^{-p} \times r^e \tag{10.28}$$

where e is the value of the unbiased exponent. In particular when $r = 2^k$, we have

$$2^{-p} \times 2^{ke} = 2^{ke-p} \tag{10.29}$$

The magnitude of the representational error equals a fraction of the above gap value and hence a function of the magnitude of the number.

The *relative error* $\boldsymbol{\delta}$ of an $m \in \mathbf{M}$ and its equivalent $x \in \mathbf{R}$ is defined as the magnitude of the representational error $x - m$ divided by x.

$$\boldsymbol{\delta} = \frac{x - m}{x} \tag{10.30}$$

If we assume a logarithmic probability distribution for normalized numbers within the range $(1/r) \leq x < 1$

$$P(x) = \frac{1}{x \times \ln r} \tag{10.31}$$

and a uniformly distributed minimum error obtained by appropriate rounding

$$Q(x) = \frac{2^{-p}}{4x} \tag{10.32}$$

then the *Average Relative Representational Error* (**ARRE**) is defined as

$$\begin{aligned}\mathbf{ARRE}(p, r) &= \int_{1/r}^{1} P(x) \times Q(x)\mathrm{d}x \\ &= \int_{1/r}^{1} \frac{2^{-p}\,\mathrm{d}x}{4x^2 \times \ln r} \\ &= \frac{(r-1)}{4 \ln r} \times 2^{-p}\end{aligned} \tag{10.33}$$

The *Maximum Relative Representational Error* (**MRRE**) over all normalized mantissas is defined as

$$\mathbf{MRRE}(p, r) = 2^{-p-1} \times r \tag{10.34}$$

The values of **ARRE** and **MRRE** for the three 32-bit floating-point systems under consideration are given in Table 10.2. The fact that the **MRRE** for binary representations is half of that for the corresponding hexadecimal representations has been used

Table 10.2 Static Characteristics of FLP Numbers in Three 32-Bit Systems with Base *r*, *p* Fraction Digits, and *q* Exponent Digits[a] (Cody [6])

r	q	p	MRRE	ARRE	Exponent Range	Number Range
2	9	22	0.5×2^{-21}	0.18×2^{-21}	$2^{2^9} - 1$	$2^{2^9} \times (1 - 2^{-22})$
4	8	23	0.5×2^{-21}	0.14×2^{-21}	$2^{2^9} - 2$	$2^{2^9} \times (1 - 2^{-23})$
16	7	24	2^{-21}	0.17×2^{-21}	$2^{2^9} - 4$	$2^{2^9} \times (1 - 2^{-24})$

[a] $p + q + 1 = 32$ where sign occupies one position.

in arguments for the superiority of binary over hex. At best this superiority is marginal, especially if the values of **ARRE** are also considered. This table shows, that among the three FLP systems the system with base $r = 4$ results in least error. The quaternary representation of a number is never less accurate than the corresponding binary representation, and the value of **ARRE** is 20 percent smaller than that of corresponding binary scheme.

Brent [3] uses a *root-mean-square* (*rms*) measure for the comparison of relative errors associated with floating-point systems of nearly identical ranges but with different bases $r = 2^k$. For the logarithmic systems, Brent has derived the following *rms Relative Error* ($\mathbf{RE}_{rms}$) function, which is defined by the ratio of the standard deviation $\boldsymbol{\delta}_{rms}$ to the mean value $\boldsymbol{\delta}_o$ of the relative error $\boldsymbol{\delta}$. The $\mathbf{RE}_{rms}$ is expressed as only a function of the power $k = \log_2 r$ of the implied base $r = 2^k$, considering only the explicitly normalized floating-point numbers with an explicit leading mantissa bit bit "one."

$$\mathbf{RE}_{rms} = \frac{\boldsymbol{\delta}_{rms}}{\boldsymbol{\delta}_o} = \sqrt{\frac{4^k - 1}{2(k \log 2)^3}} \qquad (10.35)$$

Table 10.3 lists the values of $\mathbf{RE}_{rms}$ for $k = 1, 2, \ldots, 8$ corresponding to bases $r = 2, 4, \ldots, 256$. The table shows that base-4 ($r = 4$ and $k = 2$) is best. In a worst-case study, the base-2 and base-4 systems are equally good. In this rms study, however, base-4 is definitely better. Because of the different ranges possible with base-4 and base-8, there are some choices of minimal acceptable range for which base-8 is preferable to base-4, but bases higher than 8 are always inferior to base-4 on the $\mathbf{RE}_{rms}$ criterion. Some computer systems such as the **PDP-11**, use implicitly normalized binary floating-point systems, in which the leading fraction "1" is dropped for increasing the mantissa length by one. Brent has proved that, for such systems, the result obtained in Eq. 10.34 should be halved. This shows that the implicit binary system is superior to all explicit systems described above.

Table 10.3 RMS Relative Errors for FLP Systems with Various Bases $r = 2^k$ (Brent [13])

Base r	First Fraction Bit	$\mathbf{RE}_{rms}(k)$[a]
2	Implicit	1.06
2	Explicit	2.12
4	Explicit	1.68
8	Explicit	1.87
16	Explicit	2.45
32	Explicit	3.51
64	Explicit	5.34
128	Explicit	8.47
256	Explicit	13.90

[a] $\mathbf{RE}_{rms}(k)$ was defined in Eq. 10.35 for an explicitly normalized system and equals only half of Eq. 10.35 for a normalized FLP system with an implicit most significant bit.

10.6 Error Bounds on FLP Arithmetic Operations

The number of significant digits can be used as a rough measure of the relative error contained in a number. The leading zeros and trailing zeros do not convey significant information. Ever since the universal acceptance of floating-point arithmetic in digital computers, the problems of error propagation in computations become increasingly important. Different rounding schemes may result in different degrees of accuracies. Arithmetic operations specified and implemented in previous sections should be appended with appropriate error-control mechanism. A better understanding of the error sources and error bounds will help in designing effective error-control schemes.

Errors incurred in floating-point computations have a tendency to propagate through a sequence of operations. If no appropriate control is enforced, the errors may accumulate to a degree that the final result becomes meaningless (such as deviated too far from the required precision). An error control mechanism depends heavily on the type of arithmetic system used. Design emphases associated with normalized FLP system differ from those of significance FLP design. In addition, rounding does make a difference. We introduce below some error analysis methodologies associated with floating-point arithmetic. Error bounds are derived to estimate the maximum errors associated with imposed rounding. In order to reduce the complexity, only analyses of normalized arithmetic are given. The methods can be extended to analyzing other types of FLP arithmetic systems as well.

The notations used in section 10.5 will be carried over. Let $\mathscr{E}_a(\alpha)$ be the absolute error incurred with *Case a* of machine operation (α), when comparing the machine result with that of the natural operation α, which is error-free because of no imposed truncation or rounding. Several parameters must be defined before we present the machine operations with roundings.

A small fraction η is defined below in terms of ε specified in Eq. 10.13

$$-\varepsilon \leq \eta < \varepsilon \tag{10.36}$$

where ε reflects the maximum round-off error in the $(p + 1)$th digit beyond the word length of the mantissa. We shall distinguish different fractions η_i satisfying Eq. 10.36 by indexing them differently.

Let σ be a p-digit radix-r fraction $0 \leq |\sigma| < 1$, and $\lfloor \log_r |\sigma| \rfloor$ be the largest integer smaller than or equal to the real number $\log_r |\sigma|$. It can be seen that

$$-p \leq \lfloor \log_r |\sigma| \rfloor < 0$$

or

$$r^{-p} \leq r^{\lfloor \log_r |\sigma| \rfloor} < 1 \tag{10.37}$$

Addition and Subtraction

We start our analysis on normalized FLP addition $N_1 \oplus N_2$. The machine subtraction $N_1 \ominus N_2$ can be treated as special case of addition. Assume $N_1 = (m_1, e_1)$, $N_2 = (m_2, e_2)$, and $N_3 = N_1 \oplus N_2 = (m_3, e_3)$ for $e_2 \geq e_1$

Case A. *No overflow after mantissa addition.* The machine operation $\oplus$ with rounding produces the following results

$$m_3 = \sigma \times r^{-\lfloor \log_r |\sigma| \rfloor - 1}, \tag{10.38}$$

$$e_3 = e_2 + \lfloor \log_r |\sigma| \rfloor + 1, \tag{10.39}$$

where

$$\begin{aligned} \sigma &= (m_1 \times r^{e_1 - e_2})_p^* + m_2 \\ &= m_1 \times r^{e_1 - e_2} + \eta + m_2 \end{aligned} \tag{10.40}$$

Subtracting the results of the real addition + as specified in Eq. 9.9 from the above machine results, we obtain the following error function

$$\begin{aligned} \mathscr{E}_a \oplus &= |(N_1 \oplus N_2) - (N_1 + N_2)| \\ &= |(m_1 r^{e_1 - e_2} + m_2 + \eta) \times r^{-\lfloor \log_r |\sigma| \rfloor - 1} \times r^{e_2 + \lfloor \log_r |\sigma| \rfloor + 1} \\ &\quad - (m_1 r^{e_1 - e_2} + m_2) \times r^{e_2}| \\ &= |\eta \times r^{e_2}| \\ &\leq \varepsilon \times r^{e_2} \\ &= \varepsilon \times r^{e_3 - \lfloor \log_r |\sigma| \rfloor - 1} \\ &\leq \varepsilon \times r^{e_3 + p - k} \end{aligned} \tag{10.41}$$

When the value of σ is zero, the value p is used instead of $\lfloor \log_r |\sigma| \rfloor$ in e_3, which means that there was no rounding error.

Case B. *Overflow after mantissa addition.* The machine result (m_3, e_3), are obtained as

$$\begin{aligned} m_3 &= (m_1 \times r^{e_1 - e_2 - 1})_p^* + (m_2 \times r^{-1})_p^* \\ &= (m_1 + \eta_1) \times r^{e_1 - e_2 - 1} + (m_2 + \eta_2) \times r^{-1} \end{aligned} \tag{10.42}$$

and

$$d_3 = e_2 + 1 \tag{10.43}$$

Subtracting the natural results in Eq. 9.11 from the above, we obtain

$$\begin{aligned} \mathscr{E}_b \oplus &= |(N_1 \oplus N_2) - (N_1 + N_2)| \\ &= |((m_1 + \eta_1) \times r^{e_1 - e_2 - 1} + (m_2 + \eta_2) \times r^{-1}) \times r^{e_2 + 1} \\ &\quad - (m_1 \times r^{e_1 - e_2 - 1} + m_2 \times r^{-1}) \times r^{e_2 + 1}| \\ &= |(\eta_1 \times r^{e_1 - e_2 - 1} + \eta_2 \times r^{-1}) \times r^{e_2 + 1}| \\ &\leq \varepsilon r^{-1} \times (r^{e_1 - e_2} + 1) \times r^{e_2 + 1} \\ &\leq \varepsilon r^{e_1} + \varepsilon r^{e_2} \\ &\leq 2\varepsilon r^{e_2} = 2\varepsilon r^{e_3 - 1} \end{aligned} \tag{10.44}$$

The above error bounds are expressed in terms of the known value e_3, which is available after the machine operation of $\oplus$ or $\ominus$. Both of the above two bounds satisfy the following predicted general bound for additive errors

$$\mathscr{E} \oplus = |(N_1 \oplus N_2) - (N_1 + N_2)| < \frac{r^{e_3 - 1}}{2} \tag{10.45}$$

Multiplication

Assume that a double-length accumulator formed by cascading the **A** and **Q** registers is available for mantissa multiplication. Consider $N_3 = N_1 \otimes N_2 = (m_3, e_3)$. We have for $\sigma = m_1 \times m_2$

$$\begin{aligned} m_3 &= (\sigma \times r^{-\lfloor \log_r |\sigma| \rfloor - 1})_p^* \\ &= (m_1 \times m_2 \times r^{-\lfloor \log_r |\sigma| \rfloor - 1} + \eta) \end{aligned} \tag{10.46}$$

$$e_3 = e_1 + e_2 + \lfloor \log_r |\sigma| \rfloor + 1 \tag{10.47}$$

The multiplicative error function is obtained from Eq. 9.12 and the above equations as follows

$$\begin{aligned} \mathscr{E} \otimes &= |N_1 \otimes N_2 - N_1 \times N_2| \\ &= |(m_1 - m_2 \times r^{-\lfloor \log_r |\sigma| \rfloor - 1} + \eta) \times r^{e_1 + e_2 + \lfloor \log_r |\sigma| \rfloor + 1} \\ &\quad - (m_1 \times m_2) \times r^{e_1 + e_2}| \\ &\leq \varepsilon \times r^{e_1 + e_2 + \lfloor \log_r |\sigma| \rfloor + 1} \\ &= \varepsilon \times r^{e_3} \end{aligned} \tag{10.48}$$

We have used the property

$$-2 \leq \lfloor \log_r |\sigma| \rfloor \leq -1 \tag{10.49}$$

because of Eq. 9.14, if m_1 and m_2 are nonzero. In the case where m_1 or m_2 or both are zero, there is no round-off error. Thus, the multiplicative error can be bounded in terms of known results. If the value of $\sigma = m_1 \times m_2$ is zero, the value of $\lfloor \log |\sigma| \rfloor$ is again replaced by $-p$ in e_3.

Division

It is assumed that the normalized dividend N_1 and divisor N_2 are properly aligned to avoid overflow, and therefore, result in a normalized quotient without post-operation shifting. Let the machine division be $N_3 = N_1 \oslash N_2$ with rounding. We have

$$\begin{aligned} m_3 &= \left(\frac{m_1}{m_2}\right)_p^* \times r^{-\lfloor \log_r |\sigma| \rfloor - 1} \\ &= \left(\frac{m_1}{m_2} + \eta\right) \times r^{-\lfloor \log_r |\sigma| \rfloor - 1} \\ &= \left(\frac{m_1}{m_2} + \eta\right) \times r^{+1-1} \\ &= \frac{m_1}{m_2} + \eta \end{aligned} \tag{10.50}$$

$$e_3 = e_1 - e_2 + \lfloor \log_r |\sigma| \rfloor + 1 = e_1 - e_2 \tag{10.51}$$

Table 10.4 A Summary of Error Bounds on Normalized FLP Operations (Carr [5])

FLP Operation	Arithmetic Error Function	Upper Bound on the Error Function[a]
Add or Subtract	$\lvert(N_1 \oplus N_2) - (N_1 + N_2)\rvert$ $\lvert(N_1 \ominus N_2) - (N_1 - N_2)\rvert$	*Case A*: No mantissa overflow $\varepsilon \times r^{e_3+p-1}$ *Case B*: Mantissa overflow $\varepsilon \times r^{e_3-1}$
Multiply	$\lvert(N_1 \otimes N_2) - (N_1 \times N_2)\rvert$	$\varepsilon \times r^{e_3}$
Divide	$\lvert(N_1 \oslash N_2) - (N_1/N_2)\rvert$	$2\varepsilon \times r^{e_3}$

[a] r = the base of the exponent.
p = the allowed mantissa length.
e = the value of the resulting exponent for the *sum*, *difference*, *product*, or *quotient*.
ε = the maximum roundoff error in the $(p + 1)$st digit beyond the allowed mantissa length.

where $\frac{1}{2} \le |\sigma| = |m_1/m_2| < 1$ and, therefore, $\lfloor \log_r |\sigma| \rfloor = -1$, and $-\varepsilon \le \eta < \varepsilon$. We have the error function

$$
\begin{aligned}
\mathscr{E}\oslash &= |(N_1 \oslash N_2) - (N_1/N_2)| \\
&= \left| \left(\frac{m_1}{m_2} + \eta \right) \times r^{e_1 - e_2} - \left(\frac{m_1}{m_2} \right) \times r^{e_1 - e_2} \right| \\
&\le |\eta| \times r^{e_1 - e_2} \le \varepsilon \times r^{e_1 - e_2} \\
&= \varepsilon \times r^{e_3}
\end{aligned}
\tag{10.52}
$$

A summary of the above error bounds is given in Table 10.4. These bounds are expressed as functions of the base r, the mantissa length $p = \log_r 2\varepsilon$, and the resulting exponent e_3.

10.7 A ROM-Based Rounding Scheme

In this section, we study a recently proposed rounding scheme, which can easily be implemented with ROMs or PLAs. The conventional adder-based rounding schemes involve adder hardware to handle the carry propagation caused by the "1" advanced into the low-order significant portion of a rounded FLP number. This may cause an extra time delay to complete the rounding operation. The ROM rounding to be described does not require the timing and hardware overhead associated with the adder stage. The method is considered attractive from the standpoint of both cost-effectiveness and error resistance.

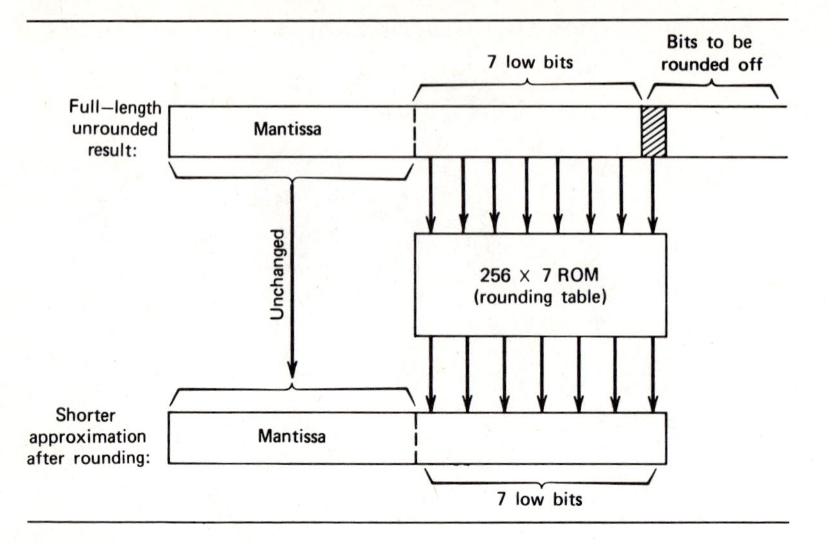

Figure 10.1 The concept of ROM-based rounding scheme (Kuck, et al. [16]).

The rounding of the low-order portion of a ROM table was discussed in section 6.10 in relation to the inexpensive ROM-based multiplication networks. Shown below is the use of ROM or PLA to implement the rounding mechanism for reducing long FLP numbers to shorter-precision approximations. The scheme is demonstrated in Fig. 10.1. The long representation corresponds to a full-length intermediate or final result in an FLP computation without the application of any rounding. The shorter approximation is the accepted machine representation after the appropriate roundoff procedure is made. Suppose that a ROM with 8 address lines and 7 output lines is used with a bit capacity of $2^8 \times 7$ bits. The 7 low-order bits of the significant part of the mantissa and the highest bit to be rounded off together form the 8-bit address input to the ROM or PLA. The output of the ROM is the 7-bit rounded value of these significant bits.

The rounding rules are microprogrammed into the ROM, with the necessary adding operations included. In case of *rounding overflow*, that is, all 7 low-order significant bits are "1," rounding is not performed by simple truncation. For the remaining 255 cases (out of 256 for the 8-bit scheme), rounding is actually performed with appropriate adding operations implemented by the ROM or PLA. The scheme is inexpensive to implement with modern memory technologies as compared with the hardwired rounding logic circuits. Thus, from a designer's viewpoint, ROM-rounding should be more cost-effective in building rounded arithmetic into an FLP processor. The error-resistance capability of this new scheme is still to be verified.

Two indicators of rounding-scheme effectiveness are used to demonstrate the error resistance of ROM-rounding applied to several existing rounding methods, similar to the *truncation* **T** (forced chopping), *proximity* **P** (nearest neighborhood),

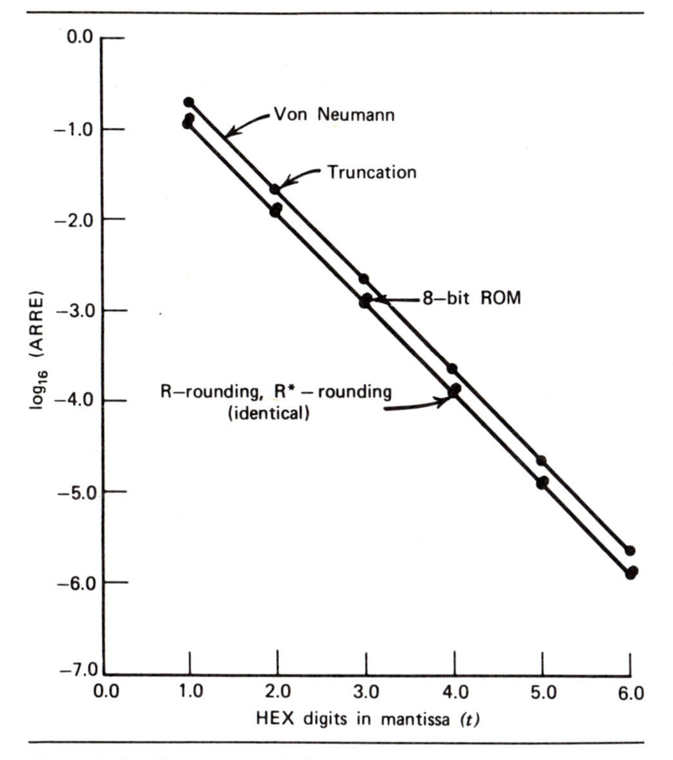

Figure 10.2 The Average Relative Representation Error (**ARRE**) for various rounding schemes.

and *augmentation* **A** (away from zero) described in previous sections. The two effectiveness indicators are:

1. The **ARRE** (average relative representational error) defined in Eq. 10.33.
2. The statistical tests with specific problem types, such as serial additions, subtractions, multiplications, and so on.

The effect of using *guard digits* to hold part of the aligned operand, which is shifted right during exponent comparison, gives the final result greater precision. This guard-digit approach depends on the accompanying *alignment* methods. By alignment we mean the use of a rounding method to reduce the aligned operand to fit in the machine word and guard digits before addition or subtraction is performed.

The **ARRE** for five existing rounding schemes are demonstrated in Fig. 10.2, in which base-16 ($r = 16$) was assumed for various mantissa lengths. The Von Neumann rounding is equivalent to truncation except that the lowest order bit of the result mantissa is forced to a "1." The **R**-rounding is the proximity rounding in a general sense as described in section 10.3. The **R***-rounding is equivalent to **R**-rounding except when the digits to be rounded off have bit values 1000, . . . , in which

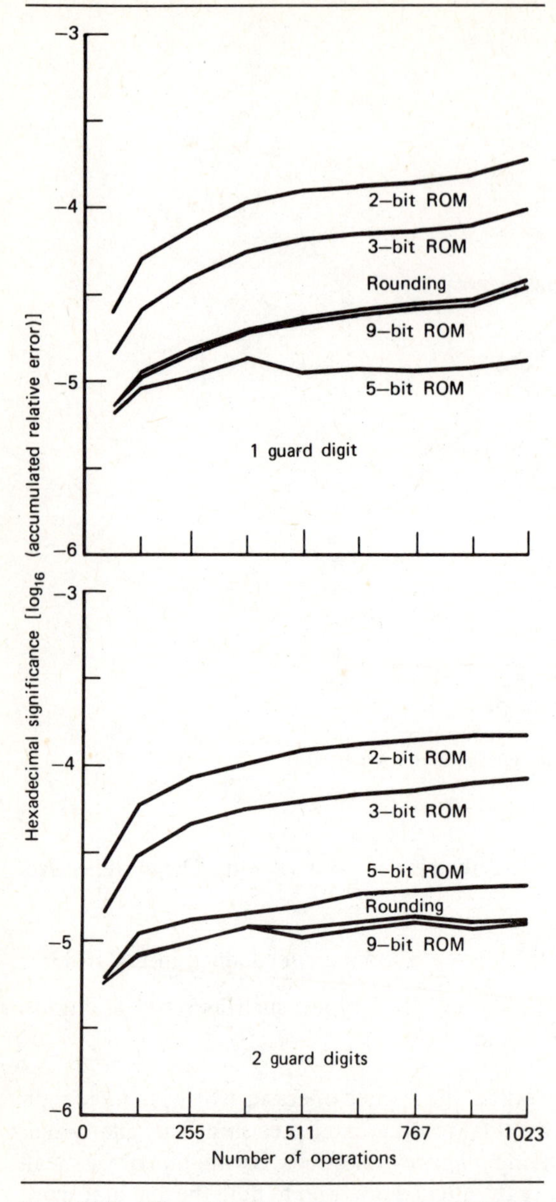

Figure 10.3 Effect of ROM length on error growth (Kuck, et al. [16]).

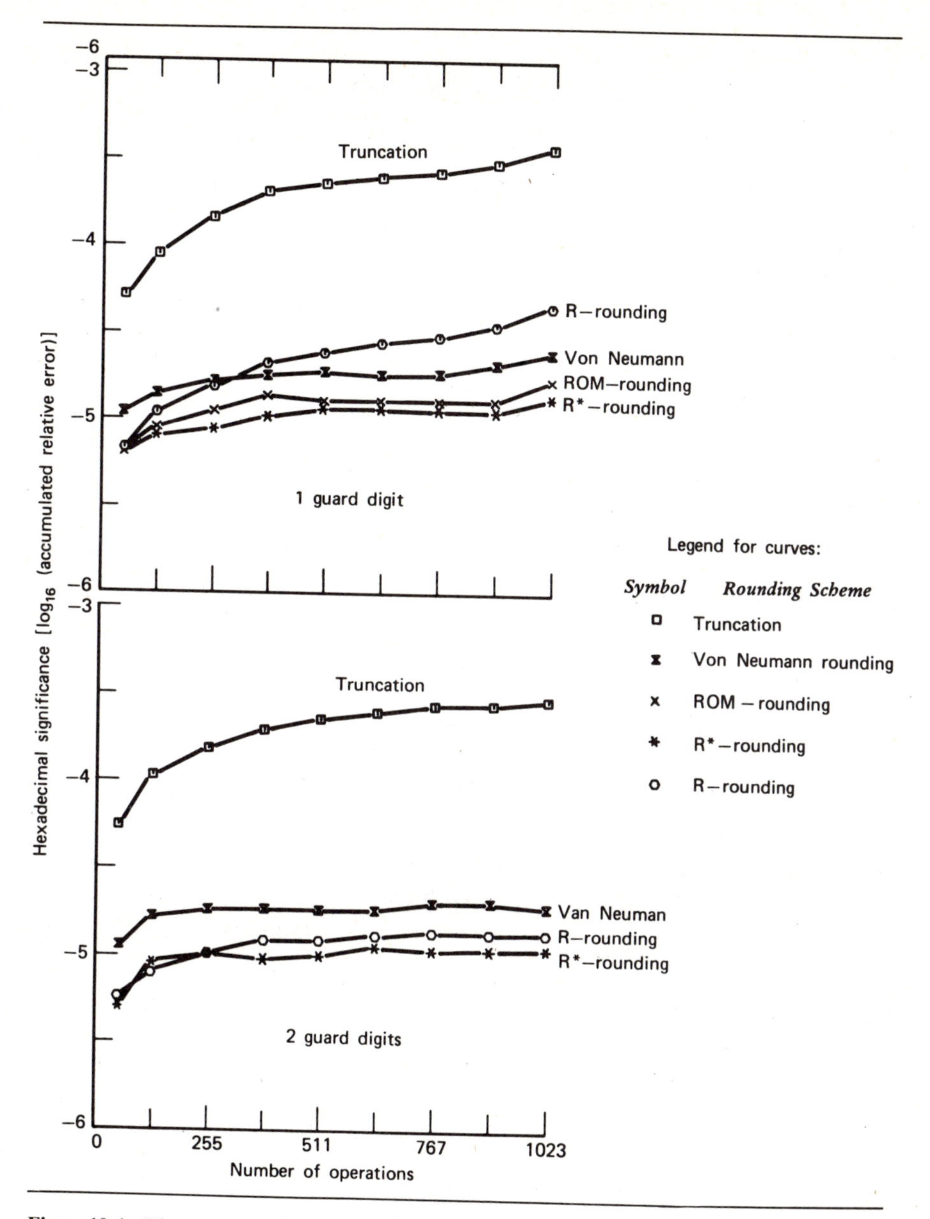

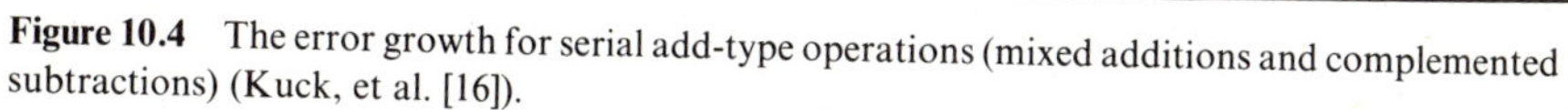

Figure 10.4 The error growth for serial add-type operations (mixed additions and complemented subtractions) (Kuck, et al. [16]).

case Von Neumann rounding is applied. The plot shows that 8-bit ROM-rounding is only marginally worse than regular rounding **R** (or **R***) for **ARRE** (this is reasonable not only because the two are identical 255 times out of 256, but also because the one case in which they differ has the lowest relative error). Figure 10.3 shows the accumulated relative error for different length ROMs in mixed add-type operations with one and two hexadecimal guard digits.

Floating-point systems with various bases (radices) and rounding schemes are frequently discriminated by applying varied sets of data to typical problems, and statistically measuring the resultant roundoff errors. Only results associated with repeated add-type operations are presented in Fig. 10.4. The plots in the figure show the error growth for the above five rounding schemes with respect to one- or two-guard digit systems. The initial bumps result from the loss of significance involved in hexadecimal postnormalization shifting. The relative errors actually decrease for most methods because the magnitude of the sum's mantissa increases faster than that of the accumulated roundoff error, but carry-out finally occurs around 256 operations; normalization of the mantissa gives it a leading digit 1 plus loss of the low-order digit and a sudden jump in relative error results. The fewer guard digits used, the more effect alignment rounding will have.

10.8 Comparison of Floating-Point Instructions

This section presents a set-theoretic study of FLP instructions. The study provides useful mathematical tools for researchers and advanced designers to formally define, analyze, compare, and evaluate FLP operations. A new concept, known as *comparison sets*, will be used to partition the space of FLP operand pairs for each of three categories of FLP algorithms: Add/Subtract, Multiply, and Divide. The theory is especially useful in predicting the effects of normalization, truncation, and rounding in most FLP operations.

In this study, a FLP number F is defined as a triple

$$F(\sigma F, \varepsilon F, \psi F) \tag{10.53}$$

where $\sigma F \in \{+, -\}$ is the *sign*, εF is the integer *exponent*, and ψF is the *fraction* (mantissa) of F. The fraction ψF is to be modeled by a *limited digit radix* (LDR) polynomial, which has positive, modulo-r coefficients with radix $r \geq 2$. The fraction length p is assumed to be $p \geq 3$; $\tilde{\psi} F$ represents the value of the fraction ψF. Zero fraction is denoted as $\emptyset$. The real arithmetic operations $+, -, *, /$ and notations for the machine arithmetic operations $\oplus$, $\ominus$, $\circledast$, and $\oslash$ are the same as before.

A radix-r polynomial is conventionally defined as

$$P = \sum_{i=-\infty}^{\infty} p_i \times r^{-i} \tag{10.54}$$

where the coefficients p_i for all i are integers (either positive or negative). When P has finite number of nonzero coefficients, we call it a *finite* polynomial. When a finite polynomial has a nonnegative value $\tilde{P} \geq 0$, we call it a *finite positive* polynomial. Furthermore, when a finite positive polynomial has all its coefficients positive within the range $\{0, 1, \ldots, r-1\}$, we call it an LDR polynomial. Formally, an LDR polynomial Q is written

$$Q = \sum_{i=-s}^{t} q_i \times r^{-i} \tag{10.55}$$

where $q_i \in \{0, 1, \ldots, r-1\}$ and $s + t + 1$ is the polynomial length. In fact, every finite positive polynomial P can be converted to an LDR polynomial Q via the following mapping

$$\lambda: P \mapsto \lambda P = Q \tag{10.56}$$

The mapping is defined, coefficient-by-coefficient, as follows: Define carries

$$C_i = 0 \quad \text{for all } i \geq t + 1, \tag{10.57}$$

$$C_i = \begin{cases} 0 & \text{if } p_{i+1} + c_{i+1} < r; \\ 1 & \text{if } p_{i+1} + c_{i+1} \geq r, \end{cases} \tag{10.58}$$

for all $-s \leq i \leq t$. The LDR coefficients are then obtained for all $-s \leq i \leq t$ as

$$q_i = (p_i + c_i)\text{Mod } r \tag{10.59}$$

In case of an LDR polynomial of length p with a positive fractional magnitude, we have the indices $s = -1$ and $t = p$ such that

$$Q = \sum_{i=1}^{p} q_i \times r^{-i} \tag{10.60}$$

is strictly less than one.

Let S_r^p be the set of all FLP numbers with base r, and $\hat{S}_r^p \subset S_r^p$ be the set of all normalized FLP numbers, both with fraction length p. The notation $\zeta_k F$ stands for the scaling of F by increasing the exponent with k and multiplying (shifting right) the fraction with r^{-k}. Consider any two FLP numbers $G, H \in S_r^p$. Denote $e = \varepsilon G - \varepsilon H$ as their exponent difference. Two predicates are required before we define FLP operations as mappings

$$G \circ H: S_r^p \times S_r^p \rightarrow S_t^p \tag{10.61}$$

Define

$$\begin{aligned} G \geq H &= \varepsilon G > \varepsilon H \text{ or } (\varepsilon G = \varepsilon H \quad \text{and} \quad \tilde{\psi} G \geq \tilde{\psi} H) \\ G < H &= \varepsilon G < \varepsilon H \text{ or } (\varepsilon G = \varepsilon H \quad \text{and} \quad \tilde{\psi} G < \tilde{\psi} H) \end{aligned} \tag{10.62}$$

FLP Addition

$$G \oplus H = \begin{cases} (\sigma G, \varepsilon G, \lambda(\psi G + \psi H * r^{-e})); & G \geq H, \sigma G = \sigma H \\ (\sigma G, \varepsilon G, \lambda(\psi G - \psi H * r^{-e})); & G \geq H, \sigma G \neq \sigma H \\ (\sigma H, \varepsilon H, \lambda(\psi H + \psi G * r^{e})); & G < H, \sigma G = \sigma H \\ (\sigma H, \varepsilon H, \lambda(\psi H - \psi G * r^{e})); & G < H, \sigma G \neq \sigma H \end{cases} \tag{10.63}$$

FLP Subtraction

$$G \ominus H = G \oplus (-1 \cdot \sigma H, \varepsilon H, \psi H) \tag{10.64}$$

FLP Multiplication

$$G \circledast H = (\sigma G \cdot \sigma H, \varepsilon G + \varepsilon H, \lambda(\psi G * \psi H)) \tag{10.65}$$

FLP Division

$$G \oslash H = (\sigma G \cdot \sigma H, \varepsilon G - \varepsilon H, \lambda(\psi G / \psi H)) \tag{10.66}$$

provided $H \neq \emptyset$.

The above operations are defined over unnormalized numbers $S_r^p \times S_r^p$. In what follows, we shall define operations over the normalized space of operand pairs as

$$S = \hat{S}_r^p \times \hat{S}_r^p \tag{10.67}$$

Normalization, truncation, and rounding in floating-point arithmetic depend on the size of the fraction in the intermediate result before normalization. The fraction value may be greater than or equal to 1. It can then be said that the intermediate result has *carry*, when it is greater than one. The intermediate result has *cancel* when its fraction has a value less than $1/r$. Two predicates are defined below to characterize the above carry and cancel conditions.

Carry Predicate

$$F \in S_r^p \quad \text{and} \quad F \neq \emptyset; \tilde{\psi} F \geq 1 \tag{10.68}$$

Cancel Predicate

$$F \in S_r^p \quad \text{and} \quad F \neq \emptyset; \tilde{\psi} F < \frac{1}{r} \tag{10.69}$$

The space of operand pairs S can be divided into subsets corresponding to different FLP operations. These subsets are called *comparison sets*. Each operation category can be characterized by a collection of comparison sets. Within each collection the subsets are mutually exclusive. The comparisons are based on the carry-borrow terms introduced in the conversion mapping λ.

Comparison Sets for FLP Add Algorithms

The carry/cancel predicates applied to the intermediate result $G \oplus \zeta_e H$ in the ideal FLP add algorithm are used to partition the space S. These predicates will be abbreviated as follows: The fraction value $(G \oplus \zeta_e H)$ is written ω^+. Then the predicates used are:

$$
\begin{aligned}
&\omega^+ \geq 1 && \text{means } G \oplus \zeta_e H \text{ has carry} \\
&\omega^+ < 1 && \text{means } G \oplus \zeta_e H \text{ has no carry} \\
&\omega^+ \geq \frac{1}{r} && \text{means } G \oplus \zeta_e H \text{ has no cancel} \\
&\omega^+ < \frac{1}{r} && \text{means } G \oplus \zeta_e H \text{ has cancel}
\end{aligned}
\tag{10.70}
$$

The union of the comparison sets defined below covers in general only those operand pairs, where $G \geq H$, because $\oplus$ operation is symmetric in G and H. Define a $\hat{S} \sqsubset S$ as

$$\hat{S} = \{(G, H) | (G, H) \in S \quad \text{and} \quad G \geq H\} \tag{10.71}$$

In the following definitions, we are considering the pair $(G, H) \in \hat{S}$. The comparison sets of FLP addition are:

$$
\begin{aligned}
\underline{S}_0 &= \{(G, H); G = \emptyset\} \\
\underline{S}_1 &= \{(G, H); G \neq \emptyset \quad \text{and} \quad \sigma G = \sigma H \quad \text{and} \quad \varepsilon G = \varepsilon H\} \\
\underline{S}_{2a} &= \{(G, H); G \neq \emptyset \quad \text{and} \quad \sigma G = \sigma H \quad \text{and} \quad 1 \leq \varepsilon G - \varepsilon H \leq p \quad \text{and} \quad \omega^+ < 1\} \\
\underline{S}_{2b} &= \{(G, H); G \neq \emptyset \quad \text{and} \quad \sigma G = \sigma H \quad \text{and} \quad \varepsilon G - \varepsilon H \geq p + 1\} \\
\underline{S}_3 &= \{(G, H); G \neq \emptyset \quad \text{and} \quad \sigma G = \sigma H \quad \text{and} \quad 1 \leq \varepsilon G - \varepsilon H \quad \text{and} \quad \omega^+ \geq 1\} \\
\underline{S}_4 &= \{(G, H); G \neq \emptyset \quad \text{and} \quad \sigma G \neq \sigma H \quad \text{and} \quad \varepsilon G = \varepsilon H\} \\
\underline{S}_{5a} &= \left\{(G, H); G \neq \emptyset \quad \text{and} \quad \sigma G \neq \sigma H \quad \text{and} \quad 1 \leq \varepsilon G - \varepsilon H \leq p \quad \text{and} \quad \omega^+ \geq \frac{1}{r}\right\} \\
\underline{S}_{5b} &= \left\{(G, H); G \neq \emptyset \quad \text{and} \quad \sigma G \neq \sigma H \quad \text{and} \quad \varepsilon G - \varepsilon H \geq p + 1 \quad \text{and} \quad \omega^+ \geq \frac{1}{r}\right\} \\
\underline{S}_6 &= \left\{(G, H); G \neq \emptyset \quad \text{and} \quad \sigma G \neq \sigma H \quad \text{and} \quad \varepsilon G - \varepsilon H = 1 \quad \text{and} \quad \omega^+ < \frac{1}{r}\right\} \\
\underline{S}_7 &= \left\{(G, H); G \neq \emptyset \quad \text{and} \quad \sigma G \neq \sigma H \quad \text{and} \quad 2 \leq \varepsilon G - \varepsilon H \quad \text{and} \quad \omega^+ < \frac{1}{r}\right\}
\end{aligned}
\tag{10.72}
$$

The union of these ten sets covers the subspace $\hat{S}$. When G and H are such that $\varepsilon G - \varepsilon H \geq p$, they cannot be a member of $\underline{S}_3$. Because $\varepsilon G - \varepsilon H \geq 2$ in $\underline{S}_7$, $\omega^+ > 1/r^2$ indicates a cancel of only one position.

Table 10.5 Percentages of FLP Operands Pairs in Comparison Sets Associated with Each Type of Operation Algorithms (Kent [14])

Operation Type	Problem	A	B	C	D	E	Total
	Percentage of FLP Add/ Subtract instructions	10.6	5.2	15.1	5.6	5.3	6.1
	$\underline{S}_0$	0.4	1.0	0	1.6	2.2	1.4
	$\underline{S}_{0a}$	23.8	27.7	1.7	12.1	0	10.5
	$\underline{S}_1$	4.6	5.2	50.9	12.2	2.3	14.4
	$\underline{S}_{2a}$	22.6	47.9	38.9	26.7	14.7	27.6
Addition or Subtraction	$\underline{S}_{2b}$	0.5	0.9	0	0.6	0	0.5
	$\underline{S}_3$	6.8	5.5	3.3	9.2	3.7	7.2
	$\underline{S}_4$	10.1	3.0	0.9	8.6	57.6	15.2
	$\underline{S}_{5a}$	15.6	4.7	2.0	15.4	6.0	11.5
	$\underline{S}_{5b}$	0.1	0	0	0.3	0	0.2
	$\underline{S}_6$	10.8	2.7	0.6	9.3	12.0	8.3
	$\underline{S}_7$	4.7	1.4	1.7	4.0	1.5	3.2
	Σ	100	100	100	100	100	100
	Problem	**A**	**B**	**C**	**D**	**E**	**Total**
Multiplication	Percentage of FLP Multiply instructions	10.4	2.2	3.2	6.9	3.0	5.7
	$\underline{S}_8$	38.6	49.3	38.9	51.1	39.7	49.1
	$\underline{S}_9$	61.4	50.7	61.1	48.9	60.3	50.9
	Σ	100	100	100	100	100	100
	Problem	**A**	**B**	**C**	**D**	**E**	**Total**
Division	Percentage of FLP Divide instructions	2.0	4.1	5.9	1.4	1.2	1.5
	$\underline{S}_{10}$	55.2	69.5	68.0	54.4	58.2	57.7
	$\underline{S}_{11}$	44.8	30.5	32.0	45.6	41.8	42.2
	Σ	100	100	100	100	100	100

Comparison Sets for FLP Multiplication

The fraction value $\tilde{\psi}(G \oplus H)$ is written ω^*. Then the predicates used are:

$$\begin{aligned} \omega^* &\geq \frac{1}{r} \qquad \text{means } G \circledast H \text{ has no cancel} \\ \omega^* &< \frac{1}{r} \qquad \text{means } G \circledast H \text{ has cancel} \end{aligned} \tag{10.73}$$

The comparison sets for FLP multiply are

$$\begin{aligned} \underline{S}_8 &= \left\{(G, H); (G, H) \in S \quad \text{and} \quad \omega^* \geq \frac{1}{r}\right\} \\ \underline{S}_9 &= \left\{(G, H); (G, H) \in S \quad \text{and} \quad \omega^* < \frac{1}{r}\right\} \end{aligned} \tag{10.74}$$

The product $G \circledast H$ will never have carry. The union $\underline{S}_8 \cup \underline{S}_9$ will cover the space S. It was noted before that when $(G, H) \in \underline{S}_9$, $(1/r^2) \leq \omega^* \leq \frac{1}{4}$, only single cancel will occur.

Comparison Sets for FLP Division

The comparison sets are

$$\begin{aligned} \underline{S}_{10} &= \{(G, H); (G, H) \in S \quad \text{and} \quad \tilde{\psi}G \geq \tilde{\psi}H\} \\ \underline{S}_{11} &= \{(G, H); (G, H) \in S \quad \text{and} \quad \tilde{\psi}G < \tilde{\psi}H\} \end{aligned} \tag{10.75}$$

It is clear that $\underline{S}_{10} \cup \underline{S}_{11} = S$ and that only carry and no carry can occur in $G \oslash H$.

Kent [14, 15] has used the above comparison sets to investigate the FLP instruction algorithms for Add/Subtract, Multiply, and Divide on the following computer series: CDC **6000**-Cyber **70**; CDC **3000**; Univac **1100**, SM3, SM4; and IBM **360-370**. These partitioned subsets of operand pairs can be used to predicate the effects of normalization and truncation in FLP instructions in most modern computers. Information on the statistical distribution of FLP operand pairs over comparison sets is found problem-dependent. Table 10.5 gives the percentages of FLP operand pairs for five programming problems in comparison sets defined above. The trace of the five programs gives 750,000 instructions of which 46,000 were FLP addition/subtraction, 43,000 were FLP multiplications, and 11,000 FLP division.

10.9 Multiple-Precision FLP Add/Subtract

In this section, we present an arithmetic algorithm for adding/subtracting *double-precision* (**DP**) FLP numbers. The presented algorithm can be implemented either in hardware or by software. In order to impress readers with real situations, we assume a machine with a 48-bit word length. A double-precision FLP operand, therefore,

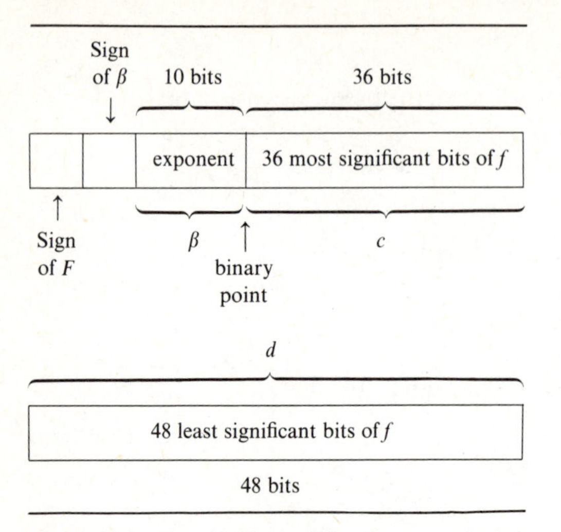

Figure 10.5 The double-precision (two word) data format for a binary machine with 48-bit basic word length. (Note that c and d are defined in Eq. 10.78.)

occupies two adjacent memory words of 96 bits in total. The data format shown in Fig. 10.5 is assumed for a **DP** FLP number

$$F = f \times 2^{\beta} \tag{10.76}$$

where $|f|$ lies in the range

$$\tfrac{1}{2} \leq |f| \leq 1 - 2^{-84} \tag{10.77}$$

The leftmost bit and the second leftmost bit of the first word represent the sign of F and the sign of the exponent β, respectively. The next 10 bits of the first word represent the magnitude of β. This leaves 84 bits for the representation of f, the high-order 36 bits in the first word and the remaining 48 low-order bits in the second word with an implied binary point as shown in the figure. We can write

$$|f| = c + d \times 2^{-36} \tag{10.78}$$

where c and d lie in the ranges

$$\begin{aligned} \tfrac{1}{2} \leq c \leq 1 - 2^{-36} \\ 0 \leq d \leq 1 - 2^{-48} \end{aligned} \tag{10.79}$$

The fraction c represents the most significant 36 bits of $|f|$ in the first word and d the least significant 48 bits of $|f|$ in the second word. A second **DP** FLP number $G = g \times 2^{\delta}$ can be similarly defined with $\frac{1}{2} \leq |g| \leq 1 - 2^{-84}$; and

$$|g| = a + b \times 2^{-36} \tag{10.80}$$

where $\frac{1}{2} \leq a \leq 1 - 2^{-36}$ and $0 \leq b \leq 1 - 2^{-48}$. Therefore, a and b represent the upper and lower portions of $|g|$, respectively. One's complement system is used in representing negative numbers. The procedure for changing the sign of a **DP** FLP number is to perform a bit-by-bit complement of the entire fraction of 85 bit (including the sign but excluding the exponent). We define the condition that F is larger than G, if

$$\beta \geq \delta \tag{10.81}$$

and F is smaller than G if otherwise. The notation L represents a pair of auxiliary adjacent memory words containing the *larger* of the two numbers F and G. On the other hand, notation S represents a pair of adjacent words containing the *smaller* of the two numbers. Now, we are ready to specify the standard **DP** arithmetic algorithms associated with **DP** FLP Add/Subtract. **DP** FLP Multiply and Divide will be described in the next section.

Double-Precision FLP Addition

$$F + G = f \times 2^{\beta} + g \times 2^{\delta} \tag{10.82}$$

Without loss of generality, assume that F is larger than G. The **DP** addition is executed by the following sequence of arithmetic operations.

1. Record δ and β and determine the sign of $\beta - \delta$ and thus determine if F is larger or smaller than G according to Eq. 10.81
2. Place f and g in L and S because F is larger than G. The following bit patterns are yielded, where "s" stands for "sign bit."

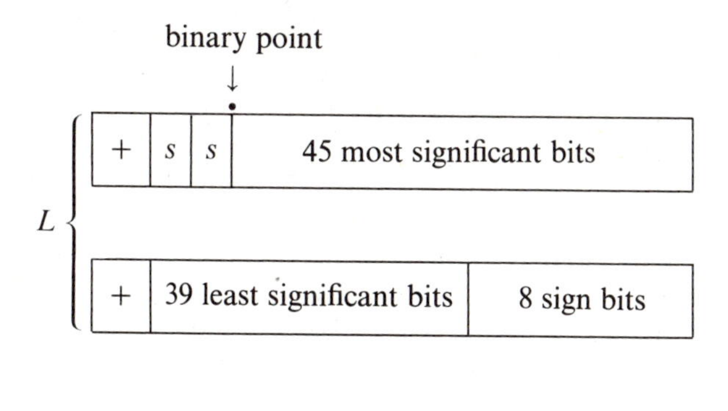

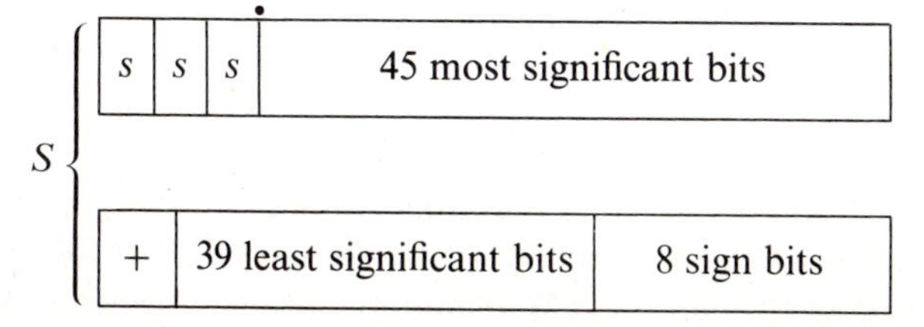

3. Shift S right $|\beta - \delta|$ places and put a "+" bit at the beginning of each of the two words. If $|\beta - \delta| \geq 84$, then there is no need to continue because all significant bits in S will be lost.

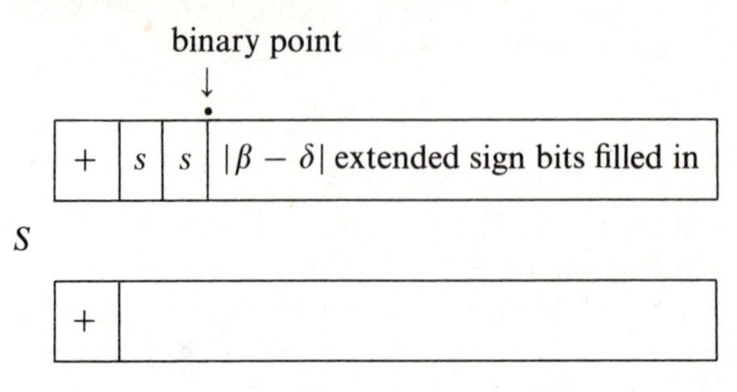

4. Add the second halves of L and S to be

If the first bit c of this sum is a "1," we actually have a carry; otherwise no carry.

5. Add the first halves of L and S and add the carry bit obtained in step 4.

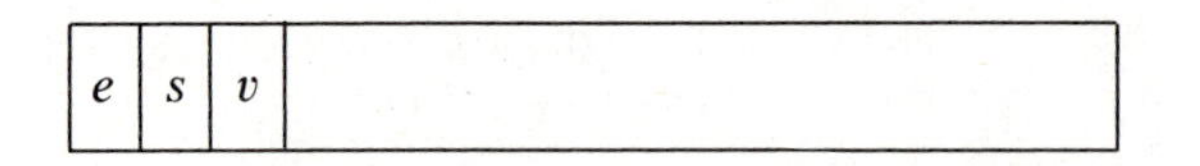

If $e = 1$, then an end-around-carry must be added to the least significant position of the word produced in step 4. The c in step 4 must be cleared to zero before the end-around-carry. If $v = s$, then v is a sign bit. However, if $v \neq s$, then there has been an overflow during the addition. An adjustment of the exponent must be made to give the correct sum.

6. Shift the second half of sum left one place to clear out the carry bit c. Then shift the double length sum left one place if $v \neq s$ and two places if $v = s$. This leaves the sum in the following form:

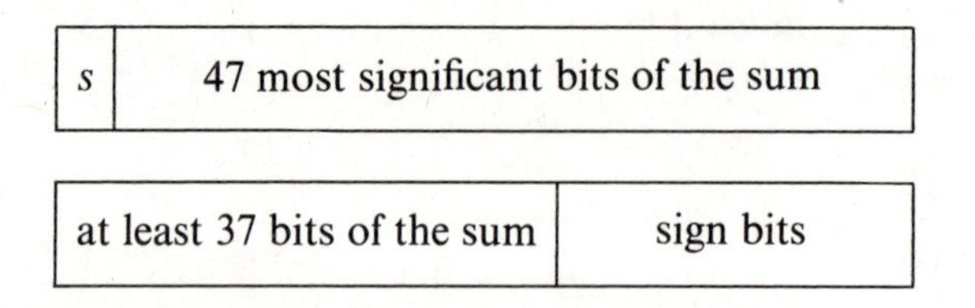

7. Normalize the sum by left shifting and adjust the exponent accordingly. If 84 left shifts are not sufficient for a normalization, then the sum should be made zero. Then pack the 84 most significant bits of the sum, together with 12 bits representing the sign and exponent, into two 48-bit words as shown

in Fig. 10.5. If the sign is negative, then the first 12 bits must be complemented before the packing takes place.

Double-Precision FLP Subtraction

$$F - G = F + (-G) \tag{10.83}$$

One merely complements G before entering the addition procedures described above. Trivial modification can be made to extend the above algorithm to triple-precision FLP system.

10.10 Multiple-Precision FLP Multiply and Divide

Following the same **DP** data format used for the 48-bit machine described in the preceding section, we present below a **DP** FLP multiplication and a **DP** FLA division algorithms. These two algorithms can be also implemented by hardware or software means. They can be also extended to handle triple-precision FLP numbers. The abbreviation "sgn F" is used to represent the sign of a FLP number F.

Double-Precision FLP Multiplication

$$\begin{aligned} F \times G &= (f \times 2^{\beta}) \times (g \times 2^{\delta}) \\ &= (\text{sgn } F \cdot G) \times |f| \times |g| \times 2^{\beta+\delta} \end{aligned} \tag{10.84}$$

The computational procedure is concerned primarily with the formation of $|f| \times |g| \times 2^{\beta+\delta}$, because the (sgn $F \cdot G$) is recorded externally. Furthermore, we record the exponents β and δ until after the fraction product $|f| \times |g|$ is formed. The following algorithm is used for **DP** FLP Multiply.

1. Record (sgn $F \cdot G$) externally, form $|F|$ and $|G|$, and then record the leftmost 12 bits of $|F|$ and $|G|$ as β and δ, respectively.
2. Shift the 84 bits of $|f|$ and $|g|$ left until each has a bit pattern:

+	47 most significant bits.

0	37 least significant bits	10 zeros

This means that $|f|$ represented in Eq. 10.78 should be adjusted to the following form

$$|f| = C + D \times 2^{-47} \tag{10.85}$$

where C and D lie in ranges

$$\frac{1}{2} \le C \le 1 - 2^{-47}$$
$$0 \le D \le 1 - 2^{-37} \qquad (10.86)$$

Likewise $|g|$ has the form

$$|g| = A + B \times 2^{-47} \qquad (10.87)$$

and A, B lie in ranges

$$\frac{1}{2} \le A \le 1 - 2^{-47}$$
$$0 \le B \le 1 - 2^{-37} \qquad (10.88)$$

3. Use FXP operations to form the fraction product

$$\begin{aligned} |f| \times |g| &= (C + D \times 2^{-47}) \times (A + B \times 2^{-47}) \\ &= CA + (CB + DA) \times 2^{-47} + DB \times 2^{-94} \\ &\cong CA + (CB + DA) \times 2^{-47} \end{aligned} \qquad (10.89)$$

Note that the term $DB \times 2^{-94}$ is dropped because only 84 most significant bits are retained. The evaluation of Eq. 10.89 is illustrated by the flow chart in Fig. 10.6.

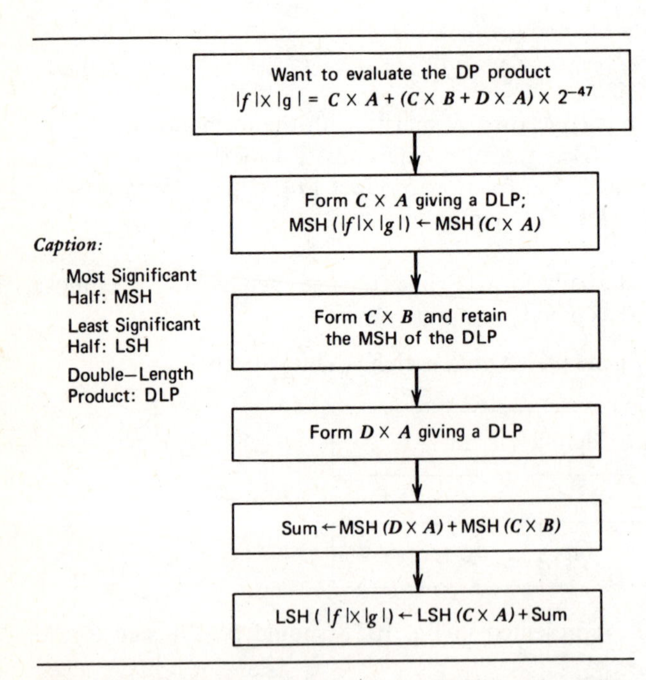

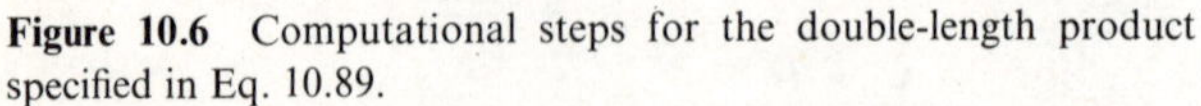

Figure 10.6 Computational steps for the double-length product specified in Eq. 10.89.

4. Round and normalize the product obtained in Eq. 10.89. Appropriate adjustment of the exponent $\beta + \delta$ due to the normalization must be performed. Then pack the 84 bits of the normalized product and the 12 bits representing the sign and adjust exponent into two 48-bit words the same as before. The final complementing step must be done if (sgn FG) is negative.

Double-Precision FLP Division

$$\frac{G}{F} = \frac{g \times 2^{\delta}}{f \times 2^{\beta}}$$

$$= (\text{sgn } G \cdot F)\left|\frac{g}{f}\right| \times 2^{\delta - \beta} \qquad \textbf{(10.90)}$$

Because we want $|g/f| < 1$, we scale down the numerator to yield

$$\frac{G}{F} = (\text{sgn } G \cdot F)\left|\frac{g/2}{f}\right| \times 2^{\delta - \beta + 1} \qquad \textbf{(10.91)}$$

The following algorithm is designed for **DP** FLP Divide

1. Repeat step 1 in **DP** multiplication algorithm described above.
2. Arrange the 84 bits of $|f|$ to give the bit pattern

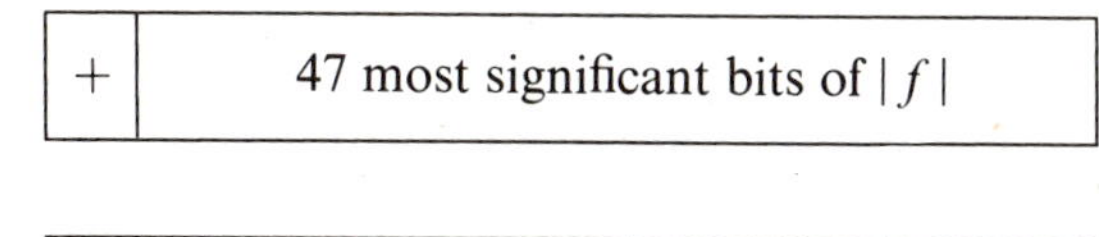

and the 84 bits of $|g|$ to give

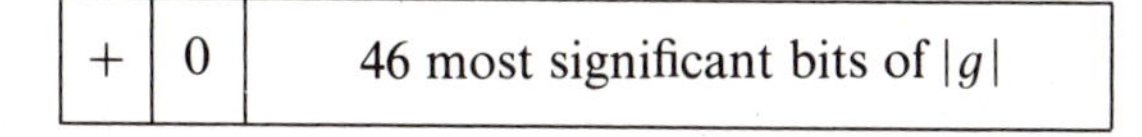

0	38 least significant bits	9 zeros

Thus $|f|$ is represented by Eqs. 10.85 and 10.86 as in the case of multiplication, We have, however,

$$\left|\frac{g}{2}\right| = A + B \times 2^{-47} \qquad \textbf{(10.92)}$$

where A and B lie in the ranges

$$\begin{aligned} \tfrac{1}{4} \le A \le \tfrac{1}{2} - 2^{-47} \\ 0 \le B \le 1 - 2^{-38} \end{aligned} \qquad \textbf{(10.93)}$$

3. Use FXP operation to form the quotient

$$\begin{aligned}
\frac{|g/2|}{|f|} &= \frac{A + B \times 2^{-47}}{C + D \times 2^{-47}} \\
&= \left[\frac{A + B \times 2^{-47}}{C}\right] \times \left[\frac{1}{1 + (D/C) \times 2^{-47}}\right] \\
&= \left[\frac{A}{C} + \frac{B}{C} \times 2^{-47}\right] \times \left[1 - \frac{D}{C} \times 2^{-47} + \frac{D^2}{C^2} 2^{-94} - \cdots\right] \\
&= \frac{A}{C} + \frac{B}{C} \times 2^{-47} - \frac{AD}{C^2} \times 2^{-47} - \frac{BD}{C^2} 2^{-94} + \cdots \\
&\cong \frac{A}{C} + \left[\frac{B}{C} - \frac{AD}{C^2}\right] \times 2^{-47} \\
&= \frac{A + [B - (AD/C)] \times 2^{-47}}{C}
\end{aligned} \tag{10.94}$$

The evaluation of Eq. 10.94 is illustrated in Fig. 10.7 by a flow chart.

4. Round and normalize the quotient obtained in the flow chart using Eq. 10.94. Any adjustment in the exponent $\delta - \beta$ must be made because of the normalization of $|g/2|/|f|$. Then repeat the packing procedures described at step 4 of the DP multiplication algorithm.

The above **DP** arithmetic algorithms on 84-bit numbers were developed by Gregory and Raney [10]. Ikebe [11] has modified these FLP **DP** Multiply and **DP** Divide algorithms to suit triple-precision numbers with $132 = 36 + 48 + 48$ bits of fraction. The data format suggested by Ikebe is shown below as an example of three-word FLP numbers.

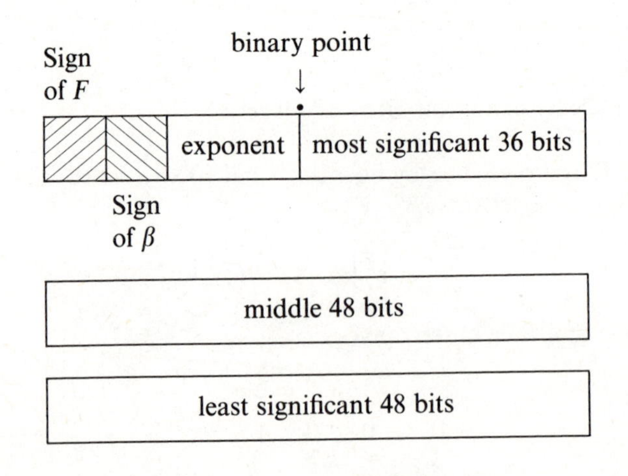

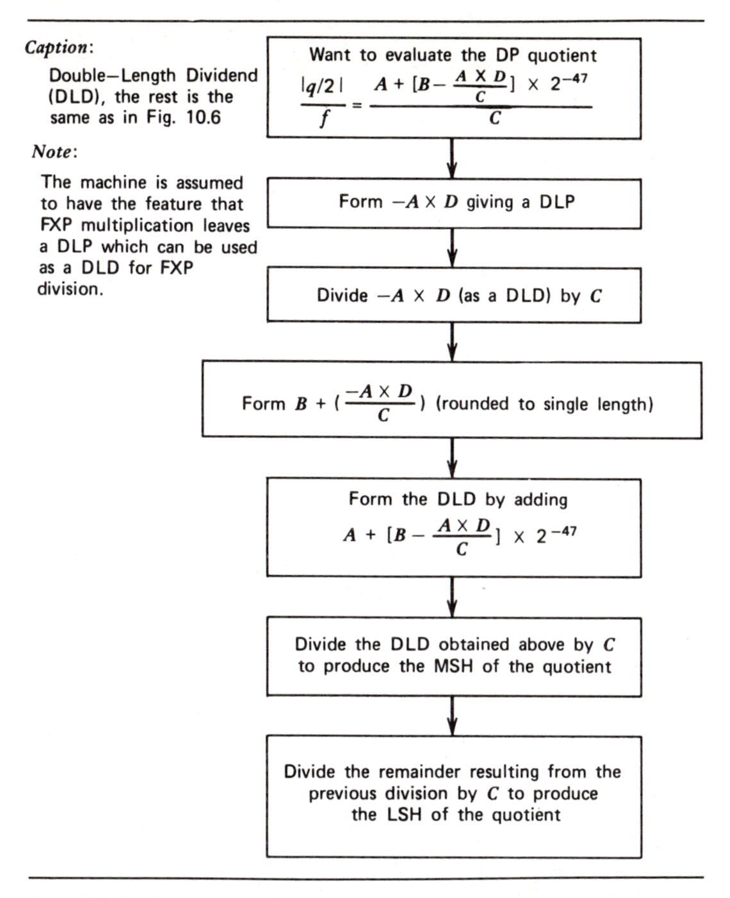

Figure 10.7 Computational steps for the double-length quotient specified in Eq. 10.94.

10.11 Bibliographic Notes

This chapter presented some advanced topics and several recent developments on FLP arithmetic systems. The unnormalized FLP arithmetic operations were studied in Ashenhurst and Metropolis [1, 22] with the development of the MANIAC III computer at University of Chicago. An excellent treatment of various rounding schemes can be found in Yohe [27, 28]. Wilkinson [26] has given a comprehensive study of rounding errors in algebraic processes. Roundoff errors in FLP computation were investigated by Caprani [4], Cody [6], Dorr and Moler [8], Keneko and Liu [13], and Tsao [25], among many others. The axiomatic rounding theory is based on the work of Kulisch [19] and Yohe [28].

Error analyses of FLP arithmetic operations were conducted by Bohlender [2], Brent [3], Carr [5], Kuki and Cody [18], among those mentioned above on roundoff errors. The choices of optimal bases for FLP arithmetic design were described in

Cody [6] and Garner [9]. Matula [21] has formalized the FLP base conversion technique. The new ROM-rounding scheme was proposed by Kuck, Parker, and Sameh [16, 17]. Kent [14, 15] has presented a set-theoretic study of FLP instructions using comparison sets to partition the space of normalized operand pairs. Lunde [20] has evaluated FLP instructions by program tracing. The algorithms on double-precision FLP arithmetic were invented by Gregory and Raney [10]. Interested readers should check details of triple-precision multiply/divide in Ikebe [11]. Sterbenz [23] has given a good treatment of double-precision calculation. Catalog information on typical FLP arithmetic instructions in existing computer systems appear in the CDC **6000** series reference manual [7], Thornton [24], and in IBM System/**370** principles of operations [12] among many newly released system manuals by leading computer manufacturers.

References

[1] Ashenhurst, R. L. and Metropolis, N., "Unnormalized Floating-Point Arithmetic," *Journal of ACM*, Vol. 6, March 1959, pp. 415–428.

[2] Bohlender, G., "Floating-Point Computation of Functions with Maximum Accuracy," *IEEE Trans. Computers*, Vol. C-26, No. 7, July 1977, pp. 621–632.

[3] Brent, R. P., "On the Precision Attainable with Various Floating-Point Number Systems," *IEEE Trans. Computers*, Vol. C-22, June 1973, pp. 601–607.

[4] Caprani, O., "Roundoff Errors in Floating-Point Summation," *BIT Vol. 15*, No. 1, January 1975, pp. 5–9.

[5] Carr, J. W. III, "Error Analysis in Floating-Point Arithmetic," *Comm. of ACM*, Vol. 2, No. 5, May 1959, pp. 10–15.

[6] Cody, W. J., Jr., "Static and Dynamic Numerical Characteristics of Floating-Point Arithmetic," *IEEE Trans. Computers*, Vol. C-22, June 1973, pp. 596–601.

[7] Control Data Corp., *CDC 6000 Series Computer Systems*, Reference Manual, Pub. No. 60100000-M, 1972.

[8] Dorr F. W. and Moler, C. B., "Roundoff Errors on the CDC 6600/7600 Computers," *SIGNUM Newsletter, Ass. for Comp. Mach.*, Vol. 8, 1973, pp. 24–26.

[9] Garner, H. L., "A Survey of Some Recent Contributions to Computer Arithmetic," *IEEE Trans. Computers*, Vol. C-25, No. 12, pp. 1277–1282.

[10] Gregory, R. T. and Raney, J. L., "Floating-Point Arithmetic with 84-bit Numbers," *Comm. of ACM*, Vol. 7, No. 1, January 1964, pp. 10–13.

[11] Ikebe, Y., "Note on Triple-Precision Floating-Point Arithmetic with 132-bit Numbers," *Comm. of ACM*, Vol. 8, No. 3, March 1965.

[12] IBM Staff, *IBM System/370, Principles of Operation*, Pub. No. GA 22-7000-2.

[13] Kaneko, T. and Liu, B., "On Local Roundoff Errors in Floating-Point Arithmetic," *Journal of ACM*, Vol. 20, No. 3, July 1973, pp. 391–398.

[14] Kent, J. G., "Highlights of A Study of Floating-Point Instructions," *IEEE Trans. Computers*, Vol. C-26, No. 7, July 1977, pp. 660–666.

[15] Kent, J. G., "Comparison Sets: A Useful Partitioning of the Space of Floating-Point Operand Pairs," *3rd Symposium on Computer Arithmetic*, IEEE Computer Society, Catalog No. 75CH1017-3C, November 1975, pp. 37–39.

[16] Kuck, D. J. et al., "ROM-Rounding: A New Rounding Scheme," *3rd Symposium on Computer Arithmetic*, IEEE Computer Society Catalog No. 75CH1017-3C, 1975, pp. 67–72.

[17] Kuck, D. J. et al., "Analysis of Rounding Methods in Floating-Point Arithmetic," *IEEE Trans. Computers*, Vol. C-26, No. 7, July 1977, pp. 643–650.

[18] Kuki, H. and Cody, W. J., "A Statistical Study of the Accuracy of Floating-Point Number System," *Comm. of ACM*, Vol. 16, No. 4, April 1973, pp. 223–230.

[19] Kulisch, U., "Mathematical Foundation of Computer Arithmetic," *IEEE Trans. on Computers*, Vol. C-26, No. 7, July 1977, pp. 610–620.

[20] Lunde, A., "Evaluation of Instruction Set Processor Architecture by Program Tracing," *Ph.D. Thesis*, Dept. of Computer Science, Carnegie-Mellon University, Pittsburgh, Pa., July 1974.

[21] Matula, D. W., "A Formalization of Floating-Point Numeric Base Conversion," *IEEE Trans. Computers*, Vol. C-19, August 1970, pp. 581–691.

[22] Metropolis, N. and Ashenhurst, R. L., "Basic Operations in an Unnormalized Arithmetic System," *IEEE Trans.*, Vol. EC-12, 1963, pp. 896–904.

[23] Sterbenz, P. H., *Floating-Point Computation*, Prentice-Hall, Englewood Cliffs, N.J., 1974.

[24] Thornton, J. E., *Design of A Computer: The CDC 6600*, Scott, Foresman and Company, Glenview, Illinois, 1970, Chapter 5.

[25] Tsao, N. K., "On the Distribution of Significant Digits and Roundoff Errors," *Comm. of ACM*, Vol. 17, No. 5, May 1974, pp. 260–271.

[26] Wilkinson, J. H., *Rounding Errors in Algebraic Processes*, Prentice-Hall, Englewood Cliffs, N.J., 1963.

[27] Yohe, J. M., "Foundations of Floating-Point Computer Arithmetic," *Tech. Rep. 1302*, Math. Res. Center, University of Wisconsin, Madison, January 1973.

[28] Yohe, J. M., "Rounding in Floating-Point Arithmetic," *IEEE Trans. Computers*, Vol. C-22, No. 6, June 1973, pp. 577–586.

Problems

Prob. 10.1 Consider floating-point numbers with n-bit word length, p-bit mantissa field, and q-bit exponent field such that $p + q + 1 = n$, where the one bit is occupied by the sign of the FLP number. Perform a statistical analysis to determine the relative frequency of addition overflow or underflow in normalized versus unnormalized FLP add-type operatons.

Prob. 10.2 Explain the following terminologies associated with floating-point arithmetic computations.

(a) Prenormalization and postnormalization
(b) Significance digit arithmetic
(c) Chopped machine arithmetic
(d) Rounded machine arithmetic
(e) Average relative representation error (ARRE)
(f) Explicity vs. implicitly normalized numbers
(g) Guard digits
(h) Comparison sets of FLP instructions.

Prob. 10.3 Consider an octal fraction in true magnitude form

$$x = (0.3725765436)_8$$

Specify the following representations and determine in each case the relative error introduced.

(a) $(x)_8$ } after chopping
(b) $(x)_6$ }
(c) $(x)_8^*$ } after rounding
(d) $(x)_6^*$ }

Prob. 10.4 Let $a = (0.10101010111000101110101011010110)_2$ be a 32-bit positive real fraction and $\bar{a}$ be the negative version of a. Consider a machine with $p = 24$ bits in its mantissa field. Answer the following questions accordingly.

(a) Determine the two consecutive members m_1 and m_2 of machine representable fractions such that $m_1 < a < m_2$.
(b) Define a rounding ρ which is a downward-directed rounding scheme, and show $\rho(a)$ for the above machine.
(c) Repeat part (b) for a upward-directed rounding scheme.
(d) Are the above rounding schemes you defined symmetrical? If not, can you figure out a symmetric one?

Prob. 10.5 Consider the same machine as in Prob. 10.4. Determine the following machine representations with respect to the five rounding schemes defined in Section 10.4.

(1) $\Delta(a) =$ ____________________,
$\nabla(a) =$ ____________________,
$T(a) =$ ____________________,
$A(a) =$ ____________________,
$P(a) =$ ____________________,

(ii) Repeat part (a) for the negative fraction a, that is, to determine

Prob. 10.6 Specify the lookup table contents of a 4-bit ROM-based rounding scheme for the same machine as specified in Prob. 10.4. The 4-bit address of the ROM is formed from 3 low bits plus the highest bit to be rounded off as shown in Fig. 10.1. The ROM should have 3-bit output. Determine $\rho_{\mathrm{ROM}}(a)$, where ρ_{ROM} is the ROM-rounding and "a" was specified in Prob. 10.4. Does $\rho_{\mathrm{ROM}}(a)$ equal any of the rounded representations in part (i) of Prob. 10.5?

Prob. 10.7 Extend the double-precision FLP addition algorithm illustrated in section 10.9 to a triple-precision FLP addition algorithm with the same data format as used by Ikebe [11].

Prob. 10.8 Use computer simulation method to verify the procedures specified in Figs. 10.6 and 10.7 by a pair of given numerical operands $|f|$ and $|g|$. What are the maximal relative errors introduced with the approximations in Eq. 10.89 and in Eq. 10.94, repsectively?

Chapter

11

Elementary Functions, Pipelined Arithmetic and Error Control

11.1 Introduction

In this chapter, we study three important aspects of arithmetic systems in digital computers. First, digital methods to evaluate elementary functions such as trigonometric, inverse-trigonometric, exponential, logarithmic, hyperbolic and inverse-hyperbolic, square root, squaring, and other transcendental function will be described. We start with the design of a cellular array square rooter and an array squaring unit. Then we show how bounded elementary functions can be calculated by polynomial approximations. The CORDIC computing techniques for elementary functions will be presented in section 11.4, based on Walther's unfied approach [35]. The convergence methods proposed by Chen [5] for the computation of exponentials, logarithms, ratios, and square roots are described with initiation, transformation, and termination rules.

Second, we study the pipelined computing principles and corresponding arithmetic processor designs. The pipelined arithmetic designs found in the Texas Instruments *Advanced Scientific Computer* (**ASC**) are described in section 11.7. We shall study a generalized arithmetic pipeline, built with cellular array logic, for the execution of multiply, divide, square root, and squaring operations. In section 11.9, we describe how pipelining can be applied to designing a special-purpose arithmetic processor for executing *Fast Fourier Transform* (**FFT**) pairs. A detailed case study of the pipelined arithmetic processors in *Control Data* **STAR-100** computer is provided in section 11.11.

Third, we give a brief introduction to the various types of logic faults that one may find in arithmetic processors, and present several methods to cope with these faults for fault-tolerant arithmetic design. A literature guide for further studies of the above topics appears at the end of the chapter.

11.2 Binary Square Rooting and Squaring

The imminence of **LSI** technology promises a reduction in logic cost in addition to speed and size improvements. It also makes the use of specialized arithmetic function generators more attractive due to the resultant emphasis on functional partitioning. In recent years, there has been considerable interest in the design and implementation of iterative arithmetic logic arrays. Iterative cellular arrays for high-speed multiplication and division were comprehensively treated in Chapters 6 and 8. In this section, we introduce the arithmetic algorithms and the corresponding cellular arrays for the extraction of square roots and for the squaring of binary numbers. These two operations can be paired together, in which case a single array with some additional control logic can be used to select either function.

Consider two positive, binary fractions A and Q such that $Q = \sqrt{A}$ or $A = Q^2$. We call Q the *square root* of A and A the *square* of Q. Denote

$$Q = \sqrt{A} = 0\,.\,q_1 q_2 \cdots q_n$$
$$A = Q^2 = 0\,.\,a_1 a_2 \cdots a_{2n-1} a_{2n} \tag{11.1}$$

A conventional nonrestoring square-root algorithm is illustrated in Fig. 11.1. The binary number A is paired off starting from the binary point as $a_1a_2, a_3a_4, \ldots, a_{2n-1}a_{2n}$. The first subtrahend $D_0 = 0.01$ is subtracted from a_1a_2. If the remainder R_1 is nonnegative, $q_1 = 1$ and the next pair a_3a_4 is appended to R_1 and $D_1 = 0\,.\,q_1 01$ is subtracted from it. If R_1 is negative, $q_1 = 0$ and a_3a_4 is appended to R_1 and $D_1 = 0\,.\,q_1 11$ is added to it. In general, the kth intermediate step requires the following operations, after the determination of the kth root bit q_k.

$$R_{k+1} \leftarrow R_k \cdot a_{2k+1}a_{2k+2} - q_1q_2 \cdots q_k 01, \quad \text{if } q_k = 1, \tag{11.2}$$

$$R_{k+1} \leftarrow R_k \cdot a_{2k+1}a_{2k+2} + q_1q_2 \cdots q_k 11, \quad \text{if } q_k = 0, \tag{11.3}$$

where the appending operation "·" can be achieved by shifting the old remainder two bits to the left with new pair entering from the right end.

A tree of possible successive subtrahend used in the nonrestoring square-rooting process is shown in Fig. 11.2, together with an 8-bit ($2n = 8$) binary square-root extractor built with a cellular array using the Controlled Add-Subtract (CAS) cells described in Fig. 2.10 (also in Fig. 8.10). Each row of the array receives a new pair of input bits. Two's complement arithmetic is used to carry out the subtraction through complemented addition. When the leftmost carry-out of a row is a "1", the remainder is positive, $q_k = 1$ and the next row should perform subtraction; otherwise, $q_k = 0$ and the next row operation is addition. This function control can be achieved by simply connecting the leftmost carry-out signal of the present row to the function-control line and the initial carry-in line of the next row. The successive root bits are generated at the left end of each row, similar to the quotient generation in Fig. 8.10.

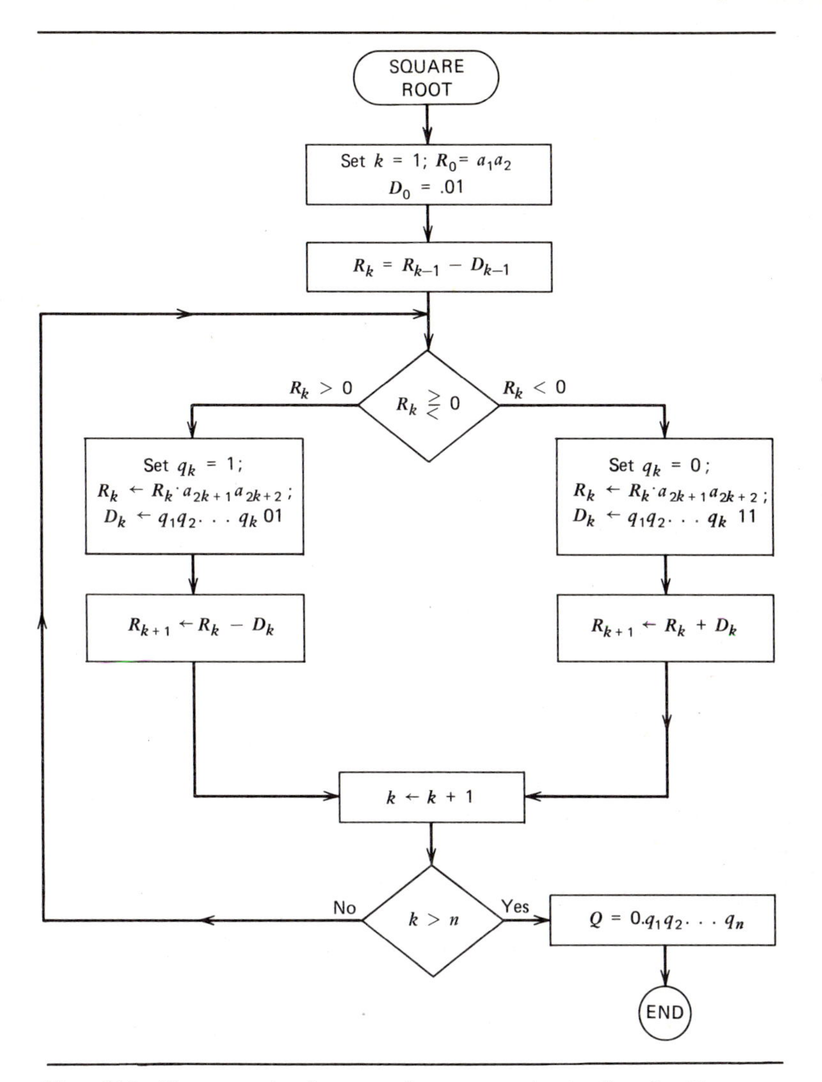

Figure 11.1 The conventional nonrestoring square-root extraction algorithm.

A numerical verification of the array is also shown with $A = (0.10101001)_2 = (169/256)_{10}$ and $Q = \sqrt{A} = \sqrt{169/256} = (13/16)_{10} = (0.1101)_2$.

What follows is an alternative approach to square rooting, which can be combined with the squaring operation to be realized. Consider the relationship specified in Eq. 11.1. If $q_1 = 1$, then $A \geq (0.1)^2 = 0.01$. Thus, a comparison is carried out between A and 0.01, if $A \geq 0.01$, then $q_1 = 1$, otherwise $q_1 = 0$. Similarly, if $A \geq (0.q_1 1)^2$,

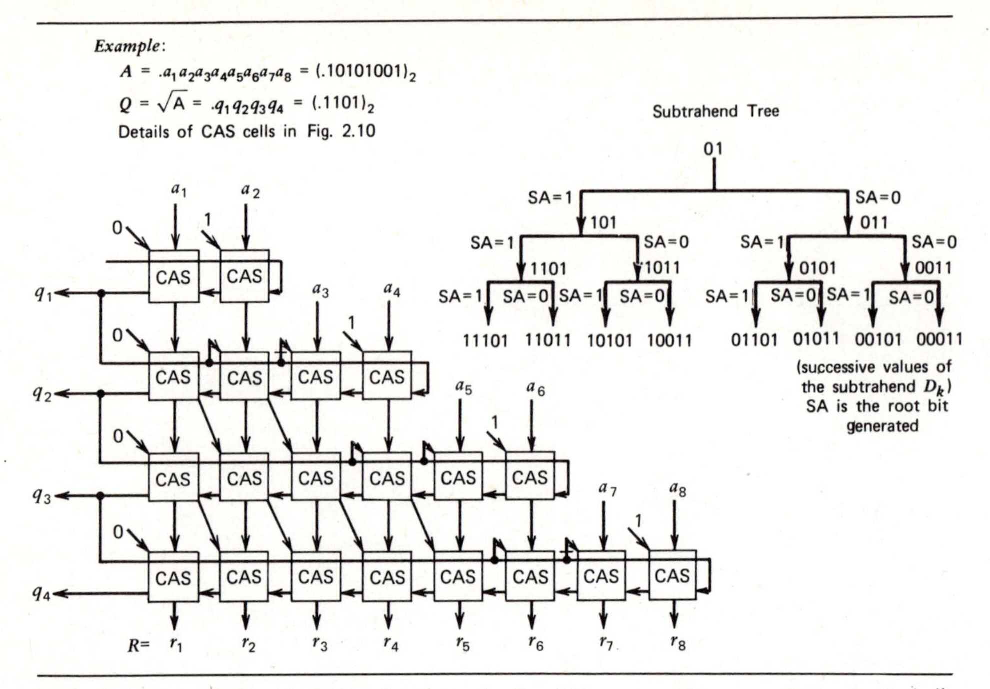

Figure 11.2 The subtrahend tree and the schematic of an 8-bit nonrestoring square-root extractor built with Controlled Add-Subtract (**CAS**) cells.

that is $A \geq (0.q_1)^2 + 0.0q_1 01$, then $q_2 = 1$, and so on. This procedure is generally formulated as follows:

$$(0.1)^2 = 0.01$$

$$(0.q_1 1)^2 = (0.q_1)^2 + 0.0q_1 01, \text{ or}$$

$$F_2 = F_1 + 0.0q_1 01 \tag{11.4}$$

where $F_1 = 0.01$ if $q_1 = 1$ and $F_1 = 0.00$ otherwise. Similarly, we have

$$(0.q_1 q_2 1)^2 = (0.q_1 q_2)^2 + 0.00q_1 q_2 01,$$

or

$$F_3 = F_2 + 0.00q_1 q_2 01$$

In general, if $q_{r+1} = 1$, then

$$F_{r+1} = F_r + D_r, \tag{11.5}$$

where $F_r = (0.q_1 q_2 \cdots q_r)^2$ is the rth *square* and $D_r = D_r = 0.\overbrace{00\cdots 0}^{r\text{ times}}q_1 q_2 \cdots q_r 01$ is called the rth *radicand*. It is obvious that $F_{r+1} = F_r$ if $q_{r+1} = 0$. The above iterative formula applies for all $r = 1, \ldots, n$.

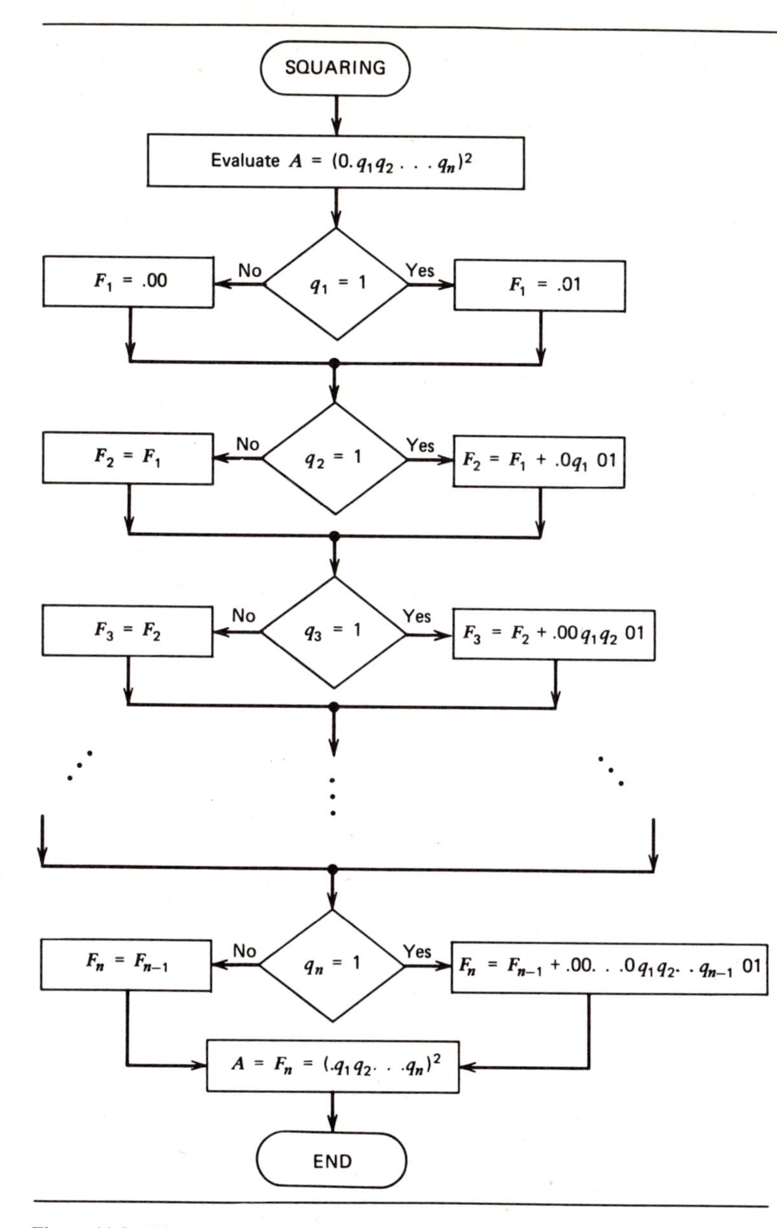

Figure 11.3 The binary squaring algorithm.

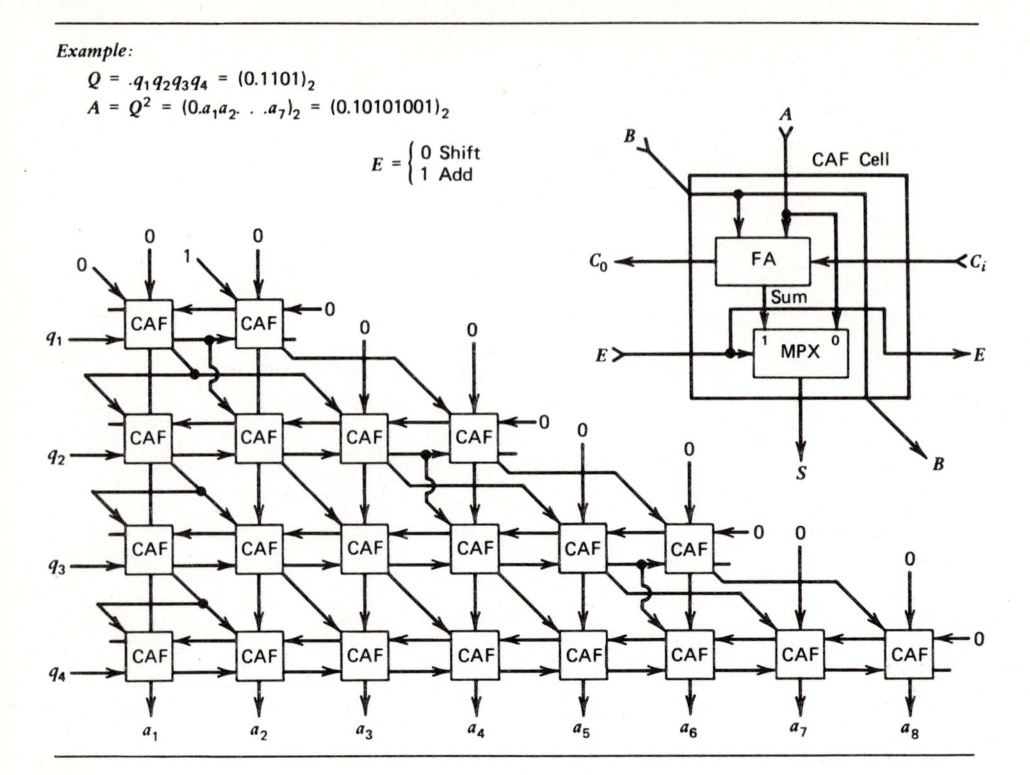

Figure 11.4 The array schematic of a 4-bit squaring unit using the Controlled Add-shift (**CAF**) cells.

The procedures for the extraction of the square root using above relations are described as follows: The first root bit q_1 is determined by subtracting the first radicand $D_1 = 0.01$ from A. If the remainder is positive, then $q_1 = 1$; otherwise $q_1 = 0$. In the latter case, the new remainder must be replaced by A. To obtain q_2, the second radicand $D_7 = 0.0q_101$ is subtracted from the new remainder (or A if $q_1 = 0$). The entire square root is revealed after n iterations in a similar manner.

Note that at each step in the above algorithm, the square of a number has been internally computed. Thus an array that computes square roots is also capable of computing squares. The binary squaring process is described in Fig. 11.3, which essentially applies the recursion defined in Eq. 11.5 repeatedly on the successive q_i digits, until the entire $2n$-bit square of the input number is produced.

Figure 11.4 shows the array realization of the squaring algorithm given in Fig. 11.3. A new type of *Controlled Add-shiFt* (CAF) cells is used in the array construction. The CAF cell is characterized by a full adder (FA) with a 2-input Multiplexer (MPX) controlled by an external Enable line to select either the old remainder input or the sum-output of the FA as the output of the cell. A binary number $Q = 0.q_1q_2q_3q_4$ is fed from the left Enable lines, one bit per each row. The initial radicand is zero from the top inputs of the array. The final square $Q^2 = A = 0.a_1a_2 \cdots a_8$ appears at the

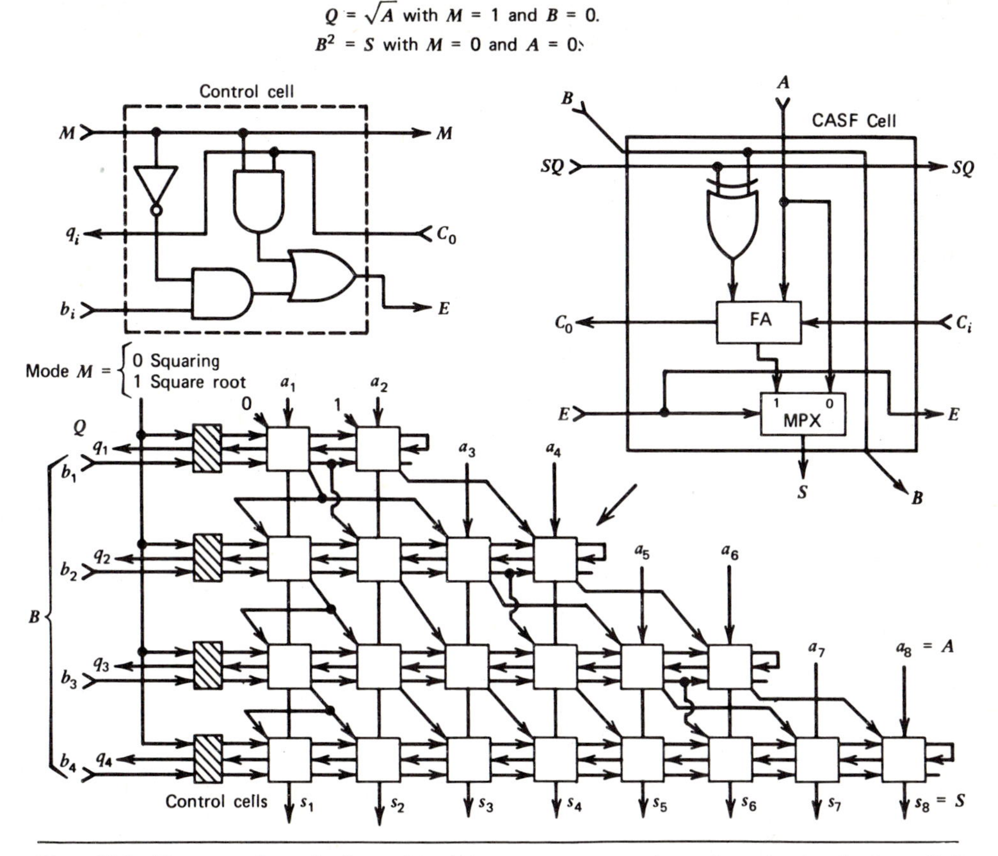

Figure 11.5 The array schematic of a combined binary square-root and squaring unit using the Controlled Add-Subtract-shift (**CASF**) cells.

bottom output lines. The numerical example used in Fig. 11.2 is also used here to verify the design.

By introducing a modified arithmetic cell as shown in Fig. 11.5, we can build a combined array unit, which can be used for either binary square rooting or binary squaring. The *Controlled Add-Subtract-shiFt* (CASF) cell is simply a combination of one CAS cell and one CAF cell. The control cell used at the left end of each row connects either the q_i input or the carry-out C_0 of the leftmost cell to the Enable input of that row. Thus, for computation of squares (mode line = 0), inputs are applied at $b_1 b_2 \cdots b_n$. The outputs are available at the sum outputs $s_1 s_2 \cdots s_8$ of the last row. Each row either adds or shifts upon the control of the q_i bit entering the row.

In the square-rooting process, the mode $M = 1$. Each row either subtracts or shifts by setting all SQ control lines "1". Again, two's complement arithmetic is assumed. The number whose square root is sought is applied at the top terminals $a_1, a_2 \cdots a_8$, whereas the 4-bit square-root answer is obtained at the left terminals

$q_1 q_2 q_3 q_4$. In general, the array requires $n(n + 1)$ CASF cells together with n control cells to compute the squares of an n-bit number or the square root of a $2n$-bit number.

11.3 Polynomial Evaluation of Elementary Functions

Some of the bounded elementary functions, such as trigonometric, exponential, logarithmic, and other bounded transcendental functions, can be evaluated with polynomial approximations within a given precision range. Mathematically, these functions can be written in the form of an infinite power series as follows:

$$f(x) = \sum_{i=0}^{\infty} a_i \times x^i \tag{11.6}$$

Listed below are examples of power-series expansions of several commonly used functions with bounded values.

$$e^{-x^2} = \sum_{j=0}^{\infty} \frac{(-1)^j}{j!} \times x^{2j} \tag{11.7}$$

$$\log_e(1 + x) = \sum_{j=1}^{\infty} \frac{(-1)^{j+1}}{j} \times x^j, \quad \text{for } -1 < x \leq 1; \tag{11.8}$$

$$\sin x = \sum_{j=0}^{\infty} \frac{(-1)^j}{(2j + 1)!} \times x^{2j+1}; \tag{11.9}$$

$$\sin^{-1} x = \sum_{j=0}^{\infty} \frac{1 \cdot 3 \cdots (2j - 1)}{2^j \cdot (j)! \cdot (2j + 1)} \times x^{2j+1}$$

$$\text{for} \quad x^2 < 1 \tag{11.10}$$

$$\cosh x = \sum_{j=0}^{\infty} \frac{1}{(2j)!} \times x^{2j} \tag{11.11}$$

This section describes a speedy hardware method using truncated power series to approximate the values of elementary functions. The truncated series expansion should be selected so that it will converge rapidly to the desired function value. The fewer the number of nonzero coefficients in the truncated series, the faster the evaluation will be. Power terms with zero coefficients do not appear in the computation. A truncated series of $k + 1$ terms takes the following form

$$f_k(x) = \sum_{j=0}^{k} a_j \times x^{m+j \cdot n} \tag{11.12}$$

where $m \geq 0$ is the initial power of the variable x and $n \geq 1$ is the increment power between adjacent power terms with nonzero coefficients. The upper limit k is usually determined by the smallest integer $j = k$ such that for a given *error tolerance* ε, the following is satisfied:

$$|a_k| \geq \varepsilon, \quad \text{but} \quad |a_{k+1}| < \varepsilon \tag{11.13}$$

Note that this method applies only to those series with monotonic decreasing coefficients, that is, $|a_i| > |a_{i+1}|$ for all i. As an example, in Eq. 11.10 we have $m = 1$, $n = 2$, and $k = 5$ for a given precision $\varepsilon = 0.018$. This means that the following truncated series

$$\sin^{-1} x = x + \frac{x^3}{6} + \frac{3}{40} \times x^5 + \frac{5}{112} \times x^7 + \frac{35}{1152} \times x^9 + \frac{63}{2816} \times x^{11} \quad \textbf{(11.14)}$$

is sufficient to approximate the value of $\sin^{-1} x$ up to a precision of $\varepsilon = 0.018$.

The way we evaluate the truncated series as in Eq. 11.12 is to rewrite it in nested-product form as follows for the general case of $k + 1$ terms:

$$\begin{aligned} f_k(x) &= x^m \times \left[\sum_{j=0}^{k} a_j \times x^{jn} \right] \\ &= x^m \times [((\ldots(((a_k \times x^n) + a_{k-1}) \times x^n + a_{k-2}) \times x^n \\ &\quad + \ldots + a_1) \times x^n + a_0)] \end{aligned} \quad \textbf{(11.15)}$$

Therefore, to evaluate $f_k(x)$ requires at most two powers of the variable x, namely, x^m and x^n. (In fact, one power of x, x^{m+n}, is sufficient, if it is available.) Note that without the nesting formulation, $k + 1$ powers of x may be required which results in longer computation time. Simple iterative multiplications and additions can be used to evaluate Eq. 11.15. In total, a $(k + 1)$-term power series may require $k + 1$ multiplications interleaved with k additions, as illustrated by the flow-chart of Fig. 11.6.

In many situations, the successive coefficients can be also obtained by iterative computations. For an example, the successive coefficients in Eq. 11.14 can be obtained by computing

$$a_j = a_{j-1} \times \frac{(2j - 1)^2}{(2j) \times (2j + 1)} \quad \textbf{(11.16)}$$

for $j = 1, 2, 3, 4, 5$ and with initial value $a_0 = 1$. Usually, all the nonzero coefficients and the integer powers used (x^m and x^n) should be predetermined and stored in a high-speed scratchpad memory or in some of the working registers, before the recursive evaluation given in Eq. 11.15 is initiated. For the example of $\sin^{-1} x$ with $(m, n, k) = (1, 2, 5)$, Eq. 11.15 takes the following form.

$$\begin{aligned} \sin^{-1} x &= x \times [(((((a_5 \times x^2) + a_4) \times x^2 \\ &\quad + a) \times x^2 + a_2) \times x^2 + a_1) \times x^2 + a_0] \\ &= x \times [(((((\tfrac{63}{2816} \times x^2) + \tfrac{35}{1152}) \times x^2 \\ &\quad + \tfrac{5}{112}) \times x^2 + \tfrac{3}{40}) \times x^2 + \tfrac{1}{6}) \times x^2 + 1] \end{aligned} \quad \textbf{(11.17)}$$

Six multiplications and five additions are required to complete the evaluation of Eq. 11.17, provided all the coefficients a_i and the powers x and x^2 are readily available from some preprocessing stage.

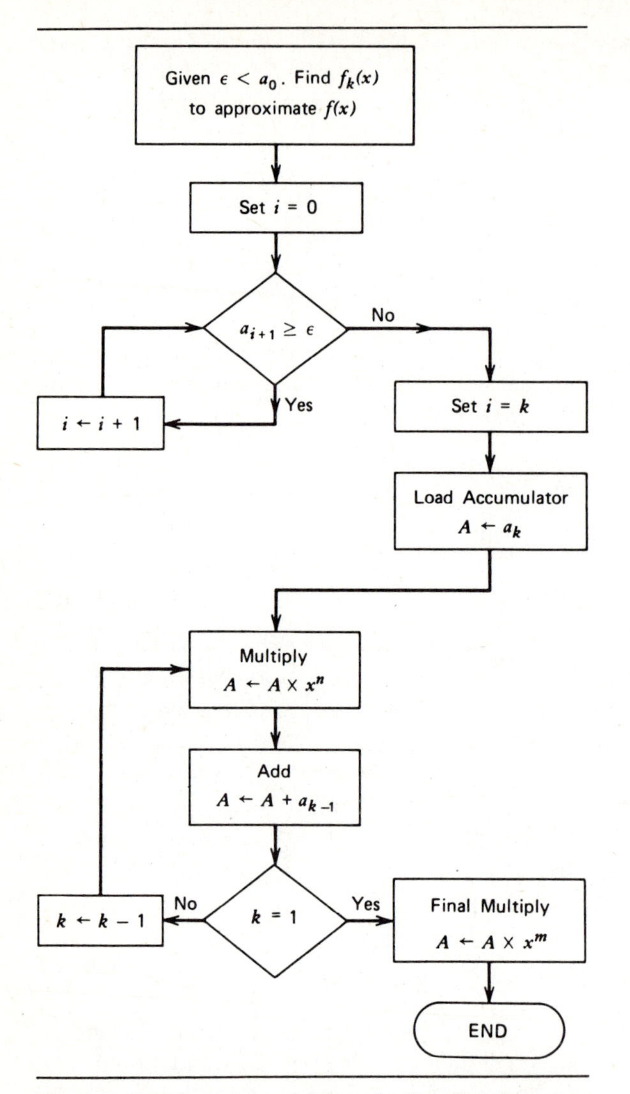

Figure 11.6 Polynomial evaluation of a bounded elementary function.

11.4 Walther's Unified CORDIC Computing Technique

The use of coordinate rotation vectors was originally proposed by Volder [34] for the development of the COordinate Rotation DIgital Computer (CORDIC) to calculate trigonometric functions, multiplication, division, and conversion between binary and mixed radix number systems. Presented below is a unified algorithm by Walther [35] for the calculation of elementary functions including *multiplication*, *division*,

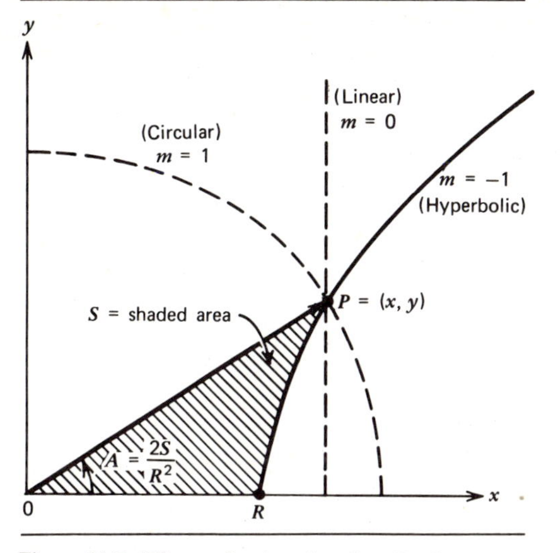

Figure 11.7 The angle A and radius R of a vector $P = (x, y)$ in three coordinate systems.

sin, cos, tan, arctan, sinh, cosh, tanh, arctanh, ln, exp, and *square root.* The primitive operations required are *shifting, adding, subtracting,* and the *recall* of prestored constants. A hardware FLP processor using the algorithm was built at Hewlett-Packard Laboratories. Possible simplifications of the algorithm for calculator designs will be discussed in section 11.12.

The basis for the algorithm is coordinate rotation in a linear, circular, of hyperbolic coordinate system, depending on which function is to be calculated. Such coordinate systems are parameterized by m, in which the radius R and angle A of the vector $P = (x, y)$ shown in Fig. 11.7 are defined as

$$R = (x^2 + m \times y^2)^{1/2}$$

$$A = \frac{1}{\sqrt{m}} \times \tan^{-1}(\sqrt{m} \times y/x) \qquad \textbf{(11.18)}$$

Three distinct values of the parameter m corresponding to the *circular* ($m = 1$), *linear* ($m = 0$), and *hyperbolic* ($m = -1$) coordinate systems are shown in the figure. A new vector $P_{i+1} = (x_{i+1}, y_{i+1})$ is obtained from $P_i = (x_i, y_i)$ according to

$$x_{i+1} = x_i + m y_i \delta_i$$

$$y_{i+1} = y_i - x_i \delta_i \qquad \textbf{(11.19)}$$

where δ_i is an arbitrary increment or decrement quantity. The angle and radius of the new vector are obtained as

$$A_{i+1} = A_i - \alpha_i,$$

$$R_{i+1} = R_i \times K_i \qquad \textbf{(11.20)}$$

Table 11.1 Equations for the Angles α_i and Radius Factors K_i in Three Coordinate Systems

Coordinate System m	Angle α_i	Radius Factor K_i
1 (circular)	$\tan^{-1} \delta_i$	$(1 + \delta_i^2)^{1/2}$
0 (linear)	δ_i	1
-1 (hyperbolic)	$\tanh^{-1} \delta_i$	$(1 - \delta_i^2)^{1/2}$

Note: δ_i is the incremental change.

where

$$\alpha_i = \frac{1}{\sqrt{m}} \times \tan^{-1} (\sqrt{m} \times \delta_i),$$

$$K_i = (1 + m \times \delta_i^2)^{1/2} \qquad \textbf{(11.21)}$$

Table 11.1 gives the equations for α_i and K_i for three coordinate systems.

After n iterations, we find

$$A_n = A_0 - \alpha$$

$$R_n = R_0 \times K \qquad \textbf{(11.22)}$$

where

$$\alpha = \sum_{i=0}^{n-1} \alpha_i$$

$$K = \prod_{i=0}^{n-1} K_i \qquad \textbf{(11.23)}$$

The total change in angle is just the sum of the incremental changes, whereas the total change in radius is the product of the incremental changes.

A third variable z is introduced for the accumulation of angular variations.

$$z_{i+1} = z_i + \alpha_i \qquad \textbf{(11.24)}$$

Solving the difference equations in Eqs. 11.19 and 11.24 for n iterations, we obtain

$$x_n = K \cdot [x_0 \cos(\alpha\sqrt{m}) + y_0 \sqrt{m} \sin(\alpha\sqrt{m})] \qquad \textbf{(11.25)}$$

$$y_n = K \cdot [y_0 \cos(\alpha\sqrt{m}) - x_0 \sqrt{m} \sin(\alpha\sqrt{m})] \qquad \textbf{(11.26)}$$

$$z_n = z_0 + \alpha$$

where α, K were defined in Eq. 11.23.

The above relations are summarized in Fig. 11.8 for various coordinate systems. The initial value x_0, y_0, z_0 are shown on the left of each block, whereas the final

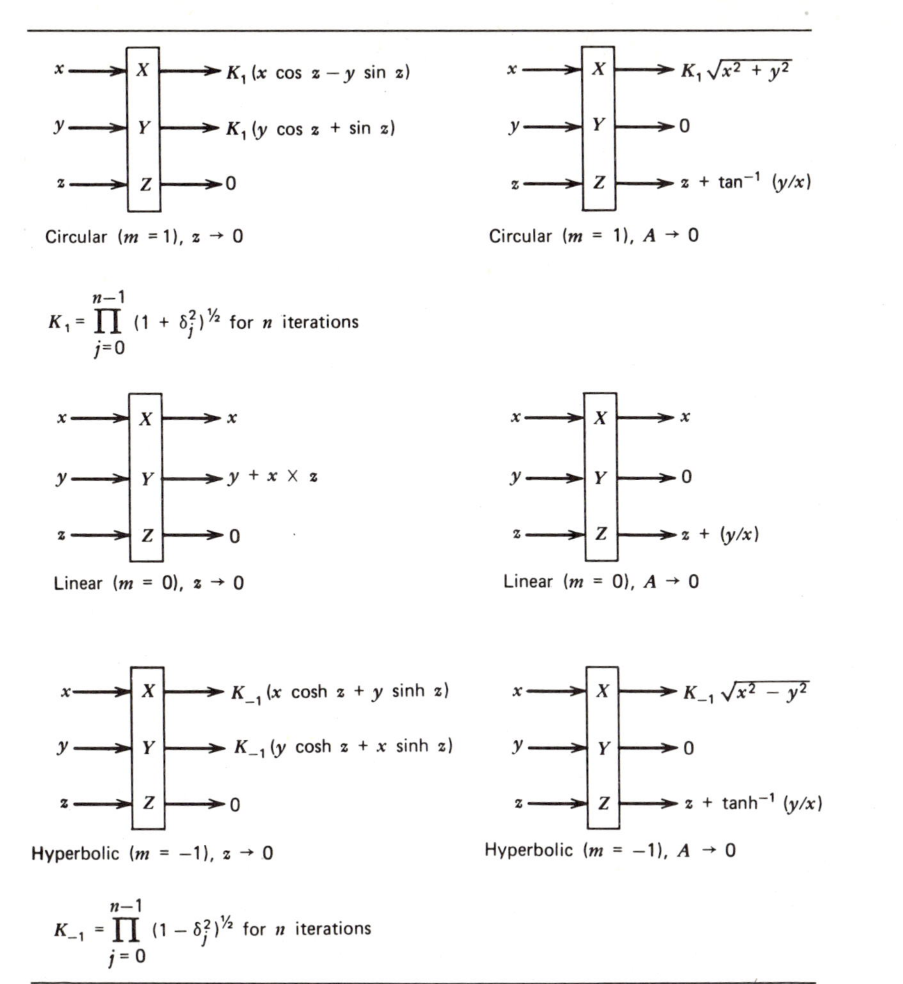

Figure 11.8 Input-output block diagram description of various CORDIC function generations using Walther's algorithm [35].

values x_n, y_n, z_n are shown on the right. By the proper choice of the initial input values, one can obtain the following functions directly at the outputs.

$$\{x \times z,\ y/x,\ \sin z,\ \cos z,\ \tan^{-1} y,\ \sinh z,\ \cosh z,\ \text{and}\ \tanh^{-1} y\} \tag{11.28}$$

In addition, the following function may be generated indirectly via the identities

$$\tan z = \sin z/\cos z, \tag{11.29}$$

$$\tanh z = \sinh z/\cosh z, \tag{11.30}$$

$$\exp z = \sinh z + \cosh z, \tag{11.31}$$

$$\ln \omega = 2 \tanh^{-1}[y/x], \quad \text{where } x = \omega + 1 \text{ and } y = \omega - 1 \tag{11.32}$$

$$\sqrt{\omega} = (x^2 - y^2)^{1/2}, \quad \text{where } x = \omega + \tfrac{1}{4} \text{ and } y = \omega - \tfrac{1}{4} \tag{11.33}$$

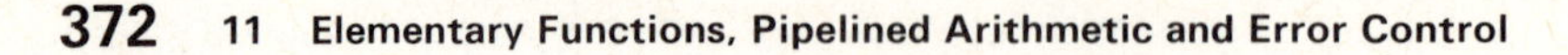

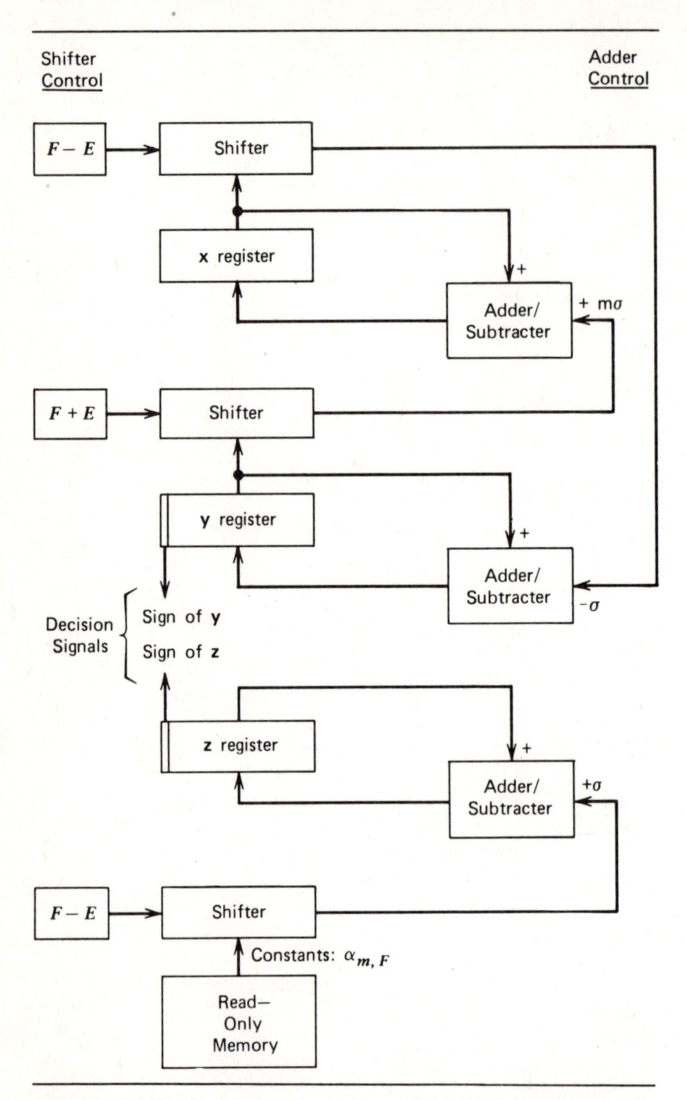

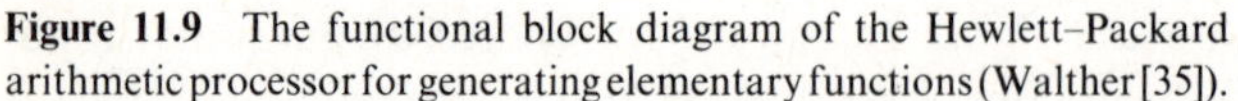
Figure 11.9 The functional block diagram of the Hewlett–Packard arithmetic processor for generating elementary functions (Walther [35]).

Figure 11.9 shows the hardware FLP processor built at Hewlett-Packard laboratories using the above unified CORDIC algorithm. The processor consists of three identical arithmetic units operated in parallel to handling the x, y, and z iterations, respectively. Each arithmetic unit contains a 64-bit register, an 8-bit parallel adder/subtractor, and an 8-out-of-48 multiplex shifter. The assembly of arithmetic units is controlled by a microprogram stored in a ROM, which also contains the angle and radius constants. The ROM has a capacity of 512 words with 48 bits per word and an access time of 200 nsec. The processor accepts three data types, 48-bit

Table 11.2 Maximum Execution Times of Various Elementary Functions in the HP Arithmetic Processor (Fig. 11.9)

Function	Execution Time (μsec)
LOAD	30
STORE	15
ADD	40
SUBTRACT	50
MULTIPLY	100
DIVIDE	100
SIN	160
COS	160
TAN	220
ATAN	90
SINH	130
COSH	130
TANH	190
ATANH	120
EXPONENTIAL	130
LOGARITHM	120
SQUARE-ROOT	100

FLP, 32-bit FXP, and 32-bit integer. All functions are calculated to 40 bits of precision (about 12 decimal digits). Table 11.2 gives the maximum execution time of the HP processor for most important functions. The speed may be slow for a general-purpose computer, but rather attractive for most calculator applications.

11.5 Chen's Convergence Computation Methods

The convergence division method described in Chapter 8 has been generalized by Chen [5] to evaluate *exponentials*, *logarithms*, *ratios*, and *square roots* of fractional arguments. The scheme is based on the co-transformation of a number pair (x, y) such that $F(x, y)$ is invariant; when x is driven approaching a known value x_ω, y is driven toward the desired result. Presented below are Chen's hardware algorithms using the convergence approach for evaluating ωe^x, $\omega + \ln x$, ω/x, and $\omega/\sqrt{x}$, where ω and x are given numbers. A brief function-theoretic background is given first, followed by algorithms for specific functions and then the hardware apparatus for carrying out the tasks.

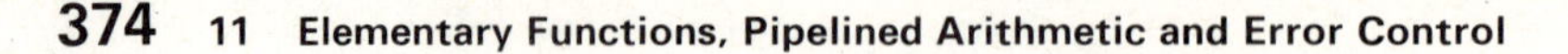

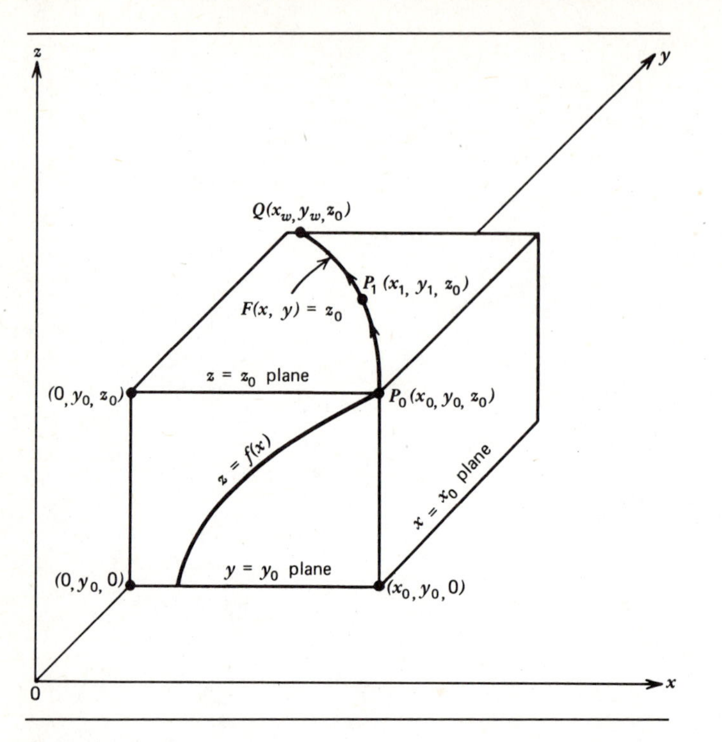

Figure 11.10 Geometrical interpretation of the convergence function for evaluating elementary functions (Chen [5]).

Consider the evaluation of a function $z_0 = f(x)|_{x=x_0}$, one can introduce a parameter y to form a two-variable *convergence function* $F(x, y)$ such that

(a) There is a known *initiation* value $y = y_0$ with $F(x_0, y_0) = z_0$.

(b) There exist convenient means to transform the pair (x_k, y_k) for $k \geq 0$ into (x_{k+1}, y_{k+1}) such that $F(x_{k+1}, y_{k+1})$ is invariant.

(c) The sequence of x-transformations converges to a known *destination* value $x = x_\omega$; the corresponding y-transformations then converges to $y = y_\omega$ such that $y_\omega = F(x_\omega, y_\omega) = z_0$.

Geometrically, the above function F can be represented by a curve on the $z = z_0$ plane in a 3-dimensional cube and passing through point $P_0(x_0, y_0, z_0)$ as shown in Fig. 11.10. We have

$$z_0 = f(x_0) = F(x, y) \tag{11.34}$$

The transformation invariance stated in (b) is merely the requirement that if the point $P_k(x_k, y_k, z_k)$ is on the curve F, then so is $P_{k+1}(x_{k+1}, y_{k+1}, z_{k+1})$. Or we may write the sequence

$$\begin{aligned} z_0 &= f(x_0) = F(x_0, y_0) = F(x_1, y_1) = \cdots = F(x_k, y_k) \\ &= \cdots = F(x_\omega, y_\omega) = y_\omega \end{aligned} \tag{11.35}$$

Condition (c) means that the curve F passes through the destination point

$$Q(x_\omega, y_\omega, z_0).$$

The iterative transformations of the number pair (x, y) in (b) present the rules to move a point along the curve F, starting from P_0, through P_1, P_2, and so on, toward the destination Q.

The successive transformation pairs can be written as two recursive functions

$$\begin{aligned} x_{k+1} &= \phi(x_k, y_k) \\ y_{k+1} &= \psi(x_k, y_k) \end{aligned} \tag{11.36}$$

The choice of the transformation rules ϕ or ψ are dictated by cost-effectiveness considerations. It is convenient to specify ϕ to be a function of x_k alone, operation over x_k being in some closed interval $[a, b]$. The x-transformation should yield an $x_{k+1} \in [a, b]$ such that

$$|x_{k+1} - x_\omega| < |x_k - x_\omega| \tag{11.37}$$

In other words, the point P_{k+1} should be a step closer than P_k to the destination $Q(x_\omega, y_\omega, z_0)$. The choice of ψ can be similarly obtained. These choices will be illustrated below for four specific elementary functions.

Convergence Algorithms for Evaluating Elementary Functions

The exponential algorithm $f(x) = \omega e^x$ for $0 \le x \le \ln 2 = 0.693\ldots$

Convergence function. $F(x, y) = y \times e^x$ **(11.38)**

with initiation $y_0 = \omega$ and destination $x_\omega = 0$

Transformation rule.
$$\begin{cases} x_{k+1} = \phi(x_k) = x_k - \ln a_k \\ y_{k+1} = \psi(y_k) = y_k \times a_k \end{cases} \tag{11.39}$$

The choice of a_k will be described in a later paragraph.

The logarithm algorithm $f(x) = \omega + \ln x$ for $\frac{1}{2} \le x < 1$.

Convergence function. $F(x, y) = y + \ln x$ **(11.40)**

with initiation $y_0 = \omega$ and destination $x_\omega = 1$.

Transformation rule.
$$\begin{cases} x_{k+1} = x_k \times a_k \\ y_{k+1} = y_k - \ln a_k \end{cases} \tag{11.41}$$

The ratio algorithm $f(x) = \omega/x$ for $1/2 \leq x < 1$.

Convergence function. $F(x, y) = y/x$ **(11.42)**

with initiation $y_0 = \omega$ and destination $x_\omega = 1$.

Transformation rule.
$$\begin{cases} x_{k+1} = x_k \times a_k \\ y_{k+1} = y_k \times a_k \end{cases} \tag{11.43}$$

The inverse square root algorithm $f(x) = \omega/\sqrt{x}$ for $1/4 \leq x M 1$.

Convergence function. $F(x, y) = y/\sqrt{x}$ **(11.44)**

with initiation $y_0 = \omega$ and destination $x_\omega = 1$.

Transformation rule.
$$\begin{cases} x_{k+1} = x_k \times a_k^2 \\ y_{k+1} = y_k \times a_k \end{cases} \tag{11.45}$$

Note that $\sqrt{x}$ can be obtained by setting $\omega = y = x$ in Eq. 11.44.

Let $k = n$ be the final step of iteration such that $x_n \to x_\omega$ and $y_n \to y_\omega$. Listed below are the *termination rules* and possible *computational errors* associated with each of the above arithmetic algorithms.

Termination Rules

Exponential algorithm

$$\begin{aligned} z_0 &= \omega \times e^{x_0} = y_0 \times e^{x_0} = (y_0 a_0) \times e^{x_0 - \ln a_0} \\ &= y_1 \times e^{x_1} = \cdots = y_n + y_n \mu \end{aligned} \tag{11.46}$$

Termination rule. $x_n = \mu \to 0, z_0 \sim y_n + y_n \times x_n$ **(11.47)**

Relative error. $0 \leq \varepsilon \leq \mu^2(1 + 2\mu/3)$ **(11.48)**

Logarithm algorithm

$$\begin{aligned} z_0 &= \omega + \ln x_0 = y_0 + \ln x_0 = (y_0 - \ln a_0) + \ln x_0 a_0 \\ &= y_1 + \ln x_1 = \cdots = y_n + \ln(1 - \mu) \sim y_n - \mu \end{aligned} \tag{11.49}$$

Termination rule. $1 - x_n = \mu \to 0; \quad z_0 \sim y_n - (1 - x_n)$ **(11.50)**

Absolute error. $0 \geq \delta \geq -\mu^2/2(1 - 2\mu/3)$ **(11.51)**

Ratio algorithm

$$\begin{aligned} z_0 &= \omega/x_0 = y_0/x_0 = (y_0 a_0)/(x_0 a_0) \\ &= y_1/x_1 = \cdots = y_n/(1 - \mu) \sim y_n + y_n \mu \end{aligned} \tag{11.52}$$

Termination rule. $1 - x_n = \mu \to 0; z_0 \sim y_n + y_n \times (1 - x_n)$ **(11.53)**

Relative error. $0 \leq \varepsilon = \mu^2/(1 - \mu^2)$ **(11.54)**

Table 11.3 Summary of Chen's Convergence Algorithms for Evaluating Elementary Functions

Range of x_0	**Function** $f(x_0)$	$F(x_k, y_k)$	x_k	x_{k+1}	y_{k+1}	x_n	**Termination Rules**
$[0, \ln_2)$	ωe^{x_0}	$y_k \times e^{x_k}$	$2^{-m} + p$	$x_k - \ln(1 + 2^{-m}) \sim p$	$y_k + 2^{-m} y_k$	$0 \le \mu$ $\mu < 2^{-N/2}$	$y_n \times e^{\mu} \sim y_n + y_n \times \mu$
$[1/2, 1)$	$\omega + \ln x_0$	$y_k + \ln x_k$	$1 - (2^{-m} + p)$	$x_k + 2^{-m} \times x_k \sim 1 - p$	$y_k - \ln(1 + 2^{-m})$	$1 - \mu$	$y_n + \ln(1 - \mu) \sim y_n - \mu$
$[1/2, 1)$	ω / x	y_k / x_k	$1 - (2^{-m} + p)$	$x_k + 2^{-m} \times x_k \sim 1 - p$	$y_k + 2^{-m} \times y_k$	$1 - \mu$	$y_n/(1 - \mu) \sim y_n + y_n \times \mu$
$[1/4, 1)$	$\omega / \sqrt{x}$	$y_k / \sqrt{x_k}$	$1 - 2(2^{-m} + p)$	$x_k \times (1 + 2^{-m})^2 \sim 1 - 2p$	$y_k + 2^{-m} \times y_k$	$1 - \mu$	$y_n/(1 - \mu)^{1/2} \sim y_n + y_n \times \mu/2$

Note: N is the word length and p satisfies Eq. 11.59.

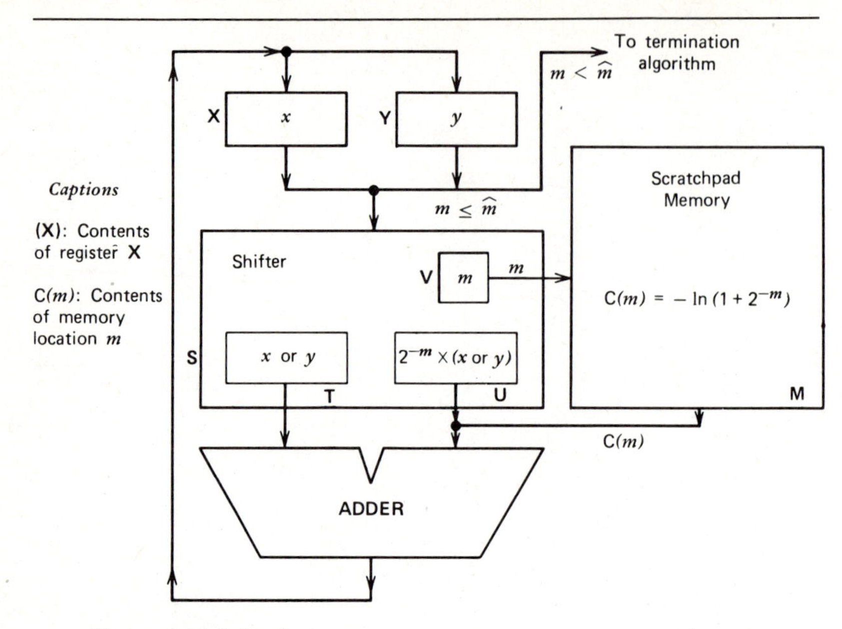

The standard choice of m:

$$\hat{m} = \begin{cases} 1 + (\text{the left-zero count of } x_k) \text{ for function } w \cdot e^x; \\ 1 + [\text{the left-zero count of } (1 - x_k)] \text{ for functions } w + \ln x;\ w/x; \\ 2 + [\text{the left-zeros count of } (1 - x_k)] \text{ for function } w/x^{1/2}. \end{cases}$$

Control microoperations

(a) Load (**X**) into **T**
(b) Deduce m from (**T**) into **V**
(c) Fetch $C(m)$ from **M**, Add $C(m)$ and (**T**)
(d) Store the sum in **X**
(e) Load (**Y**) into **T**
(f) Shift (**T**) right m places, put result in **U**; Add (**T**) and (**U**)
(g) Store the sum in **Y**
(h) Go to step (a), if $m \leq \hat{m}$, else enter the termination sequence.

Figure 11.11 The hardware arithmetic processor proposed by Chen for evaluating elementary functions (Chen [5]).

Inverse square root algorithm

$$z_0 = \omega/\sqrt{x_0} = y_0/\sqrt{x_0} = (y_0 a_0)/(x_0 a_0^2)^{1/2} = y_1/\sqrt{x_1}$$
$$= \cdots = y_n/(1 - \mu)^{1/2} \sim y_n + y_n\mu/2 \quad \textbf{(11.55)}$$

Termination rule. $1 - x_n = \mu \to 0;\ z_0 = y_n + y_n(1 - x_n)/2$ **(11.56)**

Relative error. $0 \leq \varepsilon = -1 + 1/(1 - \mu)^{1/2}(1 + \mu/2) \leq 3\mu^2/8$ **(11.57)**

We select a_k to be of the form

$$a_k = (1 + 2^{-m}) \quad \textbf{(11.58)}$$

such that multiplication by a_k can be replaced by a SHIFT and an ADD. The value m is usually chosen as the position number of the leading 1-bit in $|x_k - x_\omega|$ to the right of the binary point; but an increase of 1 is needed for the inverse square root. Table 11.3 summarizes the above algorithms with a_k replaced by Eq. 11.58, where p represents the bit pattern after the mth bit, that is,

$$0 \leq p \leq 2^{-m} \quad \textbf{(11.59)}$$

Chen has postulated a unified hardware processor to realize the above four iterative algorithms. The processor is illustrated in Fig. 11.11 with listed micro-operations. Starting with inputing x_0 in register X, $\omega(=y_0)$ in register Y, the sequences for the evaluations of four functions are:

$$\omega \times e^x\text{:}abcdefgh; \qquad \text{then } (Y) + (Y) \times (X) \sim \omega e^{x_0} \quad \textbf{(11.60)}$$

$$\omega + \ln x_0\text{:}abfdecgh; \qquad \text{then } (Y) + 1 - (X) \sim \omega + \ln x_0 \quad \textbf{(11.61)}$$

$$\omega/x_0\text{:}abfdefgh; \qquad \text{then } (Y) + (Y) \times [1 - (X)] \sim \omega/x_0 \quad \textbf{(11.62)}$$

$$\omega/\sqrt{x_0}\text{:}abfdafdefgh; \qquad \text{then } (Y) + (Y) \times [1 - (X)]/2 \sim \omega/\sqrt{x_0} \quad \textbf{(11.63)}$$

Let $T = N$ be the add-shift times for a conventional N-by-N multiplier. The expected $N/4$ iterations involve $3N/4$ add-shifts for the inverse square root and $N/2$ for the other three functions. The computation time is estimated, therefore, as $3T/4$ for square root and $T/2$ for exponential, logarithm, and ratio of N-bit fractions.

11.6 The Concept of Pipelined Computing

Two terms are frequently used to determine the computing power of a machine, one is the bandwidth and the other is latency. *Bandwidth* is the number of tasks that can be executed in a unit time interval. *Latency* is the length of time required to perform a single task. For a machine that executes one task at a time, latency is the inverse of bandwidth. In conventional design, most increases in the bandwidth of arithmetic processors have been achieved by reducing the latency with faster logic circuitry. For example, faster adders have been designed to minimize the carry propagation delay. Multipliers have been designed to allow simultaneous addition of many numbers. However, the circuit technology has almost reached its ultimate limit of light speed. Fast circuits alone will not be able to provide significant increases in computing power.

An obvious solution to this bandwidth problem is to allow simultaneous execution of many tasks by multiple arithmetic units. Parallel processing with straight hardware duplication, however, may not be economical or cost effective. Pipelined approach is an architectural design which provides significant increase in bandwidth with only a moderate increase in hardware investment.

Pipelining is an operations research technique originated from industrial assembly line processing. Pipeline computing refers to the subdivisions of the total computation workload into individual tasks, so that they can be executed in an overlapped fashion through a high-speed arithmetic pipeline under certain precedence constraints. This overlapped execution has been used in central processor design, in which the fetch, decode, effective address calculation and operand fetch of the next instruction can be overlapped with the execution of current instruction. If the instruction overhead and the execution times are nearly equal, the overlapped processor can be twice as fast as the conventional design. Here, we are mainly concerned about the overlapped executions of sequential tasks in complicated arithmetic functions.

A pipelined arithmetic unit is broadly defined as a collection of hardware resources, called *segments*, which are organized as a linear assembly line or a pipeline with synchronized timing control, such that a flow of subdivided tasks can be simultaneously executed by the successive segments of the pipeline. As illustrated in Fig. 11.12, the pipeline is characterized by a succession of segments. Each segment S_j is essentially a special-purpose combinational arithmetic logic circuit with delay τ_j, such as a complementer, an adder, a multiplier, and so on. Successive segments are

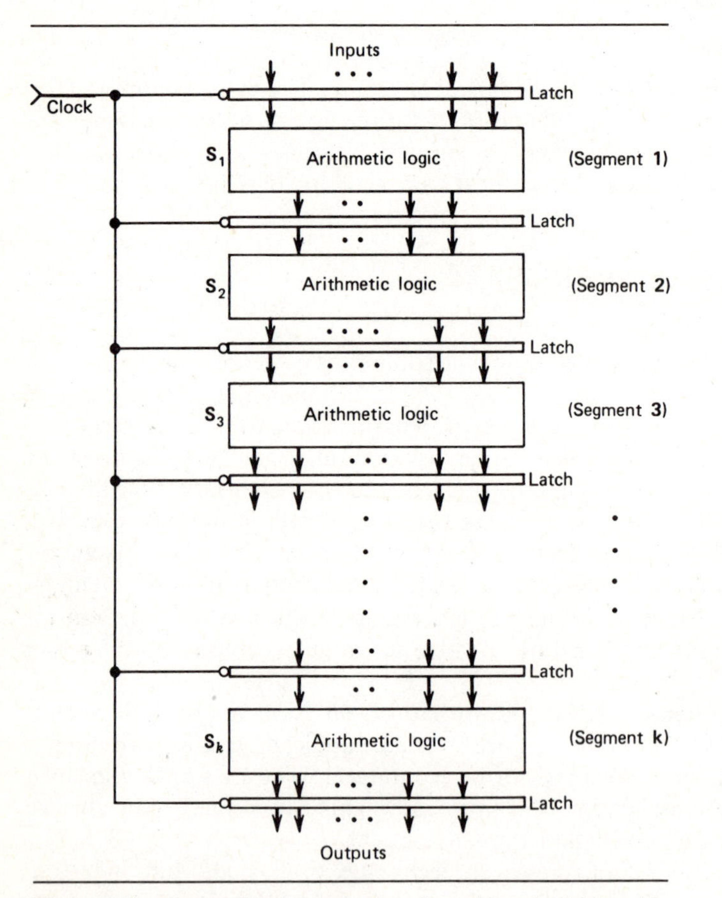

Figure 11.12 The functional organization of an arithmetic pipleine with k segments interfaced with fast latches.

interfaced with data latches, (synchronized registers) which hold the input and output bit patterns of successive segments. The b_j output bits of the segment S_j serve as the input bits a_{j+1} to segment S_{j+1}, such that the data width should match across the boundary.

$$b_j = a_{j+1} \tag{11.64}$$

Let w_j be the local word done by segment S_j within a time interval τ_j. Different segments may finish their tasks at different time intervals. In order to regulate the pipeline operation, the synchronizing clock pulse should have a clock period of

$$\tau = \text{Max}\{\tau_i\}_0^{k-1} + \tau_l \tag{11.65}$$

where τ_l is the propagate time delay of a single latch. Therefore, the rate of the pipeline system is determined by the maximum delay of one segment. Each latch releases its information to the segment on its right only when triggered by this external clock signal. The two end segments have additional latches which handle the inputs/outputs of the entire pipeline of k segments.

It takes k clock cycles to fill or to drain the pipeline. Once the pipeline is filled up in a steady state, every segment is busy executing the jth task T_j, which requires w_j amount of work. The steady-state *work rate* of the pipeline is therefore $\sum_{j=0}^{k-1} w_j$ per cycle. The *bandwidth* of the pipeline is the inverse of the maximum latency τ per

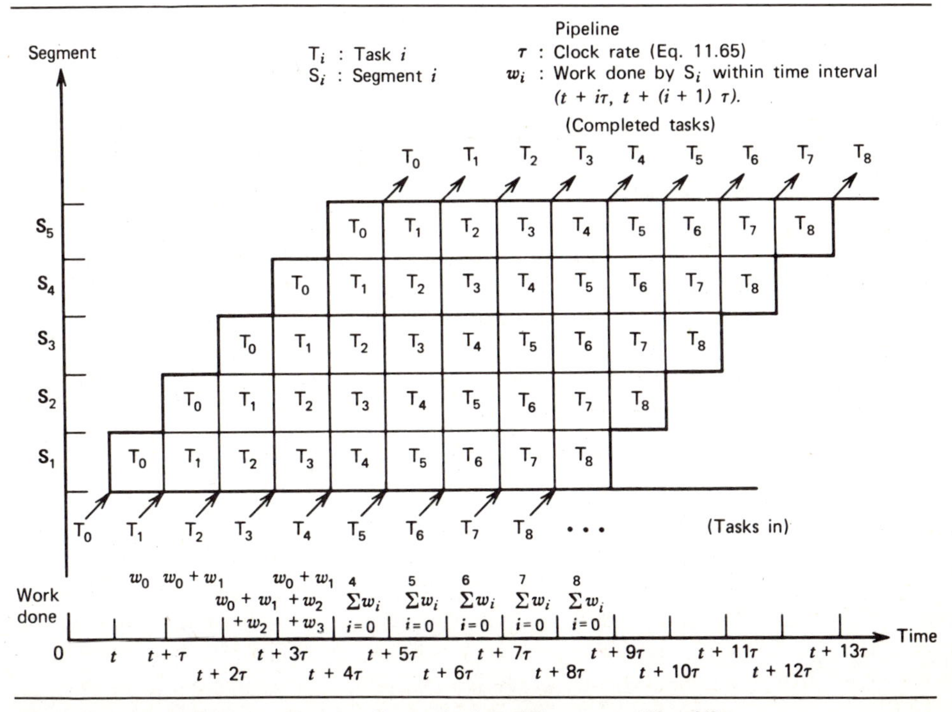

Figure 11.13 Successive tasks flow in a linear pipeline of 5 segments (Chen [4]).

segment, not the latency or duration $k \times \tau$ of the entire pipeline. Therefore, the increase in the pipeline length (number of segments) does not affect the bandwidth.

As far as task handling is concerned, the pipeline at steady state completes one task per cycle; k tasks, being simultaneously executed in the pipeline, can be done in k consecutive cycles like an automobile assembly line. Figure 11.13 shows the successive tasks through a five-segment arithmetic pipeline. Although pipelining may result in sizable increase in bandwidth, several linear pipelines can be operated in parallel to handle multiple data streams. Such parallel pipeline systems may be applied to either *Single-Instruction stream and Multiple-Data streams* (**SIMD**) or *Multiple-Instruction stream and Multiple Data stream* (**MIMD**) computer systems.

11.7 Case Study IV: Pipelined Arithmetic in TI Advanced Scientific Computer

In this section, we show how arithmetic functions can be executed by a pipeline processors. The abstract pipeline model given in the preceding section will be illustrated by the practical pipelined processors built in the Texas Instruments Advanced Scientific Computer (ASC). A fisheye view of the ASC system is shown in Fig. 11.14. All the components are labeled in the picture.

Let us start with a simple pipelined FLP Add Unit. To execute this instruction, the following operations must be performed:

1. Subtract exponents to reveal difference.
2. Shift right the smaller operand to align the mantissas.
3. Add the mantissas.
4. Normalize the resulting sum.

The execution of these steps can be performed with a pipeline unit of one to four segments depending on the goal of the design. The goal of a pipelined arithmetic design is decided primarily by the following factors:

1. Instruction repertoire such as FXP versus FLP and word sizes such as half-, single- or double-word operands.
2. The speed requirement and scalar instruction or vector instructions.
3. Control mechanism and instruction sequencing strategies.
4. Technology constraints and hardware cost.

Each of these factors must be examined because each has an influence on the total decision. Most instructions are considered *scalar instructions* which operate on only one operand or on a pair of operands to produce one numerical result. *Vector instructions* refer to vector-valued functions operated on a series of operands to produce a series of results such as the addition of two vectors; or to yield a single result such as the dot product of two vectors. The basic measurement of the processing rate of vectored arithmetic instructions, is the number of machine cycles needed per operand. The ideal arithmetic pipeline may operate upon a new set of operands for each cycle. For vectors which produce a vectored result of n components, an effective vector rate of one cycle per component operand would require n cycles plus some time

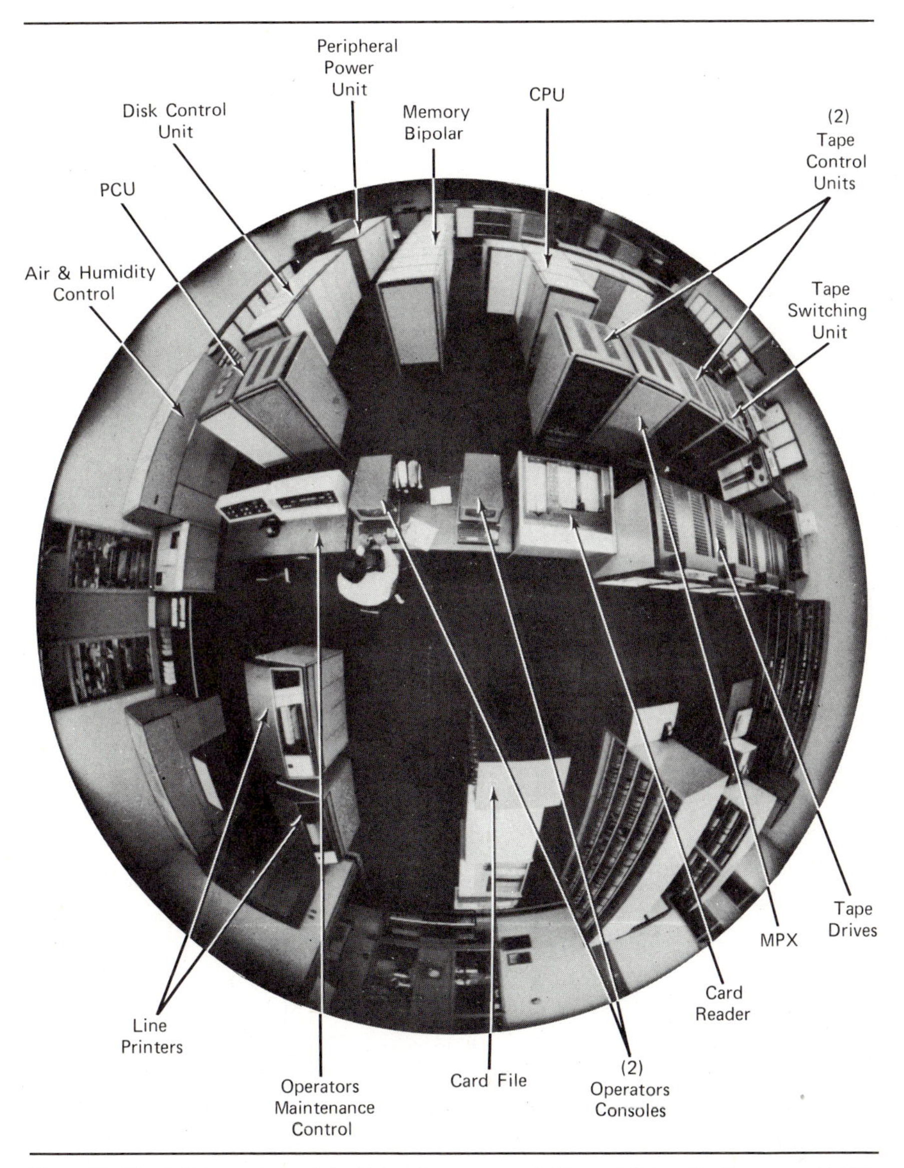

Figure 11.14 A fisheye view of the Texas Instruments Advanced Scientific Computer (**ASC**). (Courtesy of Texas Instruments, Inc. 1978).

to fill up the pipeline. If an instruction uses the same pipeline segment for two cycles without allowing the next operand to enter, the vector rate has to be raised to two cycles per component operand. Almost all scalar instructions have similar vector counterparts in **ASC** to be performed on large amounts of data. Parallel pipeline units are used to handle vector instructions, if high-speed is the primary concern.

Figure 11.15 shows the functional block diagram of the central processor in Texas Instruments **ASC**. The processor consists of three types of funcational units: The *Instruction Processing Unit* (IPU) fetches instructions, decodes the operation code, and develops the operand address. It also contains 64 addressable registers. The function of the IPU is pipelined. The *Memory Buffer Units* (MBU) fetch operands from memory and store results into memory. The MBU's have complete control during vector instructions and calculate all memory addresses required for vector

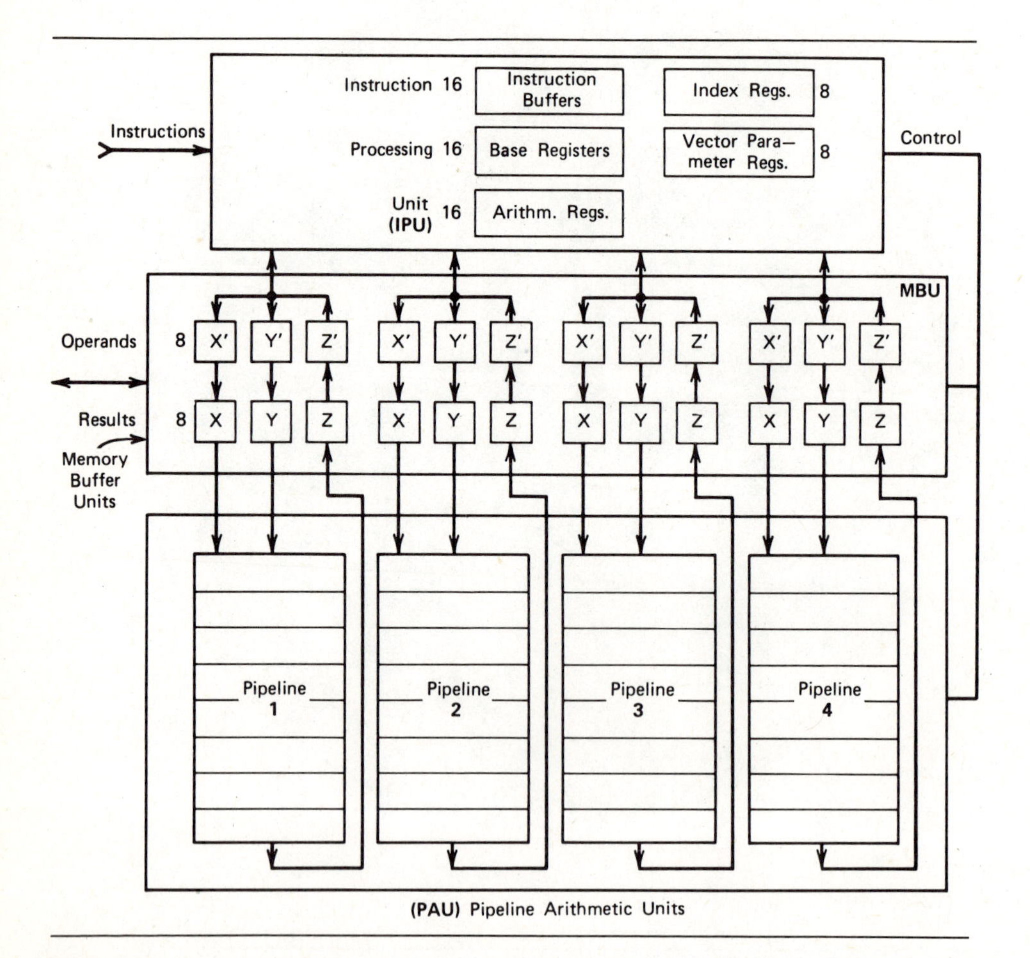

Figure 11.15 The block diagram of central processor in the TI Advanced Scientific Computer (**ASC**) (Watson [36, 37]).

instructions. The processor can have up to four *Pipelined Arithmetic Units* (PAU's) as shown. Each PAU is featured with the following requirements:

1. There are eight exclusive segments per each PAU with bypasses to execute a number of arithmetic functions in FXP or FLP formats. Each pipeline has an MBU and results of previous execution can be routed back for later uses. The basic machine cycle time is 60 nsec, which means the latency per each segment is at most 60 nsec.

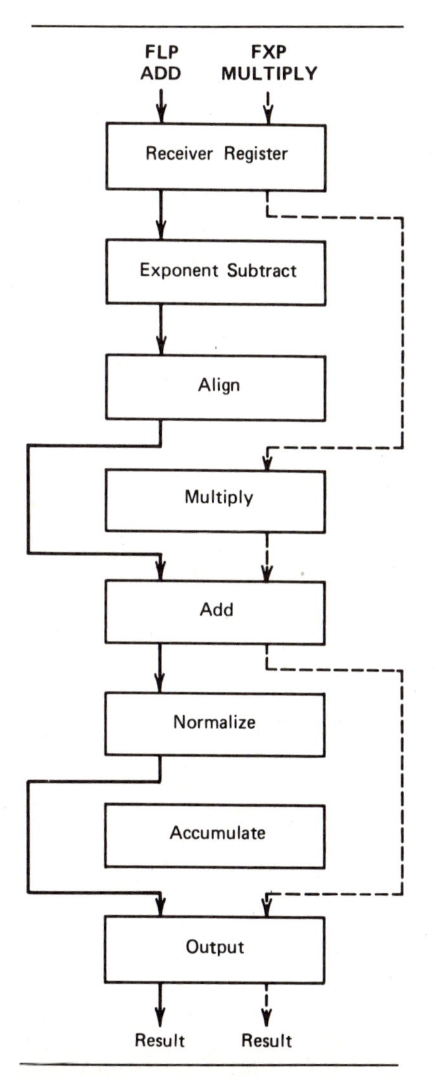

Figure 11.16 The eight segments in the Arithmetic Pipeline of the Texas Instruments **ASC**.

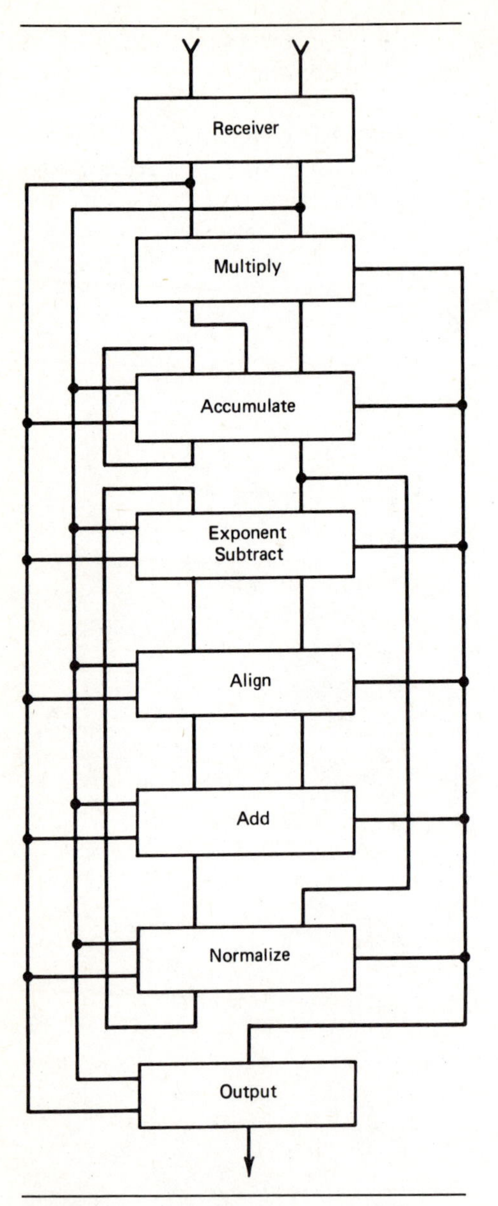

Figure 11.17 Possible segment interconnections of the **ASC** Arithmetic Pipeline for executing various instructions (Stephenson [33]).

2. The unit provides efficient scalar instruction execution.

3. The unit also provides very fast vector dot product. Each vector instruction is performed at one clock per component operand. Three-dimensional indexing is available. Each PAU is a 64-bit parallel unit which can handle 64-, 32-, and 16-bit data words.

4. The eight segments of each PAU are illustrated in Fig. 11.16 with the utilization interconnections for the execution of FLP Add and FXP Multiply shown. Most scalar and vector instructions are executable by the pipeline unit. Exceptions are double-length multiply and all types of division in which more than one clock may be required to complete the execution.

Different arithmetic instructions are allowed to take different paths through the PAU and not restrict its passage through each segment to be operated upon. Various possible pipeline cascade interconnections for executing arithmetic functions in TI ASC are shown in Fig. 11.17. The *Multiply segment* performs a full 32-by-32 multiplication. It produces two 64-bit output numbers referred to as pseudosum and pseudocarry, which are received and added by the *Accumulator segment* to produce the desired product. The Accumulator also has a provision of feeding its output back through itself to perform the double-length multiplications or divisions through repeated multiplications. This feedback path is also used to accomplish the FXP vector dot product instruction.

The *Exponent Subtract segment* determines the exponent difference and supplies the shift count to the *Align segment* to align the fractions for FLP operation, which uses the adder. The logic is shared to execute all right shifts—including logical, arithmetic, and circular shifts. The *Add segment* is a 64-bit two-level CLA Adder. *Normalize segment* normalizes all FLP results before sending them to the Output segment. All left-shift instructions share this hardware. In addition, conversion from FXP to FLP or vice versa also use the Normalize segment.

The *Output segment* provides a common point for the results of all instructions. Simple instructions such as Load or Store, logical AND, OR, and so on, use this segment only. Note that the hardware in Exponent Subtract is also shared by FXP of FLP comparison operations and the special vector instructions such as Search for Maximum or Search for Minimum. Another case study of pipelined arithmetic designs is the CDC STAR-100 computer to be given in section 11.11.

11.8 A Generalized Multifunction Arithmetic Pipeline

A multifunction cellular array arithmetic pipeline is described in this section. The unit can perform all basic arithmetic operations such as multiplication, division, square rooting, and squaring. The array uses a type of arithmetic cells modified from the CAS cells used before, each with six inputs and six outputs as shown in Fig. 11.18.

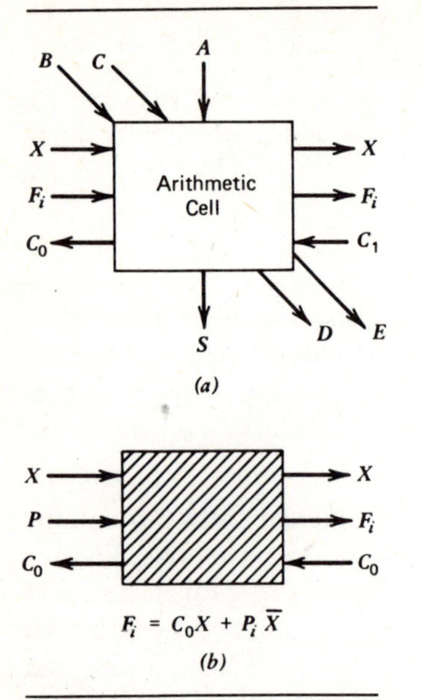

Figure 11.18 Arithmetic Cell and Control Cell used in the generalized arithmetic pipeline to be shown in Fig. 11.19. (*a*) Modified Controlled Add-Subtract Cell (Eq. 11.66) (*b*) Controll Cell.

The Boolean equations for the *Arithmetic Cell* are given below:

$$\begin{aligned} S &= [A \oplus (B \oplus X) \oplus C_1]F_i + A\bar{F}_i \\ C_o &= (B \oplus X)(A + C_1) + AC_1 \\ D &= BC + CF_i = C(B + F_i) \\ E &= B + CF_i = (B + C)(B + F_i) \end{aligned} \tag{11.66}$$

A *Control Cell* is needed at the left end of each row segment. The Boolean equation for the control cell is

$$F_i = C_o X + P_i \overline{X} \tag{11.67}$$

For multiplication and squaring operations, the control line X is set to be logical "0"; for division and square rooting, it is made logical "1". Let us briefly characterize these four operations with the use of these cells, before the complete cellular array pipeline is presented.

Multiplication and Division

The right-shift and add/subtract method is used to multiply or to divide in the array pipeline. The multiplicand or divisor is to be transmitted unchanged throughout all

Table 11.4 The Subtrahend at Different Segments for Square Rooting or Squaring in the Pipeline Arithmetic Array of Fig. 11.19

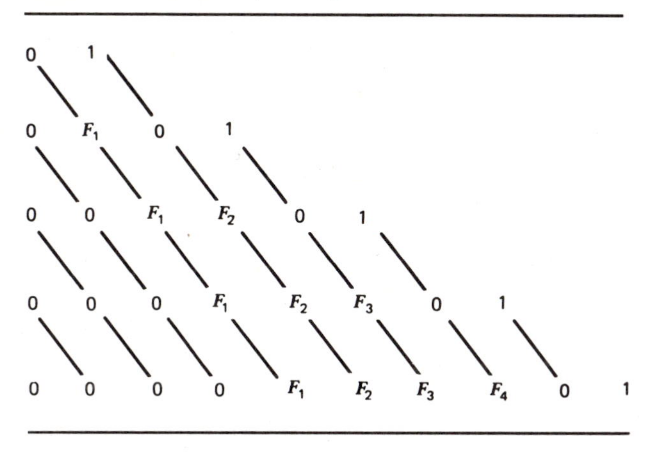

the segments in the pipeline array. This is achieved by keeping the C input bit of each individual arithmetic cell the same as the B input in Eq. 11.66.

Square Rooting

The square rooting is realized similarly to that of division, except that the subtraction changes in a particular fashion as shown in Table 11.4. Let the square root of a $2n$-bit number A be represented as

$$\sqrt{A} = F_1F_2\cdots F_i\cdots F_n \tag{11.68}$$

The first segment always subtracts 01 from the most significant digits of A. If the remainder is positive, F_1 will be "1", and otherwise "0". The old remainder will be retained as described in Eq. 11.5. when $F_i = 1$. The successive subtrahends derived this way are shown in Table 11.4.

The array pipeline requires to change the subtrahend bit of each segment in three different ways:

1. Change from "1" of any segment to "0" of the next segment by keeping the C input bit at "0".
2. Change from "0" to F_i by keeping C at "1".
3. Keep the subtrahend bit unchanged by keeping the C input bit the same as B.

Squaring

The squaring operation can be performed by utilizing Table 11.4 also. Consider the square S_n of an n-bit number $F_1F_2\cdots F_n$. If $F_1 = 1$, S_1 is taken as 01 and 00 otherwise. If F_2 is 1, the second row $0F_101$ of Table 11.4 is added to proper bits of S_1 to give S_2. The process of adding or skipping is determined by the value of F_i in Fig. 11.3. Thus, n such operations complete the squaring of a given n-bit number.

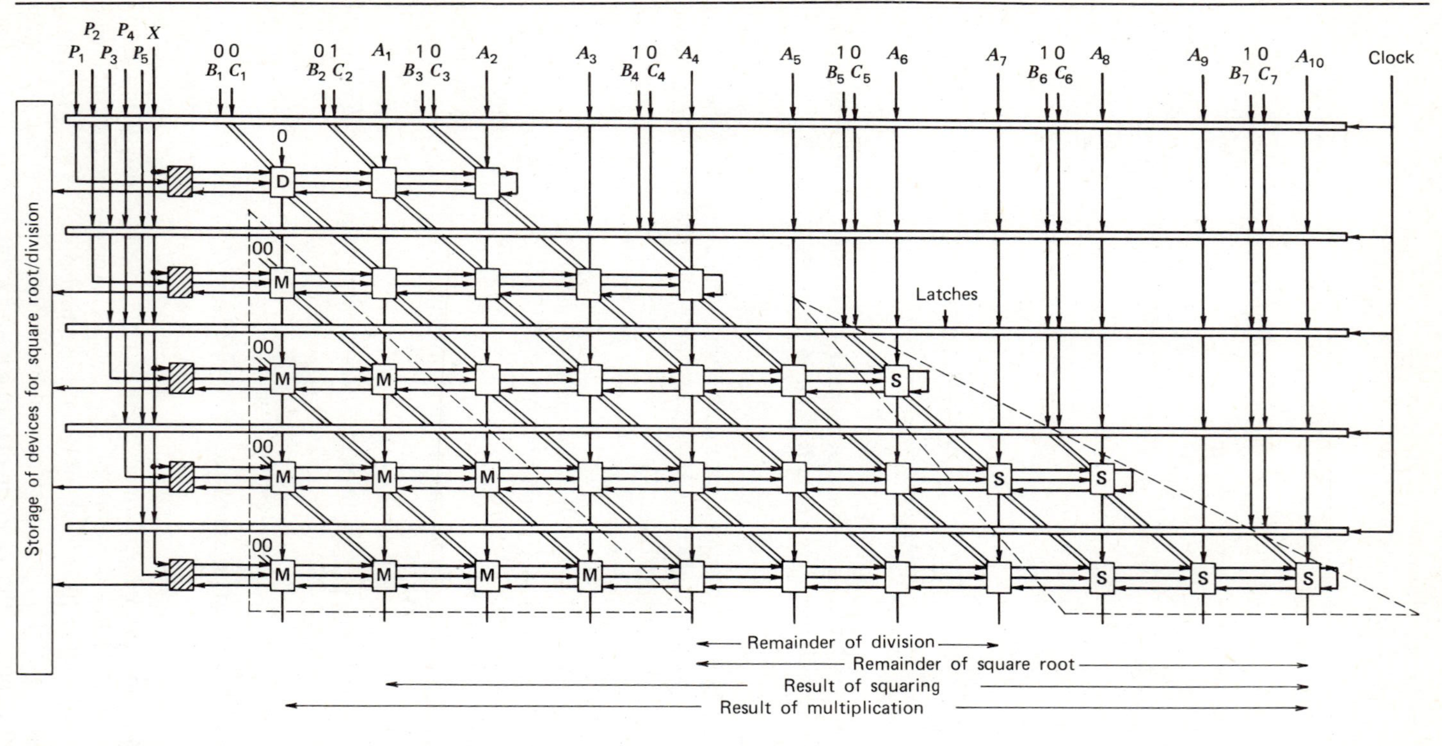

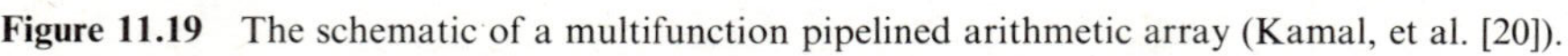

Figure 11.19 The schematic of a multifunction pipelined arithmetic array (Kamal, et al. [20]).

The pipeline array shown in Fig. 11.19 was proposed by Kamal et al [20]. The cells marked M are used only for the multiplication operation and for overflow check. The cells marked S are used for squaring and square rooting only. The array shown gives the square root of a 10-bit binary number $A = (A_1A_2 \ldots A_{10})_2$. The P inputs are made "0" and X is made "1" for getting the complement of B inputs, and to make the carry-into the lowest significant bit (the rightmost cell of each row segment). In order to generate the terms shown in Table 11.4, the B and C inputs to the latch of the first segment are given as 00, 01, 10, 10, 10, 10. The array requires $n \times (n + 2)$ arithmetic cells and n control cells for extracting the square root of a $2n$-bit number. The number whose square is to be evaluated is fed into the unit at terminals $P_1, P_2, \ldots, P_n$. When $X = 0$, the arithmetic cells act as adders, and the control cells transform P_i to F_i, thereby achieving the squaring of a binary number $P = (P_1P_2 \ldots P_n)_2$.

The array multiplies the multiplicand B by the multiplier P. X and A inputs are made zero and all C bits are made equal to the corresponding B inputs. Consider the case when B and P are four bits each. P_1 is kept to "0" to have four successive additions starting from the second segment, and the other four bits of P are given the value of the multiplier. The resulting product is obtained as the output of the last segment. The operation $A + B \times$ P can also be performed by this array, where $A = (A_1A_2 \ldots A_7)_2$. In general, the array pipeline designed for square rooting of a $2n$-bit number can multiply two numbers B and P of $(n + 3)/2$ bits each if n is add, and of $(n + 2)/2$ and $(n + 4)/2$ bits, respectively, if n is even.

The array pipeline divides a 7-bit number $A = (A_1A_2 \ldots A_7)_2$ by a 4-bit number $B = (B_1B_2B_3B_4)_2$, giving a 4-bit quotient and a 4-bit remainder. For this case, the control line X is made 1 and P inputs are made zero. The C inputs are kept the same as the B inputs. Arithmetic cells marked S are not used in the division process. Two's complement division is made possible by giving 00 inputs at B_5C_5, B_6C_6, and B_7C_7. The division process also starts from the second segment, like the multiplication. This requires the carry C_o to the control cell of the first segment be zero. In general, such array pipeline can divided an $(n + 2)$-bit number A by B. The maximum number of bits in B is $(n + 3)/2$ or $(n + 2)/2$, depending on whether n is odd or even respectively.

The clock rate of this pipeline unit is determined by the processing delay in the last segment, which uses $2n + 1$ arithmetic cells

$$\tau = (2n + 1)\tau_a + \tau_c + \tau_l \qquad \textbf{(11.69)}$$

where τ_a, τ_c, and τ_l are the delays in each arithmetic cell, control cell, and latch register, respectively.

11.9 Pipelined Fast Fourier Transform Processor

In this section, we describe possible hardware implementations of a well-known *Fast Fourier Transform* (**FFT**) algorithm applying the computing and design techniques we have learned in previous and present chapters. **FFT** plays a significant role in modern digital signal processing. The discrete **FFT** pair can be written in the following form:

$$X(n) = \frac{1}{N} \sum_{m=0}^{N-1} x(m) \times W^{mn}$$

$$x(m) = \sum_{n=0}^{N-1} X(n) \times W^{-mn} \qquad \textbf{(11.70)}$$

for $n = 0, 1, \ldots, N-1$, and $m = 0, 1, \ldots, N-1$, where W represents the Nth order complex root of unity with $j = \sqrt{-1}$.

$$W = e^{-j2\pi/N} \qquad \textbf{(11.71)}$$

An **FFT** algorithm is a sequence of computation steps for evaluating the above **FFT** pair of complex series. For simplicity in illustration, we present only radix-2 **FFT** algorithms with $N = 2^{\lambda}$ for some integer λ.

The Cooley-Tukey **FFT** algorithm is shown in Fig. 11.20 by means of signal-flow graphs for the special case of $N = 2^3 = 8$. The nodes on the graph represent Add operators on the complex inputs entering the nodes. The constants attached to the flow paths connecting the nodes indicate the rotation vector W^{ϕ}, if other than unity.

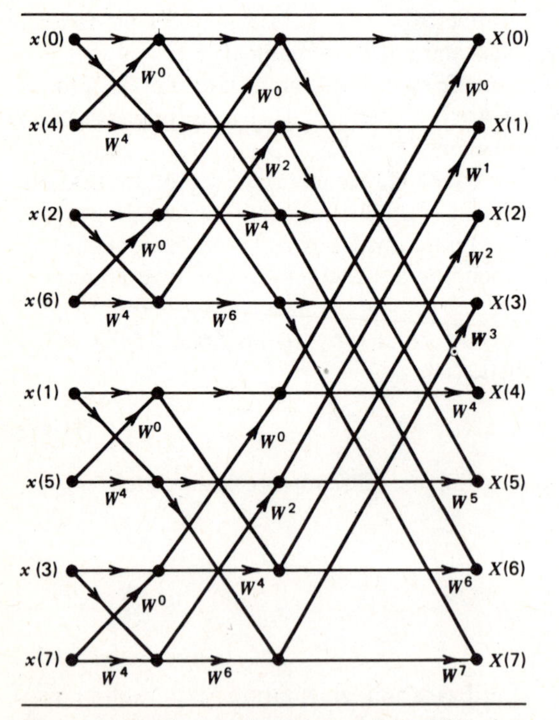

Figure 11.20 The Cooley–Tukey radix-2 **FFT** algorithm with $N = 8$ sample points.

This method is also known as the *Decimation-in-Time* algorithm. The necessary computation A_{ij} associated with each node is of the form

$$A_{ij} = a_i \times W^{\phi_i} + a_j \times W^{\phi_j} \tag{11.72}$$

where a_i and a_j are complex input variables and ϕ_i and ϕ_j are the associated rotation vectors, respectively. The initial data in this **FFT** algorithm are ordered by a bit-reversal scheme where the reverse binary index of each complex data point is used as the input order, whereas the results of the transformer are in normal order.

For the decimation-in-time formulation, there exist $N/2$ computation pairs within each of the $\log_2 N$ stages, where the two powers of W differ by exactly $N/2$. Because $W^{N/2} = -1$, the computation at each node pair can be written as

$$\begin{aligned} A' &= A + B \times W^{\phi} \\ B' &= A - B \times W^{\phi} \end{aligned} \tag{11.73}$$

where A' and B' are the results of the current transform stage in terms of those of the previous stage. A, B, and ϕ serve as indices within the calculation sequence. Figure 11.21 shows the arithmetic schematic for computing Eq. 11.73. The subscripts R and I correspond to real and imaginary components, respectively. Note that these two parts can be calculated in parallel.

The real and imaginary parts of Eq. 11.73 can be written separately as

$$\begin{aligned} A'_R &= A_R + B_R \cos \phi - B_I \sin \phi \\ B'_R &= A_R - B_R \cos \phi + B_I \sin \phi \end{aligned} \tag{11.74}$$

and

$$\begin{aligned} A'_I &= A_I + B_R \sin \phi + B_I \cos \phi \\ B'_I &= A_I - B_R \sin \phi + B_I \cos \phi, \end{aligned} \tag{11.75}$$

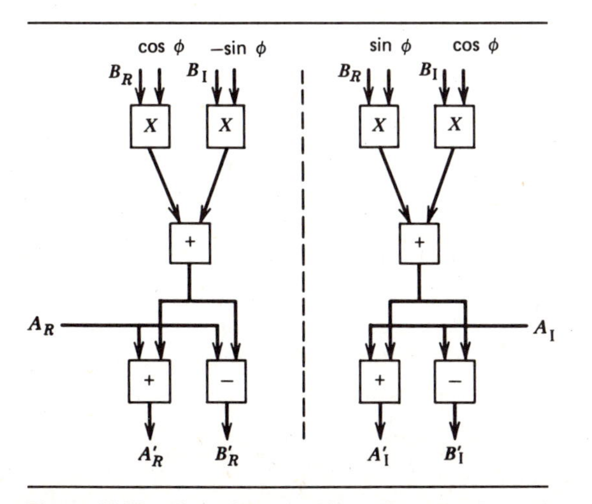

Figure 11.21 The schematic block diagram of a full Butterfly Element (**BE**) for implementing algorithm.

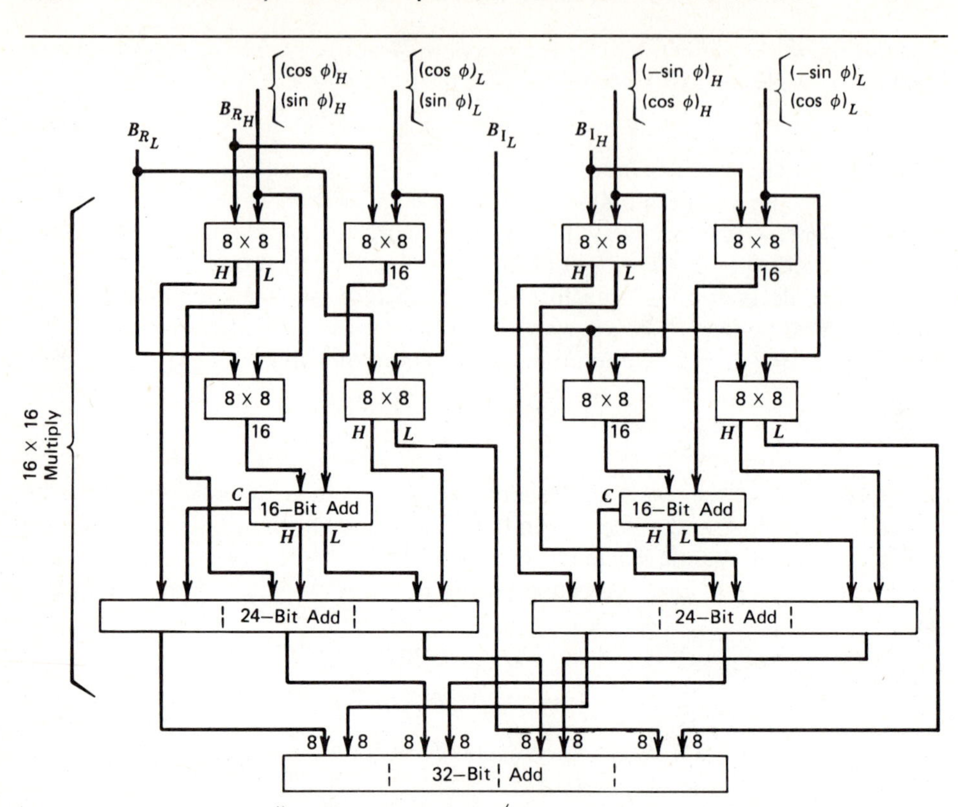

Figure 11.22 Arithmetic circuit block diagram of half butterfly element in Fig. 11.21 (Larson [20]).

where $A = A_R + jA_I$ and $B = B_R + jB_I$ are the input variables. A detailed block diagram of the arithmetic circuit for computing either Eq. 11.74 or Eq. 11.75 is shown in Fig. 11.22. Two such circuits realize the design shown in Fig. 11.21, because the right half is identical to the left half. Note that the real multiplication is performed by 8-by-8 multipliers to provide the 16-bit precision required per each operand.

The arithmetic device shown in Fig. 11.21 is called a *Butterfly Element* (BE). One wing of the butterfly is realized in Fig. 11.22. The two wings are identical except for different inputs. A complete FFT processor may require k BE's, where

$$k = \frac{N}{2} \log_2 N \qquad \textbf{(11.76)}$$

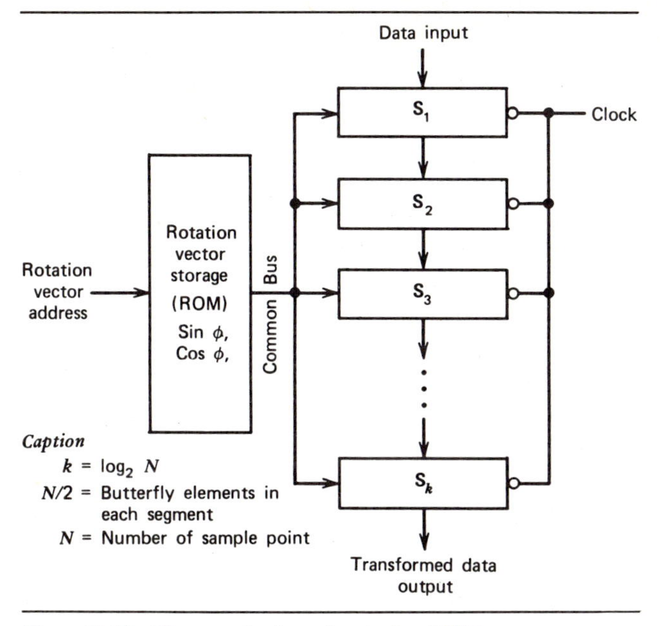

Figure 11.23 The organization of a pipelined **FFT** processor.

One can arrange these k BE's into a pipelined **FFT** processor as shown in Fig. 11.23. The pipeline consists of $\log_2 N$ segments with $N/2$ BE's per each segment. The interconnections from segment to segment have been demonstrated by the data flow graph in Fig. 11.20. The rotation vectors sin ϕ and cos ϕ for all the segments are retrieved from a lookup table in the ROM. Figure 11.20 shows that $N/2$ different rotation vectors are read to process one block of N samples. Indeed, if they are produced in the order required for the last segment, the rotation vectors required for the earlier segment may be obtained by strobing this list at the proper instant in advance of its need. Thus, the sines and cosines required for each segment may be obtained by providing a register for each arithmetic BE, all driven from a common bus. The first transform output is obtained immediately after the last data sample in the block of N is received.

An alternate approach to realize the butterfly element is to use CORDIC algorithms for generating the weighting factors cos ϕ and sin ϕ, used in each segment. Figure 11.24 shows the schematic block diagram of a CORDIC full butterfly design. The basic CORDIC algorithm for rotation of the complex quantity B by an angle $\Phi = (2\pi/N)\phi$, or $BW^{\phi} = (x + jy)W^{\phi} = (x + jy)e^{-j\Phi} = Be^{-j\Phi}$, is given by the following iterations:

$$\begin{aligned} x_{i+1} &= x_i + a_i y_i \times 2^{-i} \\ y_{i+1} &= y_i - a_i \times a_i x_i \times 2^{-i} \\ z_{i+1} &= z_i - a_i \times \theta_i \end{aligned} \tag{11.77}$$

with initial values $x_0 = x$, $y_0 = y$, $z_0 = \Phi$ and $a_i = 1$ if $z_i \geq 0$; otherwise $a_i = -1$. These 16 sign coefficients, a_i for $i = 0, 1, \ldots, 15$, are generated by the Φ cascade, where 15 input constants θ_i for $i = 0, 1, 2, \ldots, 14$ have the following values

$$\theta_{i+1} = \tan^{-1}(2^i) \tag{11.78}$$

The iteration runs μ cycles for $i = 0, 1, \ldots, \mu - 1$, where μ is the number of bits in each operand x, y, or Φ. The figure shows 16 adder stages for a word length of 16 bits. The 2^{-i} multiplier is a simple shift operation and can be implemented by hard-wired interconnections between cascade stages. Readers may refer to Larson's Doctoral thesis [21] for details of various butterfly designs.

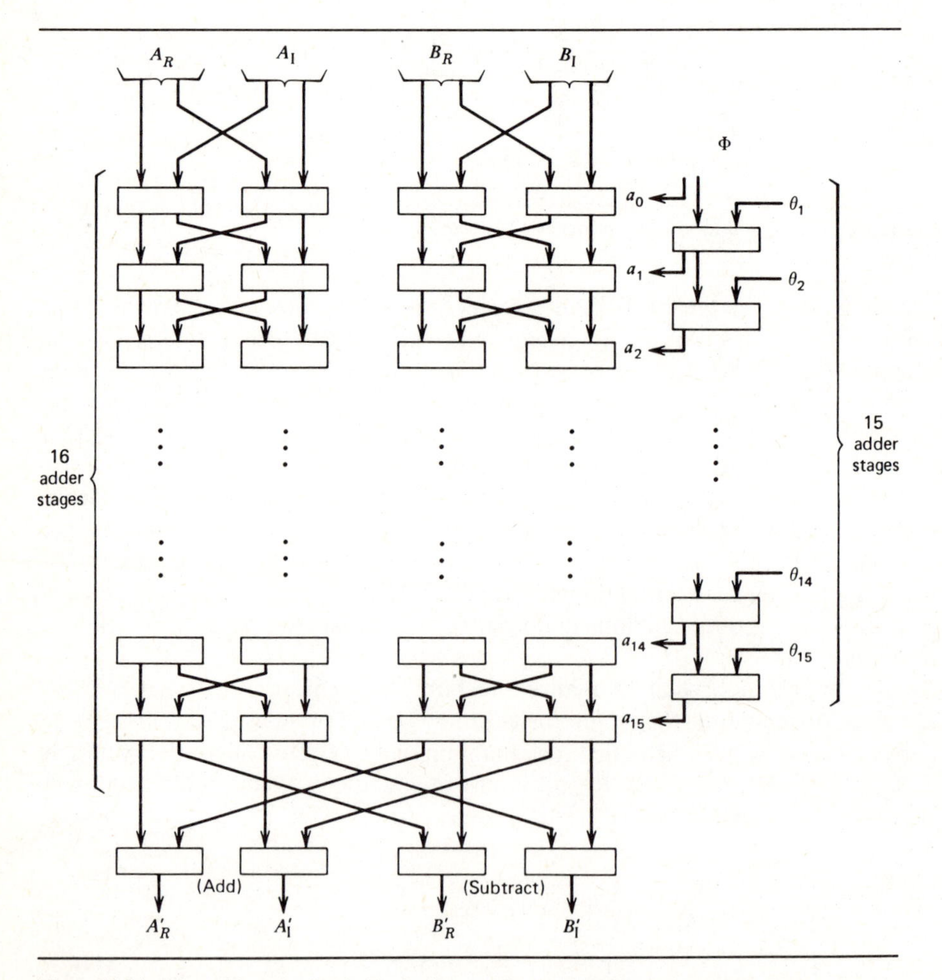

Figure 11.24 The arithmetic circuit block diagram for a CORDIC full butterfly design (Larson [20]).

11.10 Error Control in Arithmetic Processors

In this section, we briefly introduce logic fault types and possible means to cope with these logic faults in arithmetic circuits. Various causes may evoke a failure in arithmetic processors, such as circuit component malfunction due to overheating, poor contacts, stuck-types of faults, electromagnetic radiation disturbance, mechanical shocking, and so on. There are many different ways to classify the logic faults in digital circuits. Figure 11.25 presents a tree classification of logic faults.

Single fault refers to the occurrence of one incorrect *variable* (terminal value) in the circuit. Whenever there is more than one terminal variable inconsistent with its' required value, we have found *multiple faults* in the circuit. By inconsistency, we refer to the situation where logic one appears while zero is desired, and vice versa. A single fault is *local* when its immediate effect changes the value of one digit; otherwise,

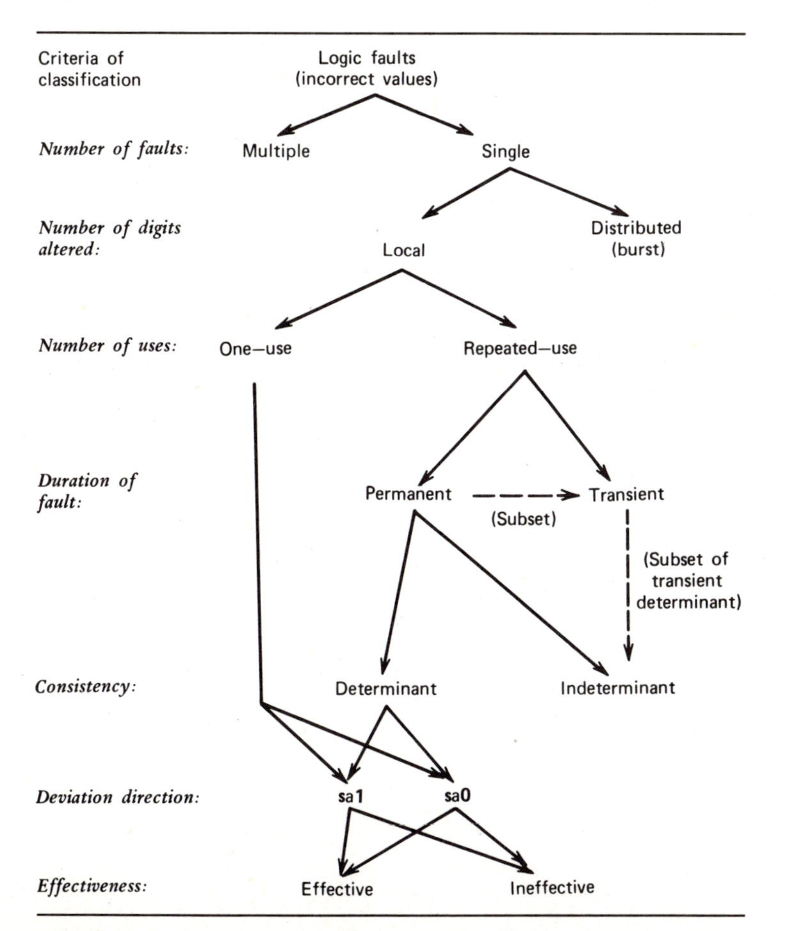

Figure 11.25 The classification of logic faults in arithmetic circuits.

distributed or *burst* faults have occurred. The effect of distributed faults may sometimes be attributed to the cumulative effect of several single faults.

In practical situations, local single fault has the highest probability of occurrence *One-use fault* refers to a faulty condition that exists only in one nonrepetitive use of the circuit during the execution of an arithmetic algorithm. *Repeated-use fault* occurs when the faulty circuit has been repeatedly used in an iterative loop of a computing process. The one-use type of faults can be further subdivided into two classes according to the direction of deviation of logic values. A *stuck-at*-0 ((*s*a**0**) and a *stuck-at*-1 (*s*a**1**) fault refers to the permanent fault-induced values of "0" and "1", respectively. The stuck-type of errors may be either *effective* or *ineffective* depending on whether or not they agree with the desired logic values at the particular time frame.

The repeated occurrence of faults may be either *permanent* or *transient*. It is transient if it does not occur at all iterative uses of the faulty unit; otherwise, it is called permanent. One can view a transient fault as a permanent fault that is ineffective during some operating cycles. Some researchers, therefore, treat transient faults as a subset of the permanent faults. Permanent sa**0** or sa**1** is called a *determinate fault* if it appears only in one form but not both. Indeterminate faults may be either sa**0** or sa**1**. Some engineers simply call *indeterminate* faults as *strick-at-x* (sa**x**) faults, in which x may be either "0" or "1". A sax fault has the same effects as two transient faults, sa**0** and

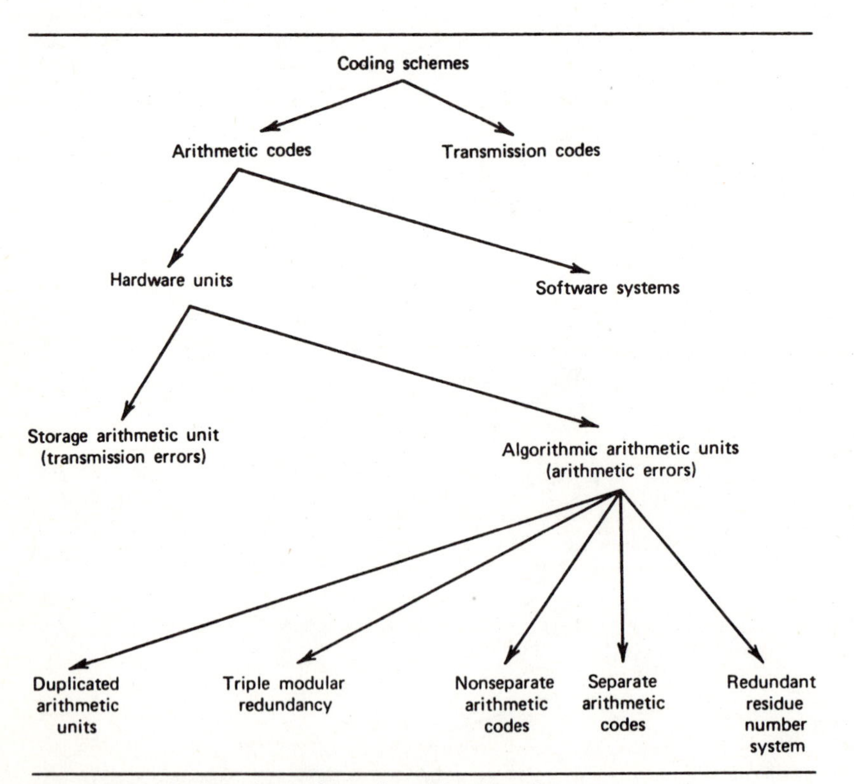

Figure 11.26 Coding schemes for coping with arithmetic errors.

sal, occurring at the same terminal variable. Therefore, indeterminant faults have been considered as a subset of transient determinant faults. The dashed lines in the classification tree show the subset relationships.

Methods to combat the above faulty conditions in arithmetic circuits are briefly described below. The defective circuit can be replaced by a spare one, after its malfunctioning is detected and before it is repaired. The replacement is usually made on a *graceful degradation* basis, in which the remaining functional components are reconfigured by switching techniques, so that faulty components will be disabled and their duties temporarily assigned to other working units until the recovery procedures are completed. The recovered system may continue computing at a degraded mode until the replacement is completed. The advantage of graceful degradation is that computing will not be terminated without completion; instead it continues computing at a reduced level during the recovery period, because the reconfigured units are usually overloaded the the job execution queues will be longer.

Another popular approach is to choose efficient arithmetic internal coding schemes to detect arithmetic faults, and then automatically correct them through sophisticated error correction procedures. These may require additional hardware such as encoders and decoders. Usually, redundancy techniques are used to mask out errors. Error codes and coding techniques for combatting arithmetic errors are illustrated in Fig. 11.26.

There are two types of internal codes used in computer systems, *arithmetic codes* and *transmission codes*. We are mainly concerned with the use of arithmetic codes. The error coding scheme can be implemented with either hardware or software. Again, we are more interested in the hardware approaches. Hardware coding schemes apply to both *algorithmic* and *storage* arithmetic units. The former produces arithmetic errors and the latter introduces transmission errors. Parity-check codes are typically storage codes.

To combat the errors in algorithmic unit usually assumes one or more of the following approaches:

1. Straight duplication of a hardware unit can provide single-error detection capability, but not correction.
2. Triple modular redundancy provides a majority vote of 2-out-of-3 to override single errors as shown in Fig. 11.27.
3. Nonseparate arithmetic codes have the checking bits immersed in the coded word. The $5N$ code shown in Table 11.5 shows an example of a nonseparate code, in which each code word, $(5N)_2$, is formed by multiplying the given 3-bit message word $(N)_2$ by 5 and expressing the resultant product in a 6-bit binary code. Note that the original message bits and check bits are not separable in the $5N$ codewords.
4. Separate arithmetic codes provide the checking capability without shuffling the giving message word. The $(N, |N|_5)$ code shown in Table 11.5 offers an example of separable code, in which the check bits are generated by appending the binary coded $N(\text{Mod } 5)$ to the original message word.

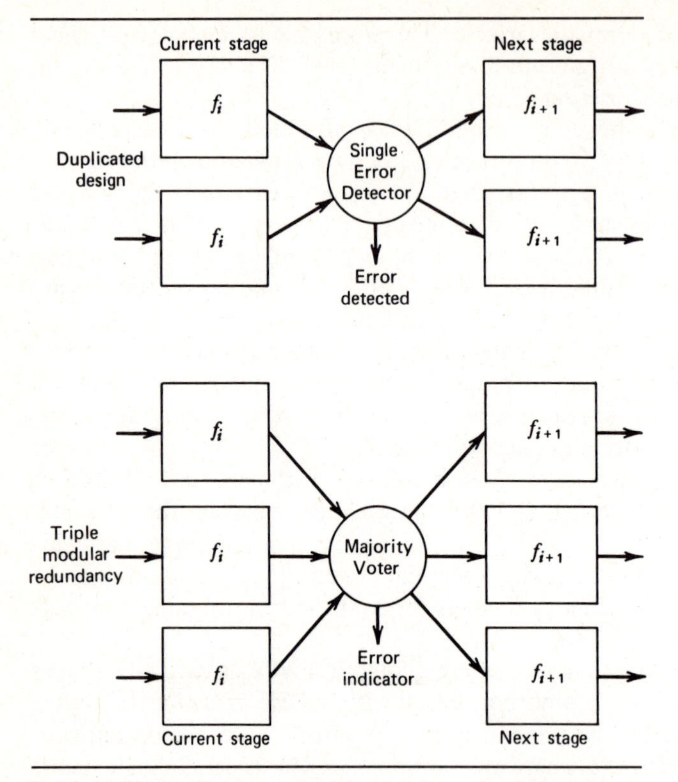

Figure 11.27 Duplication and triple modular redundancy schemes for error control in arithmetic processors.

Table 11.5 Separate vs. Nonseparate Arithmetic Codes for Error Control

Decimal N	Binary N	Binary Arithmetic Codes Representing N: Separate $[N, \|N\|_5]$ Code	Nonseparate $5N$ Code
0	0 0 0	(000,000)	0 0 0 0 0 0
1	0 0 1	(001,001)	0 0 0 1 0 1
2	0 1 0	(010,010)	0 0 1 0 1 0
3	0 1 1	(011,011)	0 0 1 1 1 1
4	1 0 0	(100,100)	0 1 0 1 0 0
5	1 0 1	(101,000)	0 1 1 0 0 1
6	1 1 0	(110,001)	0 1 1 1 1 0
7	1 1 1	(111,010)	1 0 0 0 1 1

5. The use of a redundant residue number system offers an inherent error-detecting capability. This class of arithmetic codes naturally extends the concept of residue encoding.

Interested readers are advised to study the book by Rao [30] for in-depth treatment of various classes of error coding schemes that can be used for error controls in arithmetic processors.

11.11 Case Study V: The CDC Star-100 Pipelined Arithmetic Processor

The Control Data STAR-100 (STring-ARray) computer is a pipelined processor structured around a 4-million byte (8-million byte optional) high-bandwidth memory. The STAR-100 computer contains many advanced design features such as stream processing, virtual addressing, hardware macro instructions, and a semiconductor memory register file. The pipelined arithmetic units are especially designed for sequential and parallel operations on variable length streams of data such as single bits, 8-bit bytes, and 32-bit or 64-bit FLP operands and vectors. The variable data length allows full use of the memory bandwidth and the arithmetic pipelines. Virtual addressing employs a high-speed mapping technique to convert a logical address to an absolute storage address. In a streaming mode, the system has the capability of producing 100 million 32-bit FLP results per second. Design aspects of the pipeline processors and the associated memory bus are described below.

A block diagram of the STAR-100 processor-memory configuration is shown in Fig. 11.28. The core memory has a cycle time of 1.28 μsec. It has 32 interleaved memory banks, with each bank containing 2048 512-bit words. The memory cycle is divided into 32 minor cycles with a rate of 40 nsec each, which also equals the arithmetic pipeline cycle time. Therefore, the memory system of 32 banks can deliver a total bandwidth of 512 bits of data per minor cycle. The 512 bits are subdivided into four data streams, each of 128 bits; two for the operands, one for the result, and the last for input/output requests and control vectors. This four-data bus configuration allows the pipelines to operate at their maximum rate of 100 million results per second.

The read and write buffers are used to synchronize the four active buses. The memory requests are buffered eight banks apart so that memory access conflicts can be significantly reduced. The maximum pipeline rate can therefore be sustained regardless of the distribution of addresses on the four active buses.

The floating-point arithmetic is carried out in STAR-100 by two independent pipeline processors. The Pipeline Processor 1 contains a 64-bit pipelined FLP Add unit and a 32-bit pipelined FLP Multiply unit as shown in Fig. 11.29. The 64-bit FLP Add pipeline consists of essentially four segments in cascade. The first segment compares exponents and saves the larger. The difference is used as the shift count in the second segment, where the fraction with the smaller exponent is right-shifted by the shift count. In segment three, the shifted and unshifted fractions are added and the sum and the larger exponent are gated to segment four, which selects the desired

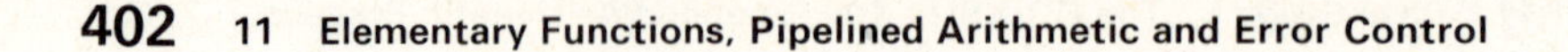

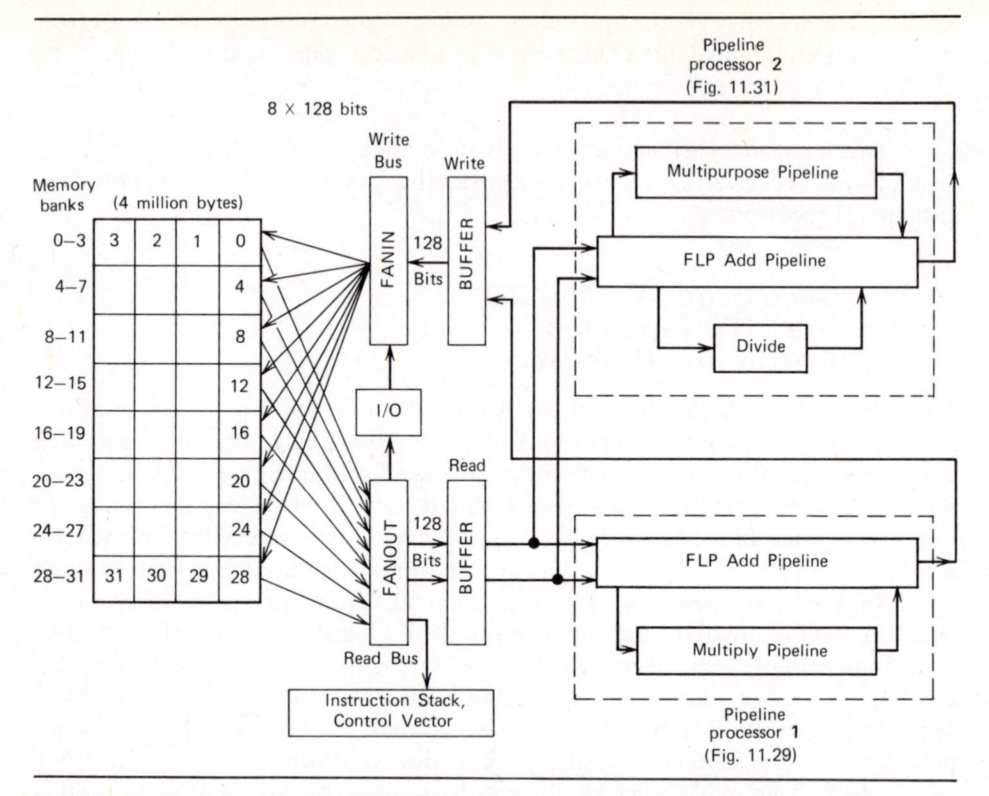

Figure 11.28 Block diagram of the CDC STAR-100 memory banks, pipeline processors, and data paths (Hintz and Tate [18]).

upper or lower half of the sum and transmits the results to the result data bus. It is possible to split the 64-bit FLP Add pipeline into two independent 32-bit pipelines with little additional hardware. The concept of pipeline splitting is used extensively in STAR-**100** to give the machine the capability of half-width (32 bits) arithmetic operations in addition to the full 64-bit operations.

Figure 11.30 shows the detailed structure of the basic 32-bit Multiply Pipeline used in the STAR-**100** computer. The two operands A and B are gated into the multiplier decode and multiplicand gating network. The multiplier is decoded into 12 two-bit groups. Multiples of the multiplicand are gated to the carry-save adders (CSA's) in the successive segments, and finally merged into a two output numbers, the partial summand the partial carry. These two numbers, when added together, form the product of the two input numbers A and B. The required normalization after FLP multiplication is carried out by the fourth segment of the Add Pipeline as shown in Fig. 11.29.

Pipeline Arithmetic Processor 2 is shown in Fig, 11.31. The processor contains a pipeline FLP Add unit similar to that in processor 1, a nonpipelined FLP register Divide unit, a 24-segment pipelined Multi-Purpose unit, and some pipelined Merge

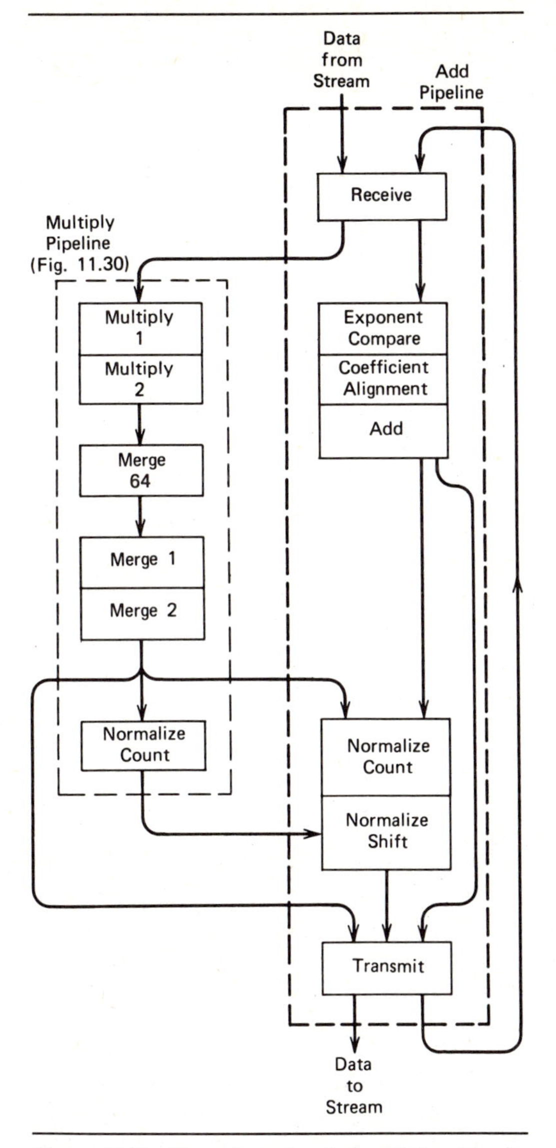

Figure 11.29 The schematic of the FLP arithmetic pipeline processor 1 in STAR-100 computer. (Reprinted with permission from "CDC STAR-100 General System Description." [7]).

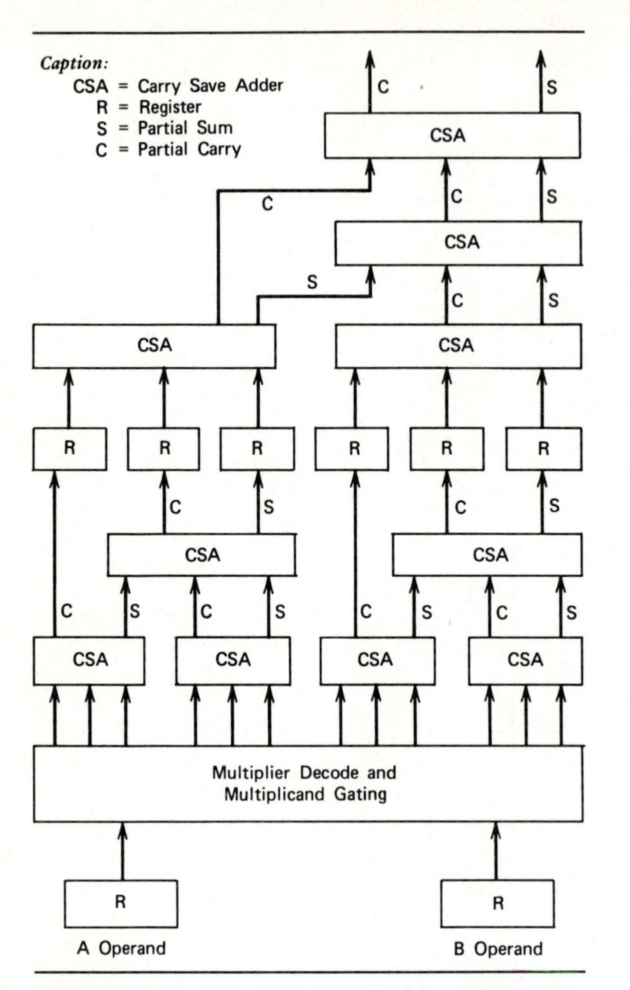

Figure 11.30 The basic 32-bit FLP multiply pipeline used in STAR-100.

units. This processor can be used for register divide, register square root, and all vector instructions. The register divide portion is a single segment divider, which can also perform binary to BCD and BCD to binary conversions. The multipurpose pipeline performs the square root, vector divide, and one-half the vector multiply instructions. There are 24 segments in the multipurpose pipeline.

Figure 11.32 shows how two of the basic Multiply Pipelines (Fig. 11.30) can be used to form a more powerful multiply pipeline processor. The processor can simultaneously execute two 32-bit multiplications or one 64-bit multiplication. When the processor is performing 32-bit operations, the left multiply pipeline multiplies inputs A and B. The partial sums and partial carries from this pipeline are merged together in Adder E. The resulting two partial numbers of the right pipeline are similarly merged by the Adder F.

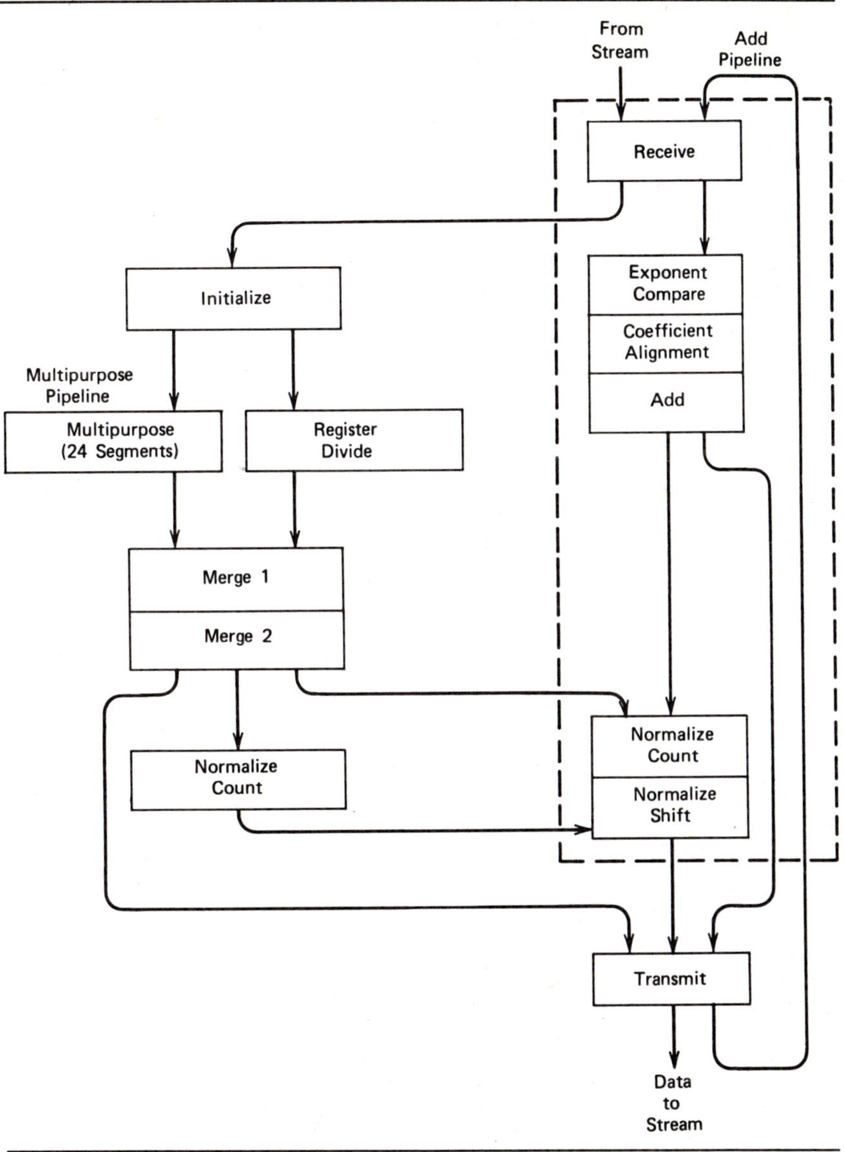

Figure 11.31 The schematic of the FLP arithmetic pipeline processor 2 in STAR-100 computer. (Reprinted with permission from "CDC STAR-100 General System Description." [7]).

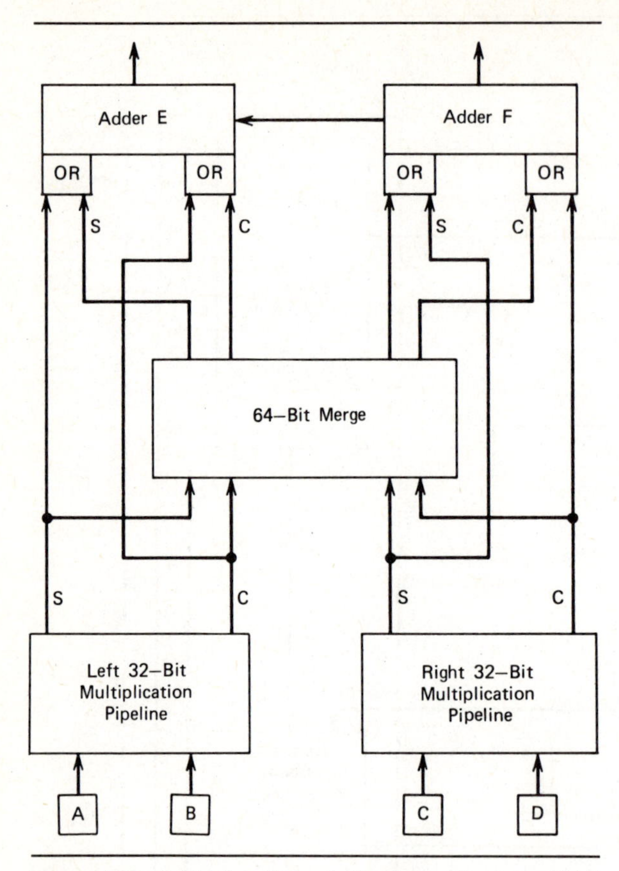

Figure 11.32 A 64-bit multiply pipeline formed from two 32-bit multiply pipelines in parallel.

To perform 64-bit multiplication, we split each of the multiplier and multiplicand into two parts such that the multiplier

$$A = A_0 + A_1 \times 2^W \tag{11.79}$$

and the multiplicand

$$B = B_0 + B_1 \times 2^W, \tag{11.80}$$

where W is the width of the basic pipeline multiplier (32 bits). Then, we have

$$A \times B = A_0 \times B_0 + (A_0 \times B_1 + A_1 \times B_0) \times 2^W + A_1 \times B_1 \times 2^{2W} \tag{11.81}$$

The 64-bit multiplication is then performed by executing the two 32-bit multiplications $A_0 \times B_0$ and $A_0 \times B_1$ during the first cycle and $A_1 \times B_0$ and $A_1 \times B_1$ on the second cycle. Four partial sums and partial carries of these splitted multiplications are then merged by the 64-bit Merge pipeline. The 64-bit outputs of the merge pipeline are then added together in the final adders to yield the 64-bit product.

Table 11.6 Maximum Numbers of Arithmetic Operations Executable in the Star-100 Computer Per Second with Respect to Word Lengths

Floating-Point Operation	32-Bit Results	64-Bit Results
Add/Subtract	100×10^6/sec	50×10^6/sec
Multiply	100×10^6/sec	25×10^6/sec
Divide/Square root	50×10^6/sec	12.5×10^6/sec

Maximum operational speeds of various arithmetic pipelines described above are summarized in Table 11.6. The entries are based on the 40 nsec pipeline cycle time. In conclusion, the STAR-100 is a versatile design which takes advantage of the pipelined arithmetic execution and the high memory bandwidth. These features allow the vectors to be streamed through the pipelines at the rate of 40 nsec.

11.12 Bibliographic Notes

The material on cellular array square-root extraction and binary squaring is based on the work of Majithia [22]. Signed-digit technique applied to minimum square rooting has been reported in Metze [24]. Ramamoorthy, et al. [29] investigated several iterative multiplication methods for square-rooting, which require no division. A convergence method for computing the inverse square root and square root of binary numbers has been proposed by Chen [5]. The polynomial approximation of bounded elementary functions presented in section 11.3 offers an improvement over the polynomial evaluator design presented in Flores [14], which requires that every power of the variable x be generated.

The CORDIC trigonometric computing technique was first proposed by Volder [34]. Later, a number of authors modified the original algorithm. The unified CORDIC approach is based on Walther [35]. DeLugish [12] described a class of algorithms for **Multiply**, **Divide**, **ln** x, e^x, **square root**, and **trigonometric** functions. His technique is based on the redundancy recoding technique in fast division algorithms. Chen's computation methods [5] are similar to the convergence transformation technique for division in IBM System **360**/Model **91**. The iterative approach, proposed by Chen for computing exponentials, logarithms, ratios, and square roots, shares most of the hardware for standard add, multiply, and divide operations. Senzig [32] has simplified the CORDIC algorithms to suit calculator architecture with word lengths on the order of 10 decimal digits. His calculator algorithms for trigonometric, hyperbolic, exponential, and logarithm functions require approximately 100 to 150 add-shift operations, a speed which is acceptable to most calculator applications. Ercegovac [13] and Sarkar, et al. [31] have studied high-radix and economic processes for obtaining elementary functions.

Chen [6] has written a chapter on overlap and pipeline processing techniques in a recent book on computer architecture edited by Stone. Original results also appear in his early work [4, 5]. The pipelined arithmetic implemented in TI **ASC** system has been reported and evaluated by Watson [36], Watson and Carr [37], Ramamoorthy and Li [27], and Stephenson [33]. Among them, Stephenson's treatment reveals most logic design considerations. The generalized pipeline array is due to Kamal, et al. [20]. The pipelined FFT processor design is based on Larson's thesis [21]. The pipeline FFT processor configuration was originally proposed by Groginsky [16], with different segment designs. The case study on CDC STAR-**100** pipelined arithmetic processors is based on description in [3, 7, 17, 18]. The schematic diagrams shown in Figs. 11.28, 11.30, and 11.32 are selected from Hintz and Tate [18]. Two high-throughout pipelined computers were recently announced. One is the **CRAY-1** described by Johnson in [19] and evaluated by Baskett and Keller in [3]; the other is the AMDAHL **470 V/6** as described in [1]. Recently, Ramamoorthy and Li [28] have provided an excellent survey of pipelined computer architecture. Other interesting articles on pipelined arithmetic design can be found in [9, 10, 11, 15, 17, 25, 26, 27].

Arithmetic error control techniques have been investigated by many authors. Avizienis [2] described how AN error-detecting codes can be used in fault-tolerant arithmetic design, such as those featured in the Jet Propulsion Laboratory **STAR** (Self-Testing And Repairing) system. A comprehensive treatment of error codes for arithmetic processors can be found in Rao [30]. The work of Chien and Hong [8] and of Massey [23] should be interesting to those who wish to further explore arithmetic error correcting codes.

References

[1] Amdahl Corp., *Amdahl 470 V/6 Machine Reference*, Sunnyvale, California, 1975.

[2] Avizienis, A., "Fault Detection in Digital Arithmetic Processors," in *Theory of Digital Computer Arithmetic*, Class Notes (Engineering 225A), University College of Los Angeles, Los Angeles, California, 1971.

[3] Baskett, F. and Keller, T. W., "An Evaluation of the CRAY-1 Computer," in *High-Speed Computer and Algorithm Organization* (Kuck, et al., eds.), Academic Press, New York, 1977, pp. 71–84.

[4] Chen, T. C., "Parallelism, Pipelining, and Computer Efficiency," *Computer Design*, January 1971, pp. 69–74.

[5] Chen, T. C. "Automatic Computation of Exponentials, Logarithms, Ratios, and Square Roots," *IBM Journal Res. and Dev.*, July 1972, pp. 380–388.

[6] Chen, T. C., "Overlap and Pipeline Processing," Chapter 9 in *Introduction to Computer Architecture*, SRA, Inc., Palo Alto, California, 1975, pp. 375–429.

[7] Control Data Corp., "Control Data STAR-100 General System Descriptions," St. Paul, Minnesota, Pub. No. 60256000-03, October 1973.

[8] Chien, R. T. and Hong, S. J., "Error Correction in High-Speed Arithmetic," *IEEE Trans. Comp.*, Vol. C-21, No. 5, May 1972.

[9] Cotton, L. W., "Circuit Implementation of High-Speed Pipeline Systems," *IEEE Trans. Comp.*, Vol. C-20, January 1971, pp. 33–38.

[10] Cotton, L. W., "Maximum-Rate Pipeline Systems," *Proc. SICC*, 1969, pp. 581–586.

[11] Davidson, E. A. et al., "Effective Control for Pipeline Computers," *IEEE 1975 Compcon Reader Digest*, Spring 1975.

[12] De Lugish, B. G., "A Class of Algorithms for Automatic Evaluation of Certain Elementary Functions in a Binary Computer," *Technical Report* No. 399, Dept. of Computer Science, University of Illinois, Urbana, Illinois, 1970.

[13] Ercegovac, M. D., "Radix-16 Evaluation of Some Elementary Functions," *IEEE Trans. Comp.*, Vol. C-20, No. 12, December 1971, pp. 1617–1618.

[14] Flores, I., *The Logic of Computer Arithmetic*, Prentice-Hall, Englewood Cliffs, New Jersey, 1963, Chapter 17.

[15] Graham, W. R., "The Parallel and the Pipeline Computers," *Datamation*, April 1970, pp. 68–71.

[16] Groginsky, H. L., et al, "A pipeline Fast Fourier Transform," *IEEE Trans. Comp.,* Vol. C-19, November 1970, pp. 1015–1019.

[17] Hallin, T. G. and Flynn, M. J., "Pipelining of Arithmetic Functions," *IEEE Trans. Comp.*, Vol. C-21, August 1972, pp. 880–886.

[18] Hintz, R. G. and Tate, D. P., "CDC STAR-100 Processor Design," *Compcon Proc.*, September 1972, pp. 1–4.

[19] Johnson, P. M., "CRAY-1 Computer System," Pub. No. 2240002A Cray Research, Inc., Minnesota, 1977.

[20] Kamal, A. K. et al., "A Generalized Pipeline Array," *IEEE Trans. Comp.*, Vol. C-23, May 1974, pp. 533–536.

[21] Larson, A. G., "Cost-Effective Processor Design with an Application to Fast Fourier Transform Computers," *Ph.D. Thesis*, Stanford University, 1973.

[22] Majithia, J. C., "Cellular Array for Extraction of Squares and Square Roots of Binary Numbers," *IEEE Trans. Comp.*, Vol. C-20, No. 12, December 1971, pp. 1617–1618.

[23] Massey, J. L. et al., "Error Correcting Codes in Computer Arithmetic," in *Advances in Information System Sciences*, Chap. 5, Plenum Press, New York, 1972.

[24] Metze, G., "Minimal Square Rooting," *IEEE Trans. Elec. Comp.*, Vol. EC-14, No. 2, 1965, pp. 181–185.

[25] Patel, J., "Improving the Throughput of Pipelines with Delays and Buffers," *Ph.D. Thesis*, Stanford University, September 1976.

[26] Peatman, J. B., *The Design of Digital Systems*, McGraw-Hill, New York, 1972.

[27] Ramamoorthy, C. V. and Kim, K. H., "Pipelining—The Generalized Concept and Sequency Strategies," *National Computer Conf.*, 1974, pp. 289–297.

[28] Ramamoorthy, C. V. and Li, H. F., "Pipeline Architecture," *ACM Computer Surveys*, Vol. 9, No. 1, March 1977, pp. 61–102.

[29] Ramamoorthy, C. V. et al., "Some Properties of Iterative Square-Rooting Methods Using High-Speed Multiplication," *IEEE Trans. Comp.*, Vol. C-21, No. 8, August 1972, pp. 837–847.

[30] Rao, T. R. N., *Error Codes for Arithmetic Processors*, Academic Press, New York, 1974.

[31] Sarakar, B. P. et al., "Economic Pseudodivision Processes for Obtaining Square-Root, Logarithm and Arc Tan," *IEEE Trans. Comp.*, Vol. C-20, No. 12, December 1971, pp. 1589–1593.

[32] Senzig, D., "Calculator Algorithms," *IEEE Compcon Reader Digest*, Spring 1975, IEEE Catalog No. 75 CH 0920-9C, pp. 139–141.

[33] Stephenson, C., "Control of a Variable Configuration Pipelined Arithmetic Unit," *Proc. 11th Allerton Conf.*, October 1973.

[34] Volder, J. E., "The CORDIC Trigonometric Computing Technique," *IEEE Trans. Elec. Comp.*, Vol. EC-9, September 1960, pp. 227–231.

[35] Walther, J. S., "A Unified Algorithm for Elementary Functions," *SJCC*, 1971, pp. 379–385.

[36] Watson, W. J., "The TI ASC—A Highly Modular and Flexible Super Computer Architecture," *FJCC*, 1972, pp. 221–228.

[37] Watson, W. J. and Carr H. M., "Operational Experiences with the TI Advanced Computer," *Proc. National Computer Conf.*, 1974, pp. 389–977.

Problems

Prob. 11.1 Design a cellular arithmetic array similar to that shown in Fig. 11.5 for both the extraction of the square roots of 12-bit numbers and the generation of squares of 6-bit numbers. Estimate the execution times required for both operations, assuming that each Controlled-Add-Subtract-shiFt (CASF) cell and each control cell has 3Δ gate delays.

Prob. 11.2 Based on the algorithm given in Fig. 11.6, explore the necessary hardware of a special function generator to evaluate the elementary functions defined by polynomials in Eqs. 11.7, 11.8, 11.9, and 11.10. Consider only 16-bit fractional arguments and allow an error tolerance of $\varepsilon = 0.0001$. Show your design with a schematic block diagram.

Prob. 11.3 Demonstrate how to generate each of the following CORDIC functions: $x \times z$, y/x, $\sin z$, $\cosh z$, $\tanh^{-1} y$, $\sinh z$, $\cosh z$, $\tan^{-1} y$, $\tan z$, $\tanh z$, $\exp z$, $\ln \omega$, and $\sqrt{\omega}$, using Walther's unified CORDIC algorithm.

Prob. 11.4 Prove that applying each pair of transformation rules defined in Eqs. 11.39, 41, 43, and 45 to the corresponding Chen's convergence functions defined in Eqs. 11.38, 40, 42, and 44, respectively, will converge to the desired function values, if the suggested initiation values and termination rules are used accordingly.

Prob. 11.5 Let τ_f and τ_l be the propagation delays of a full adder and of a latch, respectively. Express, in terms of τ_f, τ_l, and n, the maximum clock rate for each of the following n-bit adding schemes as depicted in Fig. 11.33.

(a) An n-bit ripple-carry adder with parallel input and output latches.

(b) An add pipeline with n segments and $n + 1$ latches. Each segment is simply one stage of full adder.

What is the gain on the processing speed of scheme (b) over scheme (a), assuming that the word length $n = 32$, $\tau_f = 10^{\text{nsec}}$ and $\tau_l = 2^{\text{nsec}}$. Note that the two schemes cost almost the same amount of hardware, because latches are very cheap now.

Prob. 11.6 Design an arithmetic pipeline dedicated to execute the normalized, floating-point, single-precision MULTIPLY instruction. Try to obtain a design with as many refined segments as possible, so that each segment will not exceed 10 levels of logic gating. You may assume the word length of $n = 32$ bits and an FLP data format of your own choice.

Prob. 11.7 Provide a comparative study of the pipelined arithmetic designs in Texas Instruments **ASC**, **CRAY-1**, and Control Data CDC **STAR-100** systems. Comments given in references [3, 6, 7, 15, 18, 19, 28, 33, 36, 37] should be helpful.

Prob. 11.8 Verify Kamal's generalized arithmetic pipeline in Fig. 11.19 with the following numerical operands. Show the values of all intermediate and final outputs in the pipeline array. All unused inputs should be appropriately specified.

(a) For square rooting with $A = (A_1 A_2 \cdots A_{10})_2 = (1011010110)_2$

(b) For squaring with $P = (P_1 P_2 \cdots P_5)_2 = (10111)_2$

(c) For evaluating $A + B \times P$, with $A = (0101111)_2$, $B = (1011)_2$, $P = (1010)_2$.

(d) For division with dividend $A = (0110110)_2$ and divisor $B = (1011)_2$, indicate the quotient as well as the remainder.

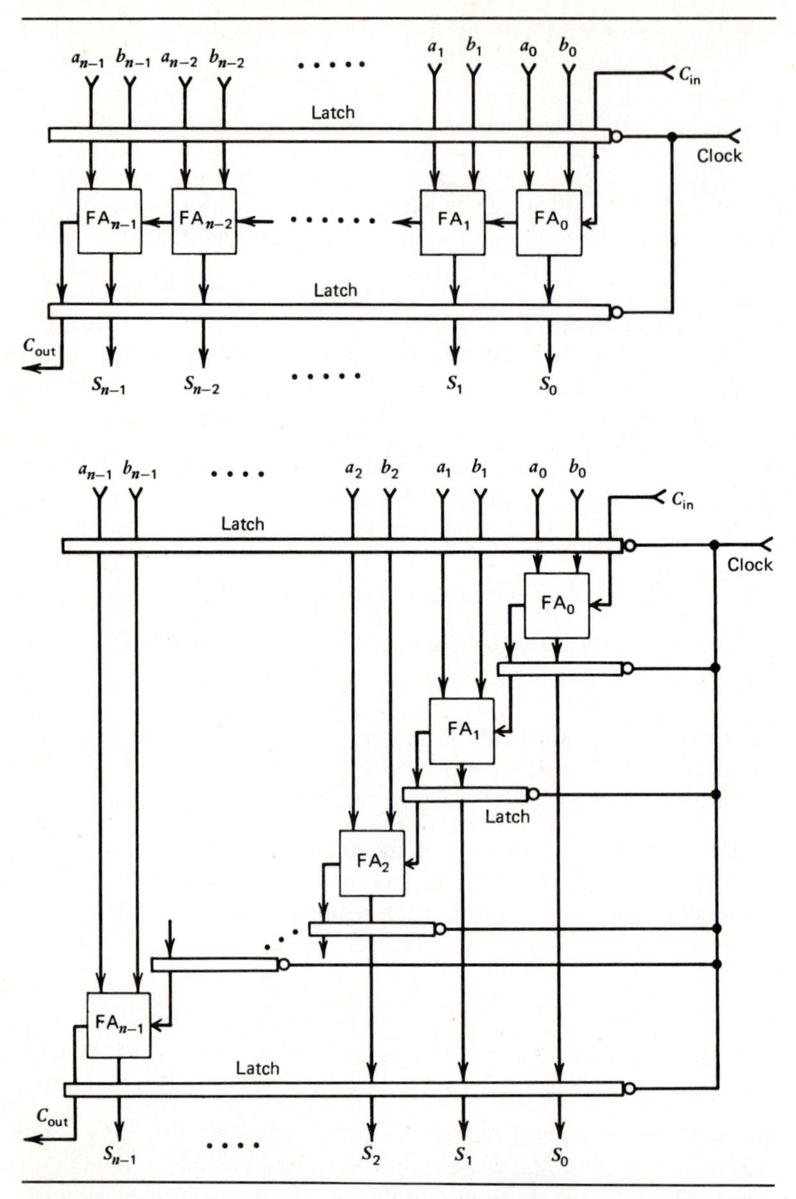

Figure 11.33 The schematic diagrams of an n-bit ripple-carry adder and of a pipelined adder with n segments (*a*) A ripple-carry adder with parallel inputs (*b*) An n-bit pipelined adder with n segments.

Prob. 11.9 Explain the functions that can be performed by each of the eight Pipeline Segments in the Pipelined Arithmetic Unit of the Texas Instruments ASC system.

Prob. 11.10 Suppose that the 16-bit butterfly elements shown in Fig. 11.22 are used to implement a **FFT** pipeline of $k = \log_2 N = \log_2 1024 = 10$ segments as shown in Fig. 11.23. Figure out the total hardware requirements in the **FFT** pipeline for $N = 1024$ sample points, in terms of the numbers of adders, 8-by-8 multipliers, latches, and so on.

Prob. 11.11 Explain the following types of logic faults and coding terminologies:

(a) Single fault vs. multiple faults

(b) Local vs. distributed faults

(c) Stuck-at-**0** vs. stuck-at-**1**

(d) Transient vs. permanent faults

(e) Graceful degradation

(f) Separate vs. nonseparate arithmetic codes

Prob. 11.12 Explain how the CDC STAR-**100** handles the execution of a 64-bit multiplication using two 32-bit Multiply Pipelines in two execution cycles. Justify the fact that their speed ratio is 4 to 1, when the word length is doubled.

Author Index

Subject Index